D0151896

PRINCIPLES OF ELECTRIC MACHINES AND POWER ELECTRONICS

PRINCIPLES OF ELECTRIC MACHINES AND POWER ELECTRONICS

P. C. SEN
Professor of Electrical Engineering
Queen's University
Kingston, Ontario, Canada

John Wiley & Sons
New York Chichester Brisbane Toronto Singapore

Library of Congress Cataloging in Publication Data:

Sen. P. C. (Paresh Chandra)
Principles of electric machines and power electronics.

Includes index.
1. Electric machinery. 2. Power electronics.
I. Title

TK2000.S44 1989 621.3 88-5725
ISBN 0-471-85084-5

Printed in the United States of America

10 9 8 7 6 5 4 3 2

To My lively children, Debashis, Priya, and Sujit;
My loving wife, Maya;
And Manidi and Sudhirda, whose affection
is always appreciated.

About the Author

Paresh C. Sen is Professor of Electrical Engineering at Queen's University, Kingston, Ontario, Canada. Dr. Sen received his Ph.D. degree from the University of Toronto in 1967. He has worked for industries in India and Canada and has been a consultant to electrical industries in Canada. He has authored or coauthored over 75 papers in the general area of power electronics and drives and is the author of the book *Thyristor DC Drives* (Wiley, 1981). He has taught electric machines, power electronics, and electric drive systems for twenty years. His fields of interest are electric machines, power electronics and drives, microcomputer control of drives, and modern control techniques for high-performance drive systems.

Dr. Sen served as an Associate Editor for the IEEE Transactions on Industrial Electronics and Control Instrumentation and as Chairman of the Technical Committee on Power Electronics. He has served on program committees of many IEEE and international conferences and has organized and chaired many technical sessions. At present, he is an active member of the Industrial Drive Committee and Static Power Converter Committee of IEEE. Dr. Sen is internationally recognized as a specialist in power electronics and drives. He received a Prize Paper Award from the Industrial Drive Committee for technical excellence at the Industry Application Society Annual Meeting in 1986.

PREFACE

Electric machines play an important role in industry as well as in our day-to-day life. They are used in power plants to generate electrical power and in industry to provide mechanical work, such as in steel mills, textile mills, and paper mills. They are an indispensable part of our daily lives. They start our cars and operate many of our household appliances. An average home in North America uses a dozen or more electric motors. Electric machines are very important pieces of equipment.

Electric machines are taught, very justifiably, in almost all universities and technical colleges all over the world. In some places, more than one semester course in electric machines is offered. This book is written in such a way that the instructor can select topics to offer one or two semester courses in electric machines. The first few sections in each chapter are devoted to the basic principles of operation. Later sections are devoted mostly to a more detailed study of the particular machine. If one semester course is offered, the instructor can select materials presented in the initial sections and/or initial portions of sections in each chapter. Later sections and/or later portions of sections can be covered in a second semester course. The instructor can skip sections, without losing continuity, depending on the material to be covered.

The book is suitable for both electrical engineering and non-electrical engineering students.

The dc machine, induction machine, and synchronous machine are considered to be basic electric machines. These machines are covered in separate chapters. A sound knowledge of these machines will facilitate understanding the operation of all other electric machines. The magnetic circuit forms an integral part of electric machines and is covered in Chapter 1. The transformer, although not a rotating machine, is indispensable in many energy conversion systems; it is covered in Chapter 2. The general principles of energy conversion are treated in Chapter 3, in which the mechanism of force and torque production in various electric machines is discussed. However, in any chapter where an individual electric machine is discussed in detail, an equivalent circuit model is used to predict the torque and other performance characteristics. This approach is simple and easily understood.

The dc machine, the three-phase induction machine, and the three-phase synchronous machine are covered extensively in Chapters 4, 5, and 6, respectively. Classical control and also solid-state control of these machines are discussed in detail. Linear induction motors (LIM) and linear synchronous motors (LSM), currently popular for application in transportation systems, are presented. Both voltage source and current source equivalent circuits for the operation of a synchronous machine are used to predict its performance. Operation of self-controlled synchronous motors for use in variable-speed drive systems is discussed. Inverter control of induction machines and the effects of time and space harmonies on induction motor operation are discussed with examples.

Comprehensive coverage of fractional horsepower single-phase motors, widely used in household and office appliances, is presented in Chapter 7. A procedure is outlined for the design of the starting winding of these motors. Special motors such as servomotors, synchro motors, and stepper motors are covered in Chapter 8. These motors play an important role in applications such as position servo systems or computer printers. The transient behavior and the dynamic behavior of the basic machines (dc, induction, and synchronous) are discussed in Chapter 9. Solid-state converters, needed for solid-state control of various electric machines, are discussed in Chapter 10.

All important aspects of electric machines are covered in this book. In the introduction to each chapter, I indicate the importance of the particular machine covered in that chapter. This is designed to stimulate the reader's interest in that machine and provide motivation to read about it. Following the introduction, I first try to provide a "physical feel" for the behavior of the machine. This is followed by analysis, derivation of the equivalent circuit model, control, application, and so forth.

A large number of worked examples are provided to aid in comprehension of the principles involved.

In present-day industry it is difficult to isolate power electronics technology from electric machines. After graduation, when a student goes into an industry as an engineer, he or she finds that in a motor drive, the motor is just a component of a complex system. Some knowledge of the solid-state control of motors is essential for understanding the functions of the motor drive system. Therefore, in any chapter where an individual motor is discussed, I present controller systems using that particular motor. This is done primarily in a qualitative and schematic manner so that the student can understand the basic operation. In the controller system the solid-state converter, which may be a rectifier, a chopper, or an inverter, is represented as a black box with defined input–output characteristics. The detailed operation of these converters is presented in a separate chapter. It is possible to offer a short course in power electronics based on material covered in Chapter 10 and controller systems discussed in other chapters.

In this book I have attempted to combine traditional areas of electric machinery with more modern areas of control and power electronics. I have presented this in as simple a way as possible, so that the student can grasp the principles without difficulty.

I thank all my undergraduate students who suggested that I write this book and, indeed, all those who have encouraged me in this venture. I acknowledge with gratitude the award of a grant from Queen's University for this purpose. I am thankful to the Dean of the Faculty of Applied Science, Dr. David W. Bacon, and to the Head of the Department of Electrical Engineering, Dr. G. J. M. Aitken, for their support and encouragement. I thank my colleagues in the power area—Drs. Jim A. Bennett, Graham E. Dawson, Tony R. Eastham, and Vilayil I. John—with whom I discussed electric machines while teaching courses on this subject. I thank Mr. Rabin Chatterjee, with whom I discussed certain sections of the manuscript. I am grateful to my graduate students, Chandra Namuduri, Eddy Ho, and Pradeep Nandam, for their assistance. Pradeep did the painful job of proofreading the final manuscript. I thank our administrative assistant, Mr. Perry Conrad, who supervised the typing of the manuscript. I thank the departmental secretaries, Sheila George, Marlene Hawkey, Marian Rose, Kendra Pople-Easton, and Jessie Griffin, for typing the manuscript at various stages. I express my profound gratitude to Chuck (Prof. C. H. R. Campling), who spent many hours reading and correcting the text. His valuable counseling and continued encouragement throughout have made it possible for me to complete this book. Finally, I appreciate the patience and solid support of my family—my wife, Maya, and my enthusiastic children, Sujit, Priya, and Debashis, who could hardly wait to have a copy of the book presented to them so that they could show it to their friends.

Queen's University
Kingston, Ontario, Canada
April 1987

P. C. SEN

CONTENTS

MAGNETIC CIRCUITS

CHAPTER one

This book is concerned primarily with the study of devices that convert electrical energy into mechanical energy or the reverse. Rotating electrical machines, such as dc machines, induction machines, and synchronous machines, are the most important ones used to perform this energy conversion. The transformer, although not an electromechanical converter, plays an important role in the conversion process. Other devices, such as actuators, solenoids, and relays, are concerned with linear motion. In all these devices, magnetic materials are used to shape and direct the magnetic fields that act as a medium in the energy conversion process. A major advantage of using magnetic material in electrical machines is the fact that high flux density can be obtained in the machine, which results in large torque or large machine output per unit machine volume. In other words, the size of the machine is greatly reduced by the use of magnetic materials.

In view of the fact that magnetic materials form a major part in the construction of electric machines, in this chapter properties of magnetic materials are discussed and some methods for analyzing the magnetic circuits are outlined.

1.1 MAGNETIC CIRCUITS

In electrical machines, the magnetic circuits may be formed by ferromagnetic materials only (as in transformers) or by ferromagnetic materials in conjunction with an air medium (as in rotating machines). In most electrical machines, except permanent magnet machines, the magnetic field (or flux) is produced by passing an electrical current through coils wound on ferromagnetic materials.

1.1.1 *i–H* RELATION

We shall first study how the current in a coil is related to the magnetic field intensity (or flux) it produces. When a

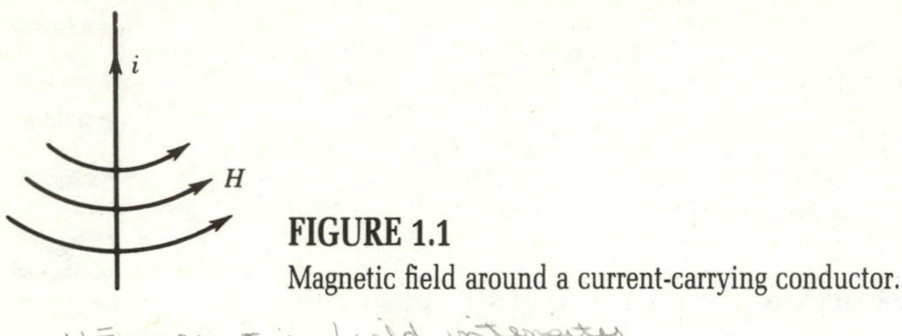

FIGURE 1.1
Magnetic field around a current-carrying conductor.

conductor carries current a magnetic field is produced around it, as shown in Fig. 1.1. The direction of flux lines or magnetic field intensity H can be determined by what is known as the *thumb rule,* which states that if the conductor is held with the right hand with the thumb indicating the direction of current in the conductor, then the fingertips will indicate the direction of magnetic field intensity. The relationship between current and field intensity can be obtained by using *Ampère's circuit law,* which states that the line integral of the magnetic field intensity H around a closed path is equal to the total current linked by the contour.

Referring to Fig. 1.2,

$$\oint \mathbf{H} \cdot \mathbf{d}l = \sum i = i_1 + i_2 - i_3 \tag{1.1}$$

where H is the magnetic field intensity at a point on the contour and dl is the incremental length at that point. If θ is the angle between vectors $\mathbf{H}$ and $\mathbf{dl}$, then

$$\oint H\, dl \cos\theta = \sum i \tag{1.2}$$

Now, consider a conductor carrying current i as shown in Fig. 1.3. To obtain an expression for the magnetic field intensity H at a distance r from the conductor, draw a circle of radius r. At each point on this circular contour, H and dl are in the same direction, that is, $\theta = 0$. Because of symmetry, H will be the same at all points on this contour. Therefore, from Eq. 1.2,

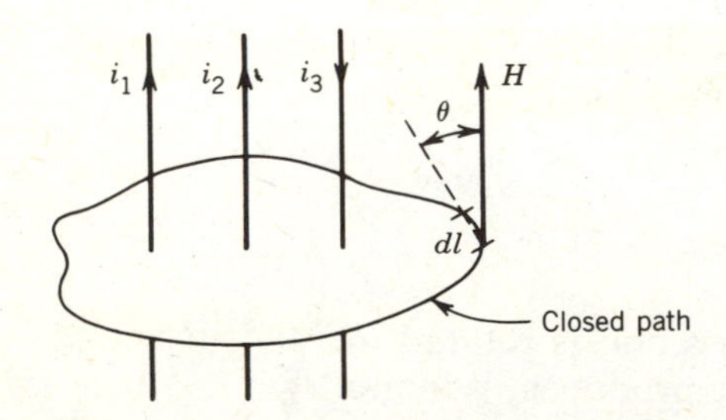

FIGURE 1.2
Illustration of Ampère's circuit law.

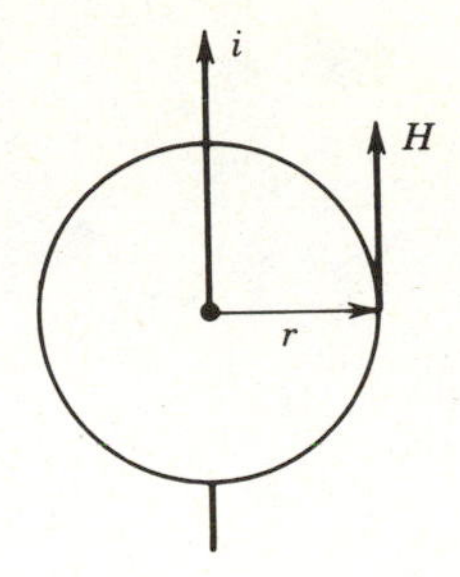

FIGURE 1.3
Determination of magnetic field intensity H due to a current-carrying conductor.

$$\oint \mathbf{H} \cdot \mathbf{dl} = i$$

$$H 2\pi r = i$$

$$H = \frac{i}{2\pi r} \text{ A/m} \tag{1.2a}$$

1.1.2 *B–H* RELATION

The magnetic field intensity H produces a magnetic flux density B everywhere it exists. These quantities are functionally related by

$$B = \mu H \text{ weber/m}^2 \quad \text{or} \quad \text{tesla} \tag{1.3}$$

$$B = \mu_r \mu_0 H \text{ Wb/m}^2 \quad \text{or} \quad \text{T} \tag{1.4}$$

where μ is a characteristic of the medium and is called the *permeability* of the medium

μ_0 is the permeability of free space and is $4\pi 10^{-7}$ henry/meter

μ_r is the *relative permeability* of the medium

For free space or electrical conductors (such as aluminum or copper) or insulators, the value of μ_r is unity. However, for ferromagnetic materials such as iron, cobalt, and nickel, the value of μ_r varies from several hundred to several thousand. For materials used in electrical machines, μ_r varies in the range of 2000 to 6000. A large value of μ_r implies that a small current can produce a large flux density in the machine.

1.1.3 MAGNETIC EQUIVALENT CIRCUIT

Figure 1.4 shows a simple magnetic circuit having a ring-shaped magnetic core, called a *toroid*, and a coil that extends around the entire circumfer-

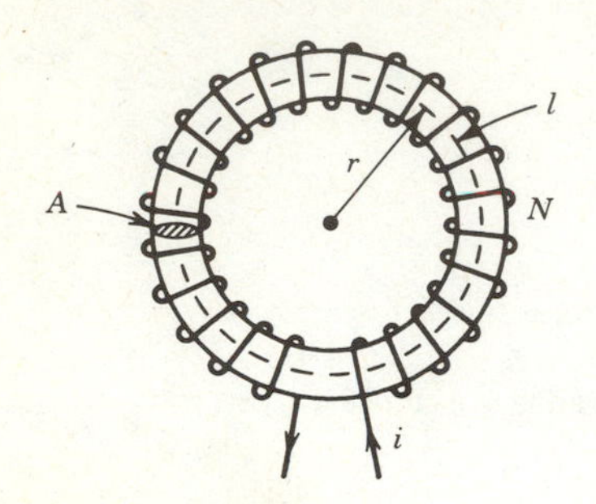

FIGURE 1.4
Toroid magnetic circuit.

ence. When current i flows through the coil of N turns, magnetic flux is mostly confined in the core material. The flux outside the toroid, called *leakage flux,* is so small that for all practical purposes it can be neglected.

Consider a path at a radius r. The magnetic intensity on this path is H and, from Ampère's circuit law,

$$\oint \mathbf{H} \cdot d\mathbf{l} = Ni \tag{1.5}$$

$$Hl = Ni \tag{1.5a}$$

$$H 2\pi r = Ni \tag{1.6}$$

The quantity Ni is called the *magnetomotive force* (*mmf*) F, and its unit is ampere-turn.

$$Hl = Ni = F \tag{1.7}$$

$$H = \frac{N}{l} i \text{ At/m} \tag{1.8}$$

From Eqs. 1.3 and 1.8

$$B = \frac{\mu N i}{l} \text{ T} \tag{1.9}$$

If we assume that all the fluxes are confined in the toroid, that is, there is no magnetic leakage, the flux crossing the cross section of the toroid is

$$\Phi = \int B \, dA \tag{1.10}$$

$$\Phi = BA \text{ Wb} \tag{1.11}$$

where B is the average flux density in the core and A is the area of cross section of the toroid. The average flux density may correspond to the path at the mean radius of the toroid. If H is the magnetic intensity for this path, then from Eqs. 1.9 and 1.11

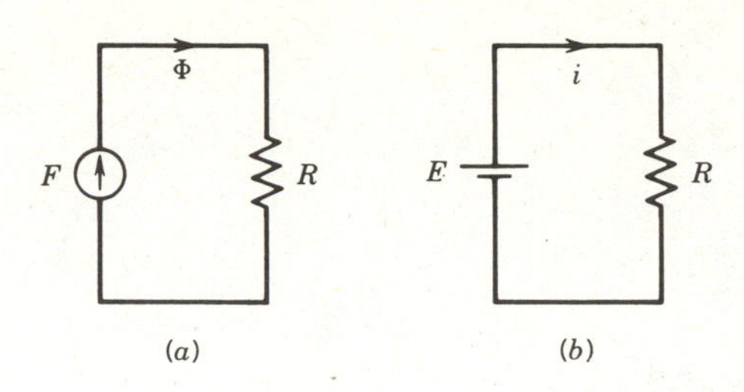

FIGURE 1.5
Analogy between (a) magnetic circuit and (b) electric circuit.

$$\Phi = \frac{\mu Ni}{l} A = \frac{Ni}{l/\mu A}$$

$$= \frac{Ni}{R} \tag{1.12}$$

$$= \frac{F}{R} \tag{1.13}$$

where

$$R = \frac{l}{\mu A} = \frac{1}{P} \tag{1.14}$$

is called the *reluctance* of the magnetic path and P is called the *permeance* of the magnetic path. Equations 1.12 and 1.13 suggest that the driving force in the magnetic circuit of Fig. 1.4 is the magnetomotive force F ($= Ni$), which produces a flux Φ against a magnetic reluctance R. The magnetic circuit of the toroid can therefore be represented by a magnetic equivalent circuit as shown in Fig. 1.5*a*. Also note that Eq. 1.13 has the form of Ohm's law for an electric circuit ($i = E/R$). The analogous electrical circuit is shown in Fig. 1.5*b*. A magnetic circuit is often looked upon as analogous to an electric circuit. The analogy is illustrated in Table 1.1.

TABLE 1.1
Electrical versus Magnetic Circuits

	Electric Circuit	Magnetic Circuit
Driving force	Emf (E)	Mmf (F)
Produces	Current ($i = E/R$)	Flux ($\Phi = F/R$)
Limited by	Resistance ($R = l/\rho A$)[a]	Reluctance ($R = l/\mu A$)[a]

[a] ρ, Conductivity; μ, permeability.

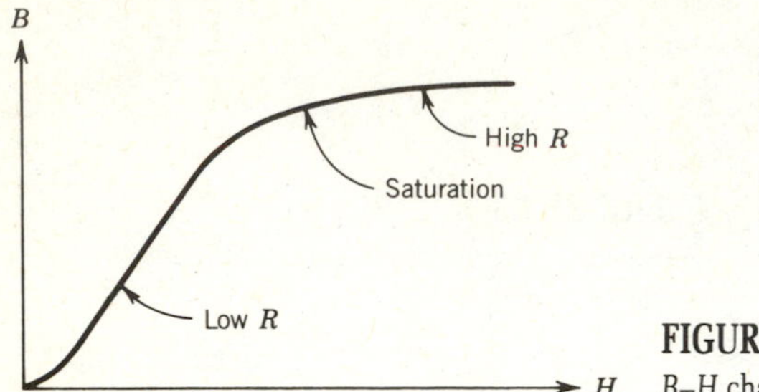

FIGURE 1.6
B–H characteristic (magnetization curve).

1.1.4 MAGNETIZATION CURVE

If the magnetic intensity in the core of Fig. 1.4 is increased by increasing current, the flux density in the core changes in the way shown in Fig. 1.6. The flux density B increases almost linearly in the region of low values of the magnetic intensity H. However, at higher values of H, the change of B is nonlinear. The magnetic material shows the effect of saturation. The B–H curve, shown in Fig. 1.6, is called the *magnetization curve*. The reluctance of the magnetic path is dependent on the flux density. It is low when B is low, high when B is high. The magnetic circuit differs from the electric circuit in this respect; resistance is normally independent of current in an electric circuit, whereas reluctance depends on the flux density in the magnetic circuit.

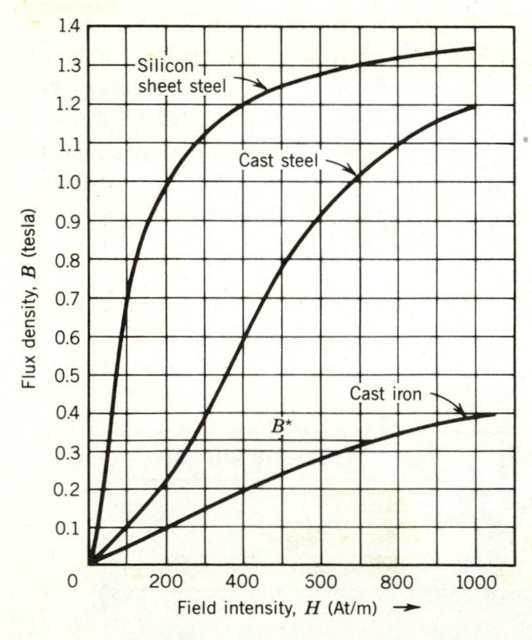

FIGURE 1.7
Magnetization curves.

The B–H characteristics of three different types of magnetic cores—cast iron, cast steel, and silicon sheet steel—are shown in Fig. 1.7. Note that to establish a certain level of flux density B^* in the various magnetic materials the values of current required are different.

1.1.5 MAGNETIC CIRCUIT WITH AIR GAP

In electric machines, the rotor is physically isolated from the stator by the air gap. A cross-sectional view of a dc machine is shown in Fig. 1.8. Practically the same flux is present in the poles (made of magnetic core) and the air gap. To maintain the same flux density, the air gap will require much more mmf than the core. If the flux density is high, the core portion of the magnetic circuit may exhibit a saturation effect. However, the air gap remains unsaturated, since the B–H curve for the air medium is linear (μ is constant).

A magnetic circuit having two or more media—such as the magnetic core and air gap in Fig. 1.8—is known as a *composite structure*. For the purpose of analysis, a magnetic equivalent circuit can be derived for the composite structure.

Let us consider the simple composite structure of Fig. 1.9*a*. The driving force in this magnetic circuit is the mmf, $F = Ni$, and the core medium and the air gap medium can be represented by their corresponding reluctances. The equivalent magnetic circuit is shown in Fig. 1.9*b*.

$$R_c = \frac{l_c}{\mu_c A_c} \tag{1.15}$$

$$R_g = \frac{l_g}{\mu_0 A_g} \tag{1.16}$$

$$\Phi = \frac{Ni}{R_c + R_g} \tag{1.17}$$

$$Ni = H_c l_c + H_g l_g \tag{1.18}$$

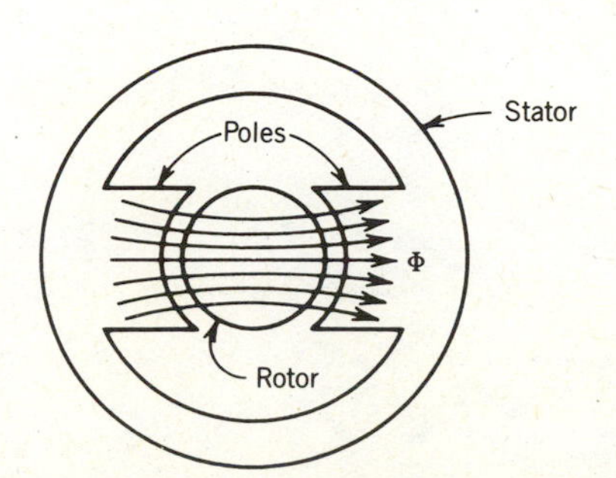

FIGURE 1.8
Cross section of a rotating machine.

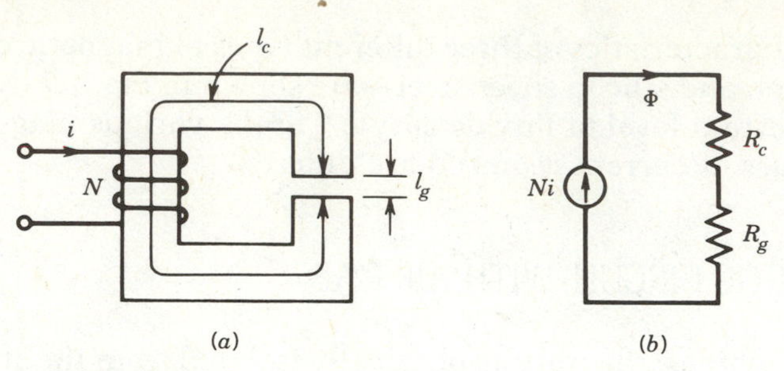

FIGURE 1.9
Composite structure. (a) Magnetic core with air gap. (b) Magnetic equivalent circuit.

where l_c is the mean length of the core
l_g is the length of the air gap

The flux densities are

$$B_c = \frac{\Phi_c}{A_c} \tag{1.19}$$

$$B_g = \frac{\Phi_g}{A_g} \tag{1.20}$$

In the air gap the magnetic flux lines bulge outward somewhat, as shown in Fig. 1.10; this is known as *fringing* of the flux. The effect of the fringing is to increase the cross-sectional area of the air gap. For small air gaps the fringing effect can be neglected. If the fringing effect is neglected, the cross-sectional areas of the core and the air gap are the same and therefore

$$A_g = A_c$$

$$B_g = B_c = \frac{\Phi}{A_c}$$

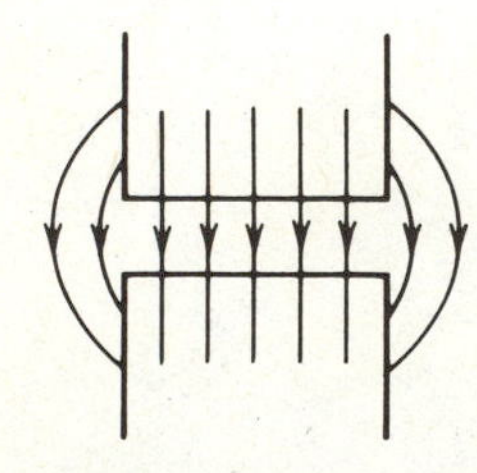

FIGURE 1.10
Fringing flux.

EXAMPLE 1.1

Figure E1.1 represents the magnetic circuit of a primitive relay. The coil has 500 turns and the mean core path is $l_c = 360$ mm. When the air gap lengths are 1.5 mm each, a flux density of 0.8 tesla is required to actuate the relay. The core is cast steel.

(a) Find the current in the coil.

(b) Compute the values of permeability and relative permeability of the core.

(c) If the air gap is zero, find the current in the coil for the same flux density (0.8 T) in the core.

Solution

(a) The air gap is small and so fringing can be neglected. Hence the flux density is the same in both air gap and core. From the B–H curve of the cast steel core (Fig. 1.7).

For

$$B_c = 0.8 \text{ T}, H_c = 510 \text{ At/m}$$

$$\text{mmf } F_c = H_c l_c = 510 \times 0.36 = 184 \text{ At}$$

For the air gap,

$$\text{mmf } F_g = H_g 2l_g = \frac{B_g}{\mu_0} 2l_g = \frac{0.8}{4\pi 10^{-7}} \times 2 \times 1.5 \times 10^{-3}$$

$$= 1910 \text{ At}$$

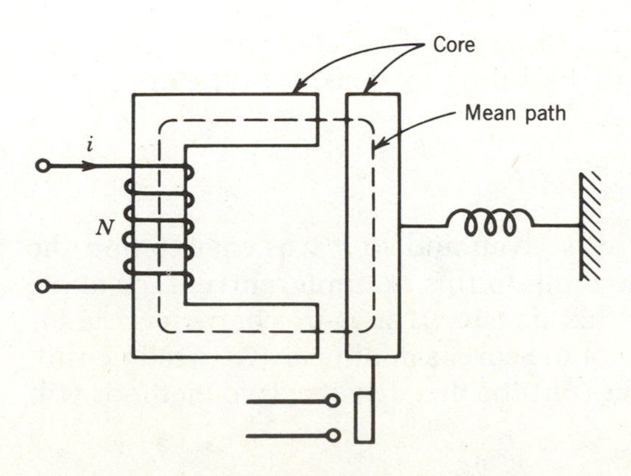

FIGURE E1.1
N = 500 turns, $l_c = 36$ cm.

Total mmf required:

$$F = F_c + F_g = 184 + 1910 = 2094 \text{ At}$$

Current required:

$$i = \frac{F}{N} = \frac{2094}{500} = 4.19 \text{ amps}$$

Note that although the air gap is very small compared to the length of the core ($l_g = 1.5$ mm, $l_c = 360$ mm), most of the mmf is used at the air gap.

(b) Permeability of core:

$$\mu_c = \frac{B_c}{H_c} = \frac{0.8}{510} = 1.57 \times 10^{-3}$$

Relative permeability of core:

$$\mu_r = \frac{\mu_c}{\mu_0} = \frac{1.57 \times 10^{-3}}{4\pi 10^{-7}} = 1250$$

(c)

$$F = H_c l_c = 510 \times 0.36 = 184 \text{ At}$$

$$i = \frac{184}{500} = 0.368 \text{ A}$$

Note that if the air gap is not present, a much smaller current is required to establish the same flux density in the magnetic circuit.

EXAMPLE 1.2

Consider the magnetic system of Example 1.1. If the coil current is 4 amps when each air gap length is 1 mm, find the flux density in the air gap.

Solution

In Example 1.1, the flux density was given and so it was easy to find the magnetic intensity and finally the mmf. In this example, current (or mmf) is given and we have to find the flux density. The B–H characteristic for the air gap is linear, whereas that of the core is nonlinear. We need nonlinear magnetic circuit analysis to find out the flux density. Two methods will be discussed.

1. *Load line method.* For a magnetic circuit with core length l_c and air gap length l_g,

$$Ni = H_g l_g + H_c l_c = \frac{B_g}{\mu_0} l_g + H_c l_c$$

Rearranging,

$$B_g = B_c = -\mu_0 \frac{l_c}{l_g} H_c + \frac{Ni\mu_0}{l_g} \tag{1.21}$$

This is in the form $y = mx + c$, which represents a straight line. This straight line (also called the *load line*) can be plotted on the B–H curve of the core. The slope is

$$m = -\mu_0 \frac{l_c}{l_g} = -4\pi 10^{-7} \frac{360}{2} = -2.26 \times 10^{-4}$$

The intersection on the B axis is

$$c = \frac{Ni\mu_0}{l_g} = \frac{500 \times 4 \times 4\pi 10^{-7}}{2 \times 10^{-3}} = 1.256 \text{ tesla}$$

The load line intersects the B–H curve (Fig. E1.2) at $B = 1.08$ tesla.

Another method of constructing the load line is as follows: If all mmf acts on the air gap (i.e., $H_c = 0$) the air gap flux density is

$$B_g = \frac{Ni}{l_g} \mu_0 = 1.256 \text{ T}$$

This value of B_g is the intersection of the load line on the B axis.

If all mmf acts on the core (i.e., $B_g = 0$),

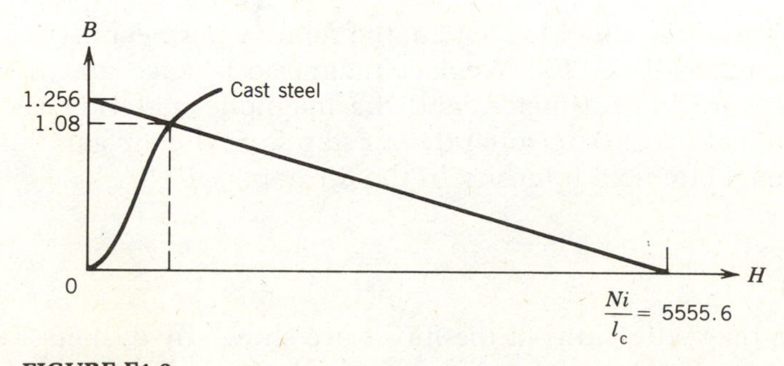

FIGURE E1.2

$$H_c = \frac{Ni}{l_c} = \frac{500 \times 4}{36 \times 10^{-2}} = 5556 \text{ At/m}$$

This value of H_c is the intersection of the load line on the H axis.

2. *Trial-and-error method.* The procedure in this method is as follows.
   (a) Assume a flux density.
   (b) Calculate H_c (from the B–H curve) and H_g $(= B_g/\mu_0)$.
   (c) Calculate F_c $(= H_c l_c)$, F_g $(= H_g l_g)$, and F $(= F_c + F_g)$.
   (d) Calculate $i = F/N$.
   (e) If i is different from the given current, assume another judicious value of the flux density. Continue this trial-and-error method until the calculated value of i is close to 4 amps.

   If all mmf acts on the air gap, the flux density is

$$B = \frac{Ni}{l_g}\mu_0 = 1.256 \text{ T}$$

   Obviously, the flux density will be less than this value. The procedure is illustrated in the following table.

B	H_c	H_g	F_c	F_g	F	i
1.1	800	8.7535×10^5	288	1750.7	2038.7	4.08
1.08	785	8.59435×10^5	282	1718.87	2000.87	4.0

EXAMPLE 1.3

In the magnetic circuit of Fig. E1.3*a*, the relative permeability of the ferromagnetic material is 1200. Neglect magnetic leakage and fringing. All dimensions are in centimeters and the magnetic material has a square cross-sectional area. Determine the air gap flux, the air gap flux density, and the magnetic field intensity in the air gap.

Solution

The mean magnetic paths of the fluxes are shown by dashed lines in Fig. E1.3*a*. The equivalent magnetic circuit is shown in Fig. E1.3*b*.

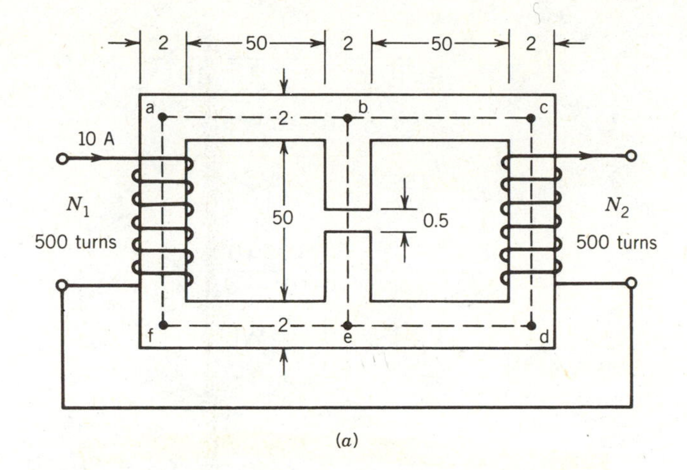

(a)

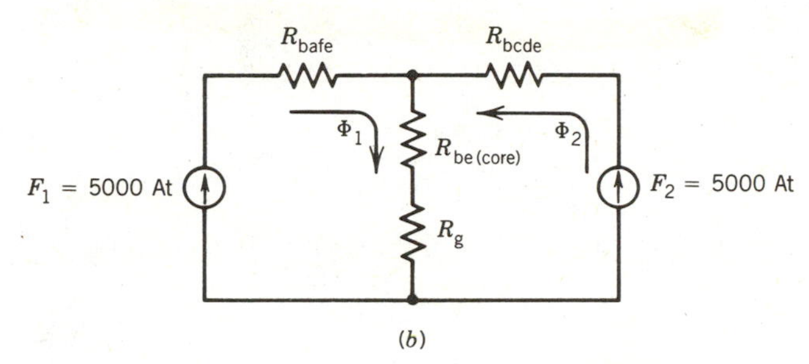

(b)

FIGURE E1.3

$$F_1 = N_1 I_1 = 500 \times 10 = 5000 \text{ At}$$

$$F_2 = N_2 I_2 = 500 \times 10 = 5000 \text{ At}$$

$$\mu_c = 1200\mu_0 = 1200 \times 4\pi 10^{-7}$$

$$\begin{aligned} R_{bafe} &= \frac{l_{bafe}}{\mu_c A_c} \\ &= \frac{3 \times 52 \times 10^{-2}}{1200 \times 4\pi 10^{-7} \times 4 \times 10^{-4}} \\ &= 2.58 \times 10^6 \text{ At/Wb} \end{aligned}$$

From symmetry

$$R_{bcde} = R_{bafe}$$

$$R_g = \frac{l_g}{\mu_0 A_g}$$

$$= \frac{5 \times 10^{-3}}{4\pi 10^{-7} \times 2 \times 2 \times 10^{-4}}$$

$$= 9.94 \times 10^{6} \text{ At/Wb}$$

$$R_{\text{be(core)}} = \frac{l_{\text{be(core)}}}{\mu_c A_c}$$

$$= \frac{51.5 \times 10^{-2}}{1200 \times 4\pi 10^{-7} \times 4 \times 10^{-4}}$$

$$= 0.82 \times 10^{6} \text{ At/Wb}$$

The loop equations are

$$\Phi_1(R_{\text{bafe}} + R_{\text{be}} + R_g) + \Phi_2(R_{\text{be}} + R_g) = F_1$$

$$\Phi_1(R_{\text{be}} + R_g) + \Phi_2(R_{\text{bcde}} + R_{\text{be}} + R_g) = F_2$$

or

$$\Phi_1(13.34 \times 10^{6}) + \Phi_2(10.76 \times 10^{6}) = 5000$$

$$\Phi_1(10.76 \times 10^{6}) + \Phi_2(13.34 \times 10^{6}) = 5000$$

or

$$\Phi_1 = \Phi_2 = 2.067 \times 10^{-4} \text{ Wb}$$

The air gap flux is

$$\Phi_g = \Phi_1 + \Phi_2 = 4.134 \times 10^{-4} \text{ Wb}$$

The air gap flux density is

$$B_g = \frac{\Phi_g}{A_g} = \frac{4.134 \times 10^{-4}}{4 \times 10^{-4}} = 1.034 \text{ T}$$

The magnetic intensity in the air gap is

$$H_g = \frac{B_g}{\mu_0} = \frac{1.034}{4\pi 10^{-7}} = 0.822 \times 10^{6} \text{ At/m}$$

1.1.6 INDUCTANCE

A coil wound on a magnetic core, such as that shown in Fig. 1.11*a*, is frequently used in electric circuits. This coil may be represented by an

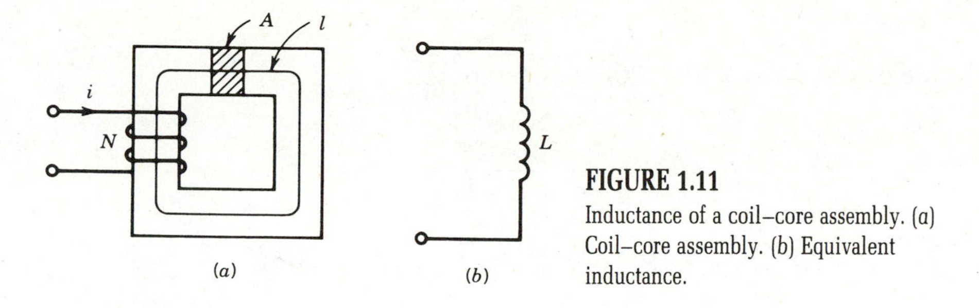

FIGURE 1.11
Inductance of a coil–core assembly. (*a*) Coil–core assembly. (*b*) Equivalent inductance.

ideal circuit element, called *inductance,* which is defined as the flux linkage of the coil per ampere of its current.

$$\text{Flux linkage} \qquad \lambda = N\Phi \tag{1.22}$$

$$\text{Inductance} \qquad L = \frac{\lambda}{i} \tag{1.23}$$

From Eqs. 1.22, 1.23, 1.11, 1.3, and 1.14,

$$L = \frac{N\Phi}{i} = \frac{NBA}{i} = \frac{N\mu HA}{i}$$

$$= \frac{N\mu HA}{Hl/N} = \frac{N^2}{l/\mu A} \tag{1.24}$$

$$L = \frac{N^2}{R} \tag{1.25}$$

Equation 1.24 defines inductance in terms of physical dimensions, such as cross-sectional area and length of core, whereas Eq. 1.25 defines inductance in terms of the reluctance of the magnetic path. Note that inductance varies as the square of the number of turns. The coil–core assembly of Fig. 1.11*a* is represented in an electric circuit by an ideal inductance as shown in Fig. 1.11*b*.

EXAMPLE 1.4

The coil in Fig. 1.4 has 250 turns and is wound on a silicon sheet steel. The inner and outer radii are 20 and 25 cm, respectively, and the toroidal core has a circular cross section. For a coil current of 2.5 A find

(a) The magnetic flux density at the mean radius of the toroid.

(b) The inductance of the coil, assuming that the flux density within the core is uniform and equal to that at the mean radius.

Solution

(a) Mean radius is $\frac{1}{2}(25 + 20) = 22.5$ cm

$$H = \frac{Ni}{l} = \frac{250 \times 2.5}{2\pi 22.5 \times 10^{-2}} = 442.3 \text{ At/m}$$

From the B–H curve for silicon sheet steel (Fig. 1.7),

$$B = 1.225 \text{ T}$$

(b) The cross-sectional area is

$$\begin{aligned} A &= \pi(\text{radius of core})^2 \\ &= \pi\left(\frac{25 - 20}{2}\right)^2 \times 10^{-4} \\ &= \pi 6.25 \times 10^{-4} \text{ m}^2 \\ \Phi &= BA \\ &= 1.225 \times \pi 6.25 \times 10^{-4} \\ &= 24.04 \times 10^{-4} \text{ Wb} \\ \lambda &= 250 \times 24.04 \times 10^{-4} \\ &= 0.601 \text{ Wb} \cdot \text{turn} \\ L &= \frac{\lambda}{i} = \frac{0.601}{2.5} = 0.2404 \text{ H} \\ &= 240.4 \text{ mH} \end{aligned}$$

Inductance can also be calculated using Eq. 1.25:

$$\begin{aligned} \mu \text{ of core} &= \frac{B}{H} = \frac{1.225}{442.3} \\ R_{\text{core}} &= \frac{l}{\mu A} = \frac{2\pi 22.5 \times 10^{-2}}{(1.225/442.3) \times \pi 6.25 \times 10^{-4}} \\ &= 2599.64 \times 10^2 \\ L &= \frac{N^2}{R} = \frac{250^2}{2599.64 \times 10^2} = 0.2404 \text{ H} \\ &= 240.4 \text{ mH} \end{aligned}$$

1.2 HYSTERESIS

Consider the coil–core assembly in Fig. 1.12*a*. Assume that the core is initially unmagnetized. If the magnetic intensity H is now increased by slowly increasing the current i, the flux density will change according to the curve $0a$ in Fig. 1.12*b*. The point a corresponds to a particular value of the magnetic intensity, say H_1 (corresponding current is i_1).

If the magnetic intensity is now slowly decreased, the B–H curve will follow a different path, such as abc in Fig. 1.12*b*. When H is made zero, the core has retained flux density B_r, known as the *residual flux density*. If H is now reversed (by reversing the current i) the flux in the core will decrease and for a particular value of H, such as $-H_c$ in Fig. 1.12*b*, the residual flux will be removed. This value of the magnetic field intensity ($-H_c$) is known as the *coercivity* or *coercive force* of the magnetic core. If H is further increased in the reverse direction, the flux density will increase in the reverse direction. For current $-i_1$ the flux density will correspond to the point e. If H is now decreased to zero and then increased to the value H_1, the B–H curve will follow the path $efga'$. The loop does not close. If H is now varied for another cycle, the final operating point is a''. The operating points a' and a'' are closer together than points a and a'. After a few cycles of magnetization, the loop almost closes, and it is called the *hysteresis loop*. The loop shows that the relationship between B and H is nonlinear and multivalued. Note that at point c the iron is magnetized, although the current in the coil is made zero. Throughout the whole cycle of magnetization, the flux density lags behind the magnetic intensity. This lagging phenomenon in the magnetic core is called *hysteresis*.

Smaller hysteresis loops are obtained by decreasing the amplitude of variation of the magnetic intensity. A family of hysteresis loops is shown in Fig. 1.12*c*. The locus of the tip of the hysteresis loop, shown dashed in Fig. 1.12*c*, is called the *magnetization curve*. If the iron is magnetized from an initial unmagnetized condition, the flux density will follow the magnetization curve. In some magnetic cores, the hysteresis loop is very narrow. If the hysteresis effect is neglected for such cores, the B–H characteristic is represented by the magnetization curve.

Deltamax Cores

Special ferromagnetic alloys are sometimes developed for special applications. The hysteresis loops for these alloys have shapes that are significantly different from those shown in Fig. 1.12. An alloy consisting of 50% iron and 50% nickel has the B–H loop shown in Fig. 1.13. Cores made of alloys having this type of almost square B–H loop are known as *deltamax cores*. A coil wound on a deltamax core can be used as a switch. Note that when the flux density is less than the residual flux density ($B < B_r$) the magnetic intensity (and hence the current) is quite low. As the flux density

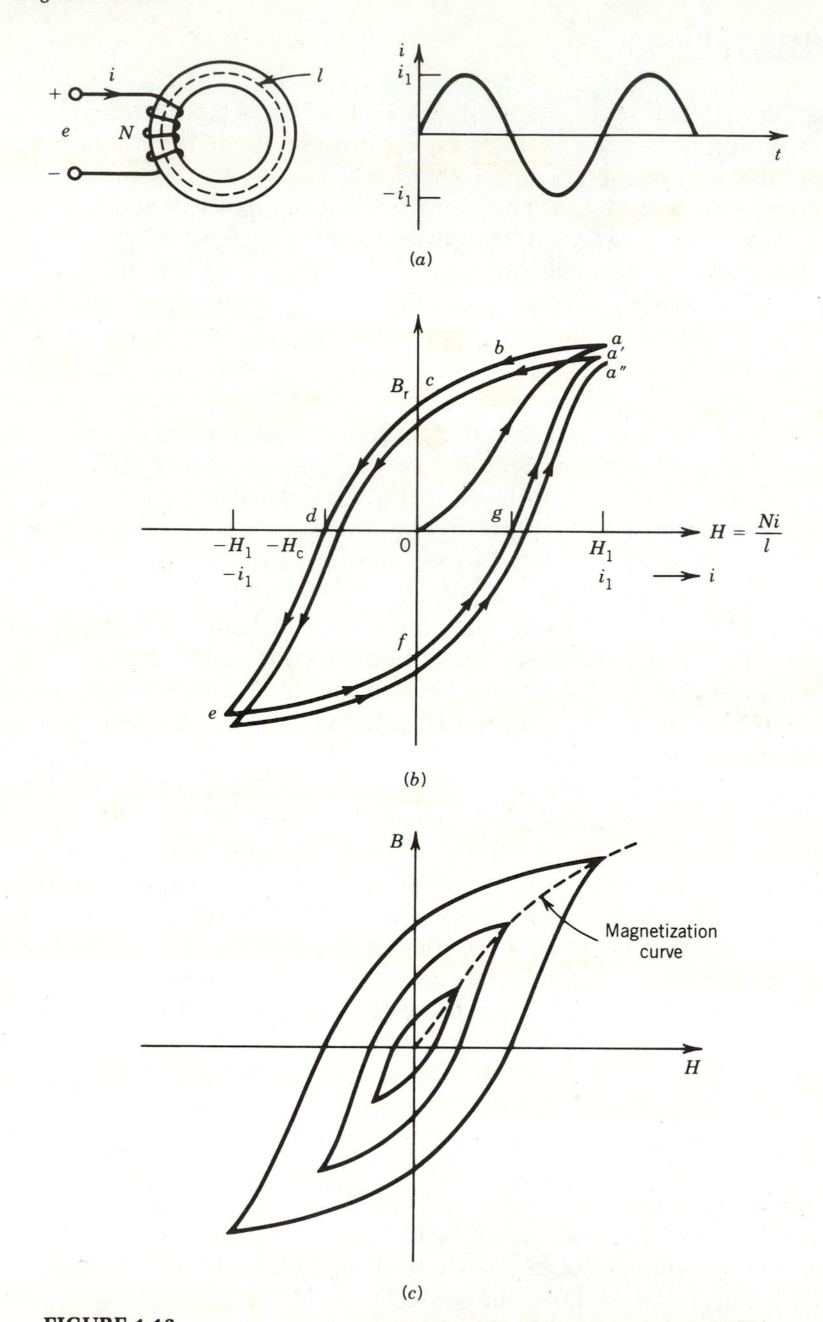

FIGURE 1.12

Magnetization and hysteresis loop. (*a*) Core–coil assembly and exciting circuit. (*b*) Hysteresis. (*c*) Hysteresis loops.

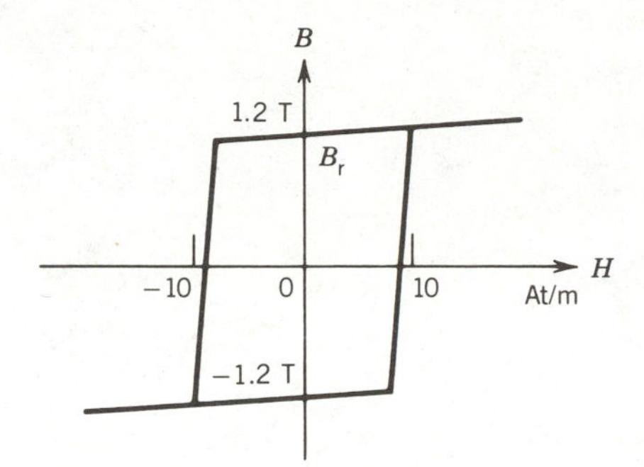

FIGURE 1.13
B–H loop for a deltamax core (50% Fe and 50% Ni).

exceeds the residual flux density ($B > B_r$), the magnetic intensity (hence the current) increases sharply. This property can be exploited to make a coil wound on a deltamax core behave as a switch (very low current when the core is unsaturated and very high current when the core is saturated).

1.2.1 HYSTERESIS LOSS

The hysteresis loops in Fig. 1.12*c* are obtained by slowly varying the current i of the coil (Fig. 1.12*a*) over a cycle. When i is varied through a cycle, during some interval of time energy flows from the source to the coil–core assembly and during some other interval of time energy returns to the source. However, the energy flowing in is greater than the energy returned. Therefore, during a cycle of variation of i (hence H), there is a net energy flow from the source to the coil–core assembly. This energy loss goes to heat the core. The loss of power in the core due to the hysteresis effect is called *hysteresis loss*. It will be shown that the size of the hysteresis loop is proportional to the hysteresis loss.

Assume that the coil in Fig. 1.12*a* has no resistance and the flux in the core is Φ. The voltage e across the coil, according to Faraday's law, is

$$e = N\frac{d\Phi}{dt} \tag{1.26}$$

The energy transfer during an interval of time t_1 to t_2 is

$$W = \int_{t_1}^{t_2} (\text{power})\, dt$$

$$= \int_{t_1}^{t_2} ei\, dt \tag{1.27}$$

From Eqs. 1.26 and 1.27

$$W = \int N \frac{d\Phi}{dt} \cdot i\, dt$$

$$= \int_{\Phi_1}^{\Phi_2} Ni\, d\Phi \qquad (1.28)$$

Now

$$\Phi = BA \quad \text{and} \quad i = \frac{Hl}{N}$$

Thus,

$$W = \int_{B_1}^{B_2} N \cdot \frac{Hl}{N} A\, dB$$

$$= lA \int_{B_1}^{B_2} H\, dB$$

$$= (V_{\text{core}}) \int_{B_1}^{B_2} H\, dB \qquad (1.29)$$

where $V_{\text{core}} = Al$, represents the volume of the core. The integral term in Eq. 1.29 represents the area shown hatched in Fig. 1.14. The energy transfer over one cycle of variation is

$$W|_{\text{cycle}} = V_{\text{core}} \oint H\, dB \qquad (1.30)$$

$$= V_{\text{core}} \times \text{area of the } B\text{–}H \text{ loop}$$

$$= V_{\text{core}} \times W_{\text{h}}$$

Where $W_{\text{h}} = \oint H\, dB$ is the energy density in the core (= area of the B–H loop). The power loss in the core due to the hysteresis effect is

$$P_{\text{h}} = V_{\text{core}} W_{\text{h}} f \qquad (1.31)$$

where f is the frequency of variation of the current i.

It is difficult to evaluate the area of the hysteresis loop because the B–H characteristic is nonlinear and multivalued and no simple mathematical expression can describe the loop. Charles Steinmetz of the General Electric Company performed a large number of experiments and found that for magnetic materials used in electric machines an approximate relation is

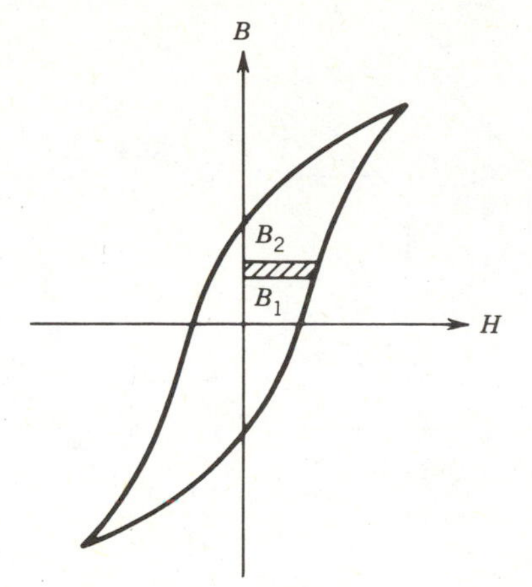

FIGURE 1.14
Hysteresis loss.

$$\text{Area of } B\text{–}H \text{ loop} = KB^{n}_{\max} \tag{1.32}$$

where $B_{\max}$ is the maximum flux density, n varies in the range 1.5 to 2.5, and K is a constant. Both n and K can be empirically determined. From Eqs. 1.31 and 1.32, the hysteresis loss is

$$P_{\mathrm{h}} = K_{\mathrm{h}}B^{n}_{\max}f \tag{1.33}$$

where K_{h} is a constant whose value depends on the ferromagnetic material and the volume of the core.

1.2.2 EDDY CURRENT LOSS

Another power loss occurs in a magnetic core when the flux density changes rapidly in the core. The cross section of a core through which the flux density B is rapidly changing is shown in Fig. 1.15*a*. Consider a path in this cross section. Voltage will be induced in the path because of the time variation of flux enclosed by the path. Consequently, a current i_{e}, known as an *eddy current,* will flow around the path. Because core material has resistance, a power loss i^2R will be caused by the eddy current and will appear as heat in the core.

The eddy current loss can be reduced in two ways.

1. A high-resistivity core material may be used. Addition of a few percent of silicon (say 4%) to iron will increase the resistivity significantly.

2. A laminated core may be used. The thin laminations are insulated from each other. The lamination is made in the plane of the flux. In trans-

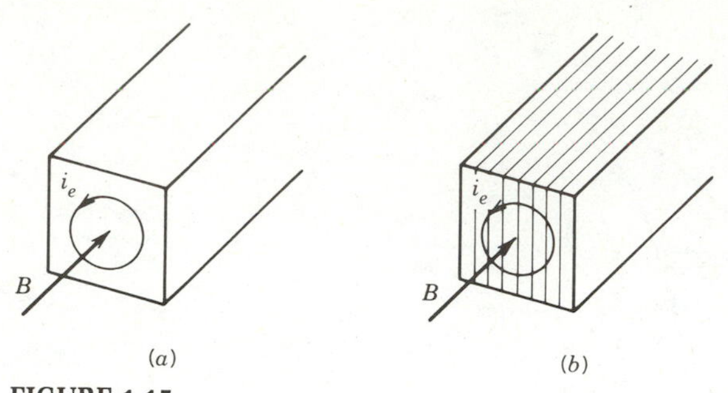

FIGURE 1.15
Eddy current in a magnetic core. (a) Solid iron core. (b) Laminated core.

formers and electric machines, the parts that are made of magnetic core and carry time-varying flux are normally laminated. The laminated core structure is shown in Fig. 1.15*b*.

The eddy current loss in a magnetic core subjected to a time-varying flux is

$$P_e = K_e B_{max}^2 f^2 \tag{1.34}$$

where K_e is a constant whose value depends on the type of material and its lamination thickness. The lamination thickness varies from 0.5 to 5 mm in electrical machines and from 0.01 to 0.5 mm in devices used in electronic circuits operating at higher frequencies.

1.2.3 CORE LOSS

The hysteresis loss and the eddy current loss are lumped together as the *core loss* of the coil–core assembly:

$$P_c = P_h + P_e \tag{1.35}$$

If the current in the coil of Fig. 1.12*a* is varied very slowly, the eddy currents induced in the core will be negligibly small. The *B–H* loop for this slowly varying magnetic intensity is called the *hysteresis loop* or *static loop.* If, however, the current through the coil changes rapidly, the *B–H* loop becomes broader because of the pronounced effect of eddy currents induced in the core. This enlarged loop is called *hystero–eddy current loop* or *dynamic loop.* The static and dynamic loops are shown in Fig. 1.16. The effect of eddy currents on the *B–H* loop can be explained as follows.

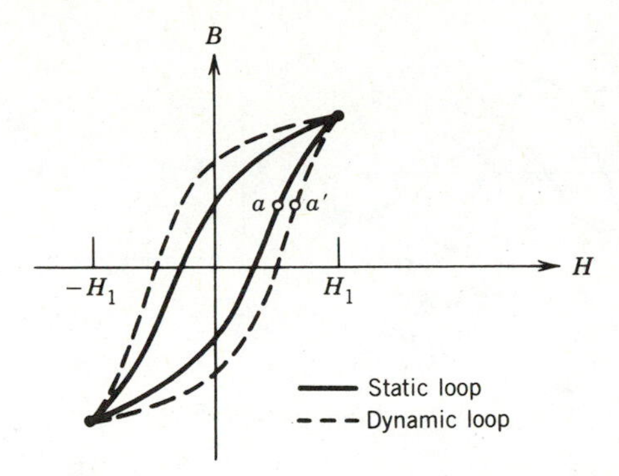

FIGURE 1.16
Static and dynamic loops.

When the coil current changes rapidly, eddy current flows in the core. The eddy current produces an mmf, which tends to change the flux. To maintain a given value of flux, the coil current must be increased by the amount necessary to overcome the effect of eddy current mmf. Therefore, a point a on the static loop will be replaced by a point a' on the dynamic loop that results from a rapidly changing coil current. This makes the dynamic loop broader than the static loop.

The core loss can be computed from the hysteresis loss and eddy current loss according to Eqs. 1.33, 1.34, and 1.35. It can also be computed from the area of the dynamic B–H loop:

$$P_c = V_{\text{core}} f \oint_{\substack{\text{dynamic}\\ \text{loop}}} H\,dB \tag{1.36}$$

$$= (\text{volume of core})(\text{frequency})(\text{area of dynamic loop})$$

Using a wattmeter, core loss can easily be measured. However, it is not easy to know how much of the loss is due to hysteresis and how much is due to eddy currents. Fortunately, it is not necessary to know the losses separately. In electrical machines that have a magnetic core and a time-varying flux, core loss occurs and the loss appears as heat in the core. This loss will be taken into account while discussing the behavior of electric machines in subsequent chapters.

1.3 SINUSOIDAL EXCITATION

In ac electric machines as well as many other applications, the voltages and fluxes vary sinusoidally with time. Consider the coil–core assembly of Fig. 1.17a. Assume that the core flux $\Phi(t)$ varies sinusoidally with time. Thus,

$$\Phi(t) = \Phi_{\max} \sin \omega t \tag{1.37}$$

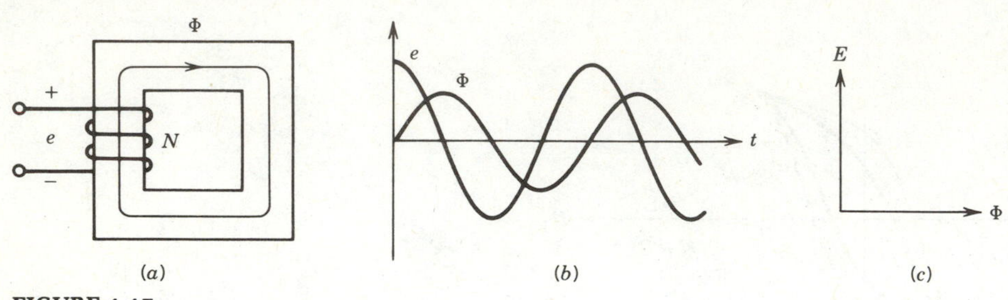

FIGURE 1.17
Sinusoidal excitation of a core. (a) Coil–core assembly. (b) Waveforms. (c) Phasor diagram.

where Φ_{max} is the amplitude of the core flux
$\omega = 2\pi f$ is the angular frequency
f is the frequency

From Faraday's law, the voltage induced in the N-turn coil is

$$e(t) = N\frac{d\Phi}{dt} \tag{1.38}$$

$$= N\Phi_{max}\omega \cos \omega t$$

$$= E_{max} \cos \omega t \tag{1.39}$$

Note that if the flux changes sinusoidally (Eq. 1.37), the induced voltage changes cosinusoidally (Eq. 1.39). The waveforms of e and Φ are shown in Fig. 1.17b, and their phasor representation is shown in Fig. 1.17c. The root-mean-square (rms) value of the induced voltage is

$$E_{rms} = \frac{E_{max}}{\sqrt{2}}$$

$$= \frac{N\omega\Phi_{max}}{\sqrt{2}}$$

$$= 4.44Nf\Phi_{max} \tag{1.40}$$

This is an important equation and will be referred to frequently in the theory of ac machines.

EXAMPLE 1.5

A square-wave voltage of amplitude E = 100 V and frequency 60 Hz is applied on a coil wound on a closed iron core. The coil has 500 turns and the cross-sectional area of the core is 0.001 m^2. Assume that the coil has no resistance.

(a) Find the maximum value of the flux and sketch the waveforms of voltage and flux as a function of time.

(b) Find the maximum value of E if the maximum flux density is not to exceed 1.2 tesla.

Solution

Refer to Fig. 1.17*a*.

(a)

$$e = N\frac{d\Phi}{dt}$$

$$N \cdot d\Phi = edt \tag{1.41}$$

$$N \cdot \Delta\Phi = E \cdot \Delta t \tag{1.42}$$

$$\text{Flux linkage change} = \text{volt–time product}$$

In the steady state, the positive volt–time area during the positive half-cycle will change the flux from negative maximum flux ($-\Phi_{max}$) to positive maximum flux ($+\Phi_{max}$). Hence the total change in flux is $2\Phi_{max}$ during a half-cycle of voltage. Also from Eq. 1.42, if E is constant, the flux will vary linearly with time. From Eq. 1.42

$$500(2\Phi_{max}) = E \times \frac{1}{120}$$

$$\Phi_{max} = \frac{100}{1000 \times 120}\ \text{Wb}$$

$$= 0.833 \times 10^{-3}\ \text{Wb}$$

The waveforms of voltage and flux are shown in Fig. E1.5.

(b)

$$B_{max} = 1.2\ \text{T}$$

$$\Phi_{max} = B_{max} \times A = 1.2 \times 0.001 = 1.2 \times 10^{-3}\ \text{Wb}$$

$$N(2\Phi_{max}) = E \times \frac{1}{120}$$

$$E = 120 \times 500 \times 2 \times 1.2 \times 10^{-3}$$

$$= 144\ \text{V}$$

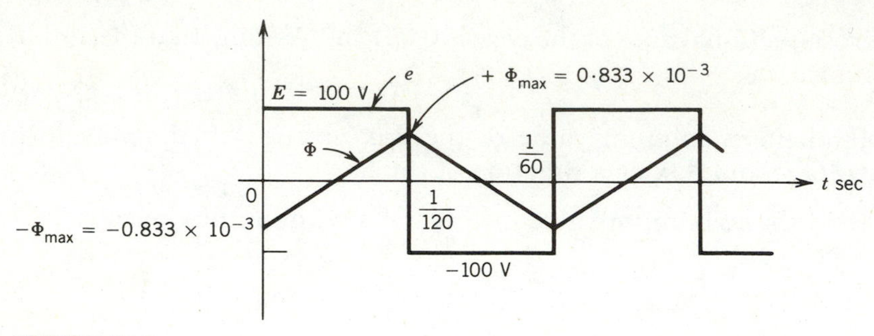

FIGURE E1.5

1.3.1 EXCITING CURRENT

If the coil of Fig. 1.17*a* is connected to a sinusoidal voltage source, a current flows in the coil to establish a sinusoidal flux in the core. This current is called the *exciting current,* i_{Φ}. If the B–H characteristic of the ferromagnetic core is nonlinear, the exciting current will be nonsinusoidal.

No Hysteresis

Let us first consider a B–H characteristic with no hysteresis loop. The B–H curve can be rescaled ($\Phi = BA$, $i = Hl/N$) to obtain the Φ–i curve for the core, as shown in Fig. 1.18*a*. From the sinusoidal flux wave and the Φ–i

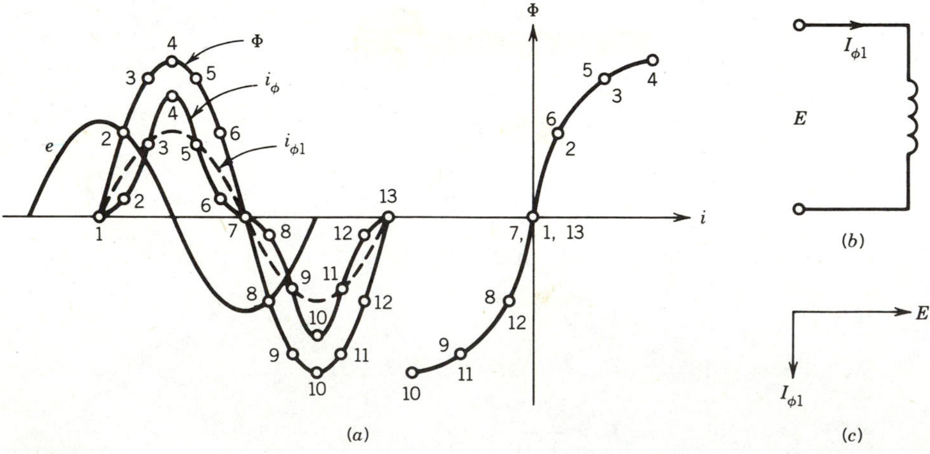

FIGURE 1.18
Exciting current for no hysteresis. (*a*) Φ–i characteristic and exciting current. (*b*) Equivalent circuit. (*c*) Phasor diagram.

curve, the exciting current waveform is obtained, as shown in Fig. 1.18*a*. Note that the exciting current i_ϕ is nonsinusoidal, but it is in phase with the flux wave and is symmetrical with respect to voltage e. The fundamental component $i_{\phi 1}$ of the exciting current lags the voltage e by 90°. Therefore no power loss is involved. This was expected, because the hysteresis loop, which represents power loss, was neglected. The excitation current is therefore a purely lagging current and the exciting winding can be represented by a pure inductance, as shown in Fig. 1.18*b*. The phasor diagram for fundamental current and applied voltage is shown in Fig. 1.18*c*.

With Hysteresis

We shall now consider the hysteresis loop of the core, as shown in Fig. 1.19*a*. The waveform of the exciting current i_ϕ is obtained from the sinusoidal flux waveform and the multivalued Φ–i characteristic of the core. The exciting current is nonsinusoidal as well as nonsymmetrical with respect to the voltage waveform. The exciting current can be split into two components, one (i_c) in phase with voltage e accounting for the core loss and the other (i_m) in phase with Φ and symmetrical with respect to e, accounting for the magnetization of the core. This magnetizing component i_m is the same as the exciting current if the hysteresis loop is neglected. The phasor diagram is shown in Fig. 1.19*b*. The exciting coil can therefore be represented by a resistance R_c, to represent core loss, and a magnetizing inductance L_m, to represent the magnetization of the core, as shown in Fig.

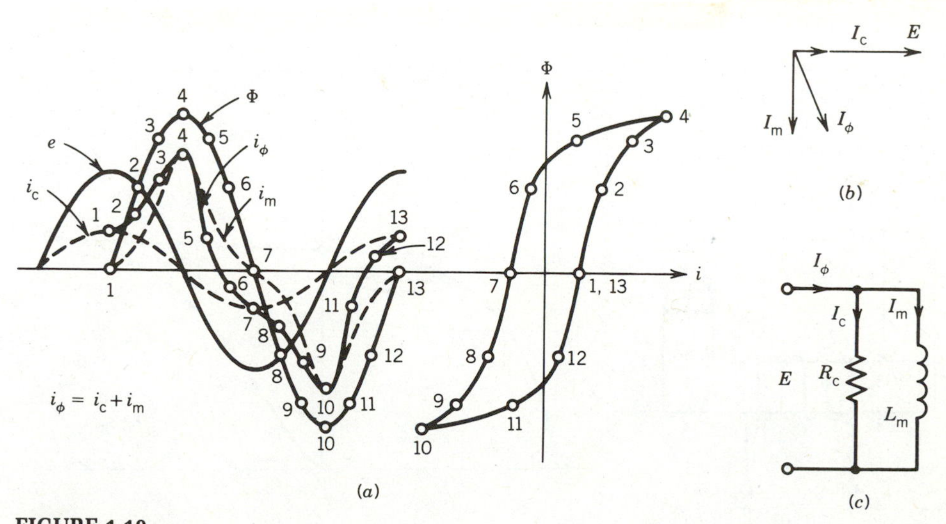

FIGURE 1.19
Exciting current with hysteresis-loop. (*a*) Φ–i loop and exciting current. (*b*) Phasor diagram. (*c*) Equivalent circuit.

1.19*c*. In the phasor diagram only the fundamental component of the magnetizing current is considered.

1.4 PERMANENT MAGNET

A permanent magnet is capable of maintaining a magnetic field without any excitation mmf provided to it. Permanent magnets are normally alloys of iron, nickel, and cobalt. They are characterized by a large B–H loop, high retentivity (high value of B_r), and high coercive force (high value of H_c). These alloys are subjected to heat treatment, resulting in mechanical hardness of the material. Permanent magnets are often referred to as *hard iron* and other magnetic materials as *soft iron*.

1.4.1 MAGNETIZATION OF PERMANENT MAGNETS

Consider the magnetic circuit shown in Fig. 1.20*a*. Assume that the magnet material is initially unmagnetized. A large mmf is applied, and on its removal the flux density will remain at the residual value B_r on the magnetization curve, point *a* in Fig. 1.20*b*. If a reversed magnetic field intensity of magnitude H_1 is now applied to the hard iron, the operating point moves to point *b*. If H_1 is removed and reapplied, the B–H locus follows a minor loop as shown in Fig. 1.20*b*. The minor loop is narrow and for all practical purposes can be represented by the straight line *bc*, known as the *recoil line*. This line is almost parallel to the tangent *xay* to the demagnetizing curve at point *a*. The slope of the recoil line is called the *recoil permeability*

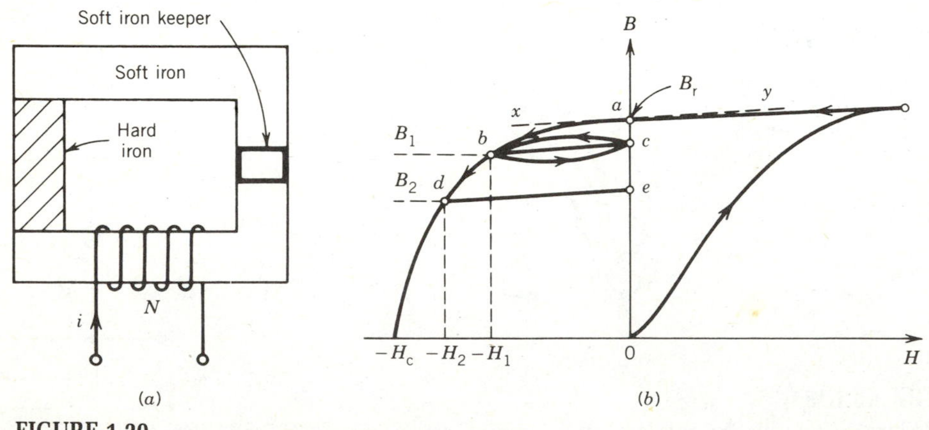

FIGURE 1.20
Permanent magnet system and its B–H locus.

μ_{rec}. For alnico magnets it is in the range of 3–5μ_0, whereas for ferrite magnets it may be as low as 1.2μ_0.

As long as the reversed magnetic field intensity does not exceed H_1, the magnet may be considered reasonably permanent. If a negative magnetic field intensity greater than H_1 is applied, such as H_2, the flux density of the permanent magnet will decrease to the value B_2. If H_2 is removed, the operation will move along a new recoil line *de*.

1.4.2 APPROXIMATE DESIGN OF PERMANENT MAGNETS

Let the permanent magnet in Fig. 1.20*a* be magnetized to the residual flux density denoted by point *a* in Fig. 1.20*b*. If the small soft iron keeper is removed, the air gap will become the active region for most applications as shown in Fig. 1.21*a*.

In order to determine the resultant flux density in the magnet and in the air gap, let us make the following assumptions.

1. There is no leakage or fringing flux.
2. No mmf is required for the soft iron.

From Ampère's circuit law,

$$H_m l_m + H_g l_g = 0 \tag{1.43}$$

$$H_m = -\frac{l_g}{l_m} H_g \tag{1.44}$$

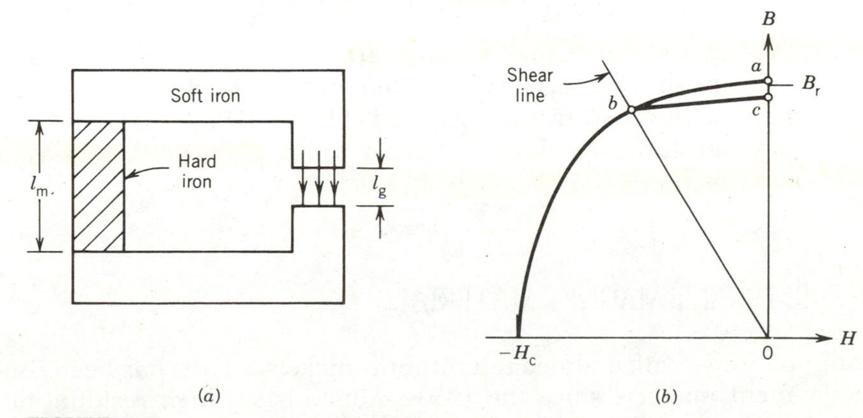

FIGURE 1.21
Permanent magnet with keeper removed and its B–H locus.

For continuity of flux,

$$\Phi = B_m A_m = B_g A_g \tag{1.45}$$

Also

$$B_g = \mu_0 H_g \tag{1.46}$$

From Eqs. 1.43, 1.45, and 1.46,

$$B_m = -\mu_0 \frac{A_g}{A_m} \frac{l_m}{l_g} H_m \tag{1.47}$$

Equation 1.47 represents a straight line through the origin, called the *shear line* (Fig. 1.21*b*). The intersection of the shear line with the demagnetization curve at point *b* determines the operating values of *B* and *H* of the hard iron material with the keeper removed. If the keeper is now reinserted, the operating point moves up the recoil line *bc*. This analysis indicates that the operating point of a permanent magnet with an air gap is determined by the demagnetizing portion of the *B–H* loop and the dimensions of the magnet and air gap.

From Eqs. 1.43, 1.45, and 1.46, the volume of the permanent magnet material is

$$\begin{aligned} V_m &= A_m l_m \\ &= \frac{B_g A_g}{B_m} \times \frac{H_g l_g}{H_m} \\ &= \frac{B_g^2 V_g}{\mu_0 B_m H_m} \end{aligned} \tag{1.48}$$

where $V_g = A_g l_g$ is the volume of the air gap.

Thus, to establish a flux density B_g in the air gap of volume V_g, a minimum volume of the hard iron is required if the final operating point is located such that the $B_m H_m$ product is a maximum. This quantity $B_m H_m$ is known as the *energy product* of the hard iron.

1.4.3 PERMANENT MAGNET MATERIALS

A family of alloys called alnico (aluminum–nickel–cobalt) has been used for permanent magnets since the 1930s. Alnico has a high residual flux density, as shown in Fig. 1.22.

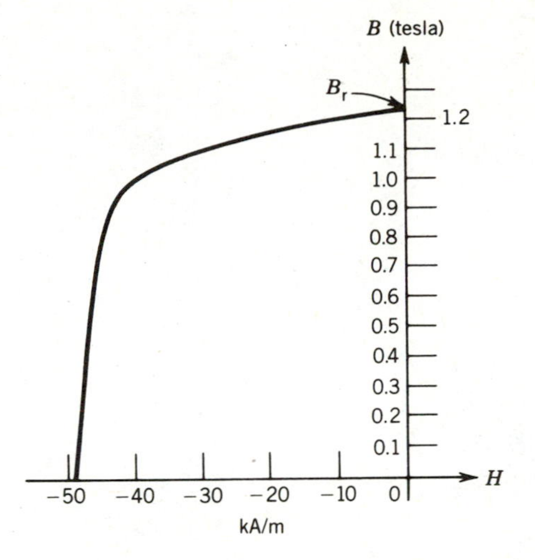

FIGURE 1.22
Demagnetization curve for alnico 5.

Ferrite permanent magnet materials have been used since the 1950s. These have lower residual flux density but very high coercive force. Figure 1.23 shows the demagnetization curve for ferrite D, which is a strontium ferrite.

Since 1960 a new class of permanent magnets known as *rare-earth* permanent magnets has been developed. The rare-earth permanent magnet materials combine the relatively high residual flux density of alnico-type materials with greater coercivity than the ferrites. These materials are compounds of iron, nickel, and cobalt with one or more of the rare-earth elements. A commonly used combination is samarium–cobalt. The demagnetization curve for this material is shown in Fig. 1.24.

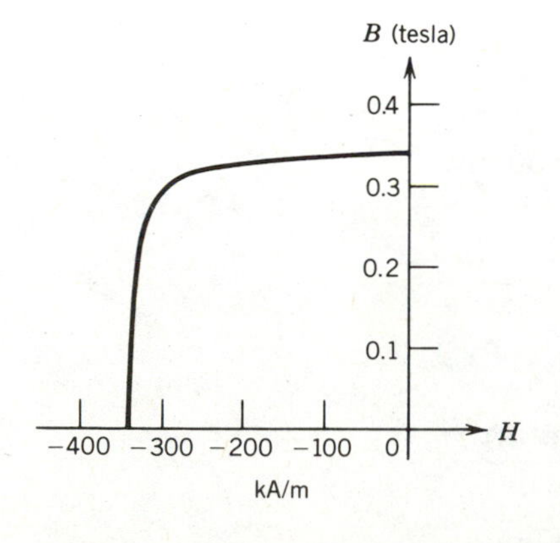

FIGURE 1.23
Demagnetization curve for ferrite D magnet.

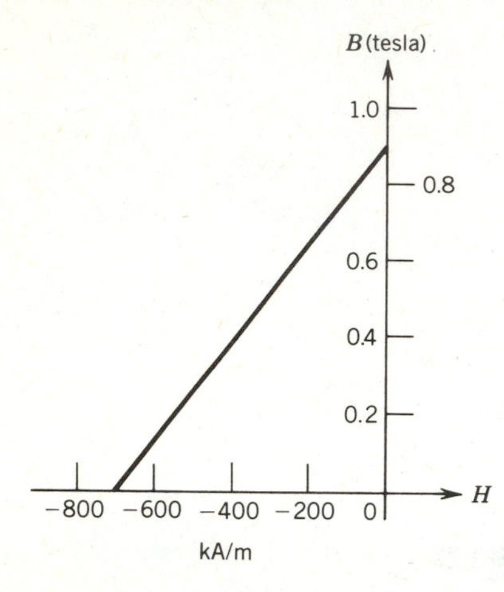

FIGURE 1.24
Demagnetization curve for samarium–cobalt magnet.

EXAMPLE 1.6

The permanent magnet in Fig. 1.21 is made of alnico 5, whose demagnetization curve is given in Fig. 1.22. A flux density of 0.8 T is to be established in the air gap when the keeper is removed. The air gap has the dimensions $A_g = 2.5\ \text{cm}^2$ and $l_g = 0.4$ cm. The operating point on the demagnetization curve corresponds to the point at which the product $H_m B_m$ is maximum, and this operating point is $B_m = 0.95$ T, $H_m = -42$ kA/m.

Determine the dimensions (l_m and A_m) of the permanent magnet.

Solution

From Eqs. 1.43 and 1.46,

$$l_m = \frac{l_g}{H_m} H_g = \frac{l_g B_g}{H_m \mu_0}$$

$$= \frac{0.4 \times 10^{-2} \times 0.8}{42 \times 10^3 \times 4\pi \times 10^{-7}}$$

$$= 0.0606 \text{ m} = 6.06 \text{ cm}$$

From Eq. 1.45,

$$A_m = \frac{B_g A_g}{B_m}$$

$$= \frac{0.8 \times 2.5 \times 10^{-4}}{0.95}$$

$$= 2.105 \text{ cm}^2$$

PROBLEMS

1.1 The long solenoid coil shown in Fig. P1.1 has 250 turns. As its length is much greater than its diameter, the field inside the coil may be considered uniform. Neglect the field outside.

(a) Determine the field intensity (H) and flux density (B) inside the solenoid ($i = 100\ A$).

(b) Determine the inductance of the solenoid coil.

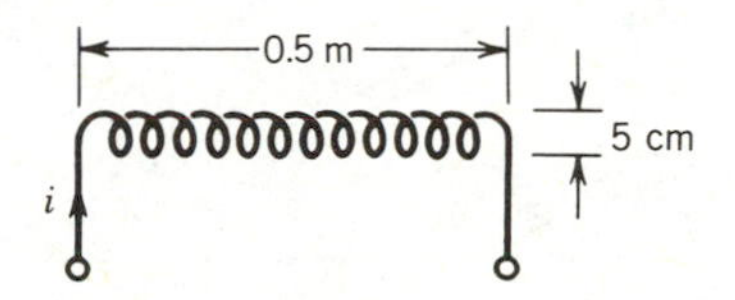

FIGURE P1.1

1.2 The magnetic circuit of Fig. P1.2 provides flux in the two air gaps. The coils ($N_1 = 700$, $N_2 = 200$) are connected in series and carry 0.5 ampere. Neglect leakage flux, reluctance of the iron (i.e., infinite permeability), and fringing at the air gaps. Determine the flux and flux density in the air gaps.

1.3 A two-pole generator, as shown in Fig. P1.3, has a magnetic circuit with the following dimensions:

Each pole (cast steel):

magnetic length = 10 cm

cross section = 400 cm^2

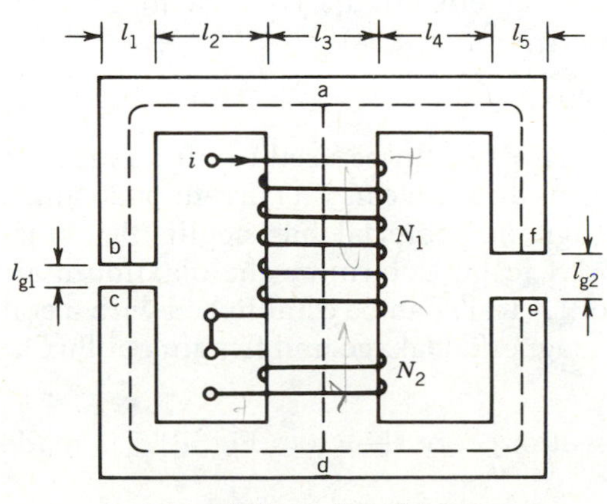

FIGURE P1.2

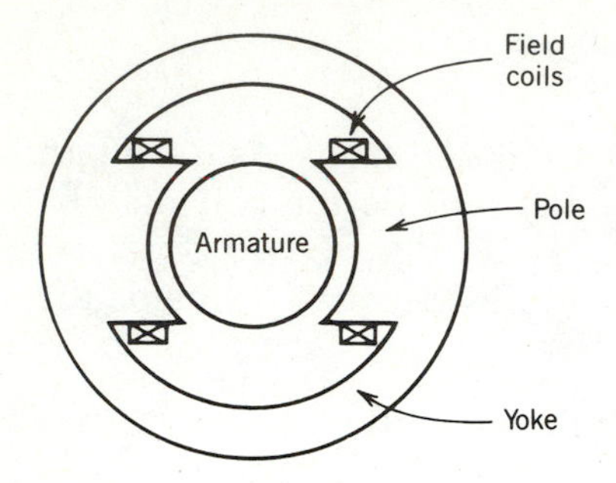

FIGURE P1.3

Each air gap:

length = 0.1 cm

cross section = 400 cm^2

Armature (Si-steel):

average length = 20 cm

average cross section = 400 cm^2

Yoke (cast steel):

mean circumference = 160 cm

average cross section = 200 cm^2

Half the exciting ampere-turns are placed on each of the two poles.

(a) Draw the magnetic equivalent circuit.

(b) How many ampere-turns per pole are required to produce a flux density of 1.1 tesla in the magnetic circuit. (Use the magnetization curves for the respective materials.)

(c) Calculate the armature flux.

1.4 The electromagnet shown in Fig. P1.4 can be used to lift a length of steel strip. The coil has 500 turns and can carry a current of 20 amps without overheating. The magnetic material has negligible reluctance at flux densities up to 1.4 tesla. Determine the maximum air gap for which a flux density of 1.4 tesla can be established with a coil current of 20 amps. Neglect magnetic leakage and fringing of flux at the air gap.

1.5 The toroidal (circular cross section) core shown in Fig. P1.5 is made from cast steel.

(a) Calculate the coil current required to produce a core flux density of 1.2 tesla at the mean radius of the toroid.

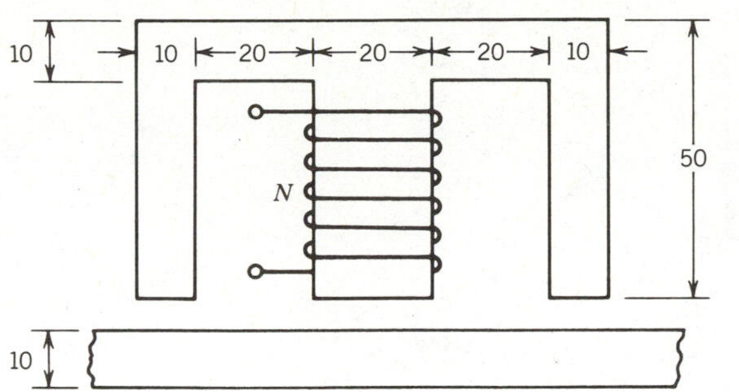

FIGURE P1.4

(b) What is the core flux, in webers?

(c) If a 2-mm-wide air gap is made in the toroid (across A–A'), determine the new coil current required to maintain a core flux density of 1.2 tesla.

1.6 An inductor is made of two coils, A and B, having 350 and 150 turns, respectively. The coils are wound on a cast steel core and in directions as shown in Fig. P1.6. The two coils are connected *in series* to a dc voltage.

(a) Determine the two possible values of current required in the coils to establish a flux density of 0.5 T in the air gap.

(b) Determine the self-inductances L_A and L_B of the two coils. Neglect magnetic leakage and fringing.

(c) If coil B is now disconnected and the current in coil A is adjusted to 2.0 A, determine the mean flux density in the air gap.

1.7 The magnetic circuit for a saturable reactor is shown in Fig. P1.7. The B–H curve for the core material can be approximated as two straight lines as in Fig. P1.7.

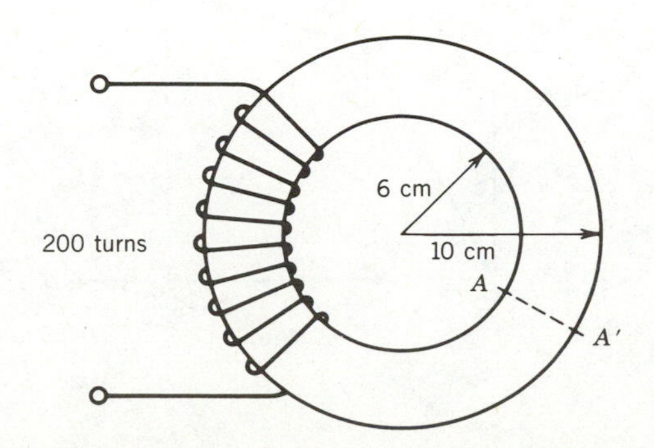

FIGURE P1.5

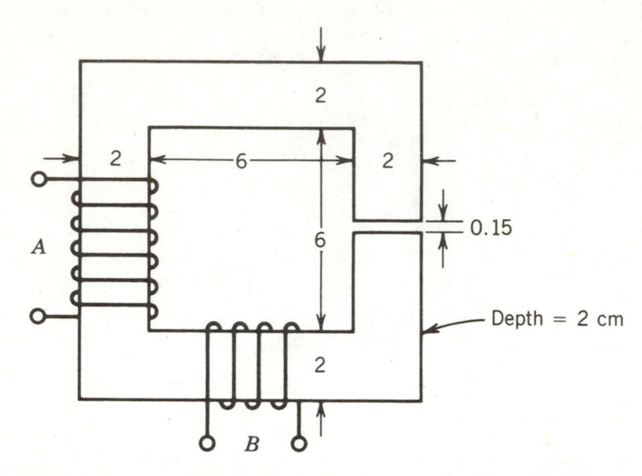

All dimensions in centimeters. $N_A = 350$, $N_B = 150$

FIGURE P1.6

(a) If $I_1 = 2.0$ A, calculate the value of I_2 required to produce a flux density of 0.6 T in the vertical limbs.

(b) If $I_1 = 0.5$ A and $I_2 = 1.96$ A, calculate the total flux in the core.

Neglect magnetic leakage.

(*Hint: Trial-and-error method.*)

1.8 A toroidal core has a rectangular cross section as shown in Fig. P1.8*a*. It is wound with a coil having 100 turns. The *B–H* characteristic of the core may be represented by the linearized magnetization curve of Fig. P1.8*b*.

(a) Determine the inductance of the coil if the flux density in any part of the core is below 1.0 Wb/m².

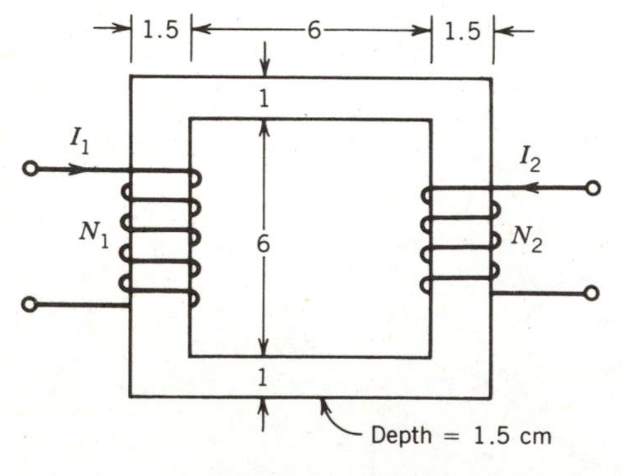

All dimensions in centimeters.
$N_1 = 200$, $N_2 = 100$

FIGURE P1.7

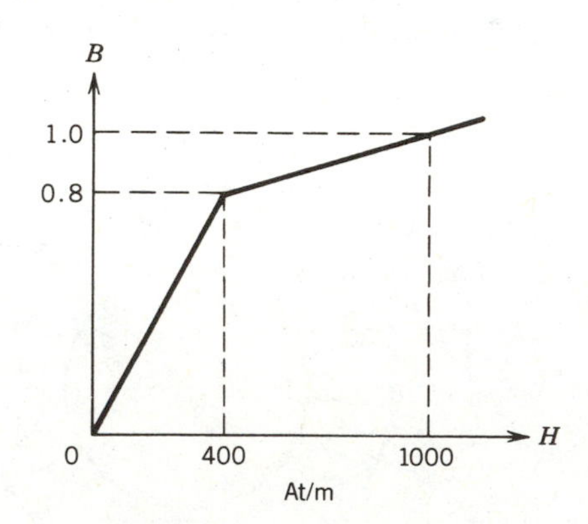

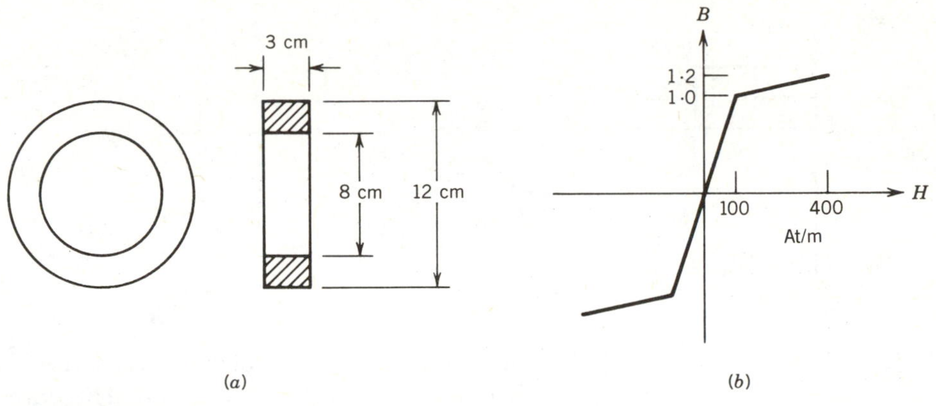

FIGURE P1.8

(b) Determine the maximum value of the current for the condition of part (a).

(c) Determine the minimum value of the current for which the complete core has a flux density of 1.0 Wb/m^2 or greater.

1.9 A coil wound on a magnetic core is excited by the following voltage sources.

(a) 100 V, 50 Hz.

(b) 110 V, 60 Hz.

Compare the hysteresis losses and eddy current losses with these two different sources. For hysteresis loss consider $n = 2$.

1.10 A toroidal core of mean length 15 cm and cross-sectional area 10 cm^2 has a uniformly distributed winding of 300 turns. The B–H characteristic of the core can be assumed to be of rectangular form, as shown in Fig. P1.10. The coil is connected to a 100 V, 400 Hz supply. Determine the hysteresis loss in the core.

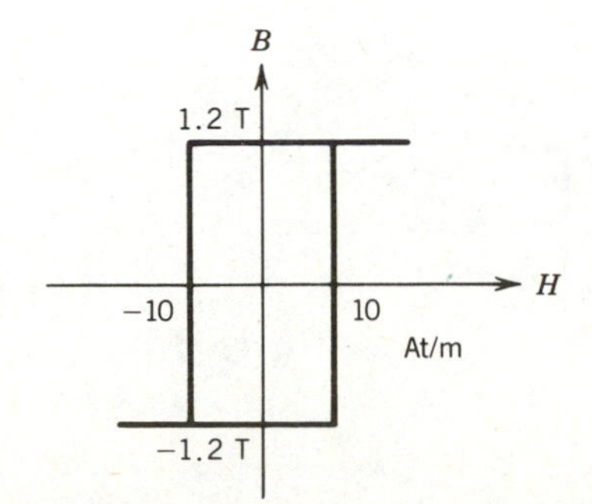

FIGURE P1.10

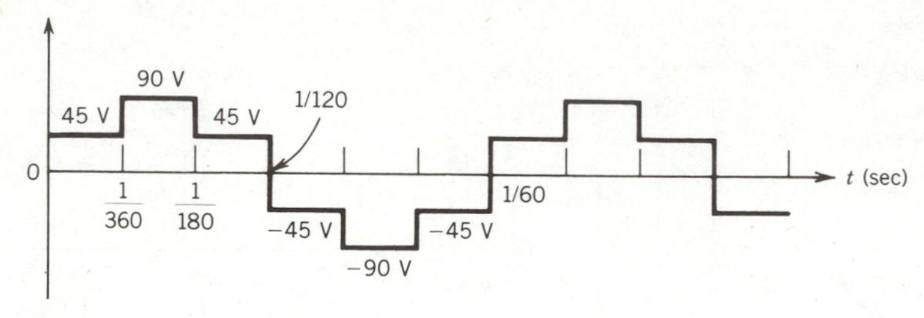

FIGURE P1.12

1.11 The core in Fig. 1.17 has the following dimensions: cross-sectional area $A_c = 5\ \text{cm}^2$, mean magnetic path length $l_c = 25$ cm. The core material is silicon sheet steel. If the coil has 500 turns and negligible resistance, determine the rms value of the 60 Hz voltage applied to the coil that will produce a peak flux density of 1.2 T.

1.12 A six-step voltage of frequency 60 Hz, as shown in Fig. P1.12, is applied on a coil wound on a magnetic core. The coil has 500 turns. Find the maximum value of the flux and sketch the waveforms of voltage and flux as a function of time.

1.13 In the circuit of Fig. P1.13*a* a resistanceless toroidal winding of 1000 turns is wound on a ferromagnetic toroid of cross-sectional area 2

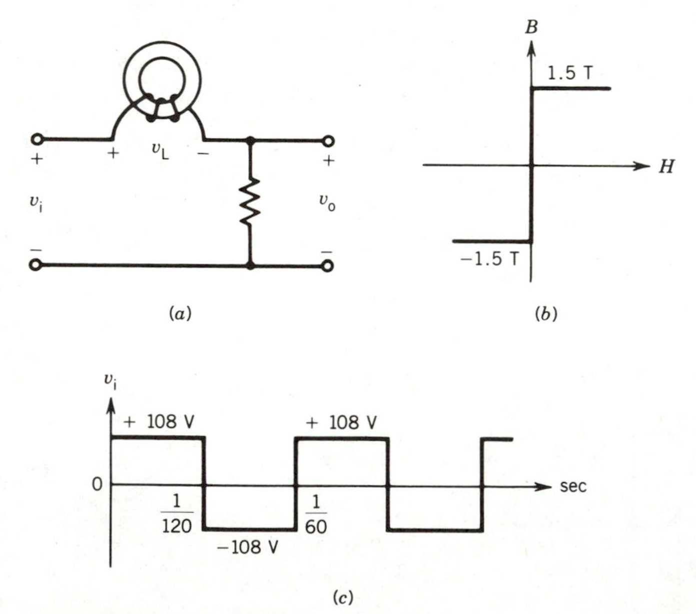

FIGURE P1.13

cm^2. The core is characterized by the ideal B–H relation shown in Fig. P1.13*b*. This circuit is excited by a 60 Hz square wave of input voltage (v_i) of amplitude 108 volts, as shown in Fig. P1.13*c*. Determine the switching instant and sketch the waveforms of the voltages v_L and v_o.

1.14 Suppose that the soft iron keeper in the permanent magnet of Example 1.6 is reinserted. Determine the flux density in the magnet if the recoil permeability (μ_{rec}) of the magnetic material is $4\mu_0$.

CHAPTER two

TRANSFORMERS

A transformer is a static machine. Although it is not an energy conversion device, it is indispensable in many energy conversion systems. It is a simple device, having two or more electric circuits coupled by a common magnetic circuit. Analysis of transformers involves many principles that are basic to the understanding of electric machines. Transformers are so widely used as electrical apparatus that they are treated along with other electric machines in all books on electric machines.

A transformer essentially consists of two or more windings coupled by a mutual magnetic field. Ferromagnetic cores are used to provide tight magnetic coupling and high flux densities. Such transformers are known as *iron core* transformers. They are invariably used in high-power applications. *Air core* transformers have poor magnetic coupling and are sometimes used in low-power electronic circuits. In this chapter we primarily discuss iron core transformers.

Two types of core constructions are normally used, as shown in Fig. 2.1. In the *core type* (Fig. 2.1*a*), the windings are wound around two legs of a magnetic core of rectangular shape. In the *shell type* (Fig. 2.1*b*), the windings are wound around the center leg of a three-legged magnetic core. To reduce core losses, the magnetic core is formed of a stack of thin laminations. Silicon-steel laminations of 0.014 inch thickness are commonly used for transformers operating at frequencies below a few hundred cycles. L-shaped laminations are used for core-type construction and E-shaped laminations are used for shell-type construction. To avoid a continuous air gap (which would require a large exciting current), laminations are stacked alternately as shown in Figs. 2.1*c* and 2.1*d*.

For small transformers used in communication circuits at high frequencies (kilocycles to megacycles) and low power levels, compressed powdered ferromagnetic alloys, known as permalloy, are used.

A schematic representation of a two-winding transformer is shown in Fig. 2.2. The two vertical bars are used to signify tight magnetic coupling between the windings. One winding is connected to an ac supply and is referred to as the *primary winding*. The other winding is connected

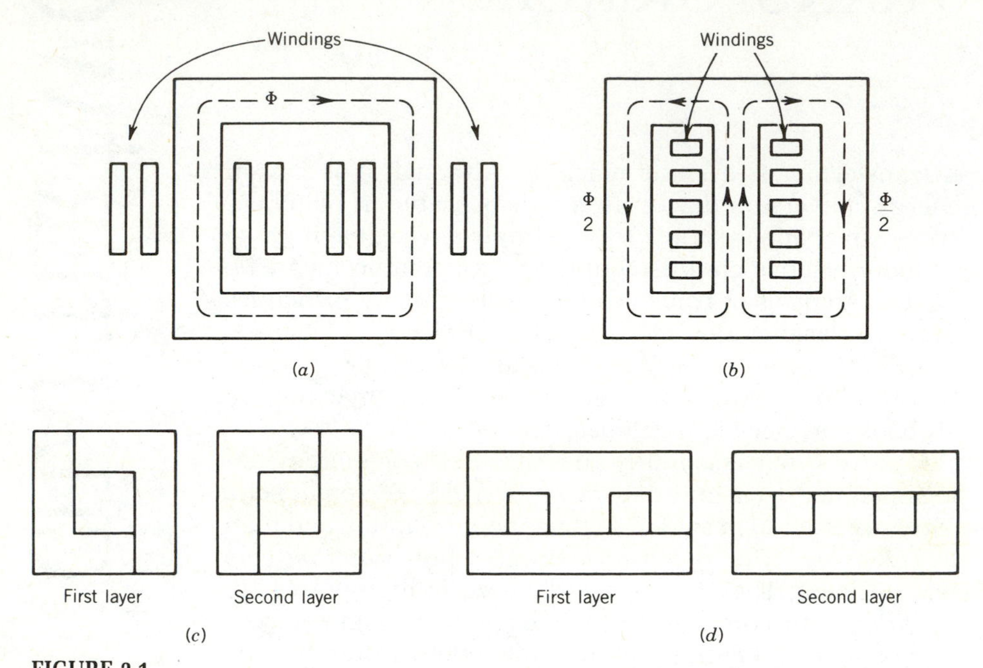

FIGURE 2.1
Transformer core construction. (*a*) Core-type. (*b*) Shell-type. (*c*) L-shaped lamination. (*d*) E-shaped lamination.

to an electrical load and is referred to as the *secondary winding*. The winding with the higher number of turns will have a high voltage and is called the high-voltage (HV) or high-tension (HT) winding. The winding with the lower number of turns is called the low-voltage (LV) or low-tension (LT) winding. To achieve tighter magnetic coupling between the windings, they may be formed of coils placed one on top of another (Fig. 2.1*a*) or side by side (Fig. 2.1*b*) in a "pancake" coil formation where primary and secondary coils are interleaved. Where the coils are placed one on top of another, the low-voltage winding is placed nearer the core and the high-voltage winding on top.

Transformers have widespread use. Their primary function is to change voltage level. Electrical power is generated in a power house at about 30,000 volts. However, in domestic houses, electric power is used at 110 or

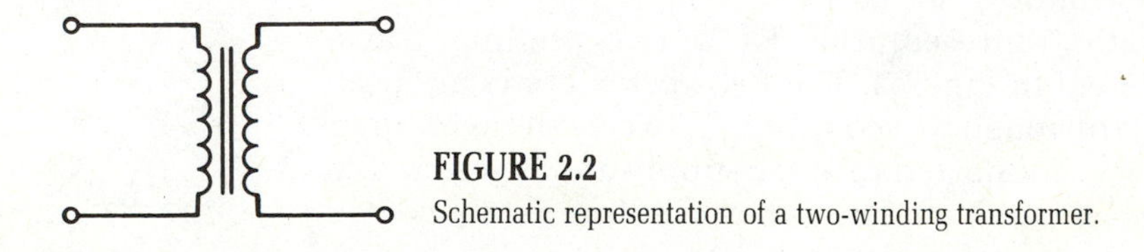

FIGURE 2.2
Schematic representation of a two-winding transformer.

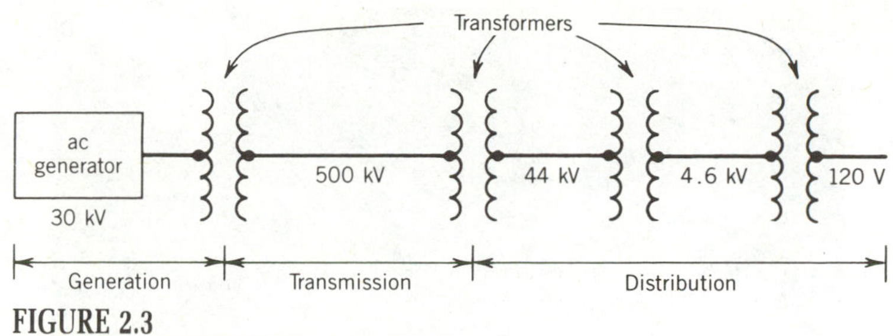

FIGURE 2.3
Power transmission using transformers.

220 volts. Electric power is transmitted from a power plant to a load center at 200,000 to 500,000 volts. Transformers are used to step up and step down voltage at various stages of power transmission, as shown in Fig. 2.3. A large power transformer used to step up generator voltage from 24 to 345 kV is shown in Fig. 2.4. A distribution transformer used in a public utility system to step down voltage from 4.6 kV to 120 V is shown in Fig. 2.5.

FIGURE 2.4
Power transformer, 24 to 345 kV. (Courtesy of Westinghouse Electric Corporation.)

FIGURE 2.5
Distribution transformer (Courtesy of Westinghouse Electric Corporation.)

Transformers are widely used in low-power electronic or control circuits to insulate one circuit from another circuit or to match the impedance of a source with its load for maximum power transfer. Transformers are also used to measure voltages and currents; these are known as *instrument transformers*.

2.1 IDEAL TRANSFORMER

Consider a transformer with two windings, a primary winding of N_1 turns and a secondary winding of N_2 turns, as shown schematically in Fig. 2.6. In a schematic diagram it is a common practice to show the two windings in

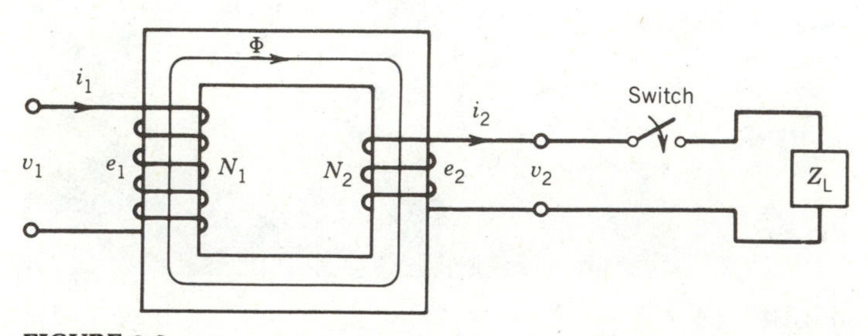

FIGURE 2.6
Ideal transformer.

the two legs of the core, although in an actual transformer the windings are interleaved. Let us consider an ideal transformer that has the following properties:

1. The winding resistances are negligible.
2. All fluxes are confined to the core and link both windings; that is, no leakage fluxes are present. Core losses are assumed to be negligible.
3. Permeability of the core is infinite (i.e., $\mu \rightarrow \infty$). Therefore, the exciting current required to establish flux in the core is negligible; that is, the net mmf required to establish a flux in the core is zero.

When the primary winding is connected to a time-varying voltage v_1, a time-varying flux Φ is established in the core. A voltage e_1 will be induced in the winding and will equal the applied voltage if resistance of the winding is neglected:

$$v_1 = e_1 = N_1 \frac{d\Phi}{dt} \tag{2.1}$$

The core flux also links the secondary winding and induces a voltage e_2, which is the same as the terminal voltage v_2:

$$v_2 = e_2 = N_2 \frac{d\Phi}{dt} \tag{2.2}$$

From Eqs. 2.1 and 2.2,

$$\frac{v_1}{v_2} = \frac{N_1}{N_2} = a \tag{2.3}$$

where a is the turns ratio.

Equation 2.3 indicates that the voltages in the windings of an ideal transformer are directly proportional to the turns of the windings.

Let us now connect a load (by closing the switch in Fig. 2.6) to the secondary winding. A current i_2 will flow in the secondary winding, and the secondary winding will provide an mmf $N_2 i_2$ for the core. This will immediately make a primary winding current i_1 flow so that a counter-mmf $N_1 i_1$ can oppose $N_2 i_2$. Otherwise $N_2 i_2$ would make the core flux change drastically and the balance between v_1 and e_1 would be disturbed. Note in Fig. 2.6 that the current directions are shown such that their mmf's oppose each other. Because the net mmf required to establish a flux in the ideal core is zero,

$$N_1 i_1 - N_2 i_2 = \text{net mmf} = 0 \tag{2.4}$$

$$N_1 i_1 = N_2 i_2 \tag{2.5}$$

$$\frac{i_1}{i_2} = \frac{N_2}{N_1} = \frac{1}{a} \tag{2.6}$$

The currents in the windings are inversely proportional to the turns of the windings. Also note that if more current is drawn by the load, more current will flow from the supply. It is this mmf-balancing requirement (Eq. 2.5) that makes the primary know of the presence of current in the secondary.

From Eqs. 2.3 and 2.6

$$v_1 i_1 = v_2 i_2 \tag{2.7}$$

that is, the instantaneous power input to the transformer equals the instantaneous power output from the transformer. This is expected, because all power losses are neglected in an ideal transformer. Note that although there is no physical connection between load and supply, as soon as power is consumed by the load, the same power is drawn from the supply. The transformer, therefore, provides a physical isolation between load and supply while maintaining electrical continuity.

If the supply voltage v_1 is sinusoidal, then Eqs. 2.3, 2.6, and 2.7 can be written in terms of rms values:

$$\frac{V_1}{V_2} = \frac{N_1}{N_2} = a \tag{2.8}$$

$$\frac{I_1}{I_2} = \frac{N_2}{N_1} = \frac{1}{a} \tag{2.9}$$

$$\underset{\substack{\uparrow \\ \text{input} \\ \text{volt-amperes}}}{V_1 I_1} = \underset{\substack{\uparrow \\ \text{output} \\ \text{volt-amperes}}}{V_2 I_2} \tag{2.10}$$

2.1.1 IMPEDANCE TRANSFER

Consider the case of a sinusoidal applied voltage and a secondary impedance Z_2, as shown in Fig. 2.7*a*.

$$Z_2 = \frac{V_2}{I_2}$$

The input impedance is

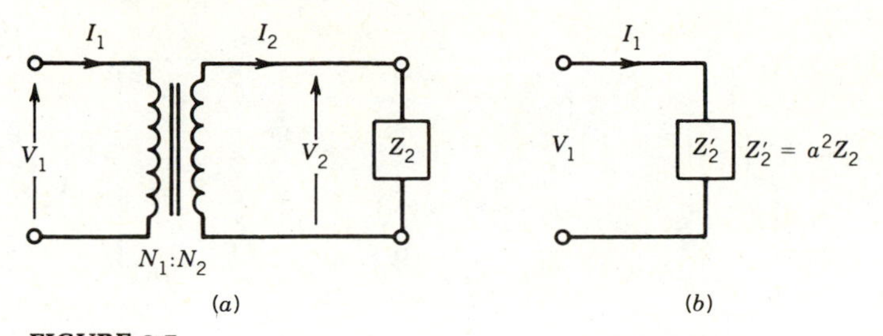

FIGURE 2.7
Impedance transfer across an ideal transformer.

$$Z_1 = \frac{V_1}{I_1} = \frac{aV_2}{I_2/a} = a^2\frac{V_2}{I_2}$$
$$= a^2Z_2 \tag{2.11}$$

so

$$Z_1 = a^2Z_2 = Z_2' \tag{2.12}$$

An impedance Z_2 connected in the secondary will appear as an impedance Z_2' looking from the primary. The circuit in Fig. 2.7*a* is therefore equivalent to the circuit in Fig. 2.7*b*. Impedance can be transferred from secondary to primary if its value is multiplied by the square of the turns ratio. An impedance from the primary side can also be transferred to the secondary side, and in that case its value has to be divided by the square of the turns ratio:

$$Z_1' = \frac{1}{a^2}Z_1 \tag{2.13}$$

This impedance transfer is very useful because it eliminates a coupled circuit in an electrical circuit and thereby simplifies the circuit.

EXAMPLE 2.1

A speaker of 9 Ω resistive impedance is connected to a supply of 10 V with internal resistive impedance of 1 Ω, as shown in Fig. E2.1*a*.

(a) Determine the power taken by the speaker.

(b) To maximize the power transfer to the speaker, a transformer of 1 : 3 turns ratio is used between source and speaker as shown in Fig. E2.1*b*. Determine the power taken by the speaker.

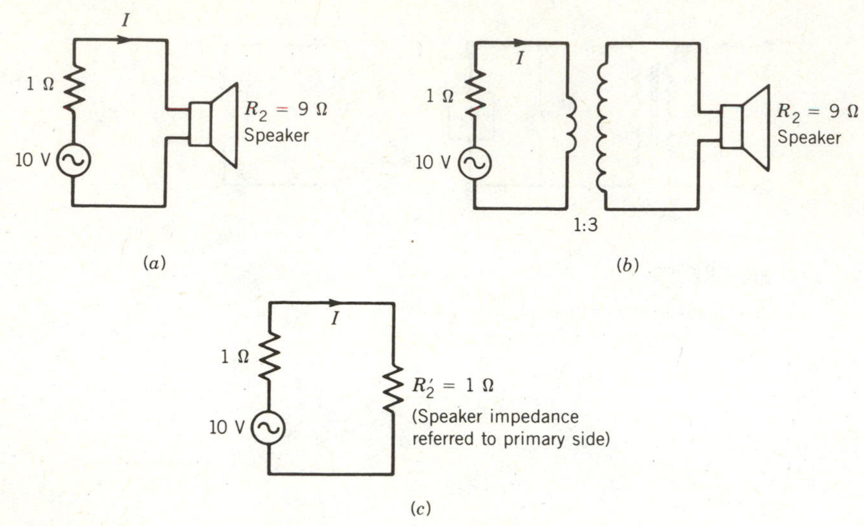

FIGURE E2.1

Solution

(a) From Fig. E2.1*a*,

$$I = \frac{10}{1 + 9} = 1 \text{ A}$$

$$P = 1^2 \times 9 = 9 \text{ W}$$

(b) If the resistance of the speaker is referred to the primary side, its resistance is

$$R_2' = a^2 R_2 = \left(\frac{1}{3}\right)^2 \times 9 = 1 \ \Omega$$

The equivalent circuit is shown in Fig. E2.1*c*.

$$\text{I} = \frac{10}{1 + 1} = 5 \text{ A}$$

$$P = 5^2 \times 1 = 25 \text{ W}$$

2.1.2 POLARITY

Windings on transformers or other electrical machines are marked to indicate terminals of like polarity. Consider the two windings shown in Fig.

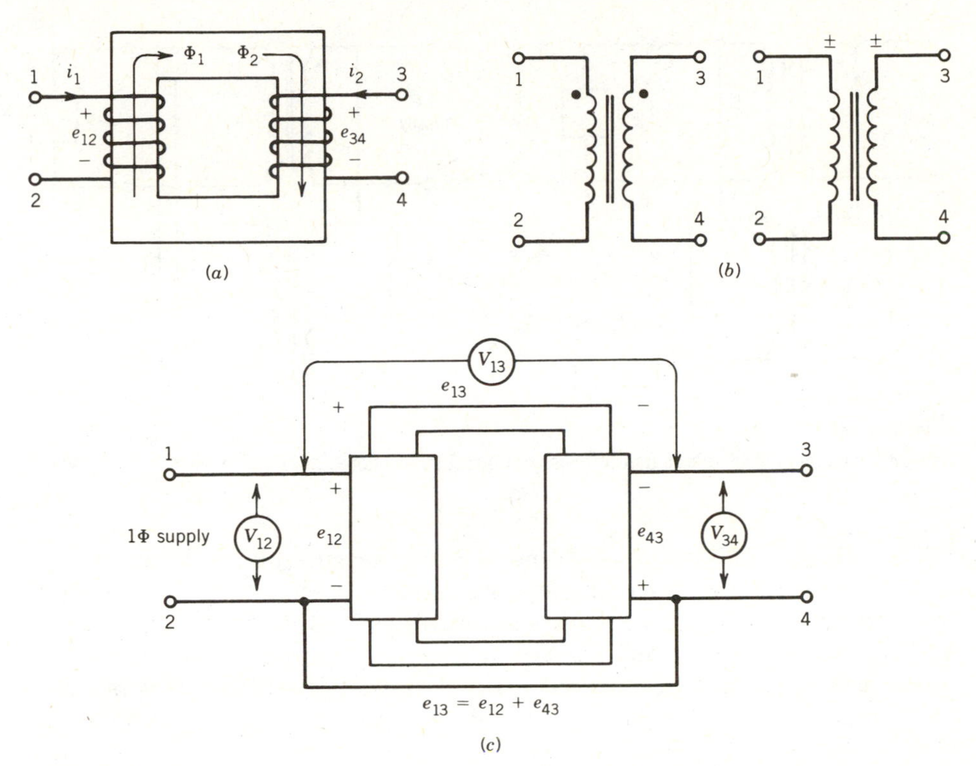

FIGURE 2.8
Polarity determination.

2.8*a*. Terminals 1 and 3 are identical, because currents entering these terminals produce fluxes in the same direction in the core that forms the common magnetic path. For the same reason, terminals 2 and 4 are identical. If these two windings are linked by a common time-varying flux, voltages will be induced in these windings such that, if at a particular instant the potential of terminal 1 is positive with respect to terminal 2, then at the same instant the potential of terminal 3 will be positive with respect to terminal 4. In other words, induced voltages e_{12} and e_{34} are in phase. Identical terminals such as 1 and 3 or 2 and 4 are sometimes marked by dots or ± as shown in Fig. 2.8*b*. These are called the polarity markings of the windings. They indicate how the windings are wound on the core.

If the windings can be visually seen in a machine, the polarities can be determined. However, usually only the terminals of the windings are brought outside the machine. Nevertheless, it is possible to determine the polarities of the windings experimentally. A simple method is illustrated in Fig. 2.8*c*, in which terminals 2 and 4 are connected together and winding 1–2 is connected to an ac supply.

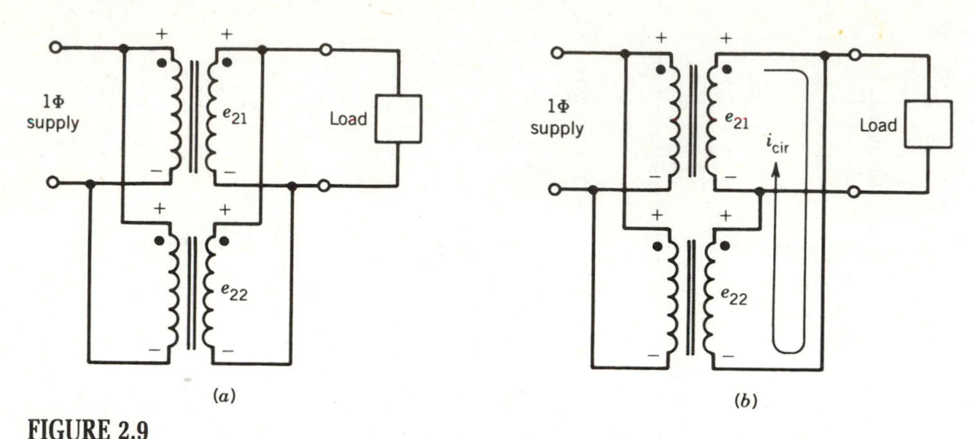

FIGURE 2.9
Parallel operation of single-phase transformers. (*a*) Correct connection. (*b*) Wrong connection.

The voltages across 1–2, 3–4, and 1–3 are measured by a voltmeter. Let these voltage readings be called V_{12}, V_{34}, and V_{13}, respectively. If a voltmeter reading V_{13} is the sum of voltmeter readings V_{12} and V_{34} (i.e., $V_{13} \simeq V_{12} + V_{34}$), it means that at any instant when the potential of terminal 1 is positive with respect to terminal 2, the potential of terminal 4 is positive with respect to terminal 3. The induced voltages e_{12} and e_{43} are in phase, as shown in Fig. 2.8*c*, making $e_{13} = e_{12} + e_{43}$. Consequently, terminals 1 and 4 are identical (or same polarity) terminals. If the voltmeter reading V_{13} is the difference between voltmeter readings V_{12} and V_{34} (i.e., $V_{13} \simeq V_{12} - V_{34}$), then 1 and 3 are terminals of the same polarity.

Polarities of windings must be known if transformers are connected in parallel to share a common load. Figure 2.9*a* shows the parallel connection of two single-phase (1ϕ) transformers. This is the correct connection because secondary voltages e_{21} and e_{22} oppose each other internally. The connection shown in Fig. 2.9*b* is wrong, because e_{21} and e_{22} aid each other internally and a large circulating current i_{cir} will flow in the windings and may damage the transformers. For three-phase connection of transformers (see Section 2.6), the winding polarities must also be known.

2.2 PRACTICAL TRANSFORMER

In Section 2.1 the properties of an ideal transformer were discussed. Certain assumptions were made which are not valid in a practical transformer. For example, in a practical transformer the windings have resistances, not all windings link the same flux, permeability of the core material is not infinite, and core losses occur when the core material is subjected to time-varying flux. In the analysis of a practical transformer, all these imperfections must be considered.

Two methods of analysis can be used to account for the departures from the ideal transformer:

1. An equivalent circuit model based on physical reasoning.
2. A mathematical model based on the classical theory of magnetically coupled circuits.

Both methods will provide the same performance characteristics for the practical transformer. However, the equivalent circuit approach provides a better appreciation and understanding of the physical phenomena involved, and this technique will be presented here.

A practical winding has a resistance, and this resistance can be shown as a lumped quantity in series with the winding (Fig. 2.10*a*). When currents flow through windings in the transformer, they establish a resultant mutual (or common) flux Φ_m that is confined essentially to the magnetic core. However, a small amount of flux known as leakage flux, Φ_l (shown in Fig. 2.10*a*), links only one winding and does not link the other winding. The leakage path is primarily in air, and therefore the leakage flux varies

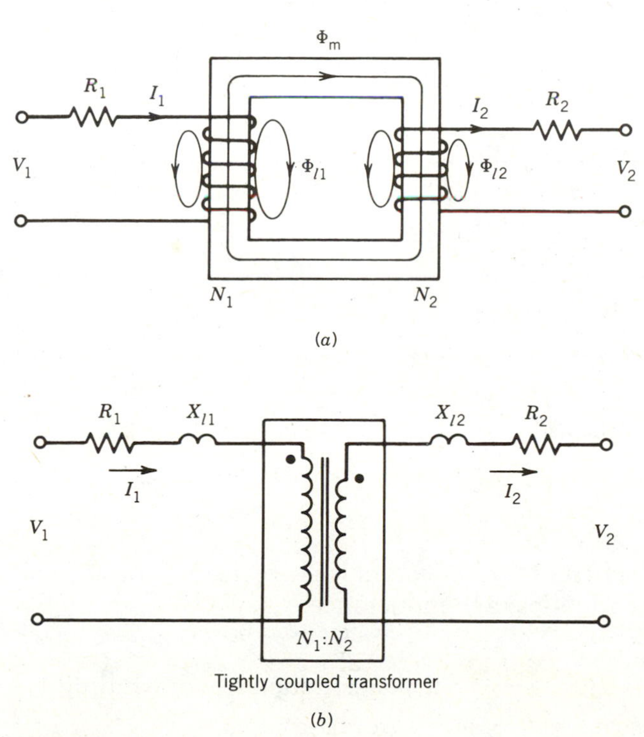

FIGURE 2.10
Development of the transformer equivalent circuits.

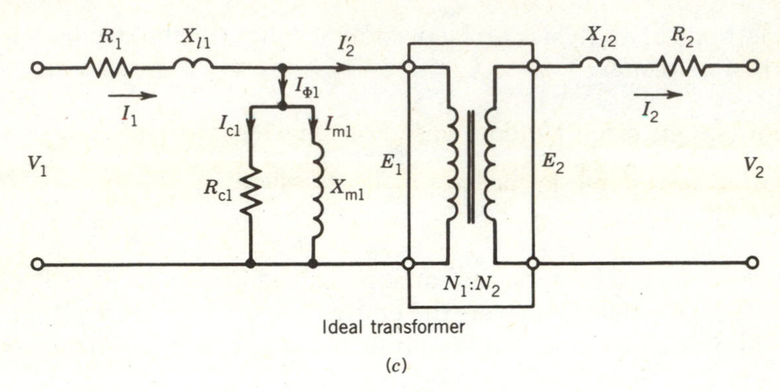

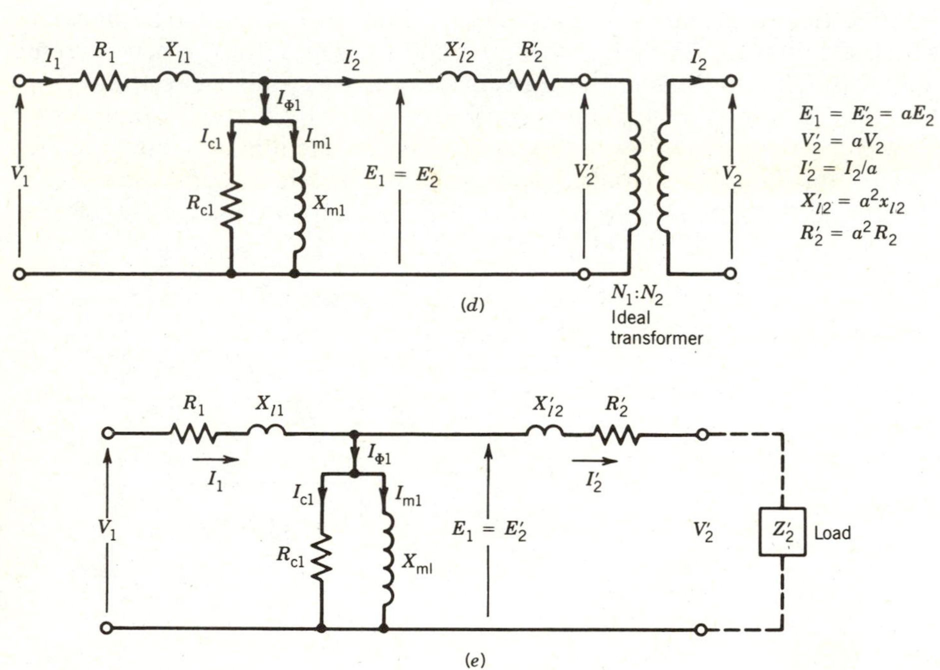

FIGURE 2.10 (*Continued*)

linearly with current. The effects of leakage flux can be accounted for by an inductance, called leakage inductance:

$$L_{l1} = \frac{N_1 \Phi_{l1}}{i_1} = \text{leakage inductance of winding 1}$$

$$L_{l2} = \frac{N_2 \Phi_{l2}}{i_2} = \text{leakage inductance of winding 2}$$

If the effects of winding resistance and leakage flux are respectively accounted for by resistance R and leakage reactance $X_l (= 2\pi f L_l)$, as shown in Fig. 2.10*b*, the transformer windings are tightly coupled by a mutual flux.

In a practical magnetic core having finite permeability, a magnetizing current I_m is required to establish a flux in the core. This effect can be represented by a magnetizing inductance L_m. Also, the core loss in the magnetic material can be represented by a resistance R_c. If these imperfections are also accounted for, then what we are left with is an ideal transformer, as shown in Fig. 2.10*c*. A practical transformer is therefore equivalent to an ideal transformer plus external impedances that represent imperfections of an actual transformer.

2.2.1 REFERRED EQUIVALENT CIRCUITS

The ideal transformer in Fig. 2.10*c* can be moved to the right or left by referring all quantities to the primary or secondary side, respectively. This is almost invariably done. The equivalent circuit with the ideal transformer moved to the right is shown in Fig. 2.10*d*. For convenience, the ideal transformer is usually not shown and the equivalent circuit is drawn, as shown in Fig. 2.10*e*, with all quantities (voltages, currents, and impedances) referred to one side. The referred quantities are indicated with primes. By analyzing this equivalent circuit the referred quantities can be evaluated, and the actual quantities can be determined from them if the turns ratio is known.

Approximate Equivalent Circuits

The voltage drops I_1R_1 and I_1X_{l1} (Fig. 2.10*e*) are normally small and $|E_1| \simeq |V_1|$. If this is true then the shunt branch (composed of R_{c1} and X_{m1}) can be moved to the supply terminal, as shown in Fig. 2.11*a*. This approximate equivalent circuit simplifies computation of currents, because both the exciting branch impedance and the load branch impedance are directly connected across the supply voltage. Besides, the winding resistances and leakage reactances can be lumped together. This equivalent circuit (Fig. 2.11*a*) is frequently used to determine the performance characteristics of a practical transformer.

In a transformer, the exciting current I_ϕ is a small percentage of the rated current of the transformer (less than 5%). A further approximation of the equivalent circuit can be made by removing the excitation branch, as shown in Fig. 2.11*b*. The equivalent circuit referred to side 2 is also shown in Fig. 2.11*c*.

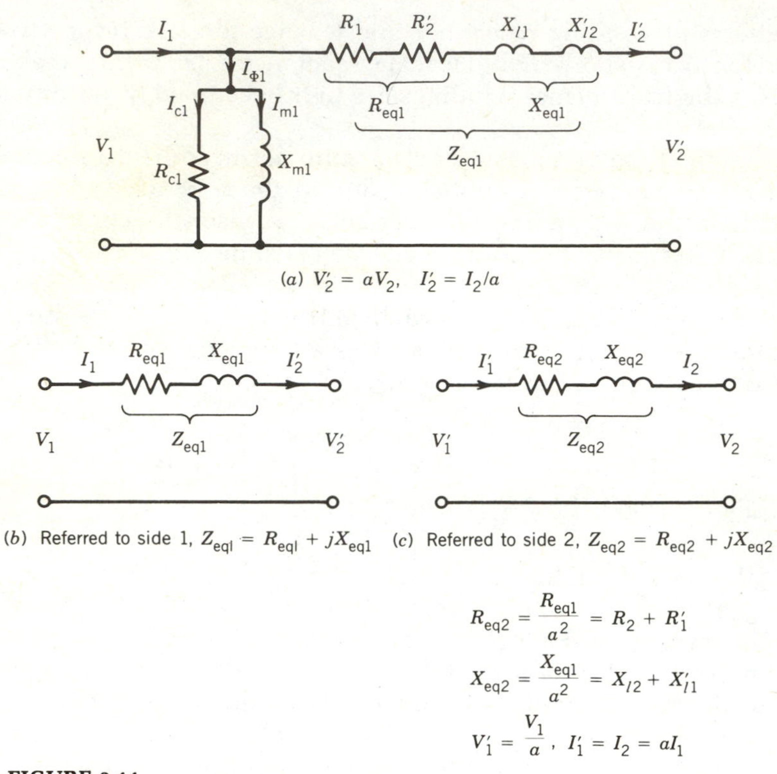

FIGURE 2.11
Approximate equivalent circuits.

2.2.2 DETERMINATION OF EQUIVALENT CIRCUIT PARAMETERS

The equivalent circuit model (Fig. 2.10*e*) for the actual transformer can be used to predict the behavior of the transformer. The parameters R_1, X_{l1}, R_{c1}, X_{m1}, R_2, X_{l2}, and a ($= N_1/N_2$) must be known so that the equivalent circuit model can be used.

If the complete design data of a transformer are available, these parameters can be calculated from the dimensions and properties of the materials used. For example, the winding resistances (R_1, R_2) can be calculated from the resistivity of copper wires, the total length, and the cross-sectional area of the winding. The magnetizing inductances L_m can be calculated from the number of turns of the winding and the reluctance of the magnetic path. The calculation of the leakage inductance (L_l) will involve accounting for partial flux linkages and is therefore complicated. However, formulas are available from which a reliable determination of these quantities can be made.

These parameters can be directly and more easily determined by performing tests that involve little power consumption. Two tests, a no-load test (or open-circuit test) and a short-circuit test, will provide information for determining the parameters of the equivalent circuit of a transformer, as will be illustrated by an example.

Transformer Rating

The kilovolt–ampere (kVA) rating and voltage ratings of a transformer are marked on its nameplate. For example, a typical transformer may carry the following information on the nameplate: 10 kVA, 1100/110 volts. What are the meanings of these ratings? The voltage ratings indicate that the transformer has two windings, one rated for 1100 volts and the other for 110 volts. These voltages are proportional to their respective numbers of turns and therefore the voltage ratio also represents the turns ratio ($a = 1100/110 = 10$). The 10 kVA rating means that each winding is designed for 10 kVA. Therefore the current rating for the high-voltage winding is $10{,}000/1100 = 9.09$ A and for the lower-voltage winding is $10{,}000/110 = 90.9$ A. It may be noted that when the rated current of 90.9 A flows through the low-voltage winding, the rated current of 9.09 A will flow through the high-voltage winding. In an actual case, however, the winding that is connected to the supply (called the primary winding) will carry an additional component of current (excitation current), which is very small compared to the rated current of the winding.

No-Load Test (or Open-Circuit Test)

This test is performed by applying a voltage to either the high-voltage side or low-voltage side, whichever is convenient. Thus, if a 1100/110 volt transformer were to be tested, the voltage would be applied to the low-voltage winding, because a power supply of 110 volts is more readily available than a supply of 1100 volts.

A wiring diagram for open-circuit test of a transformer is shown in Fig. 2.12*a*. Note that the secondary winding is kept open. Therefore, from the transformer equivalent circuit of Fig. 2.11*a* the equivalent circuit under open-circuit conditions is as shown in Fig. 2.12*b*. The primary current is the exciting current and the losses measured by the wattmeter are essentially the core losses. The equivalent circuit of Fig. 2.12*b* shows that the parameters R_c and X_m can be determined from the voltmeter, ammeter, and wattmeter readings.

Note that the core losses will be same whether 110 volts are applied to the low-voltage winding having the smaller number of turns or 1100 volts are applied to the high-voltage winding having the larger number of turns. The core loss depends on the maximum value of flux in the core, which is same in either case, as indicated by Eqs. 1.40.

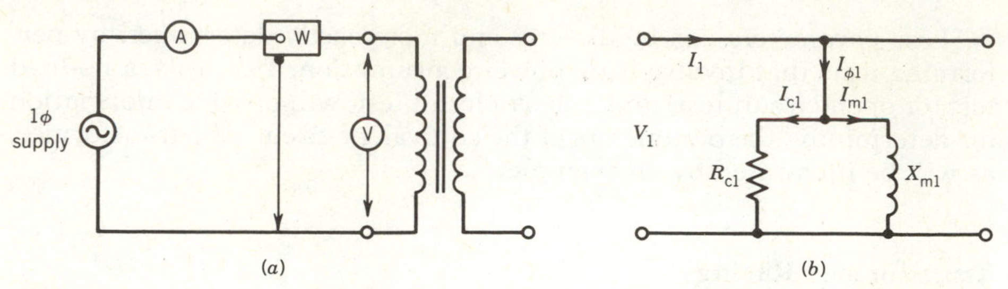

FIGURE 2.12

No-load (or open-circuit) test. (a) Wiring diagram for open-circuit test. (b) Equivalent circuit under open circuit.

Short-Circuit Test

This test is performed by short-circuiting one winding and applying a small voltage to the other winding, as shown in Fig. 2.13*a*. In the equivalent circuit of Fig. 2.11*a* for the transformer, the impedance of the excitation branch (shunt branch composed of R_c and X_m) is much larger than that of the series branch (composed of R_{eq} and X_{eq}). If the secondary terminals are shorted, the high impedance of the shunt branch can be neglected. The equivalent circuit with the secondary short-circuited can thus be represented by the circuit shown in Fig. 2.13*b*. Note that since Z_{eq} ($= R_{eq} + jX_{eq}$) is small, only a small supply voltage is required to pass rated current through the windings. It is convenient to perform this test by applying a voltage to the high-voltage winding.

As can be seen from Fig. 2.13*b*, the parameters R_{eq} and X_{eq} can be determined from the readings of voltmeter, ammeter, and wattmeter. In a well-designed transformer, $R_1 = a^2R_2 = R_2'$ and $X_{l1} = a^2X_{l2} = X_{l2}'$. Note that because the voltage applied under the short-circuit condition is small, the core losses are neglected and the wattmeter reading can be taken entirely to represent the copper losses in the windings, represented by R_{eq}.

The following example illustrates the computation of the parameters of the equivalent circuit of a transformer.

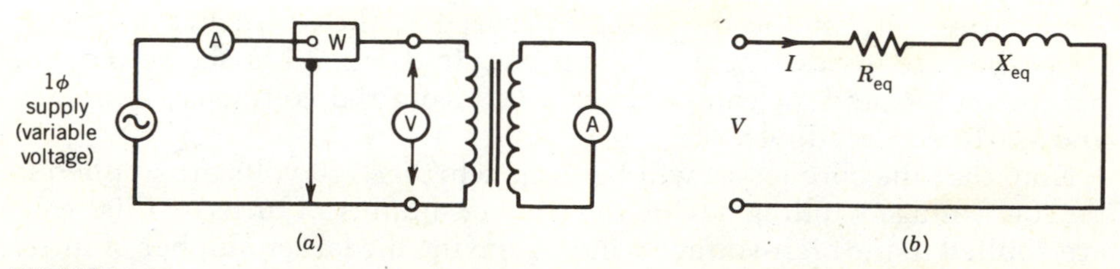

FIGURE 2.13

Short-circuit test. (a) Wiring diagram for short-circuit test. (b) Equivalent circuit at short-circuit condition.

EXAMPLE 2.2

Tests are performed on a 1ϕ, 10 kVA, 2200/220 V, 60 Hz transformer and the following results are obtained.

	Open-Circuit Test (high-voltage side open)	Short-Circuit Test (low-voltage side shorted)
Voltmeter	220 V	150 V
Ammeter	2.5 A	4.55 A
Wattmeter	100 W	215 W

(a) Derive the parameters for the approximate equivalent circuits referred to the low-voltage side and the high-voltage side.

(b) Express the excitation current as a percentage of the rated current.

(c) Determine the power factor for the no-load and short-circuit tests.

Solution

Note that for the no-load test the supply voltage (full-rated voltage of 220 V) is applied to the low-voltage winding, and for the short-circuit test the supply voltage is applied to the high-voltage winding with the low-voltage winding shorted. The subscripts H and L will be used to represent quantities for the high-voltage and low-voltage windings, respectively.

The ratings of the windings are as follows:

$$V_{\text{H(rated)}} = 2200 \text{ V}$$

$$V_{\text{L(rated)}} = 220 \text{ V}$$

$$I_{\text{H(rated)}} = \frac{1000}{2200} = 4.55 \text{ A}$$

$$I_{\text{L(rated)}} = \frac{10{,}000}{220} = 45.5 \text{ A}$$

$$V_{\text{H}} I_{\text{H(rated)}} = V_{\text{L}} I_{\text{L(rated)}} = 10 \text{ kVA}$$

(a) The equivalent circuit and the phasor diagram for the open-circuit test are shown in Fig. E2.2*a*.

$$\text{Power, } P_{\text{oc}} = \frac{V_{\text{L}}^2}{R_{\text{cL}}}$$

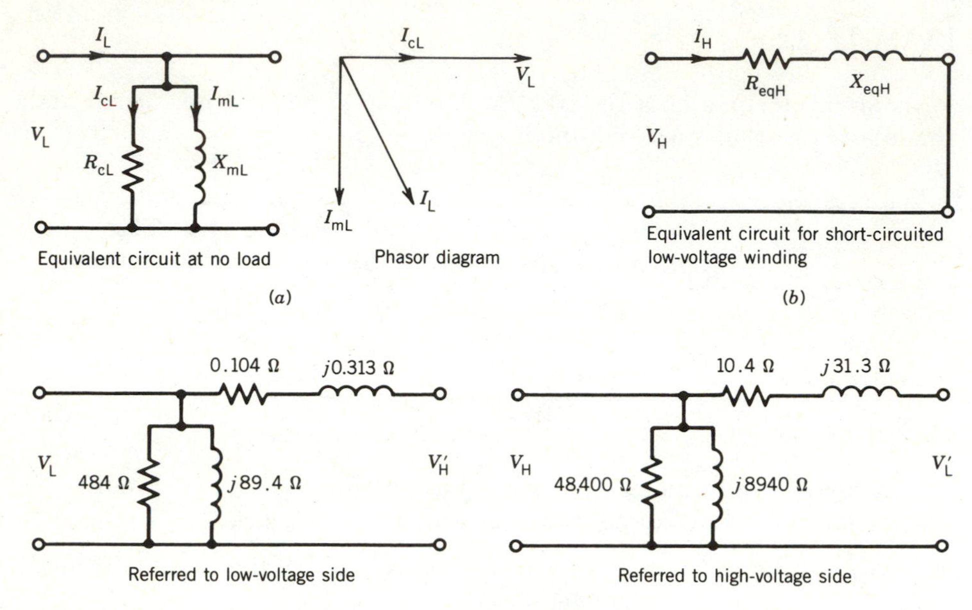

FIGURE E2.2

$$R_{cL} = \frac{220^2}{100} = 484\ \Omega$$

$$I_{cL} = \frac{220}{484} = 0.45\ \text{A}$$

$$I_{mL} = (I_L^2 - I_{cL}^2)^{1/2} = (2.5^2 - 0.45^2)^{1/2} = 2.46\ \text{A}$$

$$X_{mL} = \frac{V_L}{I_{mL}} = \frac{220}{2.46} = 89.4\ \Omega$$

The corresponding parameters for the high-voltage side are obtained as follows:

$$\text{Turns ratio} \qquad a = \frac{2200}{220} = 10$$

$$R_{cH} = a^2 R_{cL} = 10^2 \times 484 = 48{,}400\ \Omega$$

$$X_{mH} = 10^2 \times 89.4 = 8940\ \Omega$$

The equivalent circuit with the low-voltage winding shorted is shown in Fig. E2.2*b*.

Power $P_{sc} = I_H^2 R_{eqH}$

$$R_{eqH} = \frac{215}{4.55^2} = 10.4\ \Omega$$

$$Z_{eqH} = \frac{V_H}{I_H} = \frac{150}{4.55} = 32.97\ \Omega$$

$$X_{eqH} = (Z_{eqH}^2 - R_{eqH}^2)^{1/2} = (32.97^2 - 10.4^2)^{1/2} = 31.9\ \Omega$$

The corresponding parameters for the low-voltage side are as follows:

$$R_{eqL} = \frac{R_{eqH}}{a^2} = \frac{10.4}{10^2} = 0.104\ \Omega$$

$$X_{eqL} = \frac{31.3}{10^2} = 0.313\ \Omega$$

The approximate equivalent circuits referred to the low-voltage side and the high-voltage side are shown in Fig. E2.2*c*. Note that the impedance of the shunt branch is much larger than that of the series branch.

(b) From the no-load test the excitation current, with rated voltage applied to the low-voltage winding, is

$$I_\phi = 2.5\ \text{A}$$

This is $(2.5/45.5) \times 100\% = 5.5\%$ of the rated current of the winding

(c)

$$\text{Power factor at no load} = \frac{\text{power}}{\text{volt-ampere}} = \frac{100}{220 \times 2.5} = 0.182$$

$$\text{Power factor at short-circuit condition} = \frac{215}{150 \times 4.55} = 0.315$$

2.3 VOLTAGE REGULATION

Most loads connected to the secondary of a transformer are designed to operate at essentially constant voltage. However, as the current is drawn through the transformer, the load terminal voltage changes because of voltage drop in the internal impedance of the transformer. Consider Fig.

2.14*a*, where the transformer is represented by a series impedance Z_{eq}. If a load is not applied to the transformer (i.e., open-circuit or no-load condition) the load terminal voltage is

$$V_2|_{NL} = \frac{V_1}{a} \tag{2.14}$$

If the load switch is now closed and the load is connected to the transformer secondary, the load terminal voltage is

$$V_2|_L = V_2|_{NL} \pm \Delta V_2 \tag{2.15}$$

The load terminal voltage may go up or down depending on the nature of the load. This voltage change is due to the voltage drop (IZ) in the internal impedance of the transformer. A large voltage change is undesirable for many loads. For example, as more and more light bulbs are connected to the transformer secondary and the voltage decreases appreciably, the bulbs will glow with diminished illumination. To reduce the magnitude of the voltage change, the transformer should be designed for a low value of the internal impedance Z_{eq}.

A figure of merit called *voltage regulation* is used to identify this characteristic of voltage change in a transformer with loading. The voltage regu-

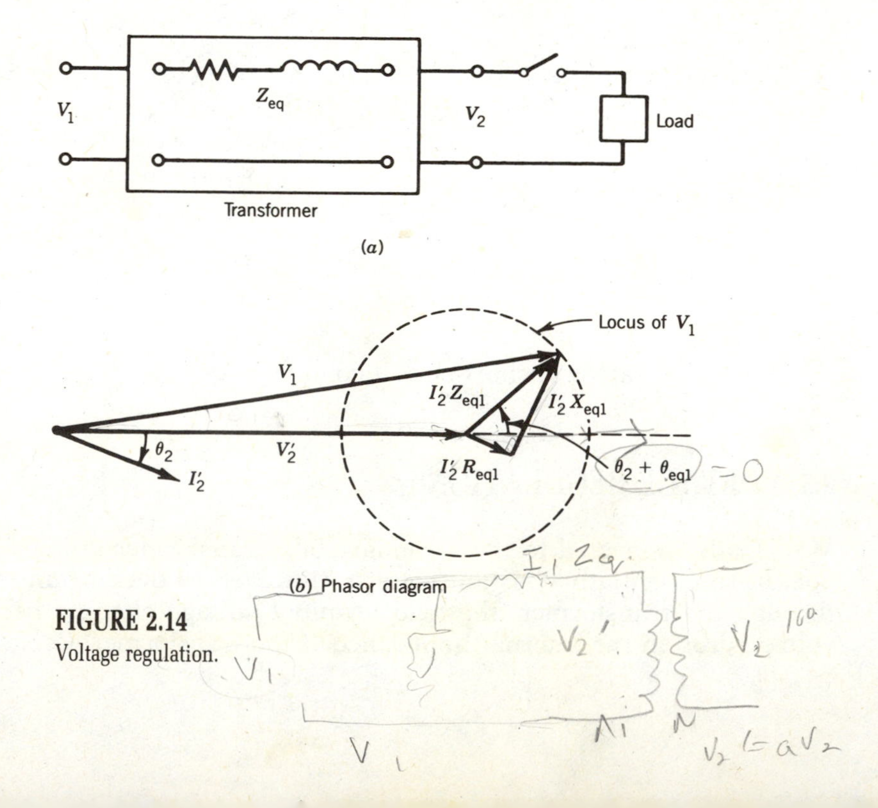

FIGURE 2.14
Voltage regulation.

lation is defined as the change in magnitude of the secondary voltage as the load current changes from the no-load to the loaded condition. This is expressed as follows:

$$\text{Voltage regulation} = \frac{|V_2|_{\text{NL}} - |V_2|_{\text{L}}}{|V_2|_{\text{L}}} \tag{2.16}$$

The absolute signs are used to indicate that it is the change in magnitudes that is important for the performance of the load. The voltages in Eq. 2.16 can be calculated by using equivalent circuits referred to either primary or secondary. Let us consider the equivalent circuit referred to the primary, shown in Fig. 2.11*b*. Equation 2.16 can also be written as

$$\text{Voltage regulation} = \frac{|V_2'|_{\text{NL}} - |V_2'|_{\text{L}}}{|V_2'|_{\text{L}}} \tag{2.17}$$

The load voltage is normally taken as the rated voltage. Therefore,

$$|V_2'|_{\text{L}} = |V_2'|_{\text{rated}} \tag{2.18}$$

From Fig. 2.11*b*,

$$V_1 = V_2' + I_2'R_{\text{eq1}} + jI_2'X_{\text{eq1}} \tag{2.19}$$

If the load is thrown off ($I_1 = I_2' = 0$), V_1 will appear as V_2'. Hence,

$$|V_2'|_{\text{NL}} = |V_1| \tag{2.20}$$

From Eqs. 2.17, 2.18, and 2.20,

$$\begin{array}{c}\text{Voltage regulation} \\ \text{(in percent)}\end{array} = \frac{|V_1| - |V_2'|_{\text{rated}}}{|V_2'|_{\text{rated}}} \times 100\% \tag{2.21}$$

The voltage regulation depends on the power factor of the load. This can be appreciated from the phasor diagram of the voltages. Based on Eq. 2.19 and Fig. 2.11*b*, the phasor diagram is drawn in Fig. 2.14*b*. The locus of V_1 is a circle of radius $|I_2'Z_{\text{eq1}}|$. The magnitude of V_1 will be maximum if the phasor $I_2'Z_{\text{eq1}}$ is in phase with V_2'. That is,

$$\theta_2 + \theta_{\text{eq1}} = 0 \tag{2.22}$$

where θ_2 is the angle of the load impedance

θ_{eq1} is the angle of the transformer equivalent impedance, Z_{eq1}.

$\theta_2 = -\theta_{eq1}$

From Eq. 2.22,

$$\theta_2 = -\theta_{eq1} \tag{2.23}$$

Therefore the maximum voltage regulation occurs if the power factor angle of the load is the same as the transformer equivalent impedance angle and the load power factor is lagging.

EXAMPLE 2.3

Consider the transformer in Example 2.2. Determine the voltage regulation in percent for the following load conditions.

(a) 75% full load, 0.6 power factor lagging.

(b) 75% full load, 0.6 power factor leading.

(c) Draw the phasor diagram for conditions (a) and (b).

Solution

Consider the equivalent circuit referred to the high-voltage side, as shown in Fig. E2.3. The load voltage is assumed to be at the rated value. The

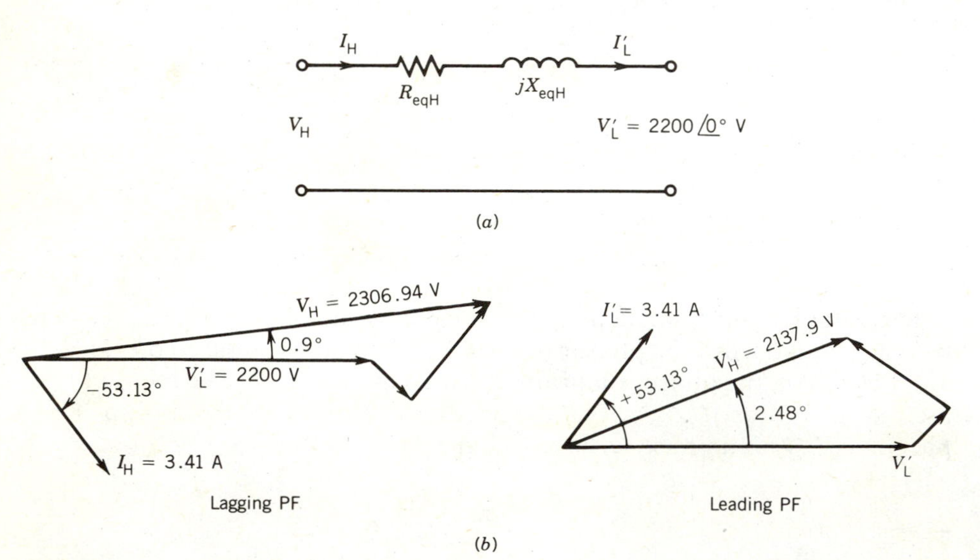

FIGURE E2.3

condition 75% full load means that the load current is 75% of the rated current. Therefore,

$$I_H = I'_L = 0.75 \times 4.55 = 3.41 \text{ A}$$

$$\text{Power factor} \qquad \text{PF} = \cos\theta_2 = 0.6$$

$$\theta_2 = \pm 53.13$$

(a) For a lagging power factor, $\theta_2 = -53.13°$

$$I_H = 3.41\underline{/-53.13°} \text{ A}$$

$$\begin{aligned} V_H &= V'_L + I'_L Z_{eqH} \\ &= 2200\underline{/0°} + 3.41\underline{/-53.13°}\,(10.4 + j31.3) \\ &= 2200 + 35.46\underline{/-53.13°} + 106.73\underline{/90° - 53.13°} \\ &= 2200 + 21.28 - j28.37 + 85.38 + j64.04 \\ &= 2306.66 + j35.67 \\ &= 2306.94\underline{/0.9°} \text{ V} \end{aligned}$$

$$\text{Voltage regulation} = \frac{2306.94 - 2200}{2200} \times 100\%$$

$$= 4.86\%$$

The meaning of 4.86% voltage regulation is that if the load is thrown off, the load terminal voltage will rise from 220 to 230.69 volts. In other words, when the 75% full load at 0.6 lagging power factor is connected to the load terminals of the transformer, the voltage drops from 230.69 to 220 volts.

(b) For leading power factor load, $\theta_2 = +53.13$

$$\begin{aligned} V_H &= 2200\underline{/0°} + 3.41\underline{/53.13°}\,(10.4 + j31.3) \\ &= 2200 + 35.46\underline{/53.13°} + 106.73\underline{/90° + 53.13°} \\ &= 2200 + 21.28 + j28.37 - 85.38 + j64.04 \\ &= 2135.9 + j92.41 \\ &= 2137.9\underline{/2.48°} \text{ V} \end{aligned}$$

$$\text{Voltage regulation} = \frac{2137.9 - 2200}{2200} \times 100\%$$

$$= -2.82\%$$

Note that the voltage regulation for this leading power factor load is negative. This means that if the load is thrown off, the load terminal voltage will decrease from 220 to 213.79 volts. To put it differently, if the leading power factor load is connected to the load terminals, the voltage will increase from 213.79 to 220 volts.

(c) The phasor diagrams for both lagging and leading power factor loads are shown in Fig. E2.3*b*.

2.4 EFFICIENCY

Equipment is desired to operate at a high efficiency. Fortunately, losses in transformers are small. Because the transformer is a static device, there are no rotational losses such as windage and friction losses in a rotating machine. In a well-designed transformer the efficiency can be as high as 99%. The efficiency is defined as follows:

$$\eta = \frac{\text{output power } (P_{out})}{\text{input power } (P_{in})} \tag{2.24}$$

$$= \frac{P_{out}}{P_{out} + \text{losses}} \tag{2.25}$$

The losses in the transformer are the core loss (P_c) and copper loss (P_{cu}). Therefore,

$$\eta = \frac{P_{out}}{P_{out} + P_c + P_{cu}} \tag{2.26}$$

The copper loss can be determined if the winding currents and their resistances are known:

$$P_{cu} = I_1^2 R_1 + I_2^2 R_2 \tag{2.27}$$

$$= I_1^2 R_{eq1} \tag{2.27a}$$

$$= I_2^2 R_{eq2} \tag{2.27b}$$

The copper loss is a function of the load current.

The core loss depends on the peak flux density in the core, which in turn depends on the voltage applied to the transformer. Since a transformer remains connected to an essentially constant voltage, the core loss is almost constant and can be obtained from the no-load test of a transformer, as shown in Example 2.2. Therefore, if the parameters of the equivalent circuit of a transformer are known, the efficiency of the transformer under any operating condition may be determined. Now,

$$P_{out} = V_2 I_2 \cos\theta_2 \tag{2.28}$$

Therefore,

$$\eta = \frac{V_2 I_2 \cos\theta_2}{V_2 I_2 \cos\theta_2 + P_c + I_2^2 R_{eq2}} \tag{2.29}$$

Normally, load voltage remains fixed. Therefore, efficiency depends on load current (I_2) and load power factor ($\cos\theta_2$).

2.4.1 MAXIMUM EFFICIENCY

For constant values of the terminal voltage V_2 and load power factor angle θ_2, the maximum efficiency occurs when

$$\frac{d\eta}{dI_2} = 0 \tag{2.30}$$

If this condition is applied to Eq. 2.29 the condition for maximum efficiency is

$$P_c = I_2^2 R_{eq2} \tag{2.31}$$

that is, core loss = copper loss. For full-load condition,

$$P_{cu,FL} = I_{2,FL}^2 R_{eq2} \tag{2.31a}$$

Let

$$X = \frac{I_2}{I_{2,FL}} = \text{per unit loading} \tag{2.31b}$$

From Eqs. 2.31, 2.31a, and 2.31b

$$P_c = X^2 P_{cu,FL}$$

$$X = \left(\frac{P_c}{P_{cu,FL}}\right)^{1/2} \tag{2.31c}$$

For constant values of the terminal voltage V_2 and load current I_2, the maximum efficiency occurs when

$$\frac{d\eta}{d\theta_2} = 0 \tag{2.32}$$

If this condition is applied to Eq. 2.29, the condition for maximum efficiency is

$$\theta_2 = 0 \tag{2.33}$$

$$\cos \theta_2 = 1 \tag{2.33a}$$

$$\text{that is, load power factor} = 1 \tag{2.33b}$$

Therefore, maximum efficiency in a transformer occurs when the load power factor is unity (i.e., resistive load) and load current is such that copper loss equals core loss. The variation of efficiency with load current and load power factor is shown in Fig. 2.15.

2.4.2 ALL-DAY (OR ENERGY) EFFICIENCY, η_{AD}

The transformer in a power plant usually operates near its full capacity and is taken out of circuit when it is not required. Such transformers are called *power transformers,* and they are usually designed for maximum efficiency occurring near the rated output. A transformer connected to the utility that supplies power to your house and the locality is called a *distribution transformer.* Such transformers are connected to the power system for 24 hours a day and operate well below the rated power output for most of the time. It is therefore desirable to design a distribution transformer for maximum efficiency occurring at the average output power.

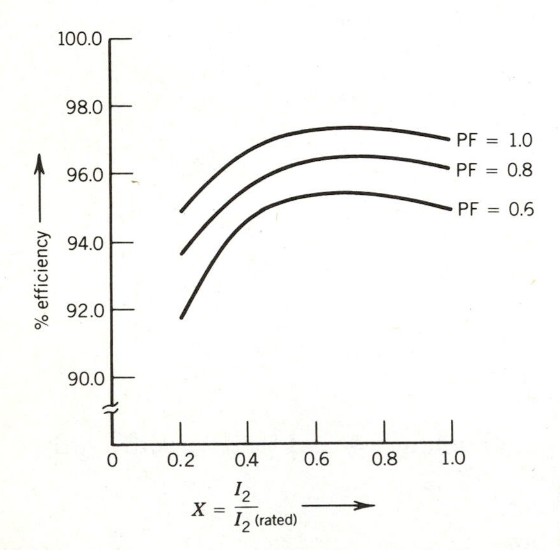

FIGURE 2.15
Efficiency of a transformer.

A figure of merit that will be more appropriate to represent the efficiency performance of a distribution transformer is the "all-day" or "energy" efficiency of the transformer. This is defined as follows:

$$\eta_{AD} = \frac{\text{energy output over 24 hours}}{\text{energy input over 24 hours}} \tag{2.34}$$

$$= \frac{\text{energy output over 24 hours}}{\text{energy output over 24 hours + losses over 24 hours}}$$

If the load cycle of the transformer is known, the all-day efficiency can be determined.

EXAMPLE 2.4

For the transformer in Example 2.2, determine

(a) Efficiency at 75% rated output and 0.6 PF.

(b) Power output at maximum efficiency and the value of maximum efficiency. At what percent of full load does this maximum efficiency occur?

Solution

(a)

$$P_{out} = V_2 I_2 \cos\theta_2$$

$$= 0.75 \times 10{,}000 \times 0.6$$

$$= 4500 \text{ W}$$

$$P_c = 100 \text{ W} \quad \text{(Example 2.2)}$$

$$P_{cu} = I_H^2 R_{eqH}$$

$$= (0.75 \times 4.55)^2 \times 10.4 \text{ W}$$

$$= 121 \text{ W}$$

$$\eta = \frac{4500}{4500 + 100 + 121} \times 100\%$$

$$= 95.32\%$$

(b) At maximum efficiency

$$P_{core} = P_{cu} \quad \text{and} \quad \text{PF} = \cos\theta = 1$$

Now, $P_{core} = 100 \text{ W} = I_2^2 R_{eq2} = P_{cu}$.

$$I_2 = \left(\frac{100}{0.104}\right)^{1/2} = 31 \text{ A}$$

$$P_{out}\big|_{\eta_{max}} = V_2 I_2 \cos\theta_2$$

$$= 220 \times 31 \times 1$$

$$= 6820 \text{ W}$$

$$\eta_{max} = \frac{6820}{6820 + \underset{\substack{\uparrow \\ P_c}}{100} + \underset{\substack{\uparrow \\ P_{cu}}}{100}} \times 100\%$$

$$= 97.15\%$$

$$\text{Output kVA} = 6.82$$

$$\text{Rated kVA} = 10$$

η_{max} occurs at 68.2% full load.

Other Method

From Example 2.2,

$$P_{cu,FL} = 215 \text{ W}$$

From Eq. 2.31c

$$X = \left(\frac{100}{215}\right)^{1/2} = 0.68$$

EXAMPLE 2.5

A 50 kVA, 2400/240 V transformer has a core loss $P_c = 200$ W at rated voltage and a copper loss $P_{cu} = 500$ W at full load. It has the following load cycle.

% Load	0.0%	50%	75%	100%	110%
Power factor		1	0.8 lag	0.9 lag	1
Hours	6	6	6	3	3

Determine the all-day efficiency of the transformer.

Solution

$$\begin{aligned}\text{Energy output over 24 hours} &= 0.5 \times 50 \times 6 + 0.75 \times 50 \times 0.8 \\ &\quad \times 6 + 1 \times 50 \times 0.9 \times 3 + 1.1 \times 50 \times 1 \\ &\quad \times 3 \text{ kWh} \\ &= 630 \text{ kWh}\end{aligned}$$

Energy losses over 24 hours:

$$\begin{aligned}\text{Core loss} &= 0.2 \times 24 = 4.8 \text{ kWh} \\ \text{Copper loss} &= 0.5^2 \times 0.5 \times 6 + 0.75^2 \times 0.5 \times 6 \\ &\quad + 1^2 \times 0.5 \times 3 + 1.1^2 \times 0.5 \times 3 \\ &= 5.76 \text{ kWh} \\ \text{Total energy loss} &= 4.8 + 5.76 = 10.56 \text{ kWh}\end{aligned}$$

$$\eta_{AD} = \frac{630}{630 + 10.56} \times 100\% = 98.35\%$$

2.5 AUTOTRANSFORMER

This is a special connection of the transformer from which a variable ac voltage can be obtained at the secondary. A common winding as shown in Fig. 2.16 is mounted on a core and the secondary is taken from a tap on the winding. In contrast to the two-winding transformer discussed earlier, the primary and secondary of an autotransformer are physically connected.

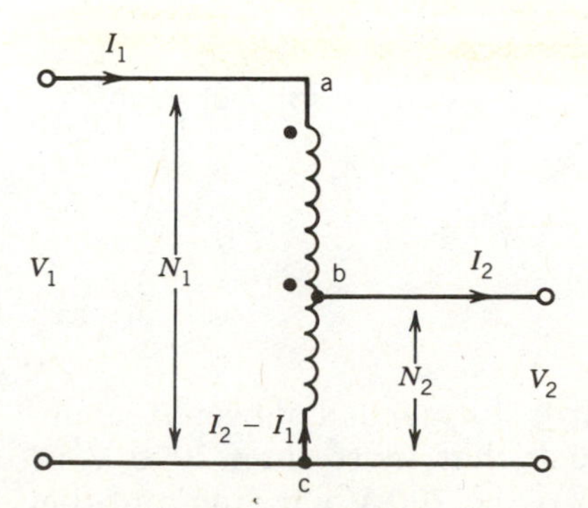

FIGURE 2.16
Autotransformer.

However, the basic principle of operation is the same as that of the two-winding transformer.

Since all the turns link the same flux in the transformer core,

$$\frac{V_1}{V_2} = \frac{N_1}{N_2} = a \tag{2.35}$$

If the secondary tapping is replaced by a slider, the output voltage can be varied over the range $0 < V_2 < V_1$.

The ampere-turns provided by the upper half (i.e., by turns between points a and b) are

$$F_U = (N_1 - N_2)I_1 = \left(1 - \frac{1}{a}\right)N_1 I_1 \tag{2.36}$$

The ampere-turns provided by the lower half (i.e., by turns between points b and c) are

$$F_L = N_2(I_2 - I_1) = \frac{N_1}{a}(I_2 - I_1) \tag{2.37}$$

For ampere-turn balance, from Eqs. 2.36 and 2.37,

$$\left(1 - \frac{1}{a}\right)N_1 I_1 = \frac{N_1}{a}(I_2 - I_1)$$

$$\frac{I_1}{I_2} = \frac{1}{a} \tag{2.38}$$

Equations 2.35 and 2.38 indicate that, viewed from the terminals of the autotransformer, the voltages and currents are related by the same turns ratio as in a two-winding transformer.

The advantages of an autotransformer connection are lower leakage reactances, lower losses, lower exciting current, increased kVA rating (see Example 2.6), and variable output voltage when a sliding contact is used for the secondary. The disadvantage is the direct connection between the primary and secondary sides.

EXAMPLE 2.6

A 1ϕ, 100 kVA, 2000/200 V two-winding transformer is connected as an autotransformer as shown in Fig. E2.6 such that more than 2000 V is obtained at the secondary. The portion ab is the 200 V winding and the

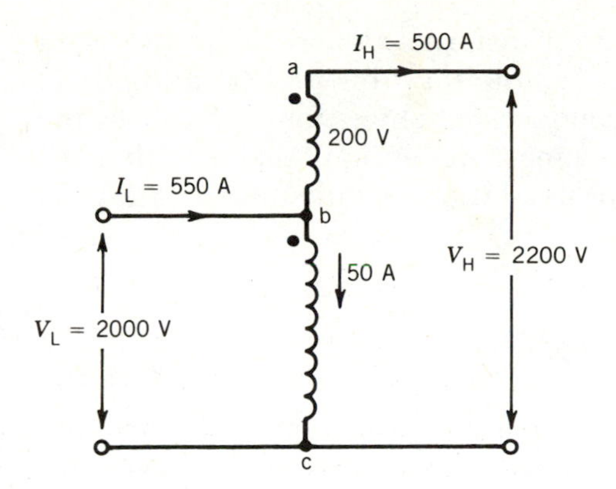

FIGURE E2.6

portion bc is the 2000 V winding. Compute the kVA rating as an autotransformer.

Solution

The current ratings of the windings are

$$I_{ab} = \frac{100{,}000}{200}\ \text{A} = 500\ \text{A}$$

$$I_{bc} = \frac{100{,}000}{2000} = 50\ \text{A}$$

Therefore, for full-load operation of the autotransformer, the terminal currents are

$$I_H = 500\ \text{A}$$

$$I_L = 500 + 50 = 550\ \text{A}$$

Now, $V_L = 2000$ V and

$$V_H = 2000 + 200 = 2200\ \text{V}$$

Therefore,

$$\text{kVA}|_L = \frac{2000 \times 550}{1000} = 1100$$

$$\text{kVA}|_H = \frac{2200 \times 500}{1000} = 1100$$

A single-phase, 100 kVA, two-winding transformer when connected as an autotransformer can deliver 1100 kVA. Note that this higher rating of an autotransformer results from the conductive connection. Not all of the 1100 kVA is transformed by electromagnetic induction. Also note that the 200 volt winding must have sufficient insulation to withstand a voltage of 2200 V to ground.

2.6 THREE-PHASE TRANSFORMERS

A three-phase system is used to generate and transmit bulk electrical energy. Three-phase transformers are required to step up or step down voltages in the various stages of power transmission. A three-phase transformer can be built in one of two ways: by suitably connecting a bank of three single-phase transformers or by constructing a three-phase transformer on a common magnetic structure.

2.6.1 BANK OF THREE SINGLE-PHASE TRANSFORMERS (THREE-PHASE TRANSFORMER BANK)

A set of three similar single-phase transformers may be connected to form a three-phase transformer. The primary and secondary windings may be connected in either wye (Y) or delta (Δ) configurations. There are therefore four possible connections for a three-phase transformer: Y–Δ, Δ–Y, Δ–Δ, and Y–Y. Figure 2.17*a* shows a Y–Δ connection of a three-phase transformer. On the primary side, three terminals of identical polarity are connected together to form the neutral of the Y connection. On the secondary side the windings are connected in series. A more convenient way of showing this connection is illustrated in Fig. 2.17*b*. The primary and secondary windings shown parallel to each other belong to the same single-phase transformer. The primary and secondary voltages and currents are also shown in Fig. 2.17*b*, where V is the line-to-line voltage on the primary side and a $(= N_1/N_2)$ is the turns ratio of the single-phase transformer. Other possible connections are also shown in Figs. 2.17*c*,*d*, and *e*. It may be noted that for all possible connections, the total kVA of the three-phase transformer is shared equally by each (phase) transformer. However, the voltage and current ratings of each transformer depend on the connections used.

Y–Δ: This connection is commonly used to step down a high voltage to a lower voltage. The neutral point on the high-voltage side can be grounded, which is desirable in most cases.

Δ–Y: This connection is commonly used to step up voltage.

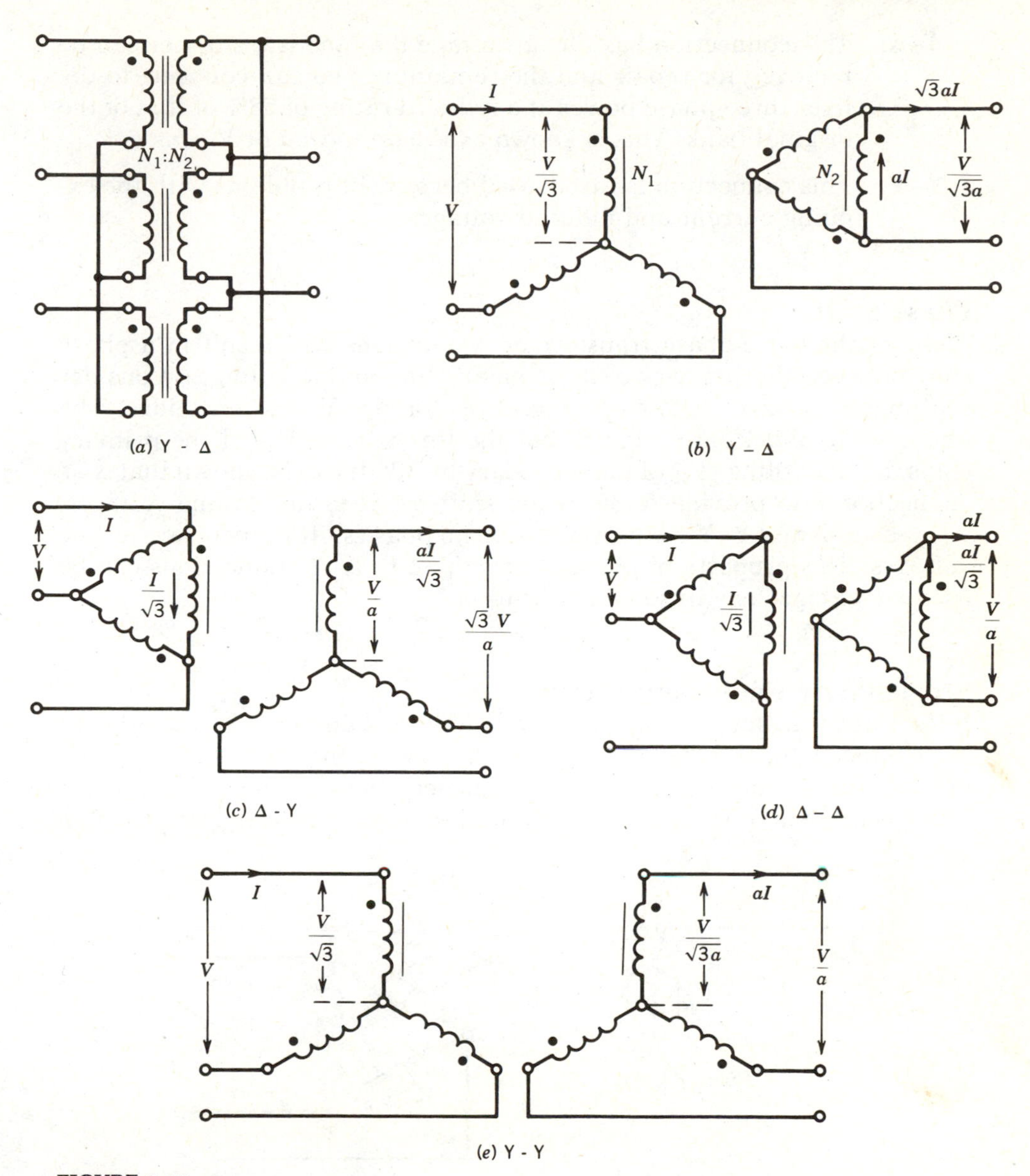

FIGURE 2.17
Three-phase transformer connections.

Δ–Δ: This connection has the advantage that one transformer can be removed for repair and the remaining two can continue to deliver three-phase power at a reduced rating of 58% of that of the original bank. This is known as the *open-delta* or *V connection.*

Y–Y: This connection is rarely used because of problems with the exciting current and induced voltages.

Phase Shift

Some of the three-phase transformer connections will result in a phase shift between the primary and secondary line-to-line voltages. Consider the phasor voltages, shown in Fig. 2.18, for the Y–Δ connections. The phasors V_{AN} and V_a are aligned, but the line voltage V_{AB} of the primary leads the line voltage V_{ab} of the secondary by 30°. It can be shown that Δ–Y connection also provides a 30° phase shift between line-to-line voltages, whereas Δ–Δ and Y–Y connections have no phase shift in their line-to-line voltages. This property of phase shift in Y–Δ or Δ–Y connections can be used advantageously in some applications.

Single-Phase Equivalent Circuit

If the three transformers are practically identical and the source and load are balanced, then the voltages and currents on both primary and secondary sides are balanced. The voltages and currents in one phase are the same as those in other phases, except that there is a phase displacement of

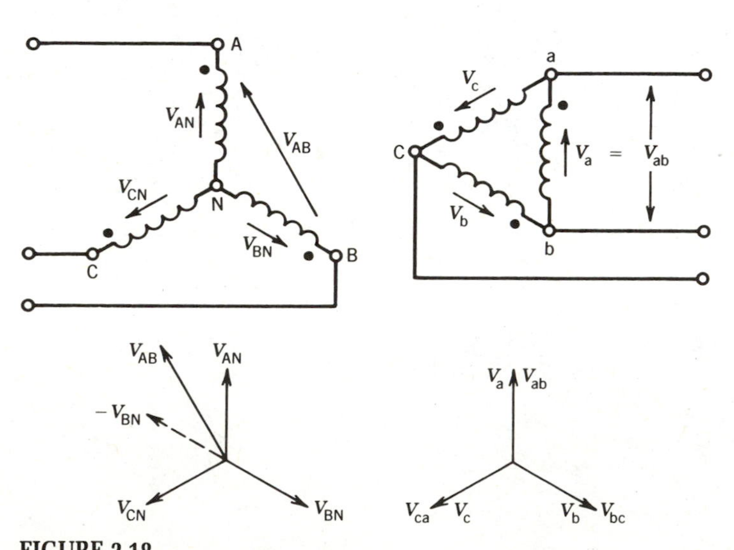

FIGURE 2.18
Phase shift in line-to-line voltages in a three-phase transformer.

120°. Therefore, analysis of one phase is sufficient to determine the variables on the two sides of the transformer. A single-phase equivalent circuit can be conveniently obtained if all sources, transformer windings, and load impedances are considered to be Y-connected. The Y load can be obtained for the Δ load by the well-known Y–Δ transformation, as shown in Fig. 2.19*b*. The equivalent Y representation of the actual circuit (Fig.

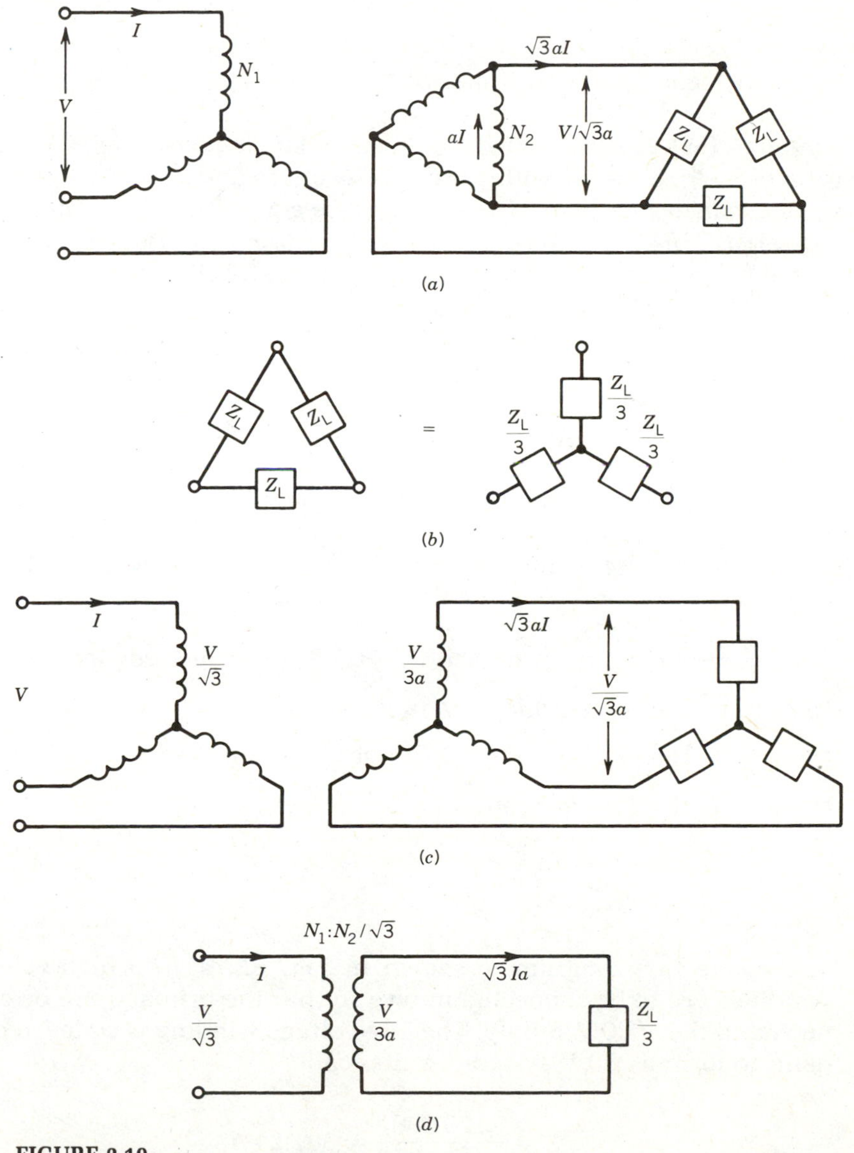

FIGURE 2.19
Three-phase transformer and equivalent circuit.

2.19*a*) is shown in Fig. 2.19*c*, in which the primary and secondary line currents and line-to-line voltages are identical to those of the actual circuit of Fig. 2.19*a*. The turns ratio a' of this equivalent Y–Y transformer is

$$a' = \frac{V/\sqrt{3}}{V/3a} = \sqrt{3}a \tag{2.39}$$

Also, for the actual transformer bank

$$\frac{\text{Primary line-to-line voltage}}{\text{Secondary line-to-line voltage}} = \frac{V}{V/\sqrt{3}\,a} = \sqrt{3}\,a \tag{2.40}$$

Therefore, the turns ratio for the equivalent single-phase transformer is the ratio of the line-to-line voltages on the primary and secondary sides of the actual transformer bank. The single-phase equivalent circuit is shown in Fig. 2.19*d*. This equivalent circuit will be useful if transformers are connected to load or power supply through feeders, as illustrated in Example 2.8.

EXAMPLE 2.7

Three 1ϕ, 50 kVA, 2300/230 V, 60 Hz transformers are connected to form a 3ϕ, 4000/230 V transformer bank. The equivalent impedance of each transformer referred to low voltage is $0.012 + j0.016\ \Omega$. The 3ϕ transformer supplies a 3ϕ, 120 kVA, 230 V, 0.85 PF (lag) load.

(a) Draw a schematic diagram showing the transformer connection.

(b) Determine the transformer winding currents.

(c) Determine the primary voltage (line-to-line) required.

(d) Determine the voltage regulation.

Solution

(a) The connection diagram is shown in Fig. E2.7*a*. The high-voltage windings are to be connected in wye so that the primary can be connected to the 4000 V supply. The low-voltage winding is connected in delta to form a 230 V system for the load.

(b)

$$I_s = \frac{120{,}000}{\sqrt{3} \times 230} = 301.24 \text{ A}$$

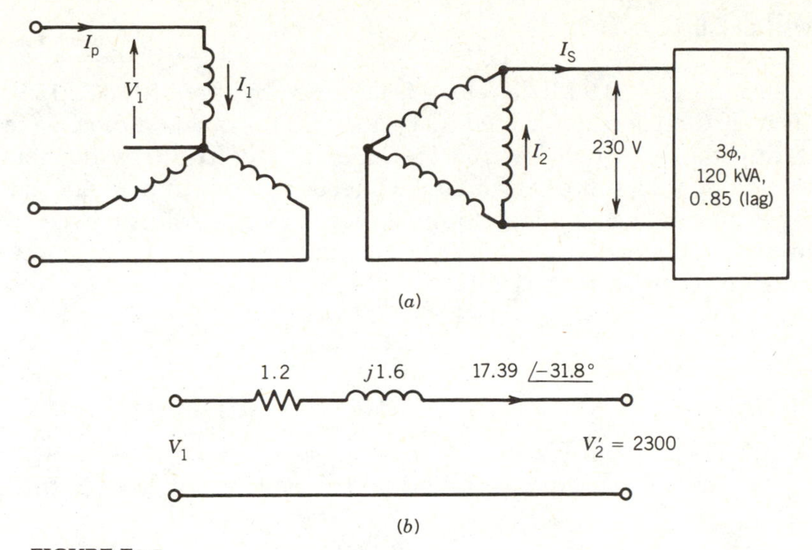

FIGURE E2.7

$$I_2 = \frac{301.24}{\sqrt{3}} = 173.92 \text{ A}$$

$$a = \frac{2300}{230} = 10$$

$$I_1 = \frac{173.92}{10} = 17.39 \text{ A}$$

(c) Computation can be carried out on a per-phase basis.

$$\begin{aligned} Z_{eq1} &= (0.012 + j0.016)10^2 \\ &= 1.2 + j1.6 \ \Omega \\ \phi &= \cos^{-1} 0.85 = 31.8^\circ \end{aligned}$$

The primary equivalent circuit is shown in Fig. E2.7*b*.

$$V_1 = 2300\underline{/0^\circ} + 17.39\underline{/-31.8^\circ}(1.2 + j1.6)$$

$$|V_1| = 2332.4 \text{ V}$$

$$\text{Primary line-to-line voltage} = \sqrt{3}\, V_1 = 4039.8 \text{ V}$$

(d)

$$\text{VR} = \frac{2332.4 - 2300}{2300} \times 100\% = 1.41\%$$

EXAMPLE 2.8

A 3ϕ, 230 V, 27 kVA, 0.9 PF (lag) load is supplied by three 10 kVA, 1330/230 V, 60 Hz transformers connected in Y–Δ by means of a common 3ϕ feeder whose impedance is $0.003 + j0.015\ \Omega$ per phase. The transformers are supplied from a 3ϕ source through a 3ϕ feeder whose impedance is $0.8 + j5.0\ \Omega$ per phase. The equivalent impedance of one transformer referred to the low-voltage side is $0.12 + j0.25\ \Omega$. Determine the required supply voltage if the load voltage is 230 V.

Solution

The circuit is shown in Fig. E2.8*a*.

The equivalent circuit of the individual transformer referred to the high-voltage side is

$$R_{\mathrm{eqH}} + jX_{\mathrm{eqH}} = \left(\frac{1330}{230}\right)^2 (0.12 + j0.25)$$

$$= 4.01 + j8.36$$

The turns ratio of the equivalent Y–Y bank is

$$a' = \frac{\sqrt{3} \times 1330}{230} = 10$$

The single-phase equivalent circuit of the system is shown Fig. E2.8*b*. All the impedances from the primary side can be transferred to the secondary side and combined with the feeder impedance on the secondary side.

$$R = (0.80 + 4.01)\frac{1}{10^2} + 0.003 = 0.051\ \Omega$$

$$X = (5 + 8.36)\frac{1}{10^2} + 0.015 = 0.149\ \Omega$$

The circuit is shown in Fig. E2.8*c*.

$$V_{\mathrm{L}} = \frac{230}{\sqrt{3}}\underline{/0^\circ} = 133\underline{/0^\circ}\ \mathrm{V}$$

$$I_{\mathrm{L}} = \frac{27 \times 10^3}{3 \times 133} = 67.67\ \mathrm{A}$$

$$\phi_{\mathrm{L}} = -\cos^{-1} 0.9 = -25.8^\circ$$

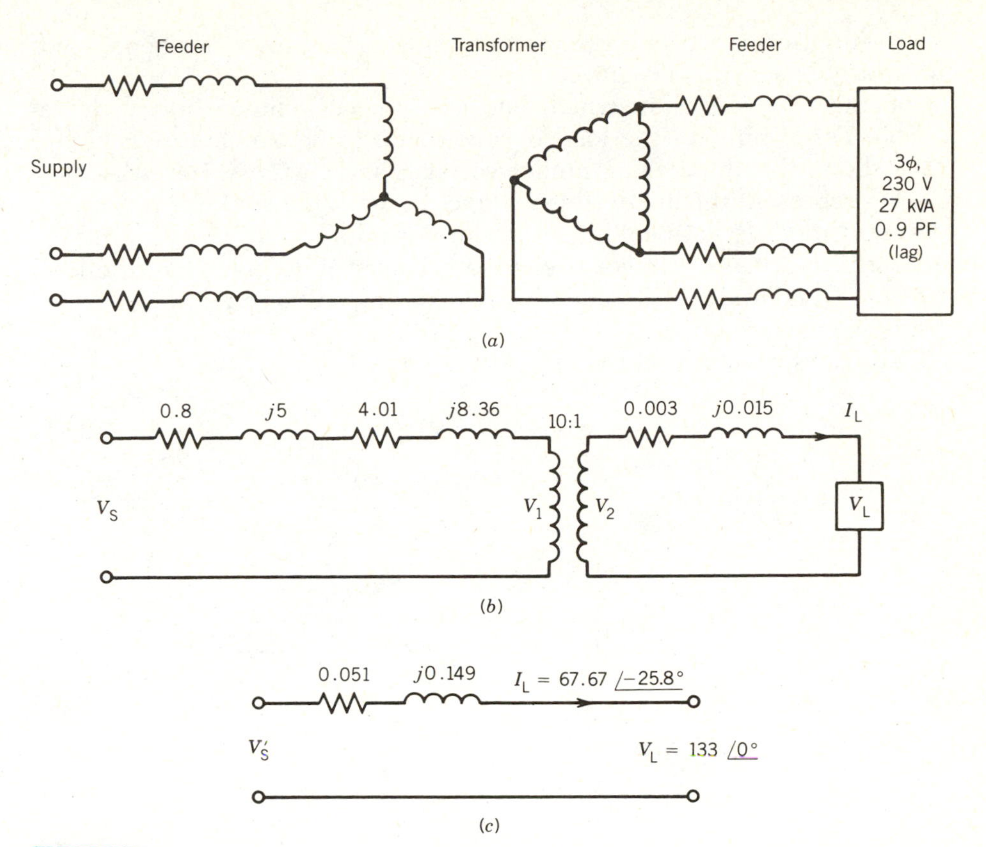

FIGURE E2.8

$$V'_s = 133\angle 0^\circ + 67.67\angle -25.8^\circ\,(0.051 + j0.149)$$
$$= 133\angle 0^\circ + 10.6571\angle 45.3^\circ$$
$$= 140.7\angle 3.1^\circ \text{ V}$$
$$V_s = 140.7 \times 10 = 1407 \text{ V}$$

The line-to-line supply voltage is

$$1407\sqrt{3} = 2437 \text{ V}$$

V Connection

It was stated earlier that in the Δ–Δ connection of three single-phase transformers, one transformer can be removed and the system can still deliver three-phase power to a three-phase load. This configuration is known as an open-delta or V connection. It may be employed in an emergency situation when one transformer must be removed for repair and continuity of service is required.

Consider Fig. 2.20*a*, in which one transformer, shown dotted, is removed. For simplicity the load is considered to be Y-connected. Figure 2.20*b* shows the phasor diagram for voltages and currents. Here V_{AB}, V_{BC}, and V_{CA} represent the line-to-line voltages of the primary; V_{ab}, V_{bc}, and V_{ca} represent the line-to-line voltages of the secondary; and V_{an}, V_{bn}, and V_{cn} represent the phase voltages of the load. For an inductive load, the load currents I_a, I_b, and I_c will lag the corresponding voltages V_{an}, V_{bn}, and V_{cn} by the load phase angle ϕ.

Transformer windings ab and bc deliver power

$$P_{ab} = V_{ab} I_a \cos(30 + \phi) \tag{2.41}$$

$$P_{bc} = V_{cb} I_c \cos(30 - \phi) \tag{2.42}$$

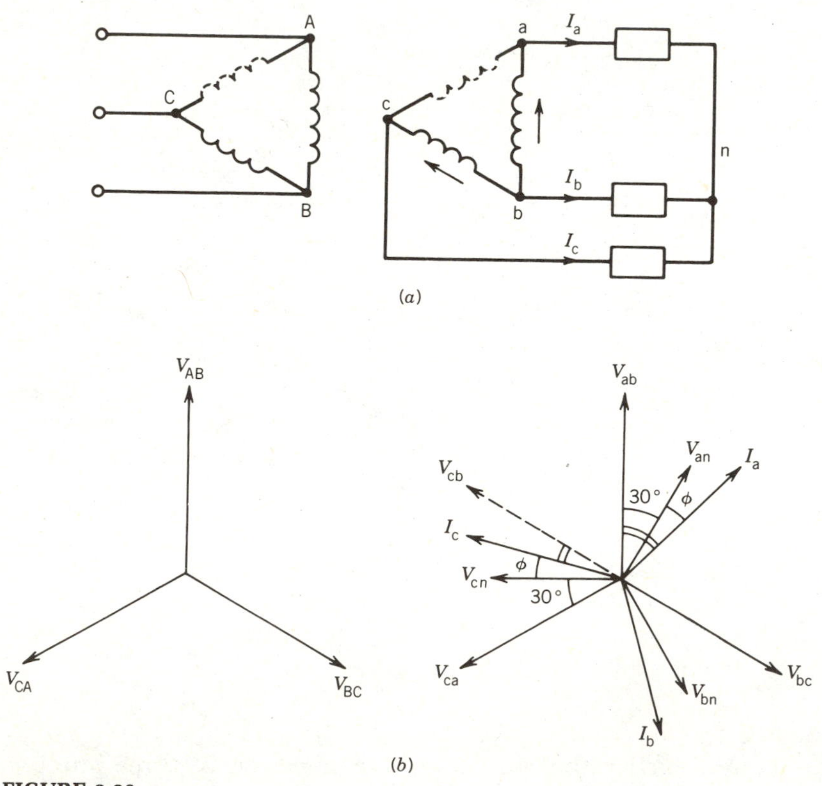

FIGURE 2.20
V connection.

Let

$|V_{ab}| = |V_{cb}| = V,$ voltage rating of the transformer secondary winding

$|I_a| = |I_c| = I,$ current rating of the transformer secondary winding

and $\phi = 0$ for a resistive load. Power delivered to the load by the V connection is

$$P_v = P_{ab} + P_{bc} = 2VI \cos 30° \quad (2.43)$$

With all three transformers connected in delta, the power delivered is

$$P_\Delta = 3VI \quad (2.44)$$

From Eqs. 2.43 and 2.44,

$$\frac{P_V}{P_\Delta} = \frac{2 \cos 30°}{3} = 0.58 \quad (2.45)$$

The V connection is capable of delivering 58% power without overloading the transformer (i.e., not exceeding the current rating of the transformer windings).

2.6.2 THREE-PHASE TRANSFORMER ON A COMMON MAGNETIC CORE (THREE-PHASE UNIT TRANSFORMER

A three-phase transformer can be constructed by having three primary and three secondary windings on a common magnetic core. Consider three single-phase core-type units as shown in Fig. 2.21*a*. For simplicity, only the primary windings have been shown. If balanced three-phase sinusoidal voltages are applied to the windings, the fluxes Φ_a, Φ_b, and Φ_c will also be sinusoidal and balanced. If the three legs carrying these fluxes are merged, the net flux in the merged leg is zero. This leg can therefore be removed as shown in Fig. 2.21*b*. This structure is not convenient to build. However, if section b is pushed in between sections a and c by removing its yokes, a common magnetic structure, shown in Fig. 2.21*c*, is obtained. This core structure can be built using stacked laminations as shown in Fig. 2.21*d*. Both primary and secondary windings of a phase are placed on the same leg. Note that the magnetic paths of legs a and c are somewhat longer than that of leg b (Fig. 2.21*c*). This will result in some im-

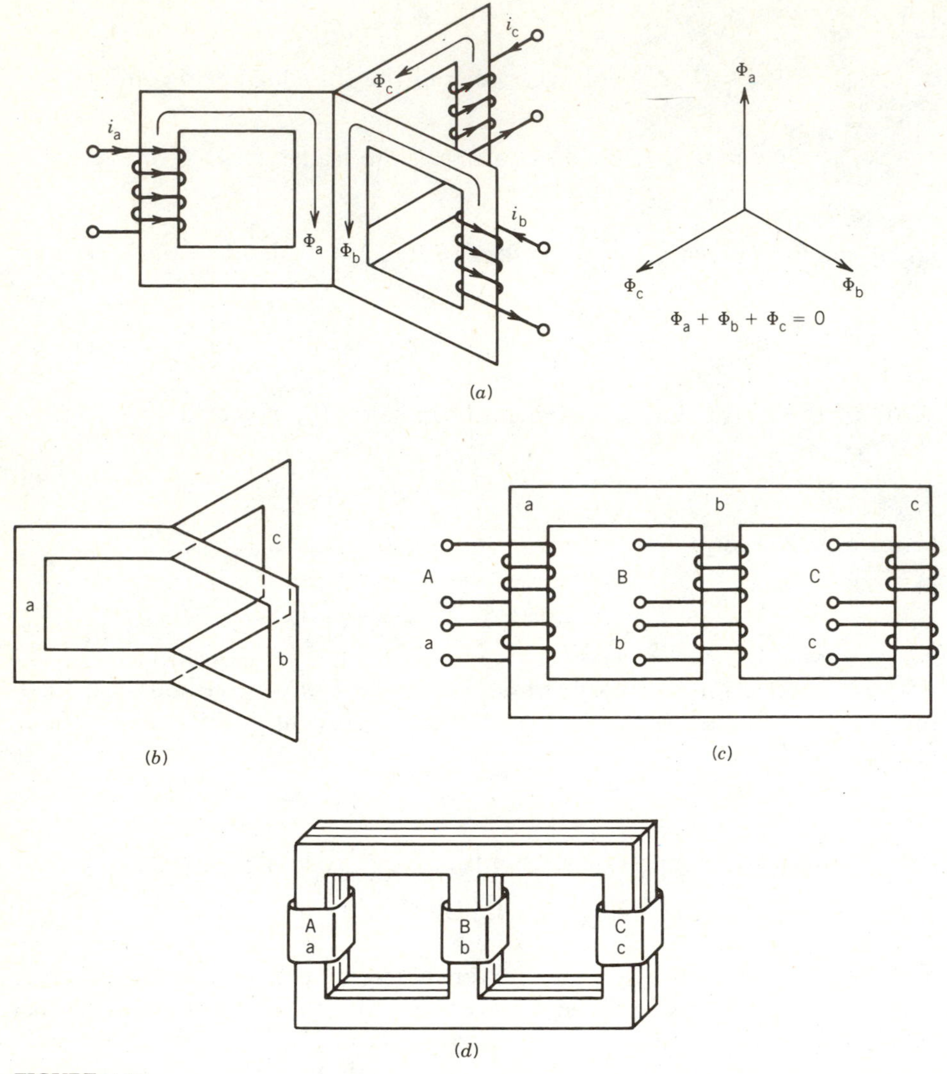

FIGURE 2.21
Development of a three-phase core-type transformer.

balance in the magnetizing currents. However, this imbalance is not significant.

Figure 2.22 shows a picture of a three-phase transformer of this type. Such a transformer weighs less, costs less, and requires less space than a three-phase transformer bank of the same rating. The disadvantage is that if one phase breaks down, the whole transformer must be removed for repair.

FIGURE 2.22
Photograph of a 3ϕ unit transformer. (Courtesy of Westinghouse Canada Inc.)

2.7 HARMONICS IN THREE-PHASE TRANSFORMER BANKS

If a transformer is operated at a higher flux density, it will require less magnetic material. Therefore, from an economic point of view, a transformer is designed to operate in the saturating region of the magnetic core. This makes the exciting current nonsinusoidal, as discussed in Chapter 1. The exciting current will contain the fundamental and all odd harmonics. However, the third harmonic is the predominant one, and for all practical purposes harmonics higher than third (fifth, seventh, ninth, etc.) can be neglected. At rated voltage the third harmonic in the exciting current can be 5 to 10% of the fundamental. At 150% rated voltage, the third harmonic current can be as high as 30 to 40% of the fundamental.

In this section we will study how these harmonics are generated in various connections of the three-phase transformers and ways to limit their effects.

Consider the system shown in Fig. 2.23*a*. The primary windings are connected in Y and the neutral point N of the supply is available. The secondary windings can be connected in Δ.

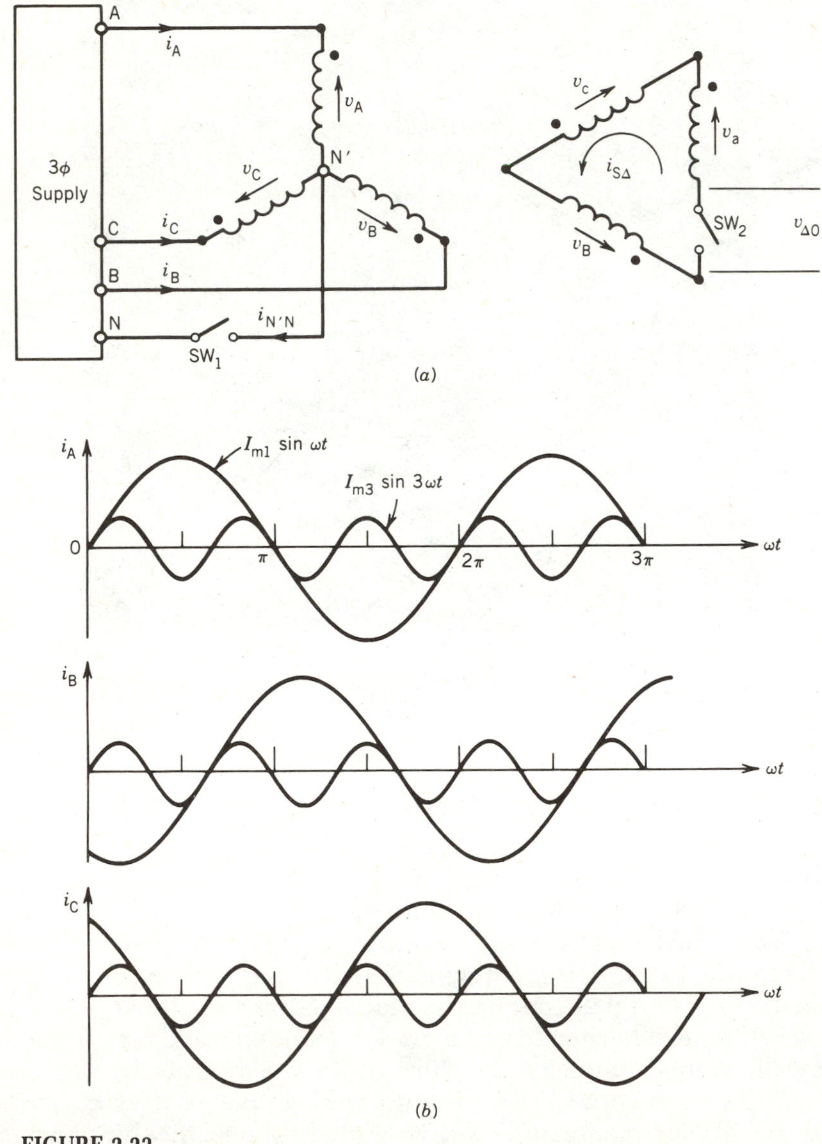

FIGURE 2.23
Harmonic current in three-phase transformer connections. (*a*) Y–Δ connection. (*b*) Waveforms of exciting currents.

Switch SW_1 Closed and Switch SW_2 Open

Because SW_2 is open, no current flows in the secondary windings. The currents flowing in the primary are the exciting currents. We assume that the exciting currents contain only fundamental and third-harmonic currents as shown in Fig. 2.23*b*. Mathematically,

$$i_A = I_{m1} \sin \omega t + I_{m3} \sin 3\,\omega t \tag{2.46}$$

$$i_B = I_{m1} \sin(\omega t - 120°) + I_{m3} \sin 3(\omega t - 120°) \tag{2.47}$$

$$i_C = I_{m1} \sin(\omega t - 240°) + I_{m3} \sin 3(\omega t - 240°) \tag{2.48}$$

The current in the neutral line is

$$i_{N'N} = i_A + i_B + i_C = 3I_{m3} \sin 3\omega t \tag{2.49}$$

Note that fundamental currents in the windings are phase-shifted by 120° from each other, whereas third-harmonic currents are all in phase. The neutral line carries only the third-harmonic current, as can be seen in the oscillogram of Fig. 2.24*a*.

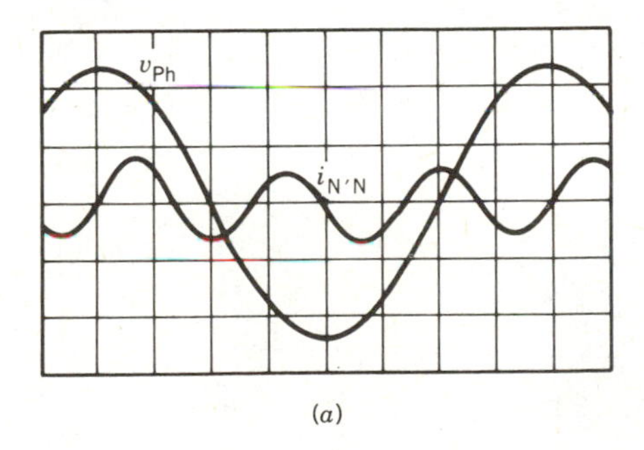

(*a*)

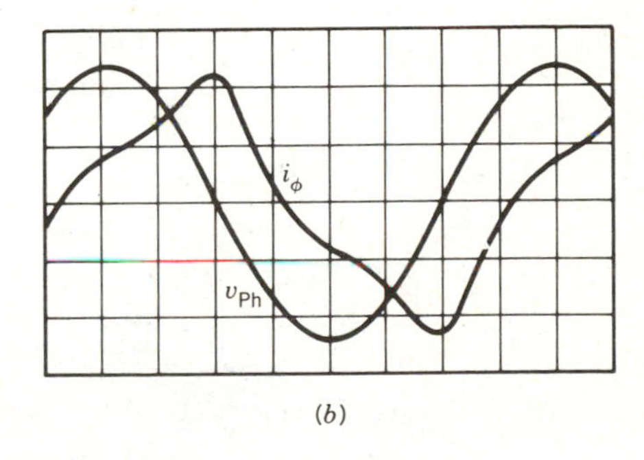

(*b*)

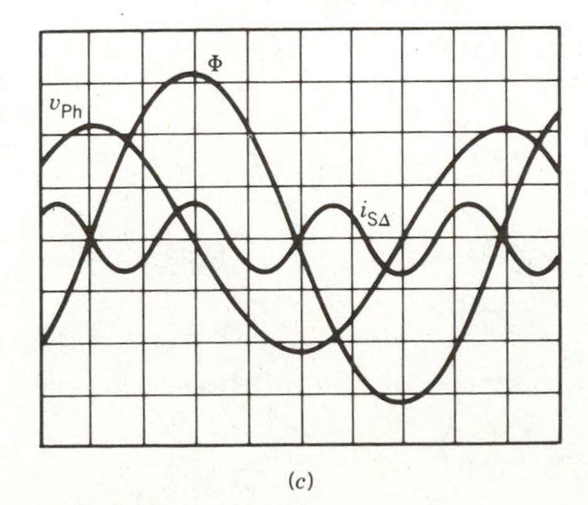

(*c*)

FIGURE 2.24

Oscillograms of currents and voltages in a Y–Δ-connected transformer.

Because the exciting current is nonsinusoidal (Fig. 2.24*b*), the flux in the core and hence the induced voltages in the windings will be sinusoidal. The secondary windings are open and therefore the voltage across a secondary winding will represent the induced voltage.

$$v_{\Delta 0} = v_a + v_b + v_c = 0 \tag{2.50}$$

Both SW_1 and SW_2 Open

In this case the third-harmonic currents cannot flow in the primary windings. Therefore the primary currents are essentially sinusoidal. If the exciting current is sinusoidal, the flux is nonsinusoidal because of nonlinear *B–H* characteristics of the magnetic core, and it contains third-harmonic components. This will induce third-harmonic voltage in the windings. The phase voltages are therefore nonsinusoidal, containing fundamental and third-harmonic voltages.

$$v_A = v_{A1} + v_{A3} \tag{2.51}$$

$$v_B = v_{B1} + v_{B3} \tag{2.52}$$

$$v_C = \underbrace{v_{C1}}_{\text{fundamental voltages}} + \underbrace{v_{C3}}_{\text{third-harmonic voltages}} \tag{2.53}$$

The line-to-line voltage is

$$v_{AB} = v_A - v_B \tag{2.54}$$

$$= v_{A1} - v_{B1} + v_{A3} - v_{B3} \tag{2.55}$$

Because v_{A3} and v_{B3} are in phase and have the same magnitude,

$$v_{A3} - v_{B3} = 0 \tag{2.56}$$

Therefore,

$$v_{AB} = v_{A1} - v_{B1} \tag{2.57}$$

Note that although phase voltages have third-harmonic components, the line-to-line voltages do not.

The open-delta voltage (Fig. 2.23*a*) of the secondary is

$$v_{\Delta 0} = v_a + v_b + v_c \tag{2.58}$$

$$= (v_{a1} + v_{b1} + v_{c1}) + (v_{a3} + v_{b3} + v_{c3}) \tag{2.58a}$$

$$= v_{a3} + v_{b3} + v_{c3} \quad (2.58b)$$

$$= 3v_{a3} \quad (2.58c)$$

The voltage across the open delta is the sum of the three third-harmonic voltages induced in the secondary windings.

Switch SW_1 Open and Switch SW_2 Closed

If switch SW_2 is closed the voltage $v_{\Delta 0}$ will drive a third-harmonic current around the secondary delta. This will provide the missing third-harmonic component of the primary exciting current and consequently the flux and induced voltage will be essentially sinusoidal, as shown in Fig. 2.24*c*.

Y–Y System with Tertiary (Δ) Winding

For high voltages on both sides, it may be desirable to connect both primary and secondary windings in Y, as shown in Fig. 2.25. In this case third-harmonic currents cannot flow either in primary or in secondary. A third set of windings, called a *tertiary winding,* connected in Δ is normally fitted on the core so that the required third-harmonic component of the exciting current can be supplied. This tertiary winding can also supply an auxiliary load if necessary.

2.8 PER-UNIT (PU) SYSTEM

Computations using the actual values of parameters and variables may be lengthy and time-consuming. However, if the quantities are expressed in a per-unit (pu) system, computations are much simplified. The pu quantity

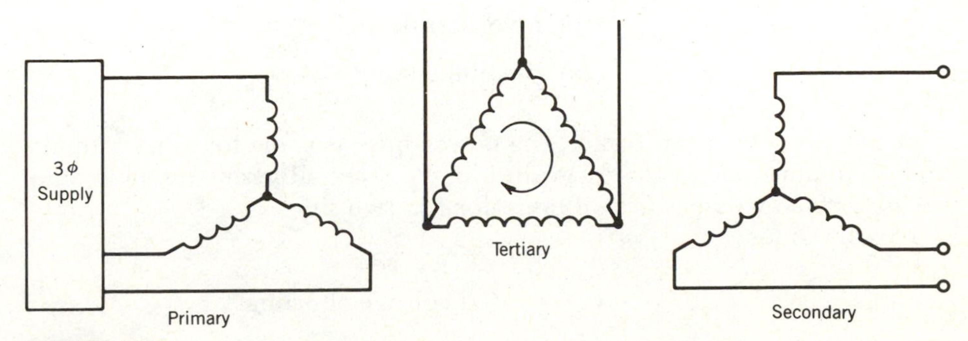

FIGURE 2.25
Y–Y system with a tertiary (Δ) transformer.

is defined as follows:

$$\text{Quantity in pu} = \frac{\text{actual quantity}}{\text{base (or reference) value of the quantity}} \tag{2.59}$$

There are two major advantages in using a per-unit system: (1) The parameters and variables fall in a narrow numerical range when expressed in a per-unit system; this simplifies computations and makes it possible to quickly check the correctness of the computed values. (2) One need not refer circuit quantities from one side to another; therefore a common source of mistakes is removed.

To establish a per-unit system it is necessary to select base (or reference) values for any two of power, voltage, current, and impedance. Once base values for any two of the four quantities have been selected, the base values for the other two can be determined from the relationship among these four quantities. Usually base values of power and voltage are selected first and base values of current and impedance are obtained as follows:

$$P_{\text{base}}, V_{\text{base}} \text{ selected}$$

$$I_{\text{base}} = \frac{P_{\text{base}}}{V_{\text{base}}} \tag{2.60}$$

$$Z_{\text{base}} = \frac{V_{\text{base}}}{I_{\text{base}}} \tag{2.61}$$

$$= \frac{V_{\text{base}}^2}{P_{\text{base}}} \tag{2.62}$$

Although base values can be chosen arbitrarily, normally the rated volt-amperes and rated voltage are taken as the base values for power and voltage, respectively.

$$P_{\text{base}} = \text{rated volt-amperes (VA)}$$

$$V_{\text{base}} = \text{rated voltage (V)}$$

In the case of a transformer, the power base is same for both primary and secondary. However, the values of V_{base} are different on each side, because rated voltages are different for the two sides.

Primary side:

$$V_{\text{base}}, V_{\text{B1}} = V_{\text{R1}} = \text{rated voltage of primary}$$

$$I_{\text{base}}, I_{\text{B1}} = I_{\text{R1}} = \text{rated current of primary}$$

$$Z_{\text{base}}, Z_{\text{B1}} = \frac{V_{\text{R1}}}{I_{\text{R1}}}$$

Let

$$Z_{\text{eq1}} = \text{equivalent impedance of the transformer referred to the primary side}$$

$$Z_{\text{eq1,pu}} = \text{per-unit value of } Z_{\text{eq1}} = Z_{\text{eq1}}/Z_{\text{B1}}$$

$$= \frac{Z_{\text{eq1}}}{V_{\text{R1}}/I_{\text{R1}}}$$

$$= Z_{\text{eq1}} \frac{I_{\text{R1}}}{V_{\text{R1}}} \tag{2.63}$$

Secondary side:

$$V_{\text{base}}, V_{\text{B2}} = V_{\text{R2}} = \text{rated voltage of secondary}$$

$$I_{\text{base}}, I_{\text{B2}} = I_{\text{R2}} = \text{rated current of secondary}$$

$$Z_{\text{base}}, Z_{\text{B2}} = \frac{V_{\text{R2}}}{I_{\text{R2}}}$$

Let

$$Z_{\text{eq2}} = \text{equivalent impedance referred to the secondary side}$$

$$Z_{\text{eq2,pu}} = \text{per-unit value of } Z_{\text{eq2}}$$

$$= \frac{Z_{\text{eq2}}}{Z_{\text{B2}}} \tag{2.64}$$

$$= \frac{Z_{\text{eq1}}/a^2}{Z_{\text{B1}}/a^2}$$

$$= \frac{Z_{\text{eq1}}}{Z_{\text{B1}}} \tag{2.65}$$

$$Z_{\text{eq2,pu}} = Z_{\text{eq1,pu}} \tag{2.66}$$

Therefore, the per-unit transformer impedance is the same referred to either side of the transformer. This is another advantage of expressing quantities in a per-unit system.

In a transformer, when voltages or currents of either side are expressed in a per-unit system, they have the same per-unit values.

$$I_{1,\mathrm{pu}} = \frac{I_1}{I_{\mathrm{B1}}} = \frac{I_1}{I_{\mathrm{R1}}} = \frac{I_2/a}{I_{\mathrm{R2}}/a} = \frac{I_2}{I_{\mathrm{R2}}} = \frac{I_2}{I_{\mathrm{B2}}} = I_{2,\mathrm{pu}} \tag{2.67}$$

$$V_{1,\mathrm{pu}} = \frac{V_1}{V_{\mathrm{B1}}} = \frac{V_1}{V_{\mathrm{R1}}} = \frac{aV_2}{aV_{\mathrm{R2}}}$$

$$= \frac{V_2}{V_{\mathrm{R2}}} = \frac{V_2}{V_{\mathrm{B2}}} = V_{2,\mathrm{pu}} \tag{2.68}$$

2.8.1 TRANSFORMER EQUIVALENT CIRCUIT IN PER-UNIT FORM

The equivalent circuit of a transformer referred to the primary side is shown in Fig. 2.26*a*. The equation in terms of actual values is

$$V_1 = I_1 Z_{\mathrm{eq1}} + V_2' \tag{2.69}$$

The equation in per-unit form can be obtained by dividing Eq. 2.69 throughout by the base value of the primary voltage.

$$\frac{V_1}{V_{\mathrm{R1}}} = \frac{I_1 Z_{\mathrm{eq1}}}{V_{\mathrm{R1}}} + \frac{V_2'}{V_{\mathrm{R1}}}$$

$$= \frac{I_1 Z_{\mathrm{eq1}}}{I_{\mathrm{R1}} Z_{\mathrm{B1}}} + \frac{\mathrm{a}V_2}{\mathrm{a}V_{\mathrm{R2}}}$$

$$V_{1,\mathrm{pu}} = I_{1,\mathrm{pu}} Z_{\mathrm{eq1,pu}} + V_{2,\mathrm{pu}} \tag{2.70}$$

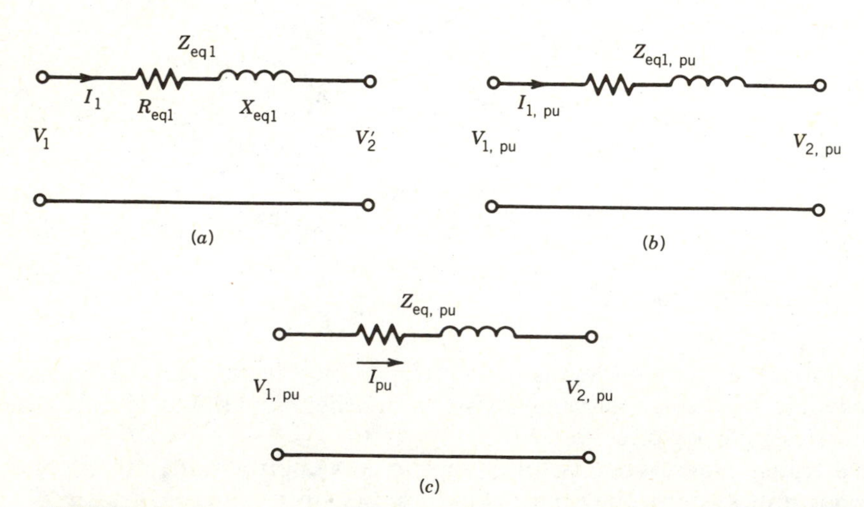

FIGURE 2.26
Transformer equivalent circuit in per-unit form.

Based on Eq. 2.70, the equivalent circuit in per-unit form is shown in Fig. 2.26*b*. It has been shown that the voltages, currents, and impedances in per-unit representation have the same values whether they are referred to primary or secondary. Hence the transformer equivalent circuit in per-unit form for either side is the one shown in Fig. 2.26*c*. Note that the values of $V_{1,\text{pu}}$ and $V_{2,\text{pu}}$ are generally close to 1 pu, and this makes the analysis somewhat easier.

2.8.2 FULL-LOAD COPPER LOSS

Let

$$\begin{aligned} P_{\text{cu,FL}} &= \text{full-load copper loss} \\ &= I_{\text{R1}}^2 R_{\text{eq1}} \end{aligned} \tag{2.71}$$

The full-load copper loss in per-unit form based on the volt-ampere rating of the transformer is

$$\begin{aligned} P_{\text{cu,FL}}\big|_{\text{pu}} &= \frac{I_{\text{R1}}^2 R_{\text{eq1}}}{P_{\text{base}}} \\ &= \frac{I_{\text{R1}}^2 R_{\text{eq1}}}{V_{\text{R1}} I_{\text{R1}}} \\ &= \frac{R_{\text{eq1}}}{V_{\text{R1}}/I_{\text{R1}}} \\ &= \frac{R_{\text{eq1}}}{Z_{\text{B1}}} \\ &= R_{\text{eq1,pu}} \end{aligned} \tag{2.72}$$

Hence the transformer resistance expressed in per-unit form also represents the full-load copper loss in per-unit form. The per-unit value of the resistance is therefore more useful than its ohmic value in determining the performance of a transformer.

EXAMPLE 2.9

The exciting current of a 1ϕ, 10 kVA, 2200/220 V, 60 Hz transformer is 0.25 A when measured on the high-voltage side. Its equivalent impedance is $10.4 + j31.3\ \Omega$ when referred to the high-voltage side. Taking the transformer rating as base,

(a) Determine the base values of voltages, currents, and impedances for both high-voltage and low-voltage sides.

(b) Express the exciting current in per-unit form for both high-voltage and low-voltage sides.

(c) Obtain the equivalent circuit in per-unit form.

(d) Find the full-load copper loss in per-unit form.

(e) Determine the per-unit voltage regulation (using the per-unit equivalent circuit from part c) when the transformer delivers 75% full load at 0.6 lagging power factor.

Solution

(a) $P_{\text{base}} = 10{,}000$ VA, $a = 10$.

Using the subscripts H and L to indicate high-voltage and low-voltage sides, the base values of voltages, currents, and impedances are

$$V_{\text{base,H}} = 2200 \text{ V} = 1 \text{ pu}$$

$$V_{\text{base,L}} = 220 \text{ V} = 1 \text{ pu}$$

$$I_{\text{base,H}} = \frac{10{,}000}{2200} = 4.55 \text{ A} = 1 \text{ pu}$$

$$I_{\text{base,L}} = \frac{10{,}000}{220} = 45.5 \text{ A} = 1 \text{ pu}$$

$$Z_{\text{base,H}} = \frac{2200}{4.55} = 483.52\ \Omega = 1 \text{ pu}$$

$$Z_{\text{base,L}} = \frac{220}{45.5} = 4.835\ \Omega = 1 \text{ pu}$$

(b)

$$I_{\phi\text{H}}\big|_{\text{pu}} = \frac{0.25}{4.55} = 0.055 \text{ pu}$$

The exciting current referred to the low-voltage side is $0.25 \times 10 = 2.5$ A. Its per-unit value is

$$I_{\phi\text{L}}\big|_{\text{pu}} = \frac{2.5}{45.5} = 0.055 \text{ pu}$$

Note that although the actual values of the exciting current are different for the two sides, the per-unit values are the same. For this transformer, this means that the exciting current is 5.5% of the rated current of the side in which it is measured.

(c)
$$Z_{eq,H}|_{pu} = \frac{10.4 + j31.3}{483.52} = 0.0215 + j0.0647 \text{ pu}$$

The equivalent impedance referred to the low-voltage side is

$$Z_{eq,L} = (10.4 + j31.3)\frac{1}{100}$$
$$= 0.104 + j0.313\ \Omega$$

Its per-unit value is

$$Z_{eq,L}|_{pu} = \frac{0.104 + j0.313}{4.835} = 0.0215 + j0.0647 \text{ pu}$$

The per-unit values of the equivalent impedances referred to the high- and low-voltage sides are the same. The per-unit equivalent circuit is shown in Fig. E2.9.

(d)
$$P_{cu,FL} = 4.55^2 \times 10.4 \text{ W}$$
$$= 215 \text{ W}$$
$$P_{cu,FL}|_{pu} = \frac{215}{10{,}000} = 0.0215 \text{ pu}$$

Note that this is same as the per-unit value of the equivalent resistance.

(e) From Fig. E2.9

$$I = 0.75\angle{-53.13^\circ} \text{ pu}$$
$$V_2 = 1\angle{0^\circ} \text{ pu}$$
$$Z_{eq,pu} = 0.0215 + j0.0647 \text{ pu}$$

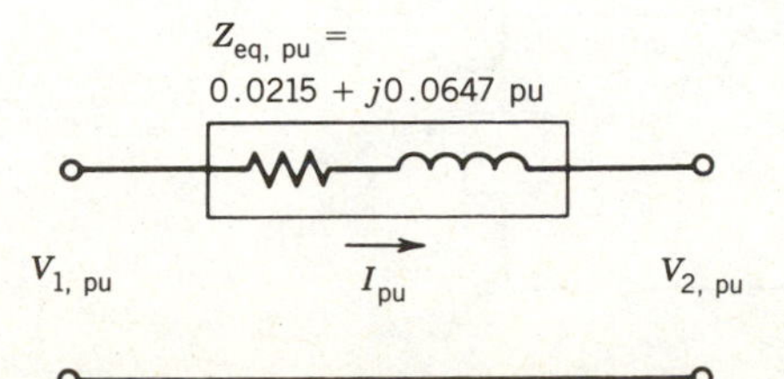

FIGURE E2.9

$$V_1 = 1\underline{/0^\circ} + 0.75\underline{/-53.13^\circ}\,(0.0215 + j0.0647)$$
$$= 1.0486\underline{/9^\circ} \text{ pu}$$

$$\text{Voltage regulation} = \frac{1.0486 - 1.0}{1.0} = 0.0486 \text{ pu}$$
$$= 4.86\% \quad \text{(see Example 2.3)}$$

Note that the computation in the per-unit system involves smaller numerical values than the computation using actual values (see Example 2.3). Also, the value of V_1 in pu form promptly gives a perception of voltage regulation.

PROBLEMS

2.1 A resistive load varies from 1 to 0.5 Ω. The load is supplied by an ac generator through an ideal transformer whose turns ratio can be changed by using different taps as shown in Fig. P2.1. The generator can be modeled as a constant voltage of 100 V (rms) in series with an inductive reactance of $j1$ Ω. For maximum power transfer to the load, the effective load resistance seen at the transformer primary (generator side) must equal the series impedance of the generator; that is, the referred value of R to the primary side is always 1 Ω.

(a) Determine the range of turns ratio for maximum power transfer to the load.

(b) Determine the range of load voltages for maximum power transfer.

(c) Determine the power transferred.

2.2 A 1ϕ, 100 kVA, 1000/100 V transformer gave the following test results:

open-circuit test (HV side open)

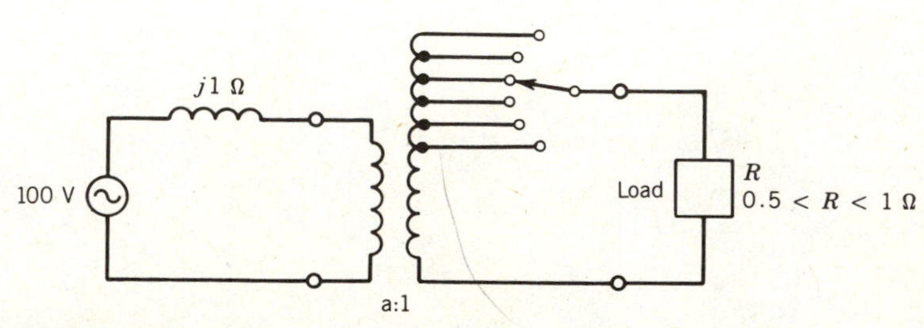

FIGURE P2.1

100 V, 6.0 A, 400 W

short-circuit test

50 V, 100 A, 1800 W

(a) Determine the rated voltage and rated current for the high-voltage and low-voltage sides.

(b) Derive an approximate equivalent circuit referred to the HV side.

(c) Determine the voltage regulation at full load, 0.6 PF leading.

(d) Draw the phasor diagram for condition (c).

2.3 A 1ϕ, 3 kVA, 240/120 V, 60 Hz transformer has the following parameters:

$$R_{HV} = 0.25\ \Omega, \qquad R_{LV} = 0.05\ \Omega$$

$$X_{HV} = 0.75\ \Omega, \qquad X_{LV} = 0.18\ \Omega$$

(a) Determine the voltage regulation when the transformer is supplying full load at 110 V and 0.9 leading power factor.

(b) If the load terminals are accidentally short-circuited, determine the currents in the high-voltage and low-voltage windings.

2.4 A single-phase, 300 kVA, 11 kV/2.2 kV, 60 Hz transformer has the following equivalent circuit parameters referred to the high-voltage side:

$$R_{c(HV)} = 57.6\ \text{k}\Omega, \qquad X_{m(HV)} = 16.34\ \text{k}\Omega$$

$$R_{eq(HV)} = 2.784\ \Omega, \qquad X_{eq(HV)} = 8.45\ \Omega$$

(a) Determine

i. No-load current as a percentage of full-load current.

ii. No-load power loss (i.e., core loss).

iii. No-load power factor.

iv. Full-load copper loss.

(b) If the load impedance on the low-voltage side is $Z_{load} = 16\angle 60°\ \Omega$ determine the voltage regulation using the approximate equivalent circuit.

2.5 A 1ϕ, 25 kVA, 2300/230 V transformer has the following parameters:

$$Z_{eq,H} = 4.0 + j5.0\ \Omega$$

$$R_{c,L} = 450\ \Omega$$

$$X_{m,L} = 300\ \Omega$$

The transformer is connected to a load whose power factor varies. Determine the worst-case voltage regulation for full-load output.

2.6 For the transformer in Problem 2.5,

(a) Determine efficiency when the transformer delivers full load at rated voltage and 0.85 power factor lagging.

(b) Determine the percentage loading of the transformer at which the efficiency is a maximum and calculate this efficiency if the power factor is 0.85 and load voltage is 230 V.

2.7 A 1ϕ, 10 kVA, 2400/240 V, 60 Hz distribution transformer has the following characteristics:

Core loss at full voltage = 100 W

Copper loss at half load = 60 W

(a) Determine the efficiency of the transformer when it delivers full load at 0.8 power factor lagging.

(b) Determine the per-unit rating at which the transformer efficiency is a maximum. Determine this efficiency if the load power factor is 0.9.

The transformer has the following load cycle:

No load for 6 hours

70% full load for 10 hours at 0.8 PF

90% full load for 8 hours at 0.9 PF

Determine the all-day efficiency of the transformer.

2.8 A 1ϕ, 10 kVA, 460/120 V, 60 Hz transformer has an efficiency of 96% when it delivers 9 kW at 0.9 power factor. This transformer is connected as an autotransformer to supply load to a 460 V circuit from a 580 V source.

(a) Show the autotransformer connection.

(b) Determine the maximum kVA the autotransformer can supply to the 460 V circuit.

(c) Determine the efficiency of the autotransformer for full load at 0.9 power factor.

2.9 Reconnect the windings of a 1ϕ, 3 kVA, 240/120 V, 60 Hz transformer so that it can supply a load at 330 V from a 110 V supply.

(a) Show the connection.

(b) Determine the maximum kVA the reconnected transformer can deliver.

2.10 Three 1ϕ, 10 kVA, 460/120 V, 60 Hz transformers are connected to form a 3ϕ, 460/208 V transformer bank. The equivalent impedance of each transformer referred to the high-voltage side is $1.0 + j2.0\ \Omega$. The transformer delivers 20 kW at 0.8 power factor (leading).

(a) Draw a schematic diagram showing the transformer connection.

(b) Determine the transformer winding current.

(c) Determine the primary voltage.

(d) Determine the voltage regulation.

2.11 Three 1ϕ, 100 kVA, 2300/460 V, 60 Hz transformers are connected to form a 3ϕ, 2300/460 V transformer bank. The equivalent impedance of each transformer referred to its low-voltage side is $0.045 + j0.16\ \Omega$. The transformer is connected to a 3ϕ source through 3ϕ feeders, the impedance of each feeder being $0.5 + j1.5\ \Omega$. The transformer delivers full load at 460 V and 0.85 power factor lagging.

(a) Draw a schematic diagram showing the transformer connection.

(b) Determine the single-phase equivalent circuit.

(c) Determine the sending end voltage of the 3ϕ source.

(d) Determine the transformer winding currents.

2.12 Two identical 250 kVA, 230/460 V transformers are connected in open delta to supply a balanced 3ϕ load at 460 V and a power factor of 0.8 lagging. Determine

(a) The maximum secondary line current without overloading the transformers.

(b) The real power delivered by each transformer.

(c) The primary line currents.

(d) If a similar transformer is now added to complete the Δ, find the percentage increase in real power that can be supplied. Assume that the load voltage and power factor remain unchanged at 460 V and 0.8 lagging respectively.

2.13 Three identical single-phase transformers, each of rating 20 kVA, 2300/230 V, 60 Hz, are connected Y–Y to form a 3ϕ transformer bank. The high-voltage side is connected to a 3ϕ, 4000 V, 60 Hz supply and the secondary is left open. The neutral of the primary is not connected to the neutral of the supply. The voltage between the primary neutral and the supply neutral is measured to be 1200 V.

(a) Describe the voltage waveform between primary neutral and supply neutral. Neglect harmonics higher than third.

(b) Determine the ratio of (i) phase voltages of the two sides and (ii) line voltages of the two sides.

(c) Determine the ratio of the rms line-to-line voltage to the rms line-to-neutral voltage on each side.

2.14 A 1ϕ, 200 kVA, 2100/210 V, 60 Hz transformer has the following characteristics. The impedance of the high-voltage winding is $0.25 + j1.5\ \Omega$ with the low-voltage winding short-circuited. The admittance (i.e., inverse of impedance) of the low-voltage winding is $0.025 - j0.075$ mhos with the high-voltage winding open-circuited.

(a) Taking the transformer rating as base, determine the base values of power, voltage, current, and impedance for both the high-voltage and low-voltage sides of the transformer.

(b) Determine the per-unit value of the equivalent resistance and leakage reactance of the transformer.

(c) Determine the per-unit value of the excitation current at rated voltage.

(d) Determine the per-unit value of the total power loss in the transformer at full-load output condition.

2.15 A single-phase transformer has an equivalent leakage reactance of 0.04 per unit. The full-load copper loss is 0.015 per unit and the no-load power loss at rated voltage is 0.01 pu. The transformer supplies full-load power at rated voltage and 0.85 lagging power factor.

(a) Determine the efficiency of the transformer.

(b) Determine the voltage regulation.

2.16 A 1ϕ, 10 kVA, 2200/220 V, 60 Hz transformer has the following characteristics:

$$\text{No-load core loss} = 100\ \text{W}$$

$$\text{Full-load copper loss} = 215\ \text{W}$$

Write a computer program to study the variation of efficiency with output kVA load and load power factor. The program should

(a) Yield the results in a tabular form showing power factor, per unit kVA load (i.e., X), and efficiency.

(b) Produce a plot of efficiency versus percent kVA load for power factors of 1.0, 0.8, 0.6, 0.4, and 0.2.

ELECTROMECHANICAL ENERGY CONVERSION

CHAPTER three

Various devices can convert electrical energy to mechanical energy and vice versa. The structures of these devices may be different depending on the functions they perform. Some devices are used for continuous energy conversion, and these are known as *motors* and *generators*. Other devices are used to produce translational forces whenever necessary and are known as actuators, such as solenoids, relays, and electromagnets. The various converters may be different structurally, but they all operate on similar principles.

This book deals with converters that use a magnetic field as the medium of energy conversion. In this chapter the basic principles of force production in electromagnetic energy conversion systems are discussed. Some general relationships are derived to tie together all conversion devices and to demonstrate that they all operate on the same basic principle.

3.1 ENERGY CONVERSION PROCESS

There are various methods for calculating the force or torque developed in an energy conversion device. The method used here is based on the principle of conservation of energy, which states that energy can neither be created nor destroyed; it can only be changed from one form to another. An electromechanical converter system has three essential parts: (1) an electric system, (2) a mechanical system, and (3) a coupling field as shown in Fig. 3.1. The energy transfer equation is as follows:

$$\begin{matrix}\text{Electrical} \\ \text{energy input} \\ \text{from source}\end{matrix} = \begin{matrix}\text{mechanical} \\ \text{energy output}\end{matrix} + \begin{matrix}\text{increase} \\ \text{in stored} \\ \text{energy in} \\ \text{coupling} \\ \text{field}\end{matrix} + \begin{matrix}\text{energy} \\ \text{losses}\end{matrix} \qquad (3.1)$$

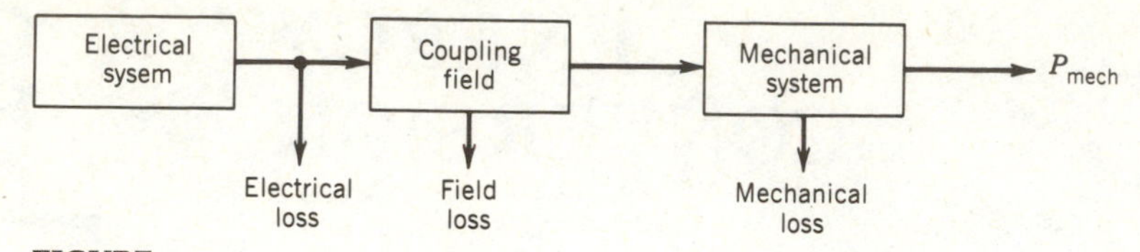

FIGURE 3.1
Electromechanical converter system.

The electrical energy loss is the heating loss due to current flowing in the winding of the energy converter. This loss is known as the i^2R loss in the resistance (R) of the winding. The field loss is the core loss due to changing magnetic field in the magnetic core. The mechanical loss is the friction and windage loss due to the motion of the moving components. All these losses are converted to heat. The energy balance equation 3.1 can therefore be written as

Electrical energy input from source − resistance loss	=	mechanical energy output + friction and windage loss	+	increase in stored field energy + core loss	(3.2)

Now consider a differential time interval dt during which an increment of electrical energy dW_e (excluding the i^2R loss) flows to the system. During this time dt, let dW_f be the energy supplied to the field (either stored or lost, or part stored and part lost) and dW_m the energy converted to mechanical form (in useful form or as loss, or part useful and part as loss). In differential forms, Eq. 3.2 can be expressed as

$$dW_e = dW_m + dW_f \tag{3.3}$$

Core losses are usually small, and if they are neglected, dW_f will represent the change in the stored field energy. Similarly, if friction and windage losses can be neglected, then all of dW_m will be available as useful mechanical energy output. Even if these losses cannot be neglected they can be dealt with separately, as done in other chapters of this book. The losses do not contribute to the energy conversion process.

3.2 FIELD ENERGY

Consider the electromechanical system of Fig. 3.2. The movable part can be held in static equilibrium by the spring. Let us assume that the movable part is held stationary at some air gap and the current is increased from zero to a value i. Flux will be established in the magnetic system.

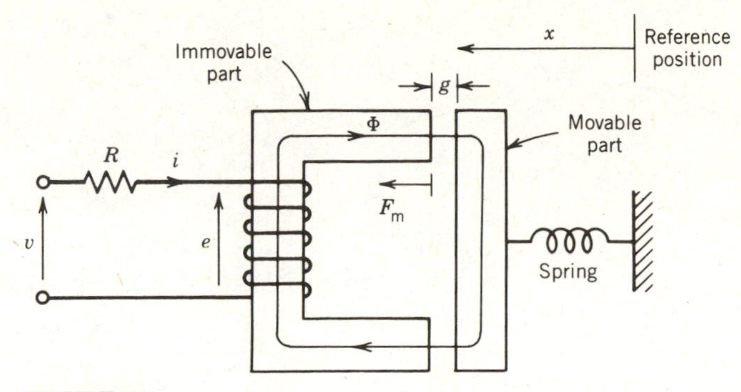

FIGURE 3.2
Example of an electromechanical system.

Obviously,

$$dW_m = 0 \tag{3.4}$$

and from Eqs. 3.3 and 3.4,

$$dW_e = dW_f \tag{3.5}$$

If core loss is neglected, all the incremental electrical energy input is stored as incremental field energy. Now,

$$e = \frac{d\lambda}{dt} \tag{3.6}$$

$$dW_e = ei\,dt \tag{3.7}$$

From Eqs. 3.5, 3.6, and 3.7,

$$dW_f = i\,d\lambda \tag{3.8}$$

The relationship between coil flux linkage λ and current i for a particular air gap length is shown in Fig. 3.3. The incremental field energy dW_f is shown as the crosshatched area in this figure. When the flux linkage is increased from zero to λ, the energy stored in the field is

$$W_f = \int_0^{\lambda} i\,d\lambda \tag{3.9}$$

This integral represents the area between the λ axis and the λ–i characteristic, the entire area shown shaded in Fig. 3.3. Other useful expressions can also be derived for the field energy of the magnetic system.

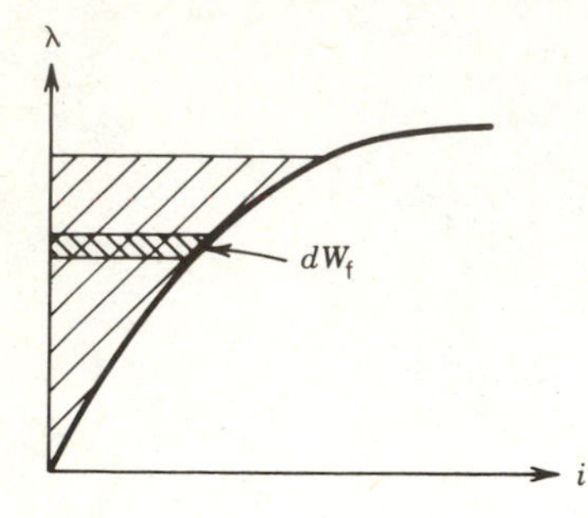

FIGURE 3.3
λ–i characteristic for the system in Fig. 3.2 for a particular air gap length.

Let

H_c = magnetic intensity in the core
H_g = magnetic intensity in the air gap
l_c = length of the magnetic core material
l_g = length of the air gap

Then

$$Ni = H_c l_c + H_g l_g \tag{3.10}$$

Also

$$\lambda = N\Phi \tag{3.11}$$

$$= NAB \tag{3.12}$$

where A is the cross-sectional area of the flux path
B is the flux density, assumed same throughout

From Eqs. 3.9, 3.10, and 3.12,

$$W_f = \int \frac{H_c l_c + H_g l_g}{N} NA\, dB \tag{3.13}$$

For the air gap,

$$H_g = \frac{B}{\mu_0} \tag{3.14}$$

From Eqs. 3.13 and 3.14,

$$W_f = \int \left(H_c l_c + \frac{B}{\mu_0} l_g\right) A\, dB \tag{3.15}$$

$$= \int \left(H_c\, dB\, A l_c + \frac{B}{\mu_0}\, dB\, l_g A\right)$$

$$= \int H_c\, dB \times \text{volume of magnetic material}$$

$$+ \frac{B^2}{2\mu_0} \times \text{volume of air gap} \tag{3.16}$$

$$= w_{fc} \times V_c + w_{fg} \times V_g \tag{3.17}$$

$$= W_{fc} + W_{fg} \tag{3.18}$$

where $w_{fc} = \int H_c\, dB_c$ is the energy density in the magnetic material

$w_{fg} = B^2/2\mu_0$ is the energy density in the air gap

V_c is the volume of the magnetic material

V_g is the volume of the air gap

W_{fc} is the energy in the magnetic material

W_{fg} is the energy in the air gap

Normally, energy stored in the air gap (W_{fg}) is much larger than the energy stored in the magnetic material (W_{fc}). In most cases W_{fc} can be neglected.

For a linear magnetic system

$$H_c = \frac{B_c}{\mu_c} \tag{3.19}$$

Therefore

$$w_{fc} = \int \frac{B_c}{\mu_c}\, dB_c = \frac{B_c^2}{2\mu_c} \tag{3.20}$$

The field energy of the system of Fig. 3.2 can be obtained by using either of Eqs. 3.9 and 3.16.

EXAMPLE 3.1

The dimensions of the actuator system of Fig. 3.2 are shown in Fig. E3.1. The magnetic core is made of cast steel whose B–H characteristic is shown in Fig. 1.7. The coil has 250 turns and the coil resistance is 5 ohms. For a

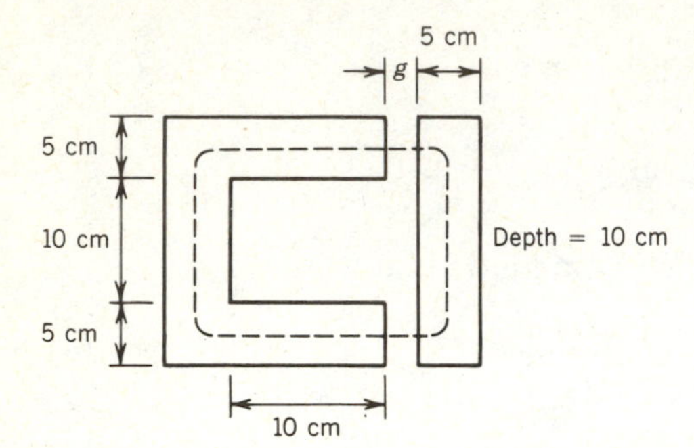

FIGURE E3.1

fixed air gap length $g = 5$ mm, a dc source is connected to the coil to produce a flux density of 1.0 tesla in the air gap.

(a) Find the voltage of the dc source.

(b) Find the stored field energy.

Solution

(a) From Fig. 1.7, magnetic field intensity in the core material (cast steel) for a flux density of 1.0 T is

$$H_c = 670 \text{ At/m}$$

Length of flux path in the core is

$$l_c \simeq 2(10 + 5) + 2(10 + 5) \text{ cm}$$
$$= 60 \text{ cm}$$

The magnetic intensity in the air gap is

$$H_g = \frac{B_g}{\mu_0} = \frac{1.0}{4\pi 10^{-7}} \text{ At/m}$$
$$= 795.8 \times 10^3 \text{ At/m}$$

The mmf required is

$$Ni = 670 \times 0.6 + 795.8 \times 10^3 \times 2 \times 5 \times 10^{-3} \text{ At}$$
$$= 402 + 7958$$
$$= 8360 \text{ At}$$

$$i = \frac{8360}{250} \text{ A}$$

$$= 33.44 \text{ A}$$

Voltage of the dc source is

$$V_{dc} = 33.44 \times 5 = 167.2 \text{ V}$$

(b) Energy density in the core is

$$w_{fc} = \int_0^{1.0} H\, dB$$

This is the energy density given by the area enclosed between the B axis and the B–H characteristic for cast steel in Fig. 1.7. This area is

$$w_{fc} \simeq \tfrac{1}{2} \times 1 \times 700$$

$$= 350 \text{ J/m}^3$$

The volume of steel is

$$V_c = 2(0.05 \times 0.10 \times 0.20) + 2(0.05 \times 0.10 \times 0.10)$$

$$= 0.0015 \text{ m}^3$$

The stored energy in the core is

$$W_{fc} = 350 \times 0.0015 \text{ J}$$

$$= 0.53 \text{ J}$$

The energy density in the air gap is

$$w_{fg} = \frac{1.0^2}{2 \times 4\pi \times 10^{-7}} \text{ J/m}^3$$

$$= 397.9 \times 10^3 \text{ J/m}^3$$

The volume of the air gap is

$$V_g = 2(0.05 \times 0.10 \times 0.005) \text{ m}^3$$

$$= 0.05 \times 10^{-3} \text{ m}^3$$

The stored energy in the air gap is

$$W_{fg} = 397.9 \times 10^3 \times 0.05 \times 10^{-3}$$
$$= 19.895 \text{ joules}$$

The total field energy is

$$W_f = 0.53 + 19.895 \text{ J}$$
$$= 20.425 \text{ J}$$

Note that most of the field energy is stored in the air gap.

3.2.1 ENERGY, COENERGY

The λ–i characteristic of an electromagnetic system (such as that shown in Fig. 3.2) depends on the air gap length and the B–H characteristics of the magnetic material. These λ–i characteristics are shown in Fig. 3.4*a* for three values of air gap length. For larger air gap length the characteristic is essentially linear. The characteristic becomes nonlinear as the air gap length decreases.

For a particular value of the air gap length, the energy stored in the field is represented by the area A between the λ axis and the λ–i characteristic, as shown in Fig. 3.4*b*. The area B between the i axis and the λ–i characteristic is known as the *coenergy* and is defined as

$$W_f' = \int_0^i \lambda \, di \tag{3.21}$$

This quantity has no physical significance. However, as will be seen later,

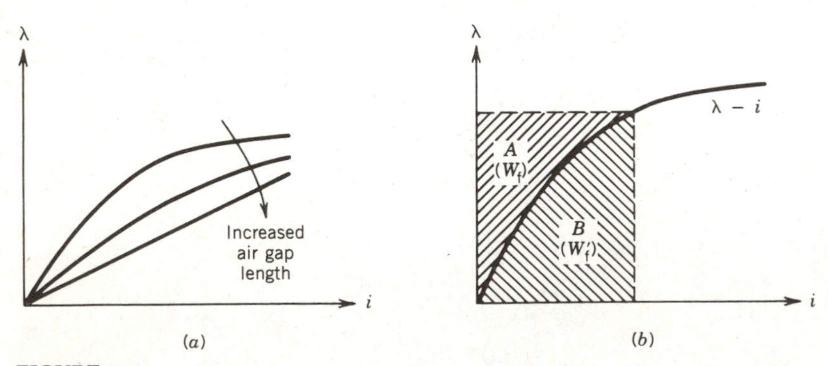

FIGURE 3.4
(*a*) λ–i characteristics for different air gap lengths. (*b*) Graphical representation of energy and coenergy.

it can be used to derive expressions for force (or torque) developed in an electromagnetic system.

From Fig. 3.4*b*,

$$W_f' + W_f = \lambda i \tag{3.22}$$

Note that $W_f' > W_f$ if the λ–i characteristic is nonlinear and $W_f' = W_f$ if it is linear.

3.3 MECHANICAL FORCE IN THE ELECTROMAGNETIC SYSTEM

Consider the system shown in Fig. 3.2. Let the movable part move from one position (say $x = x_1$) to another position ($x = x_2$) so that at the end of the movement the air gap decreases. The λ–i characteristics of the system for these two positions are shown in Fig. 3.5. The current ($i = v/R$) will remain the same at both positions in the steady state. Let the operating points be a when $x = x_1$ and b when $x = x_2$ (Fig. 3.5).

If the movable part has moved slowly, the current has remained essentially constant during the motion. The operating point has therefore moved upward from point a to b as shown in Fig. 3.5*a*. During the motion,

$$dW_e = \int ei\,dt = \int_{\lambda_1}^{\lambda_2} i\,d\lambda = \text{area } abcd \tag{3.23}$$

$$dW_f = \text{area } 0bc - 0ad \tag{3.24}$$

$$dW_m = dW_e - dW_f$$

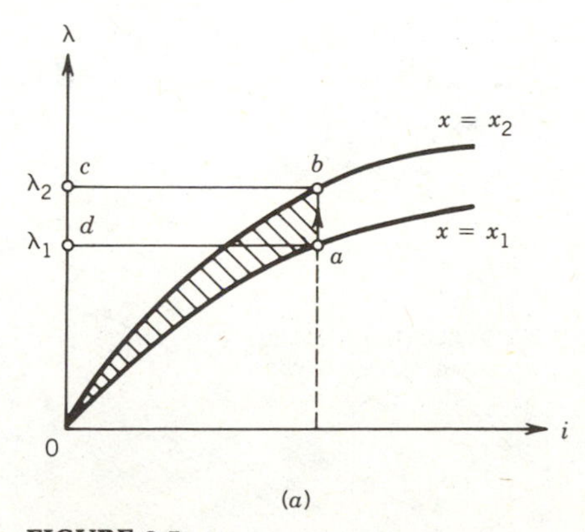

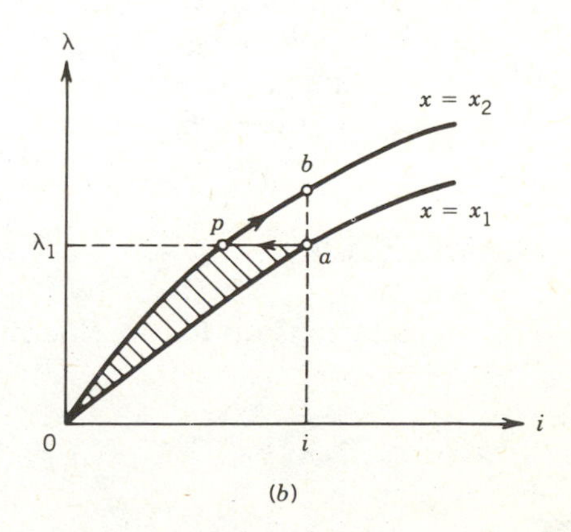

FIGURE 3.5
Locus of the operating point for motion in system of Fig. 3.2. (*a*) At constant current. (*b*) At constant flux linkage.

$$= \text{area } abcd + \text{area } 0ad - \text{area } 0bc$$

$$= \text{area } 0ab \tag{3.25}$$

If the motion has occurred under constant-current conditions, the mechanical work done is represented by the shaded area (Fig. 3.5*a*), which, in fact, is the increase in the coenergy.

$$dW_m = dW_f'$$

If f_m is the mechanical force causing the differential displacement dx,

$$f_m \, dx = dW_m = dW_f'$$

$$f_m = \left. \frac{\partial W_f'(i, x)}{\partial x} \right|_{i = \text{constant}} \tag{3.26}$$

Let us now consider that the movement has occurred very quickly. It may be assumed that during the motion the flux linkage has remained essentially constant, as shown in Fig. 3.5*b*. It can be shown that during the motion the mechanical work done is represented by the shaded area 0*ap*, which, in fact, is the decrease in the field energy. Therefore,

$$f_m \, dx = dW_m = -dW_f$$

$$f_m = -\left. \frac{\partial W_f(\lambda, x)}{\partial x} \right|_{\lambda = \text{constant}} \tag{3.27}$$

Note that for the rapid motion the electrical input is zero ($i \, d\lambda = 0$) because flux linkage has remained constant and the mechanical output energy has been supplied entirely by the field energy.

In the limit when the differential displacement dx is small, the areas 0*ab* in Fig. 3.5*a* and 0*ap* in Fig. 3.5*b* will be essentially the same. Therefore the force computed from Eqs. 3.26 and 3.27 will be the same.

EXAMPLE 3.2

The λ–i relationship for an electromagnetic system is given by

$$i = \left(\frac{\lambda g}{0.09} \right)^2$$

which is valid for the limits $0 < i < 4$ A and $3 < g < 10$ cm. For current $i = 3$ A and air gap length $g = 5$ cm, find the mechanical force on the moving part, using energy and coenergy of the field.

Solution

The λ–i relationship is nonlinear. From the λ–i relationship

$$\lambda = \frac{0.09 i^{1/2}}{g}$$

The coenergy of the system is

$$W'_f = \int_0^i \lambda \, di = \int_0^i \frac{0.09 i^{1/2}}{g} \, di$$

$$= \frac{0.09}{g} \frac{2}{3} i^{3/2} \text{ joules}$$

From Eq. 3.26

$$f_m = \left. \frac{\partial W'_f(i, g)}{\partial g} \right|_{i = \text{constant}}$$

$$= -0.09 \times \frac{2}{3} i^{3/2} \frac{1}{g^2} \bigg|_{i = \text{constant}}$$

For $g = 0.05$ m and $i = 3$ A,

$$f_m = -0.09 \times \frac{2}{3} \times 3^{3/2} \times \frac{1}{0.05^2} \text{ N} \cdot \text{m}$$

$$= -124.7 \text{ N} \cdot \text{m}$$

The energy of the system is

$$W_f = \int_0^\lambda i \, d\lambda = \int_0^\lambda \left(\frac{\lambda g}{0.09} \right)^2 d\lambda$$

$$= \frac{g^2}{0.09^2} \frac{\lambda^3}{3}$$

From Eq. 3.27

$$f_m = - \left. \frac{\partial W_f(\lambda, g)}{\partial g} \right|_{\lambda = \text{constant}}$$

$$= - \frac{\lambda^3 2g}{3 \times 0.09^2}$$

For $g = 0.05$ m and $i = 3$ A,

$$\lambda = \frac{0.09 \times 3^{1/2}}{0.05} = 3.12 \text{ Wb-turn}$$

and

$$f_m = -\frac{3.12^3 \times 2 \times 0.05}{3 \times 0.09^2}$$

$$= -124.7 \text{ N} \cdot \text{m}$$

The forces calculated on the basis of energy and coenergy functions are the same, as they should be. The selection of the energy or coenergy function as the basis for calculation is a matter of convenience, depending on the variables given. The negative sign for the force indicates that the force acts in such a direction as to decrease the air gap length.

3.3.1 LINEAR SYSTEM

Consider the electromagnetic system of Fig. 3.2. If the reluctance of the magnetic core path is negligible compared to that of the air gap path, the λ–i relation becomes linear. For this idealized system

$$\lambda = L(x)i \tag{3.28}$$

where $L(x)$ is the inductance of the coil, whose value depends on the air gap length. The field energy is

$$W_f = \int i \, d\lambda \tag{3.29}$$

From Eqs. 3.28 and 3.29

$$W_f = \int_0^\lambda \frac{\lambda}{L(x)} d\lambda = \frac{\lambda^2}{2L(x)} \tag{3.30}$$

$$= \tfrac{1}{2}L(x)i^2 \tag{3.31}$$

From Eqs. 3.27 and 3.30

$$f_m = -\frac{\partial}{\partial x}\left(\frac{\lambda^2}{2L(x)}\right)\bigg|_{\lambda = \text{constant}}$$

$$= \frac{\lambda^2}{2L^2(x)} \frac{dL(x)}{dx}$$

$$= \tfrac{1}{2} i^2 \frac{dL(x)}{dx} \tag{3.32}$$

For a linear system

$$W_\mathrm{f} = W'_\mathrm{f} = \tfrac{1}{2} L(x) i^2 \tag{3.33}$$

From Eqs. 3.26, and 3.33

$$f_\mathrm{m} = \frac{\partial}{\partial x}\left(\tfrac{1}{2} L(x) i^2\right)\Big|_{i\,=\,\text{constant}}$$

$$= \tfrac{1}{2} i^2 \frac{dL(x)}{dx} \tag{3.34}$$

Equations 3.32 and 3.34 show that the same expressions are obtained for force whether analysis is based on energy or coenergy functions. For the system in Fig. 3.2, if the reluctance of the magnetic core path is neglected,

$$Ni = H_\mathrm{g} 2g = \frac{B_\mathrm{g}}{\mu_0} 2g \tag{3.35}$$

From Eq. 3.16, the field energy is

$$W_\mathrm{f} = \frac{B_\mathrm{g}^2}{2\mu_0} \times \text{volume of air gap}$$

$$= \frac{B_\mathrm{g}^2}{2\mu_0} \times A_\mathrm{g} 2g \tag{3.36}$$

where A_g is the cross-sectional area of the air gap.

From Eqs. 3.27 and 3.36

$$f_\mathrm{m} = \frac{\partial}{\partial g}\left(\frac{B_\mathrm{g}^2}{2\mu_0} \times A_\mathrm{g} \times 2g\right)$$

$$= \frac{B_\mathrm{g}^2}{2\mu_0}(2A_\mathrm{g}) \tag{3.37}$$

The total cross-sectional area of the air gap is $2A_\mathrm{g}$. Hence, the force per unit area of air gap, called *magnetic pressure* F_m, is

$$F_\mathrm{m} = \frac{B_\mathrm{g}^2}{2\mu_0}\ \mathrm{N/m^2} \tag{3.38}$$

EXAMPLE 3.3

The lifting magnetic system shown in Fig. E3.3 has a square cross section 6 × 6 cm. The coil has 300 turns and a resistance of 6 ohms. Neglect reluctance of the magnetic core and field fringing in the air gap.

(a) The air gap is initially held at 5 mm and a dc source of 120 V is connected to the coil. Determine

(i) The stored field energy.

(ii) The lifting force.

(b) The air gap is again held at 5 mm and an ac source of 120 V (rms) at 60 Hz is connected to the coil. Determine the average value of the lift force.

Solution

(a) Current in the coil is

$$i = \frac{120}{6} = 20 \text{ A}$$

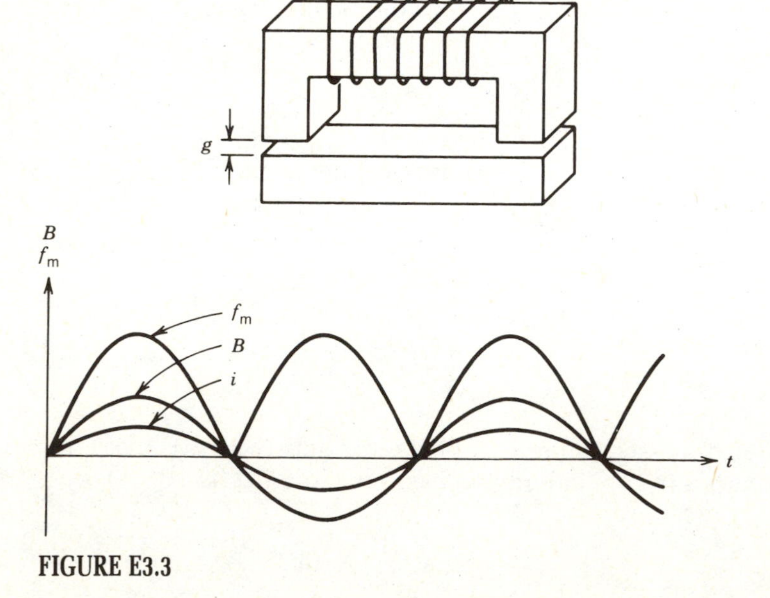

FIGURE E3.3

Because the reluctance of the magnetic core is neglected, field energy in the magnetic core is negligible. All field energy is in the air gaps.

$$Ni = H_g l_g = \frac{B_g}{\mu_0} l_g$$

$$B_g = \frac{\mu_0 Ni}{2g}$$

$$= \frac{4\pi 10^{-7} \times 300 \times 20}{2 \times 5 \times 10^{-3}}$$

$$= 0.754 \text{ tesla}$$

Field energy is

$$W_f = \frac{B_g^2}{2\mu_0} \times \text{volume of air gap}$$

$$= \frac{0.754^2}{2 \times 4\pi 10^{-7}} \times 2 \times 6 \times 6 \times 5 \times 10^{-7} \text{ J}$$

$$= 8.1434 \text{ J}$$

From Eq. 3.37 the lift force is

$$f_m = \frac{B_g^2}{2\mu_0} \times \text{air gap area}$$

$$= \frac{0.754^2}{2 \times 4\pi 10^{-7}} \times 2 \times 6 \times 6 \times 10^{-4} \text{ N}$$

$$= 1628.7 \text{ N}$$

(b) For ac excitation the impedance of the coil is

$$Z = R + j\omega L$$

Inductance of the coil is

$$L = \frac{N^2}{R_g} = \frac{N^2 \mu_0 A_g}{l_g}$$

$$= \frac{300^2 \times 4\pi 10^{-7} \times 6 \times 6 \times 10^{-4}}{2 \times 5 \times 10^{-3}}$$

$$= 40.7 \times 10^{-3} \text{ H}$$

$$Z = 6 + j377 \times 40.7 \times 10^{-3}\ \Omega$$
$$= 6 + j15.34\ \Omega$$

Current in the coil is

$$I_{\text{rms}} = \frac{120}{\sqrt{(6^2 + 15.34^2)}}$$
$$= 7.29\ \text{A}$$

The flux density is

$$B_{\text{g}} = \frac{\mu_0 N i}{2g}$$

The flux density is proportional to the current and therefore changes sinusoidally with time as shown in Fig. E3.3. The rms value of the flux density is

$$B_{\text{rms}} = \frac{\mu_0 N I_{\text{rms}}}{2g} \tag{3.38a}$$
$$= \frac{4\pi 10^{-7} \times 300 \times 7.29}{2 \times 5 \times 10^{-3}}\ \text{T}$$
$$= 0.2748\ \text{T}$$

The lift force is

$$f_{\text{m}} = \frac{B_{\text{g}}^2}{2\mu_0} \times 2A_{\text{g}} \tag{3.38b}$$
$$\propto B_{\text{g}}^2$$

The force varies as the square of the flux density as shown in Fig. E3.3.

$$f_{\text{m}}\big|_{\text{avg}} = \frac{B_{\text{g}}^2}{2\mu_0}\bigg|_{\text{avg}} \times 2A_{\text{g}}$$
$$= \frac{B_{\text{rms}}^2}{2\mu_0} \times 2A_{\text{g}}$$
$$= \frac{B_{\text{rms}}^2}{2\mu_0} \times \text{air gap area} \tag{3.38c}$$
$$= \frac{0.2748^2 \times 2 \times 6 \times 6 \times 10^{-4}}{2 \times 4\pi 10^{-7}}$$

$$= 216.3 \text{ N}$$

which is almost one-eighth of the lift force obtained with a dc supply voltage. Lifting magnets are normally operated from dc sources.

3.4 ROTATING MACHINES

The production of translational motion in an electromagnetic system has been discussed in previous sections. However, most of the energy converters, particularly the higher-power ones, produce rotational motion. The essential part of a rotating electromagnetic system is shown in Fig. 3.6. The fixed part of the magnetic system is called the *stator* and the moving part is called the *rotor*. The latter is mounted on a shaft and is free to rotate between the poles of the stator. Let us consider a general case in which both stator and rotor have windings carrying currents, as shown in Fig. 3.6. The current can be fed into the rotor circuit through fixed brushes and rotor-mounted slip rings.

The stored field energy W_f of the system can be evaluated by establishing the currents i_s and i_r in the windings keeping the system static, that is, with no mechanical output. Consequently,

$$\begin{aligned} dW_f &= e_s i_s \, dt + e_r i_r \, dt \\ &= i_s \, d\lambda_s + i_r \, d\lambda_r \end{aligned} \tag{3.39}$$

For a linear magnetic system the flux linkages λ_s of the stator winding and λ_r of the rotor winding can be expressed in terms of inductances whose values depend on the position θ of the rotor.

$$\begin{aligned} \lambda_s &= L_{ss} i_s + L_{sr} i_r \\ \lambda_r &= L_{rs} i_s + L_{rr} i_r \end{aligned} \tag{3.40}$$

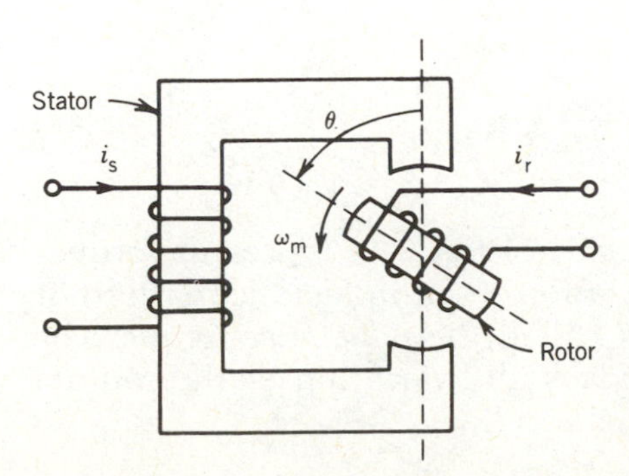

FIGURE 3.6
Basic configuration of a rotating electromagnetic system.

where L_{ss} is the self-inductance of the stator winding

L_{rr} is the self-inductance of the rotor winding

L_{sr}, L_{rs} are mutual inductances between stator and rotor windings

For a linear magnetic system $L_{sr} = L_{rs}$. Equation 3.40 can be written in the matrix form

$$\begin{vmatrix} \lambda_s \\ \lambda_r \end{vmatrix} = \begin{vmatrix} L_{ss} & L_{sr} \\ L_{sr} & L_{rr} \end{vmatrix} \begin{vmatrix} i_s \\ i_r \end{vmatrix} \tag{3.41}$$

From Eqs. 3.39 and 3.40

$$\begin{aligned} dW_f &= i_s\, d(L_{ss} i_s + L_{sr} i_r) + i_r\, d(L_{sr} i_s + L_{rr} i_r) \\ &= L_{ss} i_s\, di_s + L_{rr} i_r\, di_r + L_{sr}\, d(i_s i_r) \end{aligned}$$

The field energy is

$$\begin{aligned} W_f &= L_{ss} \int_0^{i_s} i_s\, di_s + L_{rr} \int_0^{i_r} i_r\, di_r + L_{sr} \int_0^{i_s, i_r} d(i_s i_r) \\ &= \tfrac{1}{2} L_{ss} i_s^2 + \tfrac{1}{2} L_{rr} i_r^2 + L_{sr} i_s i_r \end{aligned} \tag{3.42}$$

Following the procedure used to determine an expression for force developed in a translational actuator, it may be shown that the torque developed in a rotational electromagnetic system is

$$T = \left. \frac{\partial W_f'(i, \theta)}{\partial \theta} \right|_{i = \text{constant}} \tag{3.43}$$

In a linear magnetic system, energy and coenergy are the same, that is, $W_f = W_f'$. Therefore, from Eqs. 3.42 and 3.43,

$$T = \tfrac{1}{2} i_s^2 \frac{dL_{ss}}{d\theta} + \tfrac{1}{2} i_r^2 \frac{dL_{rr}}{d\theta} + i_s i_r \frac{dL_{sr}}{d\theta} \tag{3.44}$$

The first two terms on the right-hand side of Eq. 3.44 represent torques produced in the machine because of variation of self-inductance with rotor position. This component of torque is called the *reluctance torque.* The third term represents torque produced by the variation of the mutual inductance between the stator and rotor windings.

EXAMPLE 3.4

In the electromagnetic system of Fig. 3.6 the rotor has no winding (i.e., we have a reluctance motor) and the inductance of the stator as a function of the rotor position θ is $L_{ss} = L_0 + L_2 \cos 2\theta$ (Fig. E3.4). The stator current is $i_s = I_{sm} \sin \omega t$.

(a) Obtain an expression for the torque acting on the rotor.

(b) Let $\theta = \omega_m t + \delta$, where ω_m is the angular velocity of the rotor and δ is the rotor position at $t = 0$. Find the condition for nonzero average torque and obtain an expression for the average torque.

Solution

(a) Since $i_r = 0$, from Eq. 3.44

$$T = \tfrac{1}{2} i_s^2 \frac{dL_{ss}}{d\theta}$$

$$= \tfrac{1}{2} I_{sm}^2 \sin^2 \omega t \frac{d}{d\theta}(L_0 + L_2 \cos 2\theta)$$

$$= -I_{sm}^2 L_2 \sin 2\theta \sin^2 \omega t \text{ N} \cdot \text{m}$$

(b)

$$T = -I_{sm}^2 L_2 \sin 2(\omega_m t + \delta) \frac{(1 - \cos 2\omega t)}{2}$$

$$= -\tfrac{1}{2} I_{sm}^2 L_2 [\sin 2(\omega_m t + \delta) - \tfrac{1}{2} \sin 2\{(\omega_m + \omega)t + \delta\} - \tfrac{1}{2} \sin 2\{(\omega_m - \omega)t + \delta\}]$$

The average value of each of the three sinusoidally time-varying func-

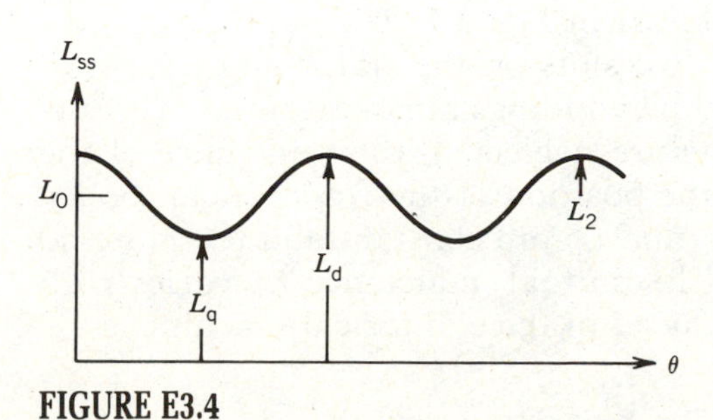

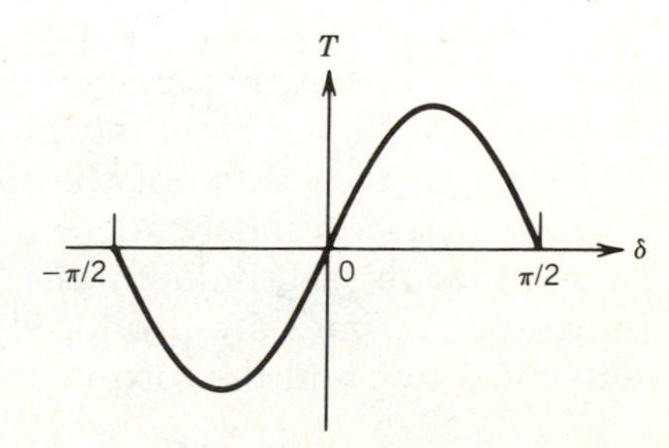

FIGURE E3.4

tions is zero unless, in one of them, the coefficient of t is zero. Average torque is produced at the following speeds:

(i) $\omega_m = 0$

The average torque at zero speed is

$$T_{avg} = -\tfrac{1}{2} I_{sm}^2 L_2 \sin 2\delta$$

From Fig. E3.4

$$L_2 = \frac{L_d - L_q}{2}$$

Therefore,

$$T_{avg} = -\tfrac{1}{4} I_{sm}^2 (L_d - L_q) \sin 2\delta$$

(ii) $\omega_m = \pm\omega$

Corresponding to this condition,

$$T_{avg} = \tfrac{1}{4} I_{sm}^2 L_2 \sin 2\delta$$
$$= \tfrac{1}{8} I_{sm}^2 (L_d - L_q) \sin 2\delta$$

The reluctance machine can therefore develop an average torque if it rotates, in either direction, at the angular velocity of the current, which is known as the *synchronous speed.* The average torque varies sinusoidally with 2δ (as shown in Fig. E3.4), where δ is the rotor position at time $t = 0$ and is known as the *power* or *torque angle.*

3.5 CYLINDRICAL MACHINES

A cross-sectional view of an elementary two-pole cylindrical rotating machine with a uniform air gap is shown in Fig. 3.7. The stator and rotor windings are shown as placed on two slots on the stator and the rotor, respectively. In an actual machine the windings are distributed over several slots. If the effects of the slots are neglected, the reluctance of the magnetic path is independent of the position of the rotor. It can be assumed that the self-inductances L_{ss} and L_{rr} are constant and therefore no reluctance torques are produced. The mutual inductance L_{sr} varies with rotor position, and the torque produced in the cylindrical machine is

$$T = i_s i_r \frac{dL_{sr}}{d\theta} \qquad (3.45)$$

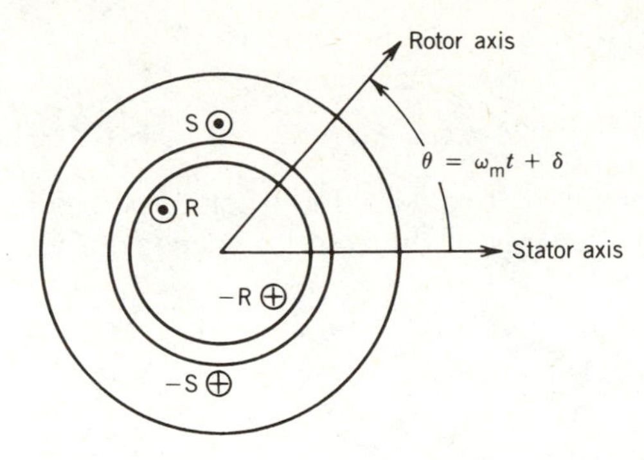

FIGURE 3.7
Cross-sectional view of an elementary two-pole cylindrical rotating machine with a uniform air gap.

Let

$$L_{sr} = M \cos \theta \tag{3.46}$$

where M is the peak value of the mutual inductance L_{sr}

θ is the angle between the magnetic axis of the stator and rotor windings

Let the currents in the two windings be

$$i_s = I_{sm} \cos \omega_s t \tag{3.47}$$

$$i_r = I_{rm} \cos(\omega_r t + \alpha) \tag{3.48}$$

The position of the rotor at any instant is

$$\theta = \omega_m t + \delta \tag{3.49}$$

where ω_m is the angular velocity of the rotor and δ is the rotor position at $t = 0$. From Eqs. 3.45 to 3.49,

$$\begin{aligned} T &= -I_{sm} I_{rm} M \cos \omega_s t \cos(\omega_r t + \alpha) \sin(\omega_m t + \delta) \\ &= -\frac{I_{sm} I_{rm} M}{4} [\sin\{(\omega_m + (\omega_s + \omega_r))t + \alpha + \delta\} \\ &\quad + \sin\{(\omega_m - (\omega_s + \omega_r))t - \alpha + \delta\} \\ &\quad + \sin\{(\omega_m + (\omega_s - \omega_r))t - \alpha + \delta\} \\ &\quad + \sin\{(\omega_m - (\omega_s - \omega_r))t + \alpha + \delta\}] \end{aligned} \tag{3.50}$$

The torque varies sinusoidally with time. The average value of each of the sinusoidal terms in Eq. 3.50 is zero, unless the coefficient of t is zero in that sinusoidal term. Therefore the average torque will be nonzero if

$$\omega_m = \pm(\omega_s \pm \omega_r) \tag{3.51}$$

The machine will develop average torque if it rotates, in either direction, at a speed that is equal to the sum or difference of the angular speeds of the stator and rotor currents:

$$|\omega_m| = |\omega_s \pm \omega_r| \tag{3.52}$$

Consider the following cases.

1. $\omega_r = 0, \alpha = 0, \omega_m = \omega_s$. The rotor current is a direct current I_R and the machine rotates at the synchronous speed. For these conditions, from Eq. 3.50, the torque developed is

$$T = -\frac{I_{sm} I_R M}{2}\{\sin(2\omega_s t + \delta) + \sin\delta\} \tag{3.53}$$

   The instantaneous torque is pulsating. The average value of the torque is

$$T_{avg} = -\frac{I_{sm} I_R M}{2}\sin\delta \tag{3.54}$$

   If the machine is brought up to its *synchronous* speed ($\omega_m = \omega_s$), it will develop an average unidirectional torque and continuous energy conversion can take place at the synchronous speed. This is the basic principle of operation of a *synchronous machine,* which normally has dc excitation in the rotor and ac excitation in the stator. Note (from Eq. 3.50) that at $\omega_m = 0$, the machine does not develop an average torque and therefore the machine is not self-starting. With one winding on the stator, the machine is called a *single-phase synchronous machine.* Although it develops an average torque, the instantaneous torque is pulsating. The pulsating torque may result in vibration, speed fluctuation, noise, and waste of energy. This may be acceptable in smaller machines but not in larger ones. The pulsating torque can be avoided in a polyphase machine, and all large machines are polyphase machines.

2. $\omega_m = \omega_s - \omega_r$. Both stator and rotor windings carry ac currents at different frequencies and the motor runs at an *asynchronous speed* ($\omega_m \neq \omega_s$, $\omega_m \neq \omega_r$). From Eq. 3.50, the torque developed is

$$T = -\frac{I_{sm} I_{rm} M}{4}[\sin(2\omega_s t + \alpha + \delta) + \sin(-2\omega_r t - \alpha + \delta)$$

$$+ \sin(2\omega_s t - 2\omega_r t - \alpha + \delta) + \sin(\alpha + \delta)] \tag{3.55}$$

The instantaneous torque is pulsating. The average value of the torque is

$$T_{avg} = -\frac{L_{sm}L_{rm}M}{4}\sin(\alpha + \delta) \tag{3.56}$$

This is the basic principle of operation of an *induction machine*, in which the stator winding is excited by an ac current and ac current is induced in the rotor winding. Note that the single-phase induction machine is also not self-starting, because at $\omega_m = 0$ no average torque is developed. The machine is brought up to the speed $\omega_m = \omega_s - \omega_r$ so that it can produce an average torque. To eliminate pulsating torque, polyphase induction machines are used for high-power applications.

The mechanism of torque production in electromagnetic systems producing both translational and rotary motions has been discussed in this chapter. In rotating machines torque can be produced by variation in the reluctance of the magnetic path or mutual inductance between the windings.

Reluctance machines are simple in construction, but torque developed in these machines is small. Cylindrical machines, although more complex in construction, produce larger torques. Most electrical machines are of the cylindrical type. The performance of the various rotating electrical machines is discussed in more detail in the following chapters.

PROBLEMS

3.1 In a translational motion actuator, the λ–i relationship is given by

$$i = \lambda^{3/2} + 2.5\lambda(x - 1)^2$$

for $0 < x < 1$ m, where i is the current in the coil of the actuator. Determine the force on the moving part at $x = 0.6$ m.

3.2 An actuator system is shown in Fig. P3.2. All dimensions are in centimeters. The magnetic material is cast steel, whose magnetization characteristic is shown in Fig. 1.7. The magnetic core and air gap have a square cross-sectional area. The coil has 500 turns and 4.0 ohms resistance.

(a) The gap is $d = 1$ mm.

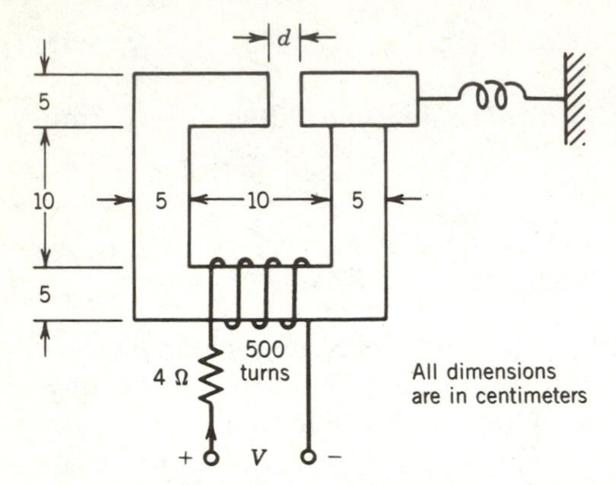

FIGURE P3.2

i. Determine the coil current and supply voltage (dc) required to establish an air gap flux density of 0.5 tesla.

ii. Determine the stored energy in the actuator system.

iii. Determine the force of attraction on the actuator arm.

iv. Determine the inductance of the coil.

(b) The actuator arm is allowed to move and finally the air gap closes.

i. For zero air gap determine the flux density in the core, force on the arm, and stored energy in the actuator system.

ii. Determine the energy transfer (excluding energy loss in the coil resistance) between the dc source and the actuator. Assume that the arm moved slowly. What is the direction of energy flow? How much mechanical energy is produced?

3.3 Fig. P3.3 shows an electromagnet system for lifting a section of steel channel. The coil has 600 turns. The reluctance of the magnetic material can be neglected up to a flux density of 1.4 tesla.

(a) For a coil current of 15 A (dc) determine the maximum air gap g for which the flux density is 1.4 tesla.

(b) For the air gap in part (a), determine the force on the steel channel.

(c) The steel channel has a mass of 1000 kg. For a coil current of 15 A, determine the largest gap at which the steel channel can be lifted magnetically against the force of gravity (9.81 m/sec^2).

3.4 The electromagnet shown in Fig. P3.4 can be used to lift a sheet of steel. The coil has 400 turns and a resistance of 5 ohms. The reluctance

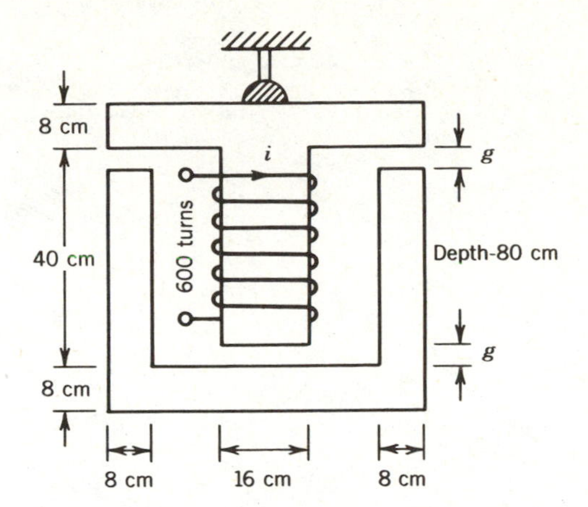

FIGURE P3.3

of the magnetic material is negligible. The magnetic core has a square cross section of 5 cm by 5 cm. When the sheet of steel is fitted to the electromagnet, air gaps, each of length $g = 1$ mm, separate them. An average force of 550 newtons is required to lift the sheet of steel.

(a) For dc supply,

i. Determine the dc source voltage.

ii. Determine the energy stored in the magnetic field.

(b) For ac supply at 60 Hz, determine the ac source voltage.

3.5 The features of a moving-iron ammeter are shown in Fig. P3.5. When current flows through the curved solenoid coil, a curved ferromagnetic rod is pulled into the solenoid against the torque of a restraining spring. The inductance of the coil is $L = 4.5 + 18\theta$ mH, where θ is angle of deflection in radians. The spring constant is 0.65×10^{-3} N · m/rad.

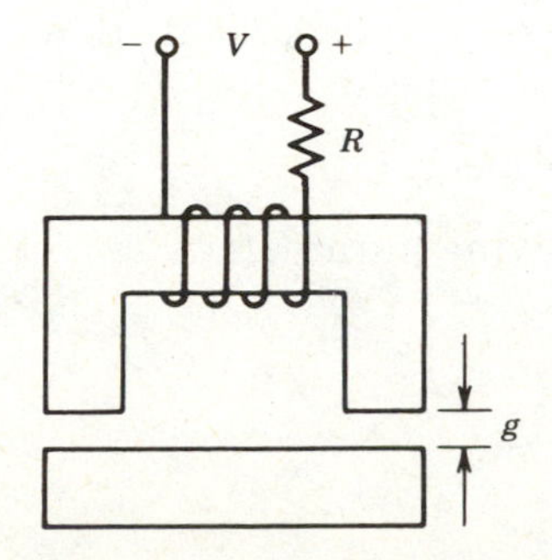

FIGURE P3.4

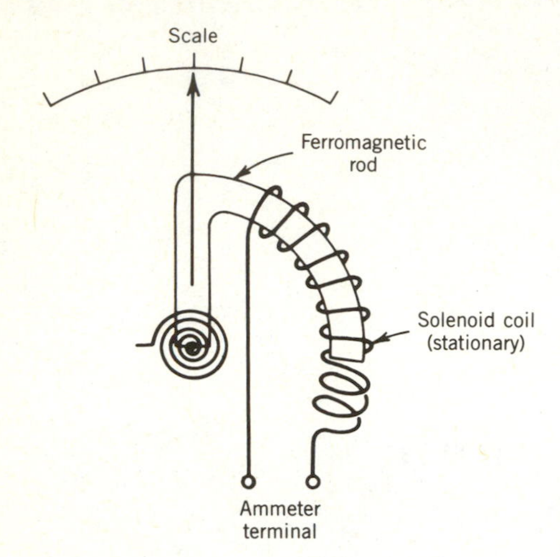

FIGURE P3.5

(a) Show that the ammeter measures the root-mean-square value of the current.

(b) Determine the deflection in degrees for a current of 10 amperes (rms).

(c) Determine the voltage drop across the ammeter terminal when 10 A (rms) at 60 Hz flows through the ammeter. The coil resistance is 0.015 Ω.

3.6 A reluctance machine of the form shown in Fig. 3.6 has no rotor winding. The inductance of the stator winding is

$$L_{ss} = 0.1 - 0.3 \cos 2\theta - 0.2 \cos 4\theta \text{ H}$$

A current of 10 A rms at 60 Hz is passed through the stator coil.

(a) Determine the values of speed (ω_m) of the rotor at which the machine will develop an average torque.

(b) Determine the maximum torque and power (mechanical) that could be developed by the machine at each speed.

(c) Determine the maximum torque at zero speed.

3.7 The rotating machine of Fig. 3.7 has the following parameters.

$$L_{ss} = 0.15 \text{ H}$$

$$L_{rr} = 0.06 \text{ H}$$

$$L_{sr} = 0.08 \cos \theta \text{ H}$$

(a) The rotor is driven at 3600 rpm. If the stator winding carries a current of 5 A rms at 60 Hz, determine the instantaneous voltage and rms voltage induced in the rotor coil. Determine the frequency of the rotor induced voltage.

(b) Suppose the stator and rotor coils are connected in series and a current of 5 A rms at 60 Hz is passed through them. Determine the speeds at which the machine will produce an average torque. Also determine the maximum torque that the machine will produce at each speed.

DC MACHINES

CHAPTER four

Applications such as light bulbs and heaters require energy in electrical form. In other applications, such as fans and rolling mills, energy is required in mechanical form. One form of energy can be obtained from the other form with the help of converters. Converters that are used to continuously translate electrical input to mechanical output or vice versa are called *electric machines*. The process of translation is known as *electromechanical energy conversion*. An electric machine is therefore a link between an electrical system and a mechanical system, as shown in Fig. 4.1. In these machines the conversion is reversible. If the conversion is from mechanical to electrical, the machine is said to act as a *generator*. If the conversion is from electrical to mechanical, the machine is said to act as a *motor*. Hence, the same electric machine can be made to operate as a generator as well as a motor. Machines are called ac machines (generators or motors) if the electrical system is ac and dc machines (generators or motors) if the electrical system is dc.

Note that the two systems in Fig. 4.1, electrical and mechanical, are different in nature. In the electrical system the primary quantities involved are *voltage* and *current*, while the analogous quantities in the mechanical system are *torque* and *speed*. The coupling medium between these different systems is the field, as illustrated in Fig. 4.2.

4.1 ELECTROMAGNETIC CONVERSION

Three electrical machines (dc, induction, and synchronous) are used extensively for electromechanical energy conversion. In these machines, conversion of energy from electrical to mechanical form or vice versa results from the following two electromagnetic phenomena:

1. When a conductor moves in a magnetic field, voltage is induced in the conductor.

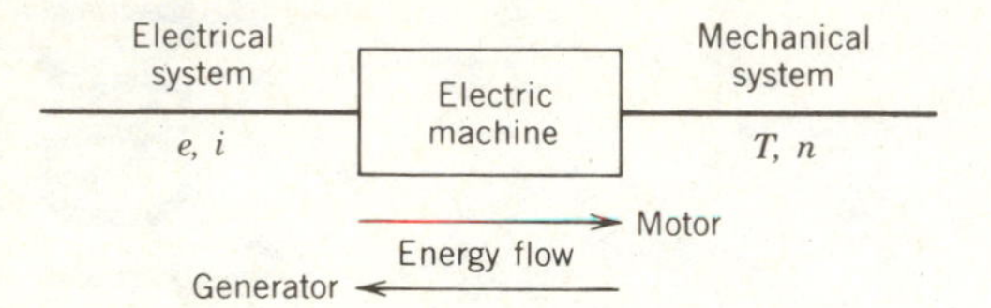

FIGURE 4.1
Electromechanical energy conversion.

2. When a current-carrying conductor is placed in a magnetic field, the conductor experiences a mechanical force.

These two effects occur simultaneously whenever energy conversion takes place from electrical to mechanical or vice versa. In motoring action, the electrical system makes current flow through conductors that are placed in the magnetic field. A force is produced on each conductor. If the conductors are placed on a structure free to rotate, an electromagnetic torque will be produced, tending to make the rotating structure rotate at some speed. If the conductors rotate in a magnetic field, a voltage will also be induced in each conductor. In generating action, the process is reversed. In this case, the rotating structure, the rotor, is driven by a prime mover (such as a steam turbine or a diesel engine). A voltage will be induced in the conductors that are rotating with the rotor. If an electrical load is connected to the winding formed by these conductors, a current i will flow, delivering electrical power to the load. Moreover, the current flowing through the conductor will interact with the magnetic field to produce a reaction torque, which will tend to oppose the torque applied by the prime mover. Note that in both motoring and generating actions, the coupling magnetic field is involved in producing a torque and an induced voltage.

The basic electric machines (dc, induction, and synchronous), which depend on electromagnetic energy conversion, are extensively used in various power ratings. The operation of these machines is discussed in detail in this and other chapters.

Motional Voltage, *e*

An expression can be derived for the voltage induced in a conductor moving in a magnetic field. As shown in Fig. 4.3*a*, if a conductor of length l

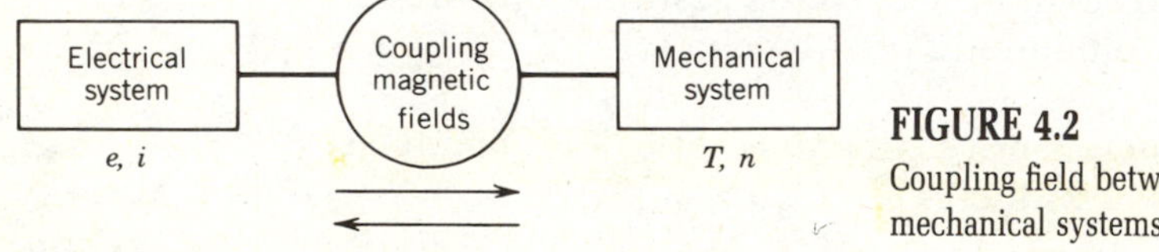

FIGURE 4.2
Coupling field between electrical and mechanical systems.

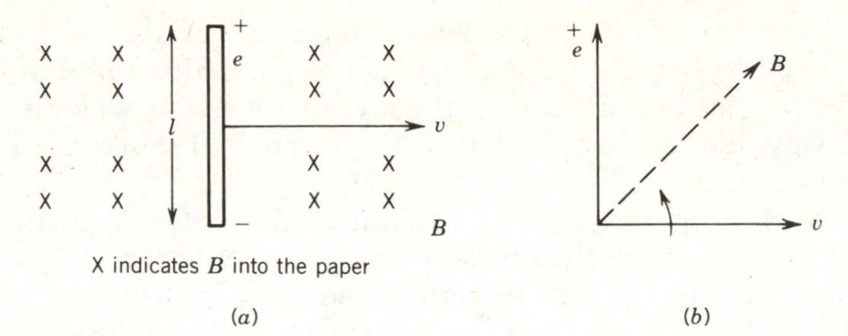

FIGURE 4.3
Motional voltage. (a) Conductor moving in the magnetic field. (b) Right-hand screw rule.

moves at a linear speed v in a magnetic field B, the induced voltage in the conductor is

$$e = Blv \tag{4.1}$$

where B, l, and v are mutually perpendicular. The polarity of the induced voltage can be determined from the so-called right-hand screw rule.

The three quantities v, B, and e are shown in Fig. 4.3b as three mutually perpendicular vectors. Turn the vector v toward the vector B. If a right-hand screw is turned in the same way the motion of the screw will indicate the direction of positive polarity of the induced voltage.

Electromagnetic Force, f

For the current-carrying conductor shown in Fig. 4.4a, the force (known as Lorentz force) produced on the conductor is

$$f = Bli \tag{4.2}$$

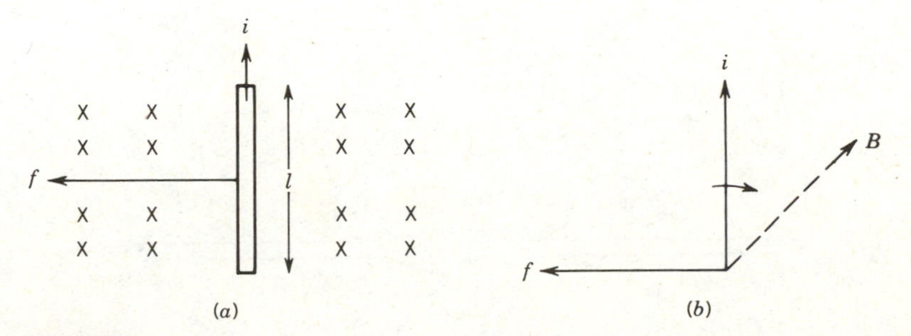

FIGURE 4.4
Electromagnetic force. (a) Current-carrying conductor moving in a magnetic field. (b) Force direction.

where B, l, and i are mutually perpendicular. The direction of the force can be determined by using the right-hand screw rule, illustrated in Fig. 4.4*b*.

Turn the current vector i toward the flux vector B. If a screw is turned in the same way, the direction in which the screw will move represents the direction of the force f.

Note that in both cases (i.e., determining the polarity of the induced voltage and determining the direction of the force) the moving quantities (v and i) are turned toward B to obtain the screw movement.

Equations 4.1 and 4.2 can be used to determine the induced voltage and the electromagnetic force or torque in an electric machine. There are, of course, other methods by which these quantities (e and f) can be determined.

Basic Structure of Electric Machines

The structure of an electric machine has two major components, stator and rotor, separated by the air gap, as shown in Fig. 4.5.

Stator: This part of the machine does not move and normally is the outer frame of the machine.

Rotor: This part of the machine is free to move and normally is the inner part of the machine.

Both stator and rotor are made of ferromagnetic materials. In most machines slots are cut on the inner periphery of the stator and outer periphery of the rotor structure, as shown in Fig. 4.5*a*. Conductors are placed in these slots. The iron core is used to maximize the coupling between the coils (formed by conductors) placed on the stator and rotor, to

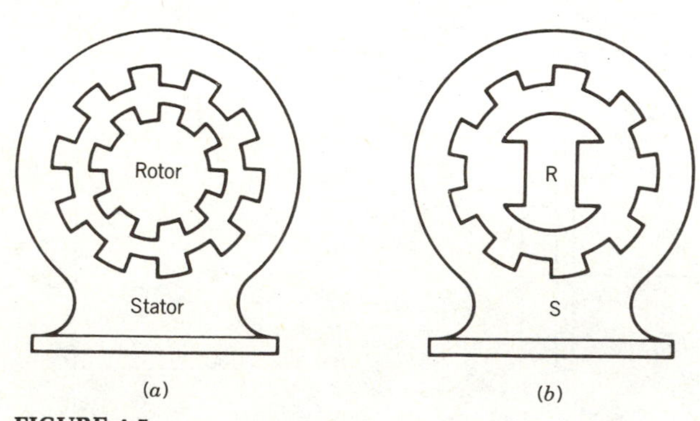

FIGURE 4.5

Structure of electric machines. (*a*) Cylindrical machine (uniform air gap). (*b*) Salient pole machine (nonuniform air gap).

increase the flux density in the machine and to decrease the size of the machine. If the stator or rotor (or both) is subjected to a time-varying magnetic flux, the iron core is laminated to reduce eddy current losses. The thin laminations of the iron core with provisions for slots are shown in Fig. 4.6.

The conductors placed in the slots of the stator or rotor are interconnected to form windings. The winding in which voltage is induced is called the *armature winding*. The winding through which a current is passed to produce the primary source of flux in the machine is called the *field winding*. Permanent magnets are used in some machines to provide the major source of flux in the machine.

Rotating electrical machines take many forms and are known by many names. The three basic and common ones are dc machines, induction machines, and synchronous machines. There are other machines, such as permanent magnet machines, hysteresis machines, and stepper machines.

DC Machine

In the dc machine, the field winding is placed on the stator and the armature winding on the rotor. These windings are shown in Fig. 4.7. A dc current is passed through the field winding to produce flux in the machine. Voltage induced in the armature winding is alternating. A mechanical

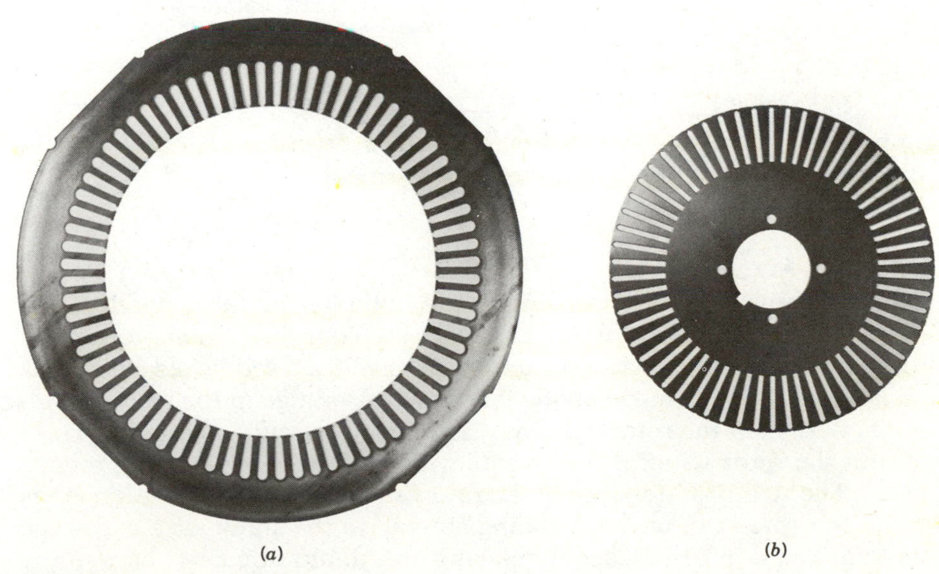

FIGURE 4.6
Laminations (Courtesy of Westinghouse Canada Inc.) (a) Stator. (b) Rotor.

(a)

(b)

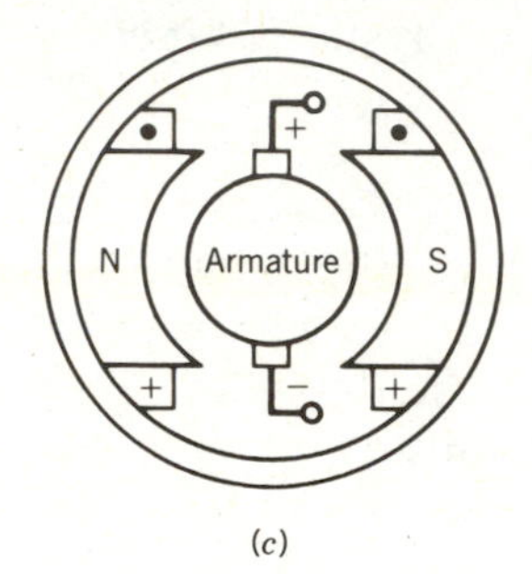

(c)

FIGURE 4.7
DC machine. (a) Stator. (b) Rotor. (Courtesy of General Electric Canada Inc.) (c) Schematic cross-sectional view.

commutator and a brush assembly function as a rectifier or inverter, making the armature terminal voltage unidirectional.

Induction Machine

In this machine the stator windings serve as both armature windings and field windings. When the stator windings are connected to an ac supply, flux is produced in the air gap and revolves at a fixed speed known as *synchronous speed.* This revolving flux induces voltage in the stator windings as well as in the rotor windings. If the rotor circuit is closed, current flows in the rotor winding and reacts with the revolving flux to produce torque. The steady-state speed of the rotor is very close to the synchronous speed. The rotor can have a winding similar to the stator or a cage-type winding. The latter is formed by placing aluminum or copper bars in the rotor slots and shorting them at the ends by means of rings. Figure 4.8 shows the structure of the induction machine.

(a)

(b)

(c)

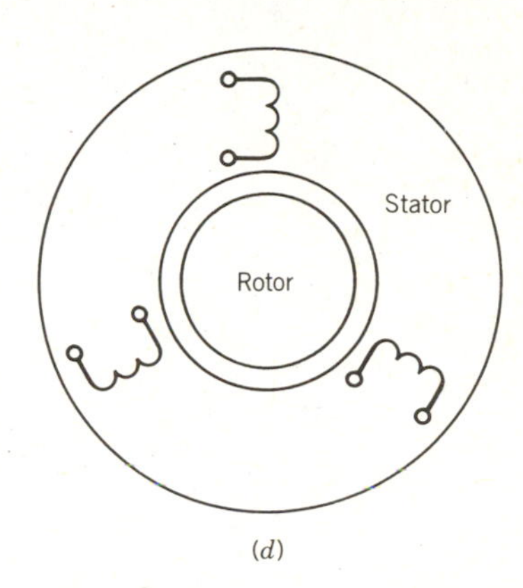

(d)

FIGURE 4.8
Induction machine. (a) Stator. (b) Rotor—cage type. (c) Rotor—wound type. (Courtesy of General Electric Canada Inc.) (d) Schematic cross-sectional view.

Synchronous Machine

In this machine, the rotor carries the field winding and the stator carries the armature winding. The structure of the synchronous machine is shown in Fig. 4.9. The field winding is excited by direct current to produce flux in the air gap. When the rotor rotates, voltage is induced in the armature winding placed on the stator. The armature current produces a revolving flux in the air gap whose speed is the same as the speed of the rotor—hence the name synchronous machine.

These three major machine types, although they differ in physical construction and appear to be quite different from each other, are in fact governed by the same basic laws. Their behavior can be explained by considering the same fundamental principles of voltage and torque production. Various analytical techniques can be used for the machines and various forms of torque or voltage equations can be derived for them, but

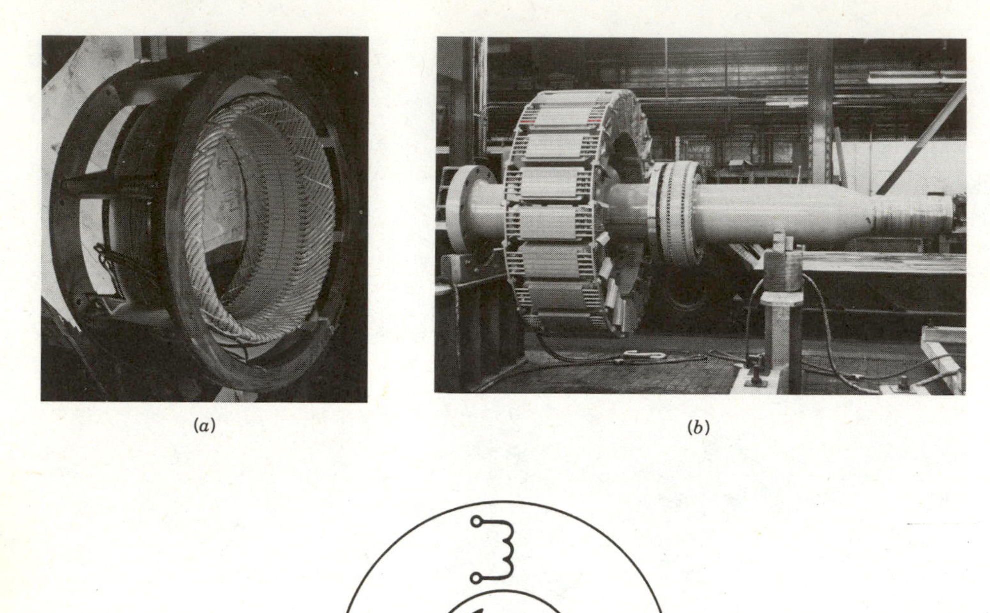

(a) (b) (c)

FIGURE 4.9
Synchronous machine. (a) Stator. (b) Rotor. (Courtesy of General Electric Canada Inc.) (c) Schematic cross-sectional view.

the forms of the equations will differ merely to reflect the difference in construction of the machines. For example, analysis will show that in dc machines the stator and rotor flux distributions are fixed in space and a torque is produced because of the tendency of these two fluxes to align. The induction machine is an ac machine and differs in many ways from the dc machine but works on the same principle. Analysis will indicate that the stator flux and the rotor flux rotate in synchronism in the air gap and the two flux distributions are displaced from each other by a torque-producing displacement angle. The torque is produced because of the tendency of the two flux distributions to align with each other. It must be emphasized at the outset that ac machines are not fundamentally different from dc machines. Their construction details are different, but the same fundamental principles underlie their operation.

The three basic and commonly used machines—dc, induction, and synchronous—are described, analyzed, and discussed in separate chapters. In this chapter the various aspects of the steady-state operation of the dc machine are studied in detail.

4.2 DC MACHINES

The dc machines are versatile and extensively used in industry. A wide variety of volt–ampere or torque–speed characteristics can be obtained from various connections of the field windings.

Although a dc machine can operate as either a generator or a motor, at present its use as a generator is limited because of the widespread use of ac power. The dc machine is extensively used as a motor in industry. Its speed can be controlled over a wide range with relative ease. Large dc motors (in tens or hundreds of horsepower) are used in machine tools, printing presses, conveyors, fans, pumps, hoists, cranes, paper mills, textile mills, rolling mills, and so forth. DC motors still dominate as traction motors used in transit cars and locomotives. Small dc machines (in fractional horsepower rating) are used primarily as control devices—such as tachogenerators for speed sensing and servomotors for positioning and tracking. The dc machine definitely plays an important role in industry.

4.2.1 CONSTRUCTION

In a dc machine, the armature winding is placed on the rotor and the field windings are placed on the stator. The essential features of a two-pole dc machine are shown in Fig. 4.10. The stator has salient poles that are excited by one or more field windings, called *shunt field windings* and *series field windings*. The field windings produce an air gap flux distribution that

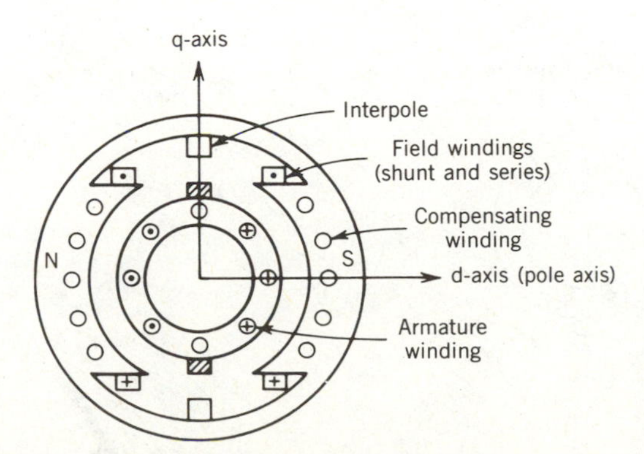

FIGURE 4.10
Schematic diagram of a dc machine.

is symmetrical about the *pole axis* (also called the *field axis, direct axis,* or *d-axis*).

The voltage induced in the turns of the armature winding is alternating. A commutator–brush combination is used as a mechanical rectifier to make the armature terminal voltage unidirectional and also to make the mmf wave due to the armature current fixed in space. The brushes are so placed that when the sides of an armature turn (or coil) pass through the middle of the region between field poles, the current through it changes direction. This makes all the conductors under one pole carry current in one direction. As a consequence, the mmf due to the armature current is along the axis midway between the two adjacent poles, called the *quadrature (or q) axis*. In the schematic diagram of Fig. 4.10, the brushes are shown placed on the q-axis to indicate that when a turn (or coil) undergoes commutation its sides are in the q-axis. However, because of the end connection, the actual brush positions will be approximately 90° from the position shown in Fig. 4.10 (see also Fig. 4.17).

Note that because of the commutator and brush assembly, the armature mmf (along the q-axis) is in quadrature with the field mmf (d-axis). This positioning of the mmf's will maximize torque production. The armature mmf axis can be changed by changing the position of the brush assembly as shown in Fig. 4.11. For improved performance, interpoles (in between two main field poles) and compensating windings (on the face of the main field poles) are required. These will be discussed in Sections 4.3.5 and 4.3.1, respectively.

4.2.2 EVOLUTION OF DC MACHINES

Consider a two-pole dc machine as shown in Fig. 4.12*a*. The air gap flux density distribution of the field poles is shown in Fig. 4.12*b*. Consider a turn a–b placed on diametrically opposite slots of the rotor. The two terminals a and b of the turn are connected to two slip rings. Two station-

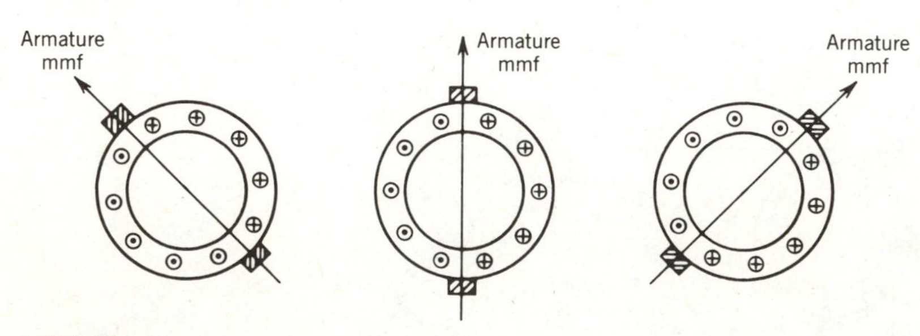

FIGURE 4.11
Shift of brush position.

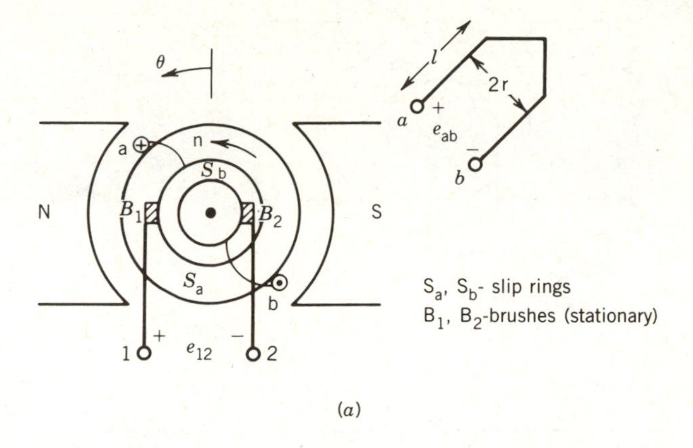

(a)

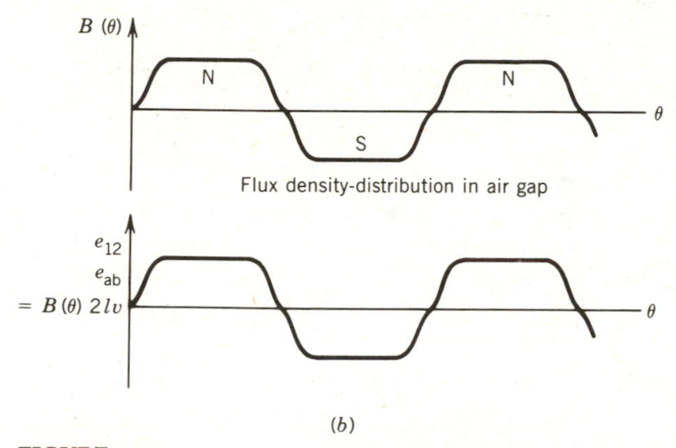

(b)

FIGURE 4.12
Induced voltage in a dc machine. (a) Two-pole dc machine. (b) Induced voltage in a turn.

ary brushes pressing against the slip rings provide access to the revolving turn a–b.

The voltage induced in the turn is due primarily to the voltage induced in the two sides of the turn under the poles. Using the concept of "conductor cutting flux" (Eq. 4.1), these two voltages are in series and aid each other. The voltage induced in the turn, e_{ab} (same as the voltage e_{12} across the brushes), is alternating in nature, and its waveform is the same as that of the flux density distribution wave in space.

Let us now replace the two slip rings by two commutator segments (which are copper segments separated by insulating materials) as shown in Fig. 4.13*a*. Segment C_a is connected to terminal a of the turn and segment C_b to terminal b of the turn. For counterclockwise motion of the rotor the terminal under the N pole is positive with respect to the terminal under the S pole. Therefore, brush terminal B_1 is always connected to the

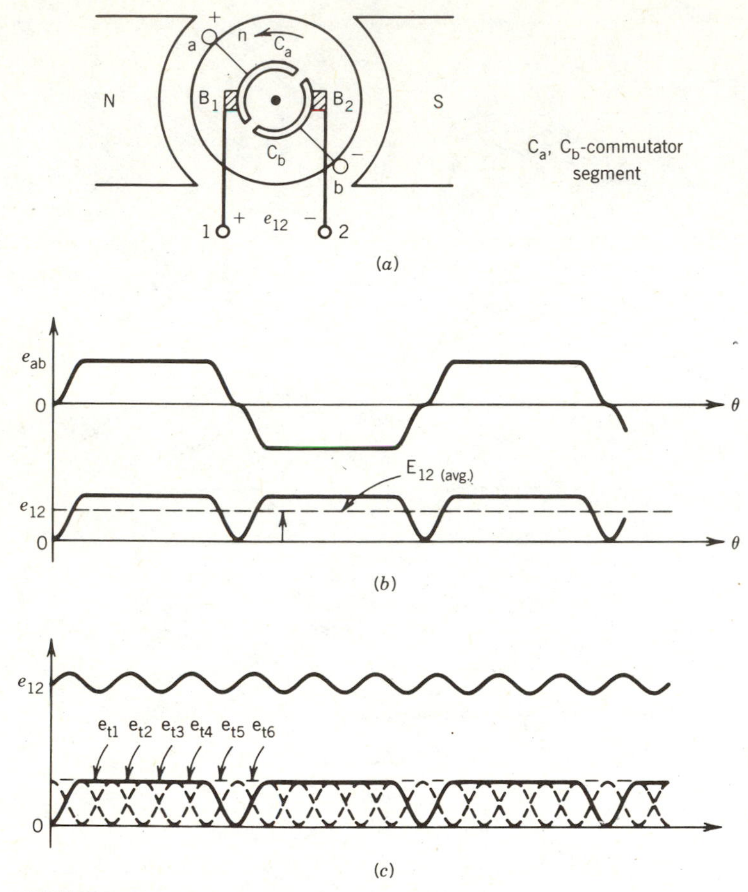

FIGURE 4.13

Voltage rectification by commutators and brushes. (*a*) DC machine with commutator segments. (*b*) Single turn machine. (*c*) Multiturn machine.

positive end of the turn (or coil) and brush terminal B_2 to the negative end of the turn (or coil). Consequently, although the voltage induced in the turn, e_{ab}, is alternating, the voltage at the brush terminals, e_{12}, is unidirectional as shown in Fig. 4.13*b*. This voltage contains a significant amount of ripple. In an actual machine a large number of turns are placed in several slots around the periphery of the rotor. By connecting these in series through the commutator segments (to form an armature winding) a good dc voltage (having a small amount of ripple) can be obtained across the brushes of the rotor armature as shown in Fig. 4.13*c*.

Note that turn a–b is short-circuited by the brushes when its sides pass midway between the field poles (i.e., the q-axis). In the case of a dc motor, current will be fed into the armature through the brushes. The current in the turn will reverse when the turn passes the interpolar region and the commutator segments touch the other brushes. This phenomenon is illustrated by the three positions of the turn in Fig. 4.14.

4.2.3 ARMATURE WINDINGS

As stated earlier, in the dc machine the field winding is placed on the stator to excite the field poles, and the armature winding is placed on the rotor so that the commutator and brush combination can rectify the voltage. There are various ways to construct an armature winding. Before these are discussed, some basic components of the armature winding and terms related to it are defined.

A *turn* consists of two conductors connected to one end by an end connector.

A *coil* is formed by connecting several turns in series.

A *winding* is formed by connecting several coils in series.

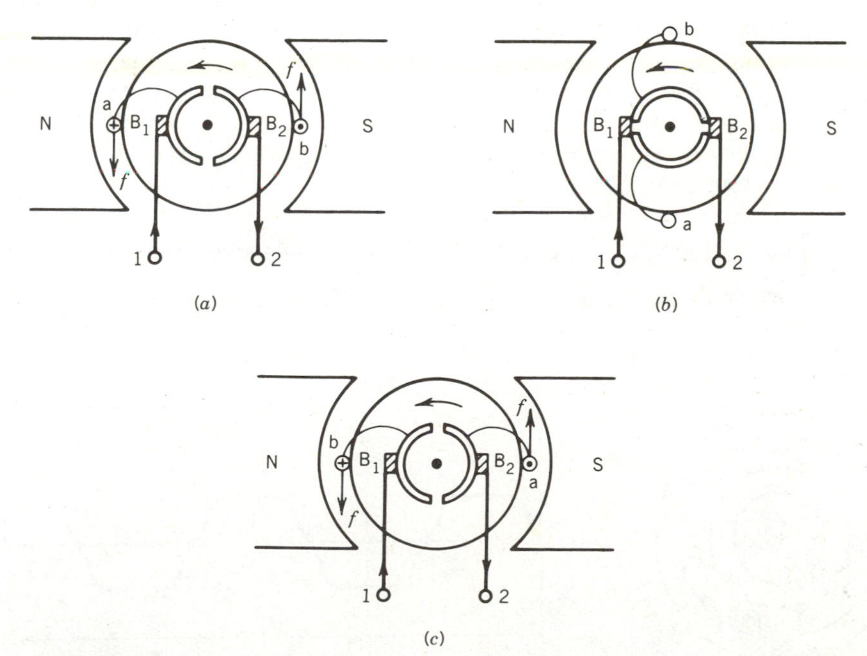

FIGURE 4.14
Current reversal in a turn by commutators and brushes. (a) End a touches brush B_1; current flows from a to b. (b) The turn is shorted; turn is in interpolar region. (c) End a touches brush B_2; current flows from b to a.

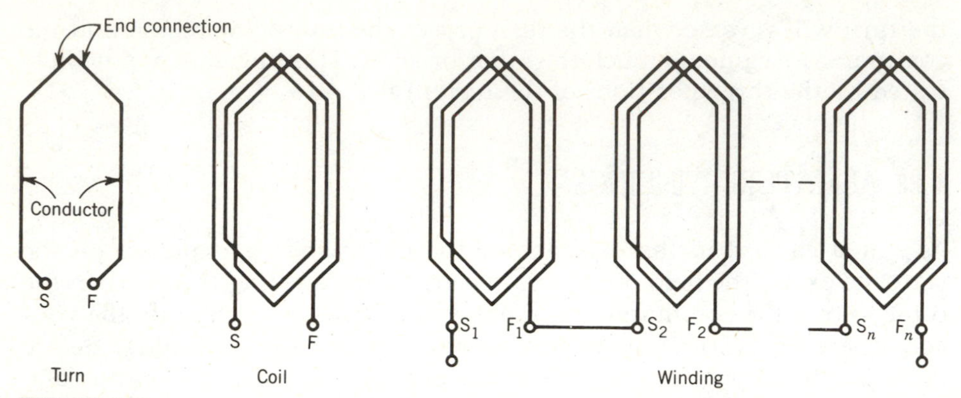

FIGURE 4.15
Turn, coil and winding.

The turn, coil, and winding are shown schematically in Fig. 4.15. The beginning of the turn, or coil, is identified by the symbol S and the end of the turn or coil by the symbol F.

Most dc machines, particularly larger ones, have more than two poles, so most of the armature conductors can be in the region of high air gap flux density. Figure 4.16 shows the stator of a dc machine with four poles. This calls for an armature winding that will also produce four poles on the rotor. The air gap flux density distribution due to the stator poles is shown in Fig. 4.16*b*. Note that, for the four-pole machine, in going around the air gap once (i.e., one mechanical cycle) two cycles of variation of the flux density distribution are encountered. If we define

$$\theta_{md} = \text{mechanical degrees or angular measure in space}$$

$$\theta_{ed} = \text{electrical degrees or angular measure in cycles}$$

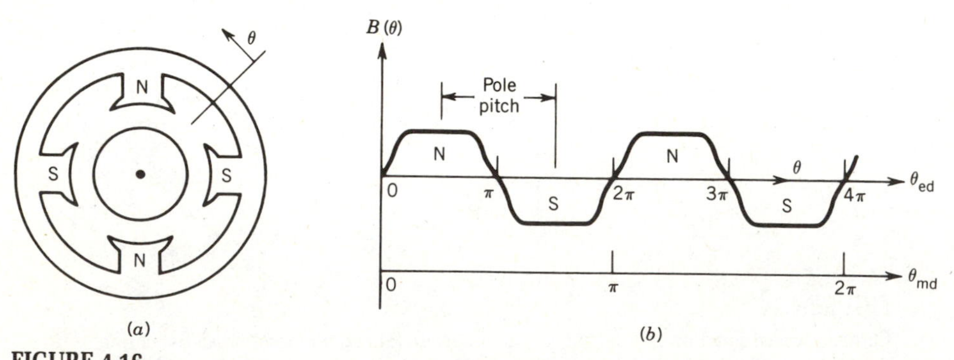

FIGURE 4.16
Mechanical and electrical degrees. (*a*) Four-pole dc machine. (*b*) Flux density distribution.

then, for a p-pole machine,

$$\theta_{ed} = \frac{p}{2}\theta_{md} \tag{4.3}$$

The distance between the centers of two adjacent poles is known as *pole pitch* or pole span. Obviously,

$$\text{One pole pitch} = 180°_{ed} = \frac{360°_{md}}{p}$$

The two sides of a coil are placed in two slots on the rotor surface. The distance between the two sides of a coil is called the *coil pitch.* If the coil pitch is one pole pitch, it is called a *full-pitch* coil. If the coil pitch is less than one pole pitch, the coil is known as a *short-pitch* (or *fractional-pitch*) coil. Short-pitch coils are desirable in ac machines for various reasons (see Appendix A). The dc armature winding is mostly made of full-pitch coils.

There are a number of ways in which the coils of the armature windings of a dc machine can be interconnected. Two kinds of interconnection, *lap* and *wave,* are very common. These are illustrated in Figs. 4.17 and 4.18, respectively.

Lap Winding

Figure 4.17 illustrates an unrolled lap winding of a dc armature, along with the commutator segments (bars) and stationary brushes. The brushes are located under the field poles at their centers. Consider the coil shown by dark lines with one end connected to the commutator bar numbered 2. The coil is placed in slots 2 and 7 such that the coil sides are placed in similar positions under adjacent poles. The other end of the coil is connected to the commutator bar numbered 3. The second coil starts at commutator 3 and finishes at the next commutator, numbered 4. In this way all the coils are added in series and the pattern is continued until the end of the last coil joins the start of the first coil. This is called a lap winding because as the winding progresses the coil *laps* back on itself. It progresses in a continuous loop fashion.

Note that there is one coil between two adjacent commutator bars. Also note that $1/p$ of the total coils of the winding are connected in series between two adjacent brushes and the total voltage induced in these series-connected coils will appear across these two brushes. The brushes making up the positive set are connected together, as are the brushes in the negative set. In a four-pole machine, therefore, there are four parallel paths between the positive and negative terminals of the armature as shown in Fig. 4.17*b*.

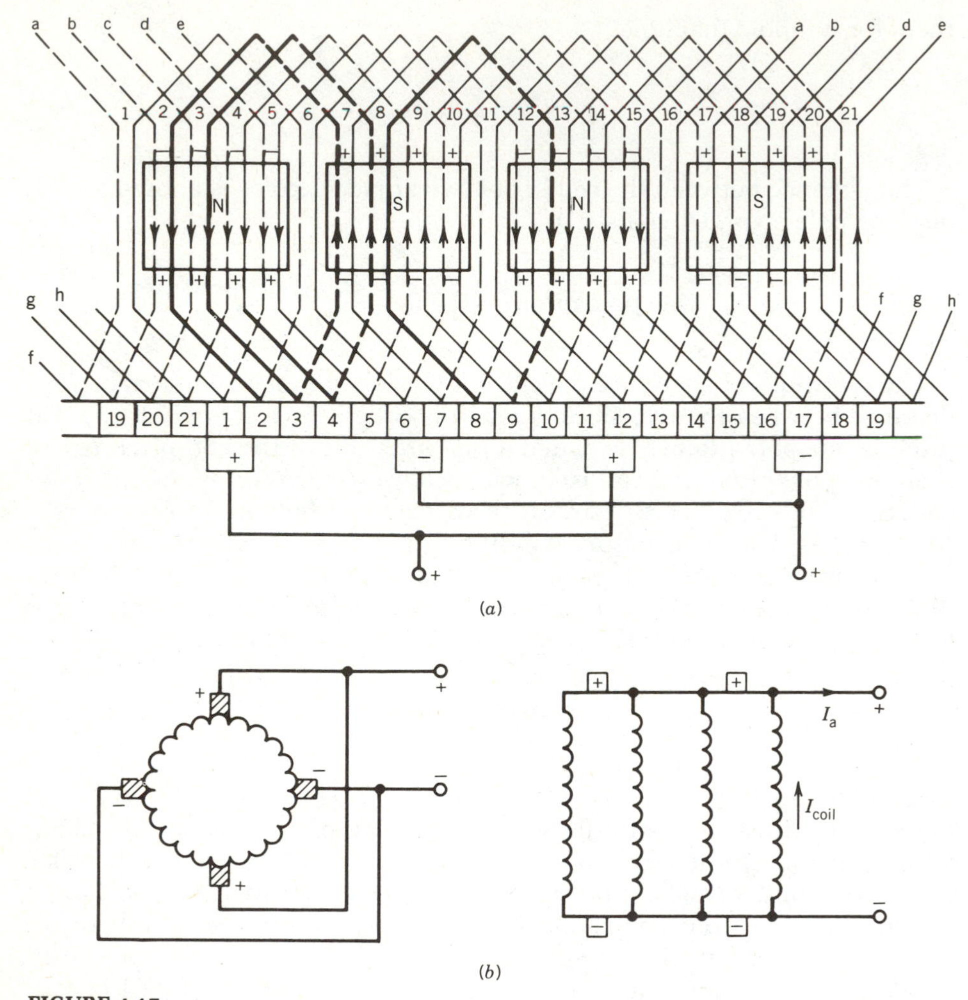

FIGURE 4.17
Lap winding. (a) Unrolled winding. (b) Equivalent coil representation.

In a lap-winding, the number of parallel paths (a) *is always equal to the number of poles* (p) *and also to the number of brushes.*

Wave Winding

The layout of a wave-wound armature winding is shown in Fig. 4.18*a*. The coil arrangement and the end connections are illustrated by the dark lines shown in Fig. 4.18*a* for two coils. One end of the coil starts at commutator bar 2 and the coil sides are placed in slots 7 and 12. The other end of the coil is connected to commutator bar 13. The second coil starts at this

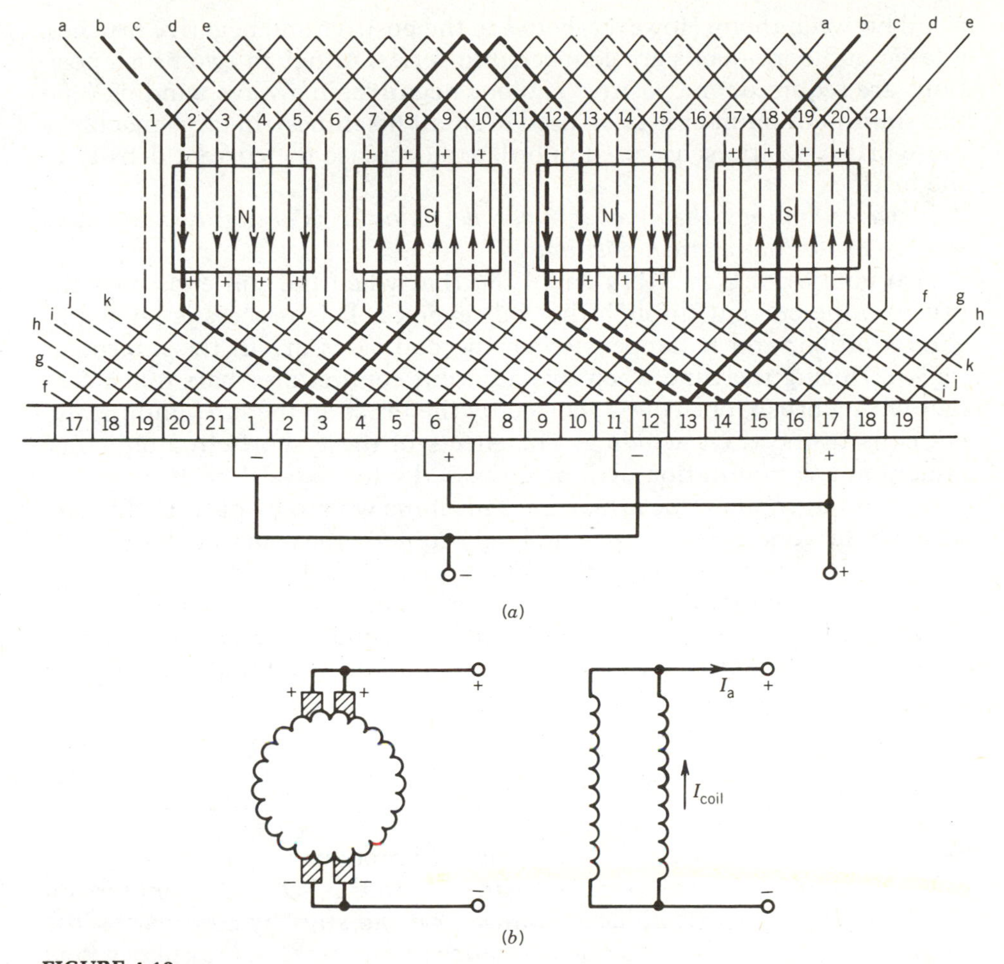

FIGURE 4.18
Wave winding. (a) Unrolled winding. (b) Equivalent coil representation.

commutator bar and is placed in slots 18 and 2 and ends on commutator bar 3. The coil connections are continued in this fashion. The winding is called a wave winding because the coils are laid down in a wave pattern.

Note that between two adjacent commutator bars there are $p/2$ coils connected in series, as opposed to a single coil in the lap winding. Between two adjacent brushes there are $1/p$ of the total commutator bars. Between two adjacent brushes, therefore, there are $(p/2)(1/p)$ or $\frac{1}{2}$ of all the coils. This indicates that in the wave winding the coils are arranged in two parallel paths, irrespective of the number of poles, as illustrated in Fig. 4.18*b*. Note also in Fig. 4.18*a* that the two brushes of the same polarity are connected essentially to the same point in the winding, except that there is a coil between them. However, between the positive and negative brushes, a large number of coils are connected in series. Although two brush positions are required, one positive and one negative, in a wave winding (and this minimum number is often used in small machines), in large machines more brush positions are used in order to decrease the current density in the brushes.

In wave windings, the number of parallel paths (a) *is always two and there may be two or more brush positions.*

Also note from Figs. 4.17*a* and 4.18*a* that when the coil ends pass the brushes, the current through the coil reverses. This process is known as commutation, and it happens when the coil sides are in the interpolar region. During the time when two adjacent commutator bars make contact with a brush, one coil is shorted by the brush in the lap winding and $p/2$ coils in the wave winding. The effects of these short-circuited coils, undergoing commutation, will be discussed later.

In small dc motors, the armature is machine wound by putting the wire into the slots one turn at a time. In larger motors, the armature winding is composed of prefabricated coils that are placed in the slots.

Because many parallel paths can be provided with a lap winding, it is suitable for high-current, low-voltage dc machines, whereas wave windings having only two parallel paths are suitable for high-voltage, low-current dc machines.

4.2.4 ARMATURE VOLTAGE

As the armature rotates in the magnetic field produced by the stator poles, voltage is induced in the armature winding. In this section an expression will be derived for this induced voltage. We can start by considering the induced voltage in the coils due to change of flux linkage (Faraday's law) or by using the concept of "conductor cutting flux." Both approaches will provide the same expression for the armature voltage.

The waveform of the voltage induced in a turn is shown in Fig. 4.12*b*, and because a turn is made of two conductors, the induced voltage in a turn a–b (Fig. 4.12) from Eq. 4.1 is

$$e_{\mathrm{t}} = 2B(\theta)l\omega_{\mathrm{m}}r \tag{4.4}$$

where l is the length of the conductor in the slot of the armature

ω_{m} is the mechanical speed

r is the distance of the conductor from the center of the armature, that is, the radius of the armature

The average value of the induced voltage in the turn is

$$\bar{e}_t = 2\overline{B(\theta)}l\omega_m r \tag{4.5}$$

Let

$$\Phi = \text{flux per pole}$$

$$\text{A} = \text{area per pole} = \frac{2\pi r l}{p}$$

Then

$$\overline{B(\theta)} = \frac{\Phi}{A} = \frac{\Phi p}{2\pi r l} \tag{4.6}$$

From Eqs. 4.5 and 4.6,

$$\bar{e}_t = \frac{\Phi p}{\pi}\omega_m \tag{4.7}$$

The voltages induced in all the turns connected in series for one parallel path across the positive and negative brushes will contribute to the average terminal voltage E_a. Let

$$N = \text{total number of turns in the armature winding}$$

$$a = \text{number of parallel paths}$$

Then

$$E_a = \frac{N}{a}\bar{e}_t \tag{4.8}$$

From Eqs. 4.7 and 4.8,

$$E_a = \frac{Np}{\pi a}\Phi\omega_m$$

$$E_a = K_a\Phi\omega_m \tag{4.9}$$

where K_a is known as the machine (or armature) constant and is given by

$$K_a = \frac{Np}{\pi a} \tag{4.10}$$

or

$$K_a = \frac{2Zp}{\pi a} \tag{4.11}$$

where Z is the total number of conductors in the armature winding. In the MKS system, if Φ is in webers and ω_m in radians per second, then E_a is in volts.

This expression for induced voltage in the armature winding is independent of whether the machine operates as a generator or a motor. In the case of generator operation it is known as a *generated voltage,* and in motor operation it is known as *back emf.*

4.2.5 DEVELOPED (OR ELECTROMAGNETIC) TORQUE

There are various methods by which an expression can be derived for the torque developed in the armature (when the armature winding carries current in the magnetic field produced by the stator poles). However, a simple method is to use the concept of Lorentz force, as illustrated by Eq. 4.2.

Consider the turn aa′b′b shown in Fig. 4.19, whose two conductors aa′ and bb′ are placed under two adjacent poles. The force on a conductor (placed on the periphery of the armature) is

$$f_c = B(\theta) l\, i_c = B(\theta) l \frac{I_a}{a} \tag{4.12}$$

where i_c is the current in the conductor of the armature winding

I_a is the armature terminal current

The torque developed by a conductor is

$$T_c = f_c r \tag{4.13}$$

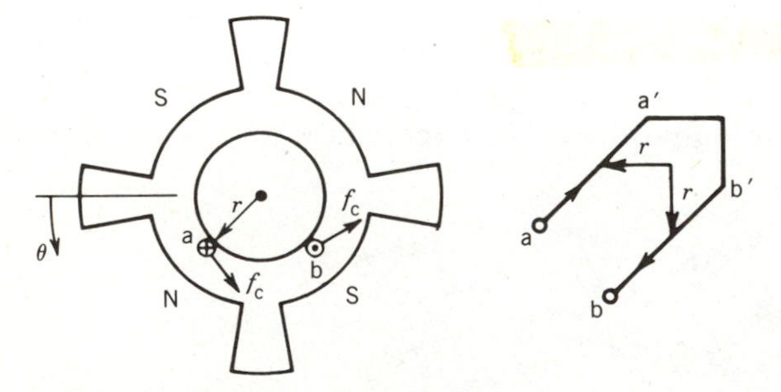

FIGURE 4.19
Torque production in dc machine.

The average torque developed by a conductor is

$$\bar{T}_c = \overline{B(\theta)} l \frac{I_a}{a} r \tag{4.14}$$

From Eqs. 4.6 and 4.14

$$\bar{T}_c = \frac{\Phi p I_a}{2\pi a} \tag{4.15}$$

All the conductors in the armature winding develop torque in the same direction and thus contribute to the average torque developed by the armature. The total torque developed is

$$T = 2N\bar{T}_c \tag{4.16}$$

From Eqs. 4.15 and 4.16

$$T = \frac{N\Phi p}{\pi a} I_a = K_a \Phi I_a \tag{4.17}$$

In the case of motor action, the electrical power input ($E_a I_a$) to the magnetic field by the electrical system must be equal to the mechanical power ($T\omega_m$) developed and withdrawn from the field by the mechanical system. The converse is true for generator action. This is confirmed from Eqs. 4.9 and 4.17. Electrical power,

$$E_a I_a = K_a \Phi \omega_m I_a = T\omega_m, \qquad \text{mechanical power} \tag{4.17a}$$

EXAMPLE 4.1

A four-pole dc machine has an armature of radius 12.5 cm and an effective length of 25 cm. The poles cover 75% of the armature periphery. The armature winding consists of 33 coils, each coil having seven turns. The coils are accommodated in 33 slots. The average flux density under each pole is 0.75T.

1. If the armature is lap-wound,
   (a) Determine the armature constant K_a.
   (b) Determine the induced armature voltage when the armature rotates at 1000 rpm.

(c) Determine the current in the coil and the electromagnetic torque developed when the armature current is 400 A.

(d) Determine the power developed by the armature.

2. If the armature is wave-wound, repeat parts (a) to (d) above. The current rating of the coils remains the same as in the lap-wound armature.

Solution

1. Lap-wound dc machine

(a)
$$K_a = \frac{Np}{\pi a} = \frac{Z}{2a}\frac{p}{\pi}$$

$$Z = 2 \times 33 \times 7 = 462, \qquad a = p = 4$$

$$K_a = \frac{462 \times 4}{2 \times 4 \times \pi} = 73.53$$

(b)
$$\text{Pole area,} \quad A_p = \frac{2\pi \times 0.125 \times 0.25 \times 0.75}{4}$$

$$= 36.8 \times 10^{-3}\ \text{m}^2$$

$$\Phi = A_p \times B = 36.8 \times 10^{-3} \times 0.75$$

$$= 0.0276\ \text{Wb}$$

$$E_a = K_a\Phi\omega_m = 73.53 \times 0.0276 \times \frac{1000}{60}$$

$$\times 2\pi = 212.5\ \text{V}$$

(c)
$$I_{coil} = \frac{I_a}{a} = \frac{400}{4} = 100\ \text{A}$$

$$T = K_a\Phi I_a = 73.53 \times 0.0276 \times 400 = 811.8\ \text{N}\cdot\text{m}$$

(d)
$$P_a = E_a I_a = 212.5 \times 400 = 85.0\ \text{kW}$$

or

$$= T\omega_m = 811.8 \times \frac{1000}{60} \times 2\pi = 85.0\ \text{kW}$$

2. Wave-wound dc machine

$$p = 4, \qquad a = 2, \qquad Z = 462$$

(a)
$$K_a = \frac{462 \times 4}{2 \times 2 \times \pi} = 147.06;$$

$$\omega_m = \frac{1000}{60} \times 2\pi = 104.67 \text{ rad/sec}$$

(b)
$$E_a = 147.06 \times 0.0276 \times 104.67 = 425 \text{ V}$$

(c)
$$I_{coil} = 100 \text{ A}$$
$$I_a = 2 \times 100 = 200 \text{ A}$$
$$T = 147.06 \times 0.0276 \times 200 = 811.8 \text{ N} \cdot \text{m}$$

(d)
$$P_a = 425 \times 200 = 85.0 \text{ kW}$$

4.2.6 MAGNETIZATION (OR SATURATION) CURVE OF A DC MACHINE

A dc machine has two distinct circuits, a field circuit and an armature circuit. The mmf's produced by these two circuits are at quadrature—the field mmf is along the direct axis and the armature mmf is along the quadrature axis. A simple circuit representation of the dc machine is shown in Fig. 4.20.

The flux per pole of the machine will depend on the ampere turns F_p provided by one or more field windings on the poles and the reluctance R of the magnetic path. The magnetic circuit of a two-pole dc machine is shown in Fig. 4.21*a*. The flux passes through the pole, air gap, rotor teeth, rotor core, rotor teeth, air gap, and opposite pole and returns through the yoke of the stator of the machine. The magnetic equivalent circuit is shown in Fig. 4.21*b*, where different sections of the magnetic system in which the flux density can be considered reasonably uniform are represented by separate reluctances.

The magnetic flux Φ that crosses the air gap under each pole depends on the magnetomotive force F_p (hence the field current) of the coils on each pole. At low values of flux Φ the magnetic material may be considered to have infinite permeability, making the reluctances for magnetic core sections zero. The magnetic flux in each pole is then

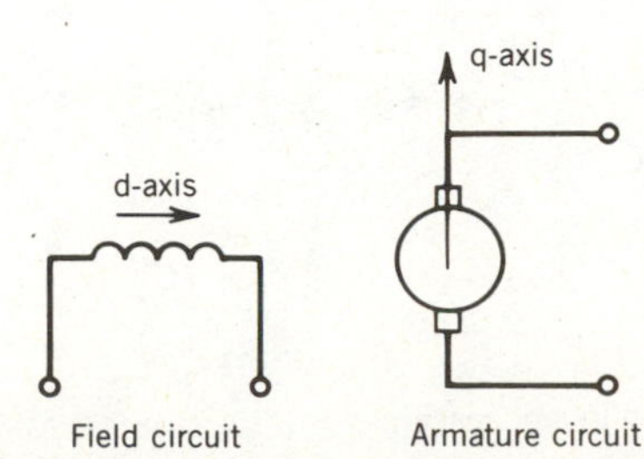

FIGURE 4.20
DC machine representation.

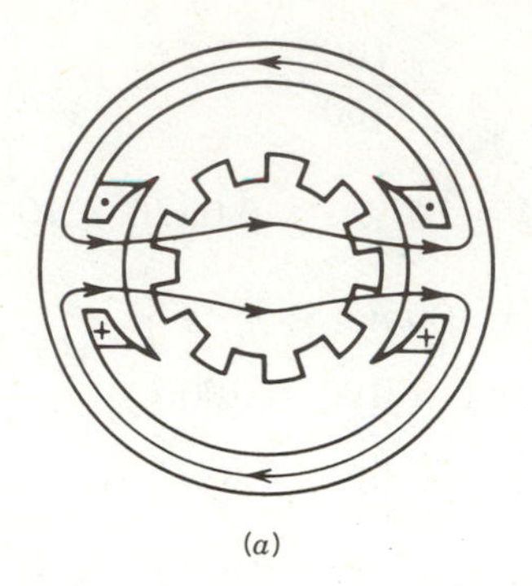

(a)

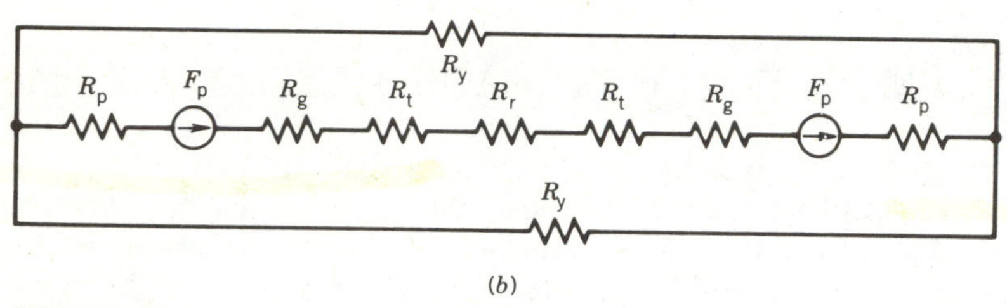

(b)

FIGURE 4.21
Magnetic circuit. (a) Cross-sectional view. (b) Equivalent circuit.

$$\Phi = \frac{2F_p}{2R_g} = \frac{F_p}{R_g} \tag{4.18}$$

If F_p is increased, flux Φ will increase and saturation will occur in various parts of the magnetic circuit, particularly in the rotor teeth. The relationship between field excitation mmf F_p and flux Φ in each pole is shown in Fig. 4.22. With no field excitation, the flux in the pole is the residual flux left over from the previous operation. As the field excitation is increased, the flux increases linearly, as long as the reluctance of the iron core is

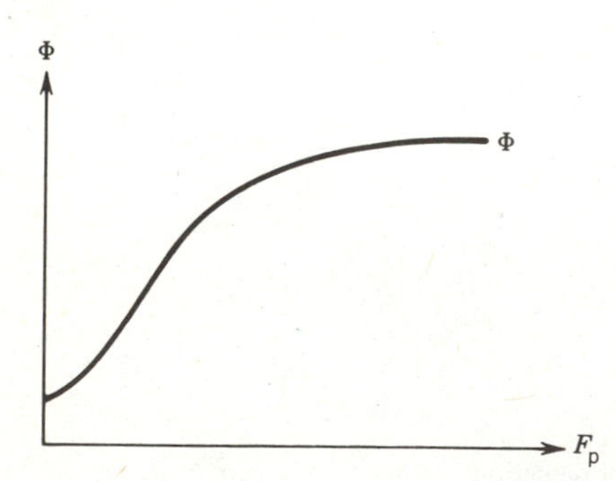

FIGURE 4.22
Flux–mmf relation in a dc machine.

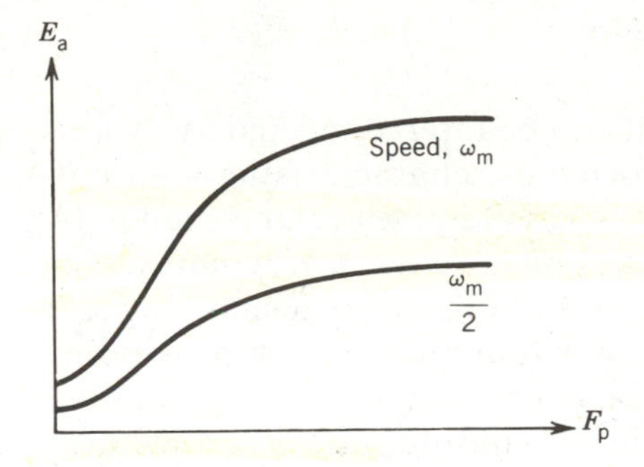

FIGURE 4.23
Magnetization curve.

negligible compared with that of the air gap. Further increase in the field excitation will result in saturation of the iron core, and the flux increase will no longer be linear with the field excitation. It is assumed here that the armature mmf has no effect on the pole flux (d-axis flux) because the armature mmf acts along the q-axis. We shall reexamine this assumption later on.

The induced voltage in the armature winding is proportional to flux times speed (Eq. 4.9). It is more convenient if the magnetization curve is expressed in terms of armature induced voltage E_a at a particular speed. This is shown in Fig. 4.23. This curve can be obtained by performing tests on a dc machine. Figure 4.24 shows the magnetization curve obtained experimentally by rotating the dc machine at 1000 rpm and measuring the open-circuit armature terminal voltage as the current in the field winding is changed. This magnetization curve is of great importance because it represents the saturation level in the magnetic system of the dc machine for various values of the excitation mmf.

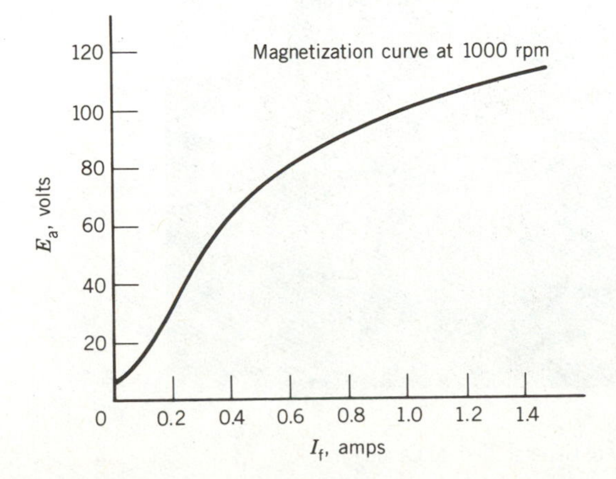

FIGURE 4.24
Test result: magnetization curve.

4.2.7 CLASSIFICATION OF DC MACHINES

The field circuit and the armature circuit can be interconnected in various ways to provide a wide variety of performance characteristics—an outstanding advantage of dc machines. Also, the field poles can be excited by two field windings, a *shunt field winding* and a *series field winding*. The shunt winding has a large number of turns and takes only a small current (less than 5% of the rated armature current). A picture of a shunt winding is shown in Fig. 4.25. This winding can be connected across the armature (i.e., parallel with it), hence the name shunt winding. The series winding has fewer turns but carries a large current. It is connected in series with the armature, hence the name series winding. If both shunt and series windings are present, the series winding is wound on top of the shunt winding, as shown in Fig. 4.26.

The various connections of the field circuit and armature circuit are shown in Fig. 4.27. In the *separately excited* dc machine (Fig. 4.27*a*), the field winding is excited from a separate source. In the *self-excited* dc machine, the field winding can be connected in three different ways. The field winding may be connected in series with the armature (Fig. 4.27*b*), resulting in a series dc machine; it may be connected across the armature (i.e., in shunt), resulting in a *shunt machine* (Fig. 4.27*c*); or both shunt and series windings may be used (Fig. 4.27*d*), resulting in a compound machine. If the shunt winding is connected across the armature, it is known as *short-shunt* machine. In an alternative connection, the shunt winding is connected across the series connection of armature and series winding, and

FIGURE 4.25
Shunt field winding. (Courtesy of General Electric Canada Inc.)

FIGURE 4.26
Series winding on top of shunt winding. (Courtesy of General Electric Canada Inc.)

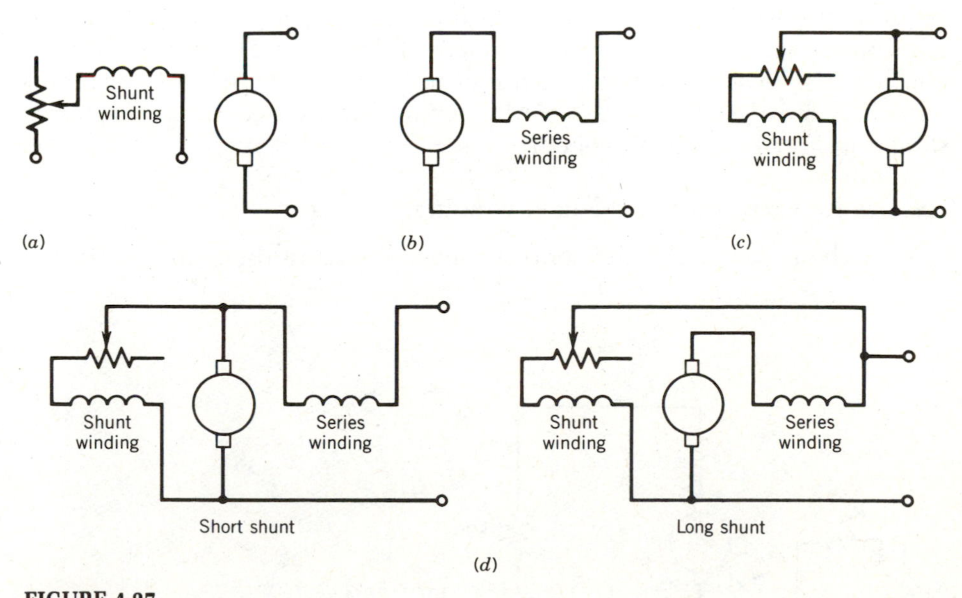

FIGURE 4.27
Different connections of dc machines. (*a*) Separately excited dc machine. (*b*) Series dc machine. (*c*) Shunt dc machine. (*d*) Compound dc machine.

the machine is known as *long-shunt* machine. There is no significant difference between these two connections, which are shown in Fig. 4.27*d*. In the compound machine, the series winding mmf may aid or oppose the shunt winding mmf, resulting in different performance characteristics.

A rheostat is normally included in the circuit of the shunt winding to control the field current and thereby to vary the field mmf.

Field excitation may also be provided by permanent magnets. This may be considered as a form of separately excited machine, the permanent magnet providing the separate but constant excitation.

In the following sections the operation of the various dc machines, first as generators and then as motors, will be studied.

4.3 DC GENERATORS

The dc machine operating as a generator is driven by a prime mover at a constant speed and the armature terminals are connected to a load. In many applications of dc generators, knowledge of the variation of the terminal voltage with load current, known as the *external* or (*terminal*) characteristic, is essential.

4.3.1 SEPARATELY EXCITED DC GENERATOR

As stated in Section 4.2.7, in the separately excited dc generator, the field winding is connected to a separate source of dc power. This source may be another dc generator, a controlled rectifier, or a diode rectifier, or a battery. The steady-state model of the separately excited dc generator is shown in Fig. 4.28. In this model

R_{fw} is the resistance of the field winding.

R_{fc} is the resistance of the control rheostat used in the field circuit.

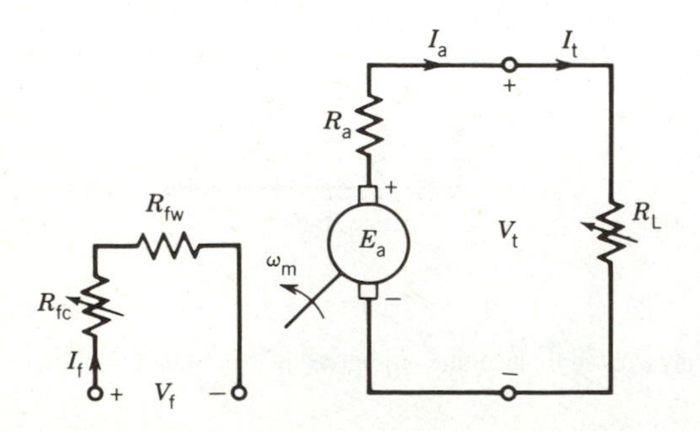

FIGURE 4.28
Steady-state model of a separately excited dc generator.

$R_f = R_{fw} + R_{fc}$ is the total field circuit resistance.

R_a is the resistance of the armature circuit, including the effects of the brushes. Sometimes R_a is shown as the resistance of the armature winding alone; the brush–contact voltage drop is considered separately and is usually assumed to be about 2 V.

R_L is the resistance of the load.

In the steady-state model, the inductances of the field winding and armature winding are not considered.

The defining equations are the following:

$$V_f = R_f I_f \quad (4.19)$$

$$E_a = V_t + I_a R_a \quad (4.20)$$

$$E_a = K_a \Phi \omega_m \quad (4.21)$$

$$V_t = I_t R_L \quad (4.22)$$

$$I_a = I_t \quad (4.23)$$

From Eq. 4.20

$$V_t = E_a - R_a I_a \quad (4.24)$$

Equation 4.20 defines the terminal or external characteristic of the separately excited dc generator; the characteristic is shown in Fig. 4.29. As the

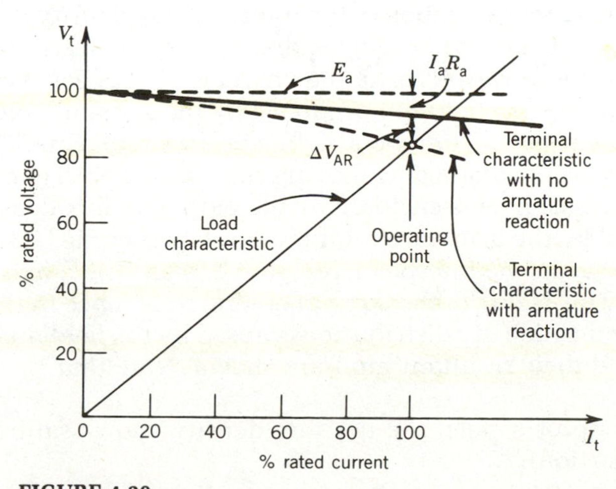

FIGURE 4.29
Terminal characteristic of a separately excited dc generator.

terminal (i.e., load) current I_t increases, the terminal voltage V_t decreases linearly (assuming E_a remains constant) because of the voltage drop across R_a. This voltage drop I_aR_a is small, because the resistance of the armature circuit R_a is small. A separately excited dc generator maintains an essentially constant terminal voltage.

At high values of the armature current a further voltage drop (ΔV_{AR}) occurs in the terminal voltage; that is known as *armature reaction* (or the *demagnetization effect*) and causes a divergence from the linear relationship. This effect can be neglected for armature currents below the rated current. It will be discussed in the next section.

The load characteristic, defined by Eq. 4.22, is also shown in Fig. 4.29. The point of intersection between the generator external characteristic and the load characteristic determines the operating point, that is, the operating values of the terminal voltage V_t and the terminal current I_t.

Armature Reaction (AR)

With no current flowing in the armature, the flux in the machine is established by the mmf produced by the field current, as shown in Fig. 4.30*a*. However, if the current flows in the armature circuit it produces its own mmf (hence flux) acting along the q-axis. Therefore, the original flux distribution in the machine due to the field current is disturbed. The flux produced by the armature mmf opposes flux in the pole under one half of the pole and aids under the other half of the pole, as shown in Fig. 4.30*b*. Consequently, flux density under the pole increases in one half of the pole and decreases under the other half of the pole. If the increased flux density causes magnetic saturation, the net effect is a reduction of flux per pole. This is illustrated in Fig. 4.30*c*.

To have a better appreciation of the mmf and flux density distribution in a dc machine, consider the developed diagram of Fig. 4.31*a*. The armature mmf has a sawtooth waveform as shown in Fig. 4.31*b*. For the path shown by the dashed line, the net mmf produced by the armature current is zero because it encloses equal numbers of dot and cross currents. The armature mmf distribution is obtained by moving this dashed path and considering the dot and cross currents enclosed by the path. The flux density distribution produced by the armature mmf is also shown in Fig. 4.31*b* by a solid curve. Note that in the interpolar region (i.e., near the q-axis), this curve shows a dip. This is due to the large magnetic reluctance in this region. In Fig. 4.31*c* the flux density distributions caused by the field mmf, the armature mmf, and their resultant mmf are shown. Note that

- Near one tip of a pole, the net flux density shows saturation effects (dashed portion).
- The zero flux density region moves from the q-axis when armature current flows.

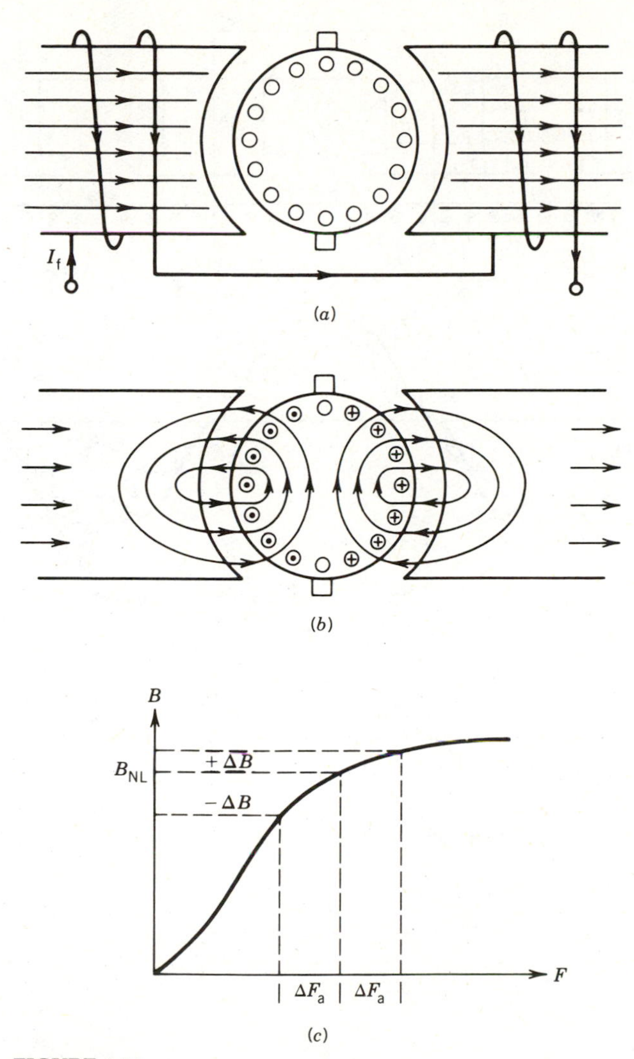

FIGURE 4.30
Armature reaction effects.

- If saturation occurs, the flux per pole decreases. This demagnetizing effect of armature current increases as the armature current increases.

At no load ($I_a = I_t = 0$) the terminal voltage is the same as the generated voltage ($V_{t0} = E_{a0}$). As the load current flows, if the flux decreases because of armature reaction, the generated voltage will decrease (Eq. 4.21). The terminal voltage will further decrease because of the I_aR_a drop (Eq. 4.24).

In Fig. 4.32, the generated voltage for an actual field current $I_{f(actual)}$ is E_{a0}. When the load current I_a flows the generated voltage is $E_a = V_t + I_aR_a$.

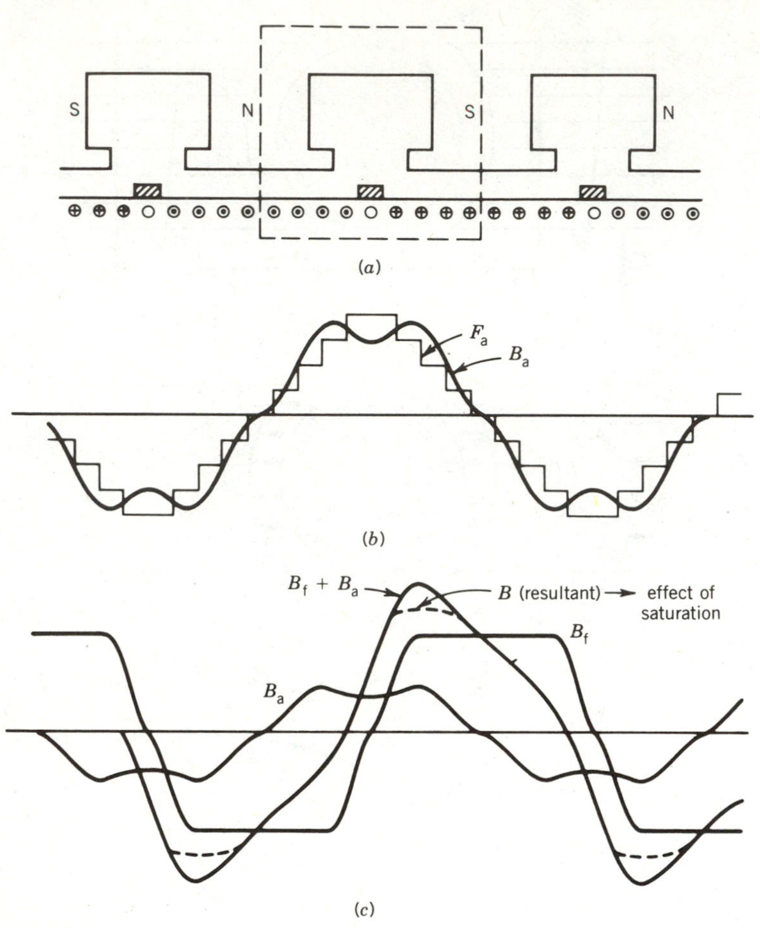

FIGURE 4.31
MMF and flux density distribution.

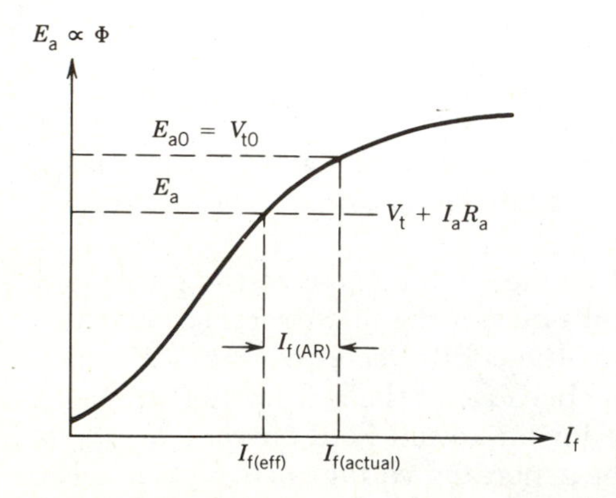

FIGURE 4.32
Effect of armature reaction.

If $E_a < E_{a0}$, the flux has decreased (assuming the speed remains unchanged) because of armature reaction, although the actual field current $I_{f(actual)}$ in the field winding remains unchanged. In Fig. 4.32, the generated voltage E_a is produced by an effective field current $I_{f(eff)}$. The net effect of armature reaction can therefore be considered as a reduction in the field current. The difference between the actual field current and effective field current can be considered as armature reaction in equivalent field current. Hence,

$$I_{f(eff)} = I_{f(actual)} - I_{f(AR)} \tag{4.25}$$

where $I_{f(AR)}$ is the armature reaction in equivalent field current.

Compensating Winding

The armature mmf distorts the flux density distribution and also produces the demagnetizing effect known as armature reaction. The zero flux density region shifts from the q-axis because of armature mmf (Fig. 4.31), and this causes poor commutation leading to sparking (Section 4.3.5). Much of the rotor mmf can be neutralized by using a compensating winding, which is fitted in slots cut on the main pole faces. These pole face windings are so arranged that the mmf produced by currents flowing in these windings opposes the armature mmf. This is shown in the developed diagram of Fig. 4.33*a*. The compensating winding is connected in series with the armature winding so that its mmf is proportional to the armature mmf. Figure 4.33*b* shows a schematic diagram and Fig. 4.33*c* shows the stator of a dc machine having compensating windings. These pole face windings are expensive. Therefore they are used only in large machines or in machines that are subjected to abrupt changes of armature current. The dc motors used in steel rolling mills are large as well as subjected to rapid changes in speed and current. Such dc machines are always provided with compensating windings.

EXAMPLE 4.2

A 12 kW, 100 V, 1000 rpm dc shunt generator has armature resistance $R_a = 0.1\ \Omega$, shunt field winding resistance $R_{fw} = 80\ \Omega$, and $N_f = 1200$ turns per pole. The rated field current is 1 ampere. The magnetization characteristic at 1000 rpm is shown in Fig. 4.24.

The machine is operated as a separately excited dc generator at 1000 rpm with rated field current.

(a) Neglect the armature reaction effect. Determine the terminal voltage at full load.

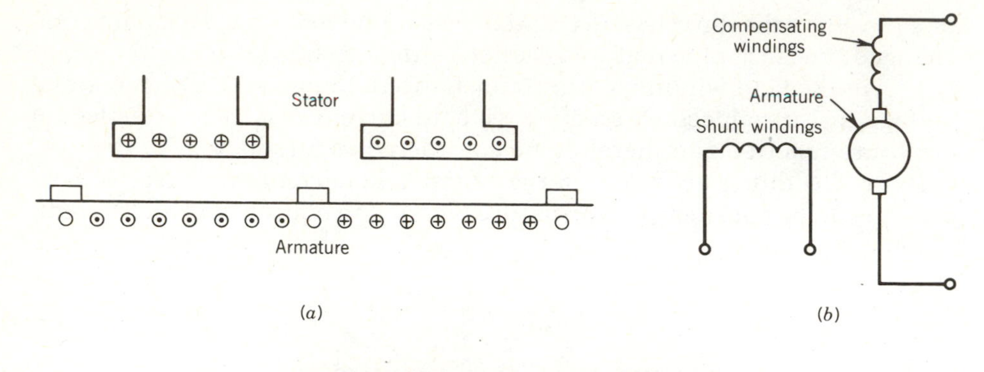

(c)

FIGURE 4.33
Compensating winding. (a) Developed diagram. (b) Schematic diagram. (c) Photograph. (Courtesy of General Electric Canada Inc.)

(b) Consider that armature reaction at full load is equivalent to 0.06 field amperes.

(i) Determine the full-load terminal voltage.

(ii) Determine the field current required to make the terminal voltage $V_t = 100$ V at full-load condition.

Solution

In a dc shunt generator, the main field winding is the shunt field winding. Also, from the data on the machine,

$$\text{Rated terminal voltage } V_t|_{\text{rated}} = 100 \text{ V}$$

Rated armature power (or full load) = 12 kW

Rated armature current (or full load) $I_a|_{rated} = 12{,}000/100 = 120$ A

Rated speed = 1000 rpm

Rated field current $I_f|_{rated} = 1$ A

(a)

$$V_t = E_a - I_aR_a = 100 - 120 \times 0.1 = 88 \text{ V}$$

(b) (i) From Eq. 4.25

$$I_{f(eff)} = 1 - 0.06 = 0.94 \text{ A}$$

From Fig. 4.24, at this field current

$$E_a = 98 \text{ V}$$

$$V_t = E_a - I_aR_a = 98 - 120 \times 0.1 = 86 \text{ V}$$

(ii)

$$E_a = V_t + I_aR_a = 100 + 120 \times 0.1 = 112 \text{ V}$$

From Fig. 4.24, the effective field current required is

$$I_{f(eff)} = 1.4 \text{ A}$$

From Eq. 4.25,

$$I_{f(actual)} = 1.4 + 0.06 = 1.46 \text{ A}$$

4.3.2 SHUNT (SELF-EXCITED) GENERATOR

In the shunt or self-excited generator the field is connected across the armature so that the armature voltage can supply the field current. Under

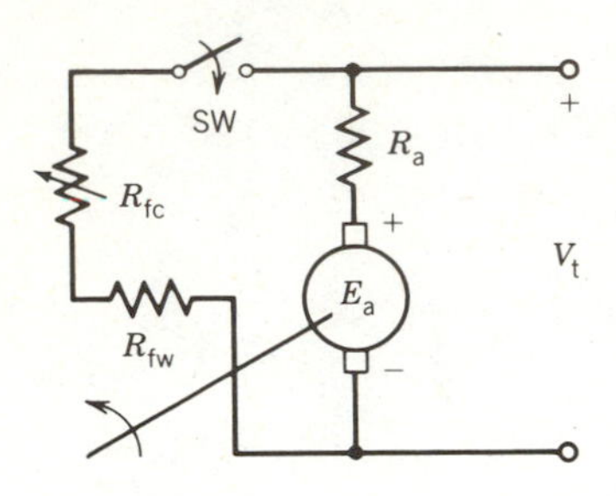

FIGURE 4.34
Schematic of a shunt or self-excited dc machine.

certain conditions, to be discussed here, this generator will build up a desired terminal voltage.

The circuit for the shunt generator under no-load conditions is shown in Fig. 4.34. If the machine is to operate as a self-excited generator, some residual magnetism must exist in the magnetic circuit of the generator. Figure 4.35 shows the magnetization curve of the dc machine. Also shown in this figure is the *field resistance line,* which is a plot of $R_f I_f$ versus I_f. A simplistic explanation of the voltage buildup process in the self-excited dc generator is as follows.

Assume that the field circuit is initially disconnected from the armature circuit and the armature is driven at a certain speed. A small voltage, E_{ar}, will appear across the armature terminals because of the residual magnetism in the machine. If the switch SW is now closed (Fig. 4.34) and the field circuit is connected to the armature circuit, a current will flow in the field winding. If the mmf of this field current aids the residual magnetism, eventually a current I_{f1} will flow in the field circuit. The buildup of this current will depend on the time constant of the field circuit. With I_{f1} flowing in the field circuit, the generated voltage is E_{a1}—from the magnetiza-

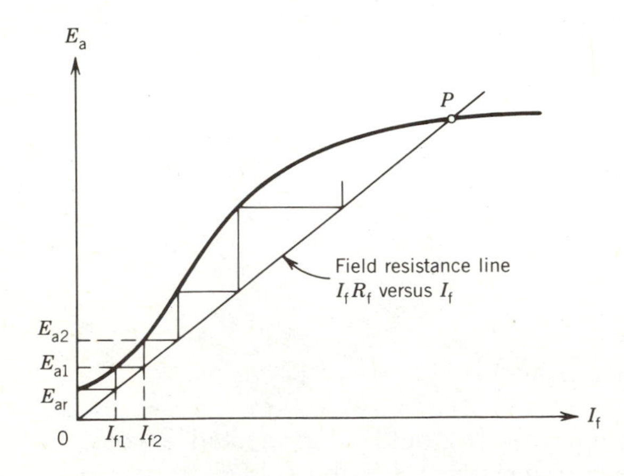

FIGURE 4.35
Voltage buildup in a self-excited dc generator.

tion curve—but the terminal voltage is $V_t = I_{f1}R_f < E_{a1}$. The increased armature voltage E_{a1} will eventually increase the field current to the value I_{f2}, which in turn will build up the armature voltage to E_{a2}. This process of voltage buildup continues. If the voltage drop across R_a is neglected (i.e., $R_a << R_f$), the voltage builds up to the value given by the crossing point (P in Fig. 4.35) of the magnetization curve and the field resistance line. At this point, $E_a = I_f R_f = V_t$ (assume R_a is neglected), and no excess voltage is available to further increase the field current. In the actual case, the changes in I_f and E_a take place simultaneously and the voltage buildup follows approximately the magnetization curve, instead of climbing the flight of stairs.

Figure 4.36 shows the voltage buildup in the self-excited dc generator for various field circuit resistances. At some resistance value R_{f3}, the resistance line is almost coincident with the linear portion of the magnetization curve. This coincidence condition results in an unstable voltage situation. This resistance is known as the *critical field circuit resistance.* If the resistance is greater than this value, such as R_{f4}, buildup (V_{t4}) will be insignificant. On the other hand, if the resistance is smaller than this value, such as R_{f1} or R_{f2}, the generator will build up higher voltages (V_{t1}, V_{t2}). To sum up, three conditions are to be satisfied for voltage buildup in a self-excited dc generator:

1. Residual magnetism must be present in the magnetic system.
2. Field winding mmf should aid the residual magnetism.
3. Field circuit resistance should be less than the critical field circuit resistance.

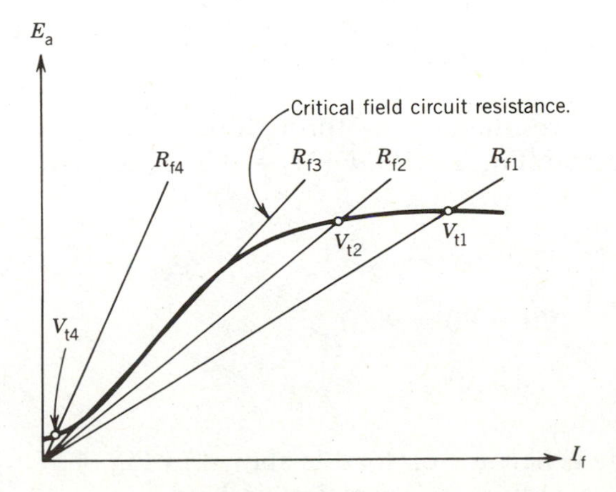

FIGURE 4.36
Effect of field resistance.

EXAMPLE 4.3

The dc machine in Example 4.2 is operated as a self-excited (shunt) generator at no load.

(a) Determine the maximum value of the generated voltage.

(b) Determine the value of the field circuit control resistance (R_{fc}) required to generate rated terminal voltage.

(c) Determine the value of the critical field circuit resistance.

Solution

(a) The maximum voltage will be generated at the lowest value of the field circuit resistance, $R_{fc} = 0$. Draw a field resistance line (Fig. E4.3*b*) for $R_f = R_{fw} = 80\ \Omega$. The maximum generated voltage is

$$E_a = 111 \text{ volts}$$

(b)

$$\begin{aligned} V_t &= E_a - I_a R_a \\ &\simeq E_a \\ &= 100 \text{ V} \end{aligned}$$

Draw a field resistance line that intersects the magnetization curve at 100 V (Fig. E4.3*b*). For this case,

$$I_f = 1 \text{ A}$$

$$R_f = \frac{100}{1} = 100\ \Omega = R_{fw} + R_{fc}$$

$$R_{fc} = 100 - 80 = 20\ \Omega$$

(c) Draw the critical field resistance line passing through the linear portion of the magnetization curve (Fig. E4.3*b*). For $I_f = 0.5$, E_a is 85 V.

$$R_{f(crit)} = \frac{85}{0.5} = 170\ \Omega$$

$$R_{fc} = 170 - 80 = 90\ \Omega$$

Voltage–Current Characteristics

The circuit of the self-excited dc generator on load is shown in Fig. 4.37. The equations that describe the steady-state operation on load are

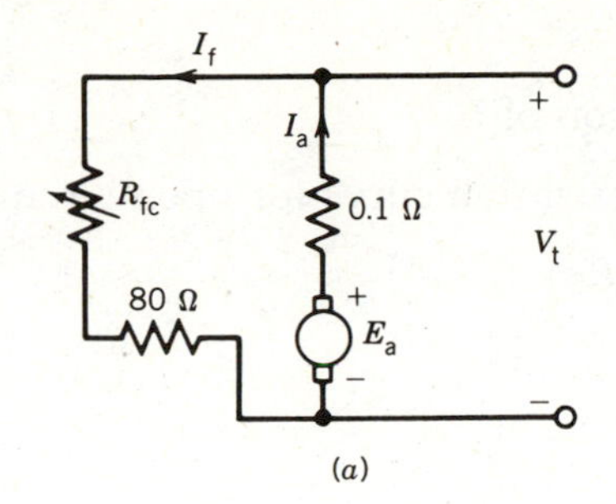

(a)

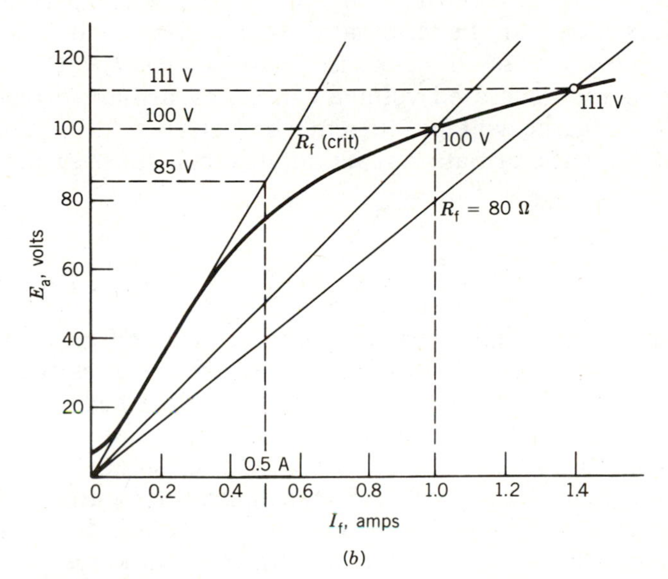

(b)

FIGURE E4.3

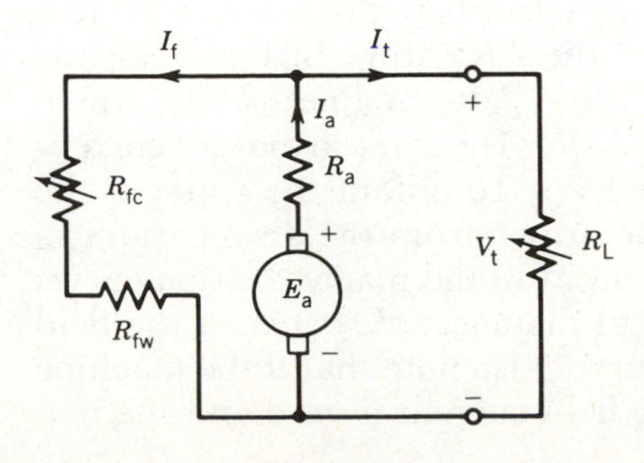

FIGURE 4.37
Self-excited dc generator with load.

$$E_a = V_t + I_a R_a \tag{4.25a}$$

$$E_a = K_a \Phi \omega_m = \text{function of } I_f \tag{4.25b}$$

$\rightarrow$ magnetization curve (or open-circuit saturation curve)

$$V_t = I_f R_f = I_f(R_{f\omega} + R_{fc}) \tag{4.25c}$$

$$V_t = I_L R_L \tag{4.25d}$$

$$I_a = I_f + I_L \tag{4.25e}$$

The terminal voltage (V_t) will change as the load draws current from the machine. This change in the terminal voltage with current (also known as voltage regulation) is due to the internal voltage drop $I_a R_a$ (Eq. 4.25a) and the change in the generated voltage caused by armature reaction (Eq. 4.25b). In finding the voltage–current characteristics (V_t versus I_a) we shall first neglect the armature reaction and then subsequently consider its effects.

Without Armature Reaction

The voltage–current characteristic of the self-excited generator can be obtained from the magnetization curve and the field resistance line, as illustrated in Fig. 4.38. Note that the vertical distance between the magnetization curve and the field resistance line represents the $I_a R_a$ voltage drop. Consider the various points on the field resistance line, which also represents the terminal voltage V_t. For each terminal voltage, such as V_{t1}, compute the armature current I_{a1} from the $I_a R_a$ voltage drop, which is the vertical distance between V_{t1} and E_{a1}. If this calculation is performed for various terminal voltages, the voltage–current characteristic of the dc generator, shown in Fig. 4.38*b*, is obtained. Note that (Fig. 4.37) at $I_t = 0$, $I_a = I_f$, and therefore the actual no-load voltage, V_{t0}, is not the voltage given by the crossing point P of the magnetization curve and the field resistance line, as predicted earlier because of neglecting R_a. However, for all practical purposes $V_{to} = V_p$.

A convenient way to construct the voltage–current characteristic from the magnetization curve and field resistance line is to draw a vertical line at point P. This vertical line represents the $I_a R_a$ drop. In Fig. 4.38, the vertical line pq represents the voltage drop $I_a R_a$. A line qbn is drawn parallel to $0p$. Therefore $pq = ab = mn = I_{a1} R_a$. The same armature current results in two terminal voltages, V_{t1} and V_{t2}. To obtain the value of the maximum armature current that can be drawn from the dc generator, a line rs is drawn parallel to $0p$ and tangential to the magnetization curve. This will result in the maximum vertical distance, sk, between the field resistance line and the magnetization curve. Also note that if the machine terminals are shorted (i.e., $R_L \rightarrow 0$), the field current is zero and the ma-

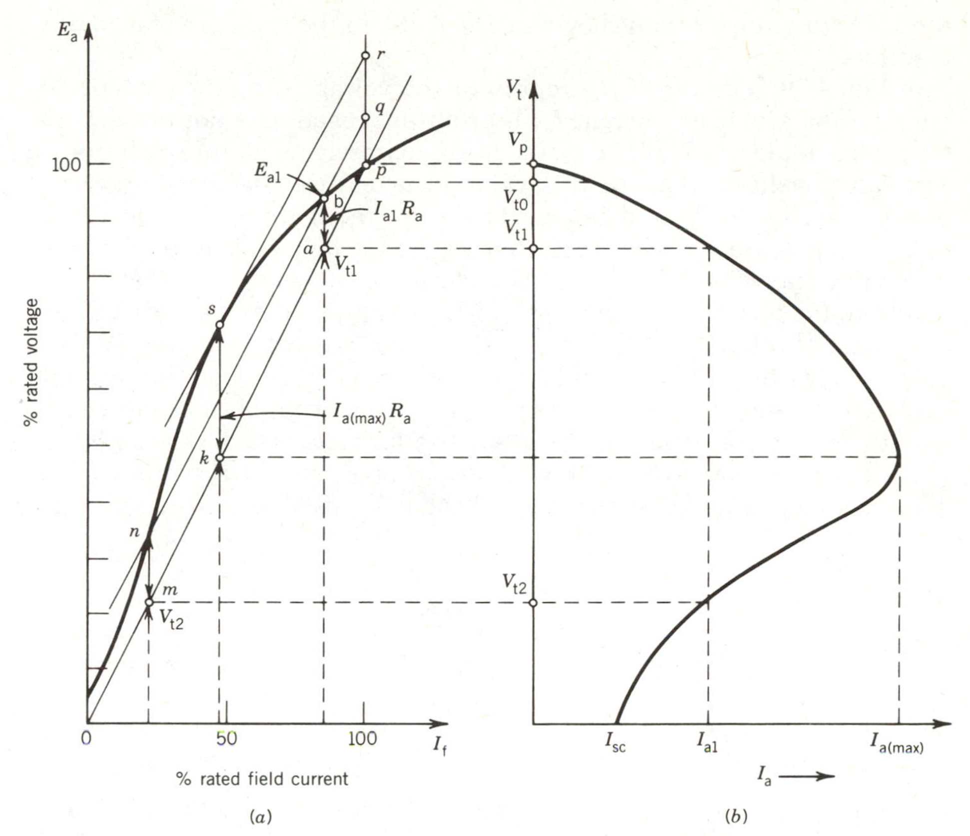

FIGURE 4.38
Terminal characteristic of a self-excited dc generator.

chine currents ($I_a = I_t = I_L = E_a/R_a$) are not very high. However, before R_L is reduced to zero, the armature current may be large enough (such as the current $I_{a(max)}$ in Fig. 4.38*b*) to cause damage to the machine.

From Figs. 4.29 and 4.38*b* it is apparent that the terminal voltage drops faster with the armature current in the self-excited generator. The reason is that, as the terminal voltage decreases with load in the self-excited generator, the field current also decreases, resulting in less generated voltage, whereas in the separately excited generator the field current and hence the generated voltage remain unaffected.

With Armature Reaction

When armature current flows it produces an internal voltage drop I_aR_a. If the armature produces demagnetizing effects on the pole, there will be a further voltage drop in the terminal voltage. The terminal voltage will therefore drop faster than shown in Fig. 4.38*b* in the presence of armature reaction.

In Fig. 4.39, let *pq* (= I_aR_a) represent the voltage drop for a particular value of the armature current I_a. If armature reaction is not present, the terminal voltage is V_{t1}. Let *qr* (= $I_{f(AR)}$) represent armature reaction in equivalent field current for this value of armature current. A line *rc* is drawn parallel to 0*p* and intersects the magnetization curve at *c*. The triangle *pqr* is drawn as *abc* such that *a* is on the field resistance line and *c* is on the magnetization curve. Therefore, in the presence of armature reaction, the terminal voltage is V_{ta}, which is lower than V_{t1}, where V_{t1} is the terminal voltage if armature reaction is not present. Note that $V_t = V_{ta}$, $E_a = V_t + I_aR_a = V_{ta} + ab$ and $I_{f(eff)} = I_f - I_{f(AR)} = I_f - bc$. The terminal voltage corresponding to any other value of the armature current can be determined by constructing a triangle similar to *pqr*, such that *pq* is proportional to I_aR_a and *qr* is proportional to $I_{f(AR)}$, and fitting this triangle between the magnetization curve and the field resistance line.

EXAMPLE 4.4

The dc machine in Example 4.2 is operated as a self-excited generator.

(a) The no-load terminal voltage is adjusted to 100 V. Determine the full-load terminal voltage. Neglect armature reaction effects.

(b) Repeat (a), assuming that the effect of armature reaction at full load is equivalent to 0.06 field amperes, that is, $I_{f(AR)} = 0.06$ A.

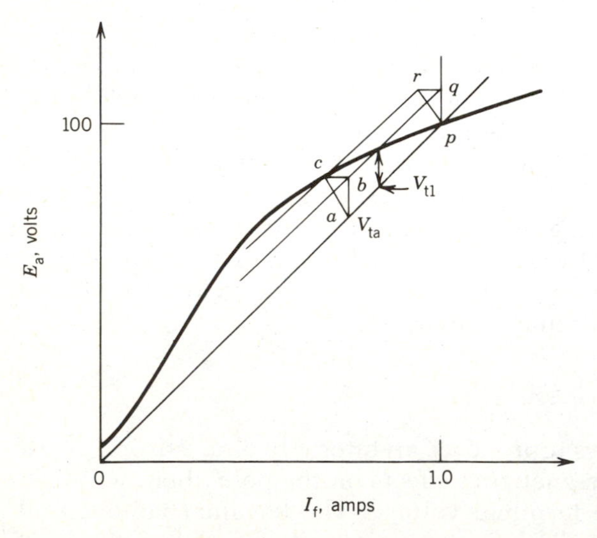

FIGURE 4.39
Determination of terminal voltage.

(c) Determine the maximum value of the armature current that the generator can supply and the corresponding value of the terminal voltage. Assume that $I_{f(AR)}$ is proportional to I_a.

(d) Determine the short-circuit current of the generator.

Solution

(a) Draw the field resistance line $0p$ such that it intersects the magnetization curve at 100 V (Fig. E4.4*a*).

$$I_a|_{FL} = 120 \text{ A}$$

$$I_aR_a = 120 \times 0.1 = 12 \text{ V} = pq$$

Fit $I_aR_a = 12 \text{ V} = a'b'$ between the magnetization curve and the field resistance line (Fig. E4.4).

$$V_t = 80 \text{ V}$$

(b) Construct the triangle pqr with $pq = 12$ V and $qr = 0.06$ A, and fit this triangle as abc between the magnetization curve and the field resistance line (Fig. E4.4).

$$V_t = 75 \text{ V}$$

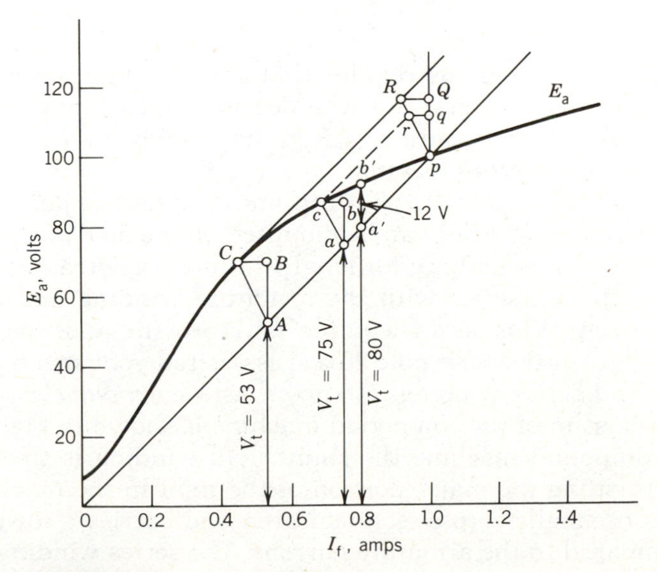

FIGURE E4.4

(c) Draw a tangent CR parallel to $0p$. Extend the triangle pqr to touch the tangent at R. Draw AC parallel to pR and construct the triangle $pQR = ABC$. Note that $pQ = AB$ represents I_aR_a, $QR = BC$ represents $I_{f(AR)}$, and triangle ABC is the largest triangle that will fit between the magnetization curve and the field resistance line.

$$I_aR_a = AB = 17 \text{ V}$$

$$I_a = \frac{17}{0.1} = 170 \text{ A}$$

$$V_t = 53 \text{ V}$$

(d) With the generator terminals short-circuited, $V_t = 0$ and so $I_f = 0$ (Fig. E4.3*a*). The generated voltage is due to residual magnetism and

$$E_a = E_r = 6 \text{ V}$$

$$I_aR_a = 6 \text{ V}$$

$$I_a = \frac{6}{0.1} = 60 \text{ A}$$

Note that because $I_f = 0$, the machine operates at a low flux level in the linear region of the magnetization curve and so there will be no demagnetizing effect due to armature reaction.

4.3.3 COMPOUND DC MACHINES

Many practical applications require that the terminal voltage remains constant when load changes. But when dc machines deliver current, the terminal voltage drops because of I_aR_a voltage drop and decrease in pole fluxes caused by armature reaction.

To overcome the effects of I_aR_a drop and decrease of pole fluxes with armature current, a winding can be mounted on the field poles along with the shunt field winding. This additional winding, known as a series winding, is connected in series with the armature winding and carries the armature current. This series winding may provide additional ampere-turns to increase or decrease pole fluxes, as desired. A dc machine that has both shunt and series windings is known as a *compound dc machine.* A schematic diagram of the compound machine is shown in Fig. 4.40. Note that in a compound machine the shunt field winding is the main field winding, providing the major portion of the mmf in the machine. It has many turns of smaller cross-sectional area and carries a lower value of current compared to the armature current. The series winding has fewer turns, larger cross-sectional area, and carries the armature current. It

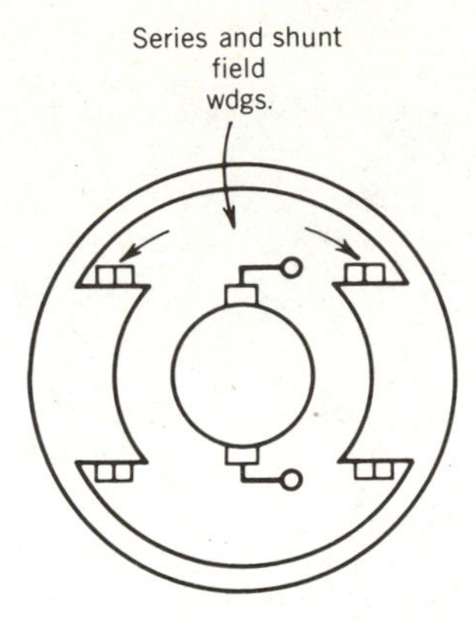

FIGURE 4.40
Compound dc machine.

provides mmf primarily to compensate the voltage drops caused by I_aR_a and armature reaction.

Figure 4.41 shows the two connections for the compound dc machine. In the *short-shunt* connection the shunt field winding is connected across the armature, whereas in the *long-shunt* connection the shunt field winding is connected across the series combination of armature and series winding. The equations that govern the steady-state performance are as follows.

Short Shunt

$$V_t = E_a - I_aR_a - I_tR_{sr} \tag{4.26}$$

$$I_t = I_a - I_f \tag{4.27}$$

where R_{sr} is the resistance of the series field windings.

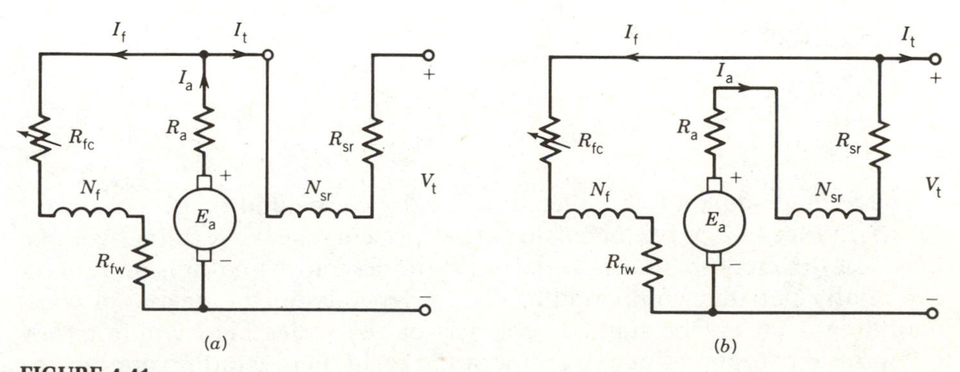

FIGURE 4.41
Equivalent circuits of compound dc machines. (a) Short shunt. (b) Long shunt.

Long Shunt

$$V_t = E_a - I_a(R_a + R_{sr}) \tag{4.28}$$

$$I_t = I_a - I_f \tag{4.29}$$

$$I_f = \frac{V_t}{R_{f\omega} + R_{fc}} \tag{4.30}$$

For either connection, assuming magnetic linearity, the generated voltage is

$$E_a = K_a(\Phi_{sh} \pm \Phi_{sr})\omega_m \tag{4.31}$$

where Φ_{sh} is the flux per pole produced by the mmf of the shunt field winding

Φ_{sr} is the flux per pole produced by the mmf of the series field winding

When these two fluxes aid each other the machine is called a *cumulative compound machine,* and when they oppose each other the machine is called a *differential compound machine.*

Note that both shunt field mmf and series field mmf act on the same magnetic circuit. Therefore, the total effective mmf per pole is

$$F_{eff} = F_{sh} \pm F_{sr} - F_{AR} \tag{4.32}$$

$$N_f I_{f(eff)} = N_f I_f \pm N_{sr} I_{sr} - N_f I_{f(AR)} \tag{4.33}$$

where N_f is the number of turns per pole of the shunt field winding

N_{sr} is the number of turns per pole of the series field winding

F_{AR} is the mmf of the armature reaction

From Eq. 4.33,

$$I_{f(eff)} = I_f \pm \frac{N_{sr}}{N_f} I_{sr} - I_{f(AR)} \tag{4.34}$$

The voltage–current characteristics of the compound dc generator are shown in Fig. 4.42. With increasing armature current the terminal voltage may rise (overcompounding), decrease (undercompounding), or remain essentially flat (flat compounding). This depends on the degree of compounding, that is, the number of turns of the series field winding. For differential compounding (i.e., mmf of the series field winding opposed to that of the shunt field winding) the terminal voltage drops very quickly

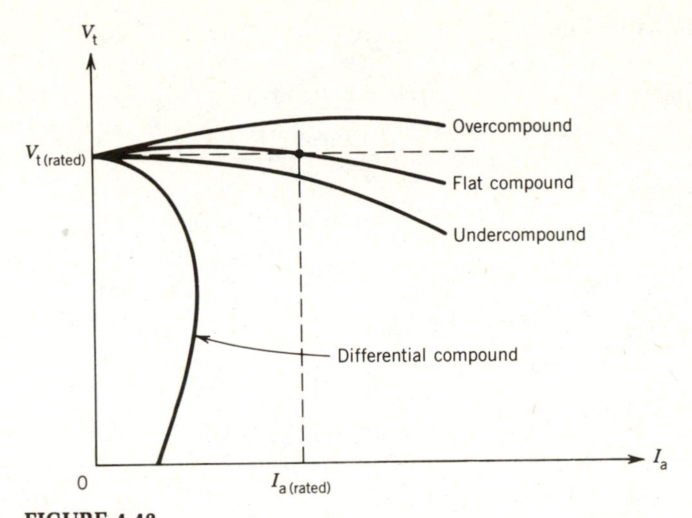

FIGURE 4.42
V–I characteristics of compound dc generators.

with increasing armature current. In fact, the armature current remains essentially constant. This current-limiting feature of the differentially compounded dc generator makes it useful as a welding generator.

EXAMPLE 4.5

The dc machine in Example 4.2 is provided with a series winding so that it can operate as a compound dc machine. The machine is required to provide a terminal voltage of 100 V at no load as well as at full load (i.e., zero voltage regulation) by cumulatively compounding the generator. If the shunt field winding has 1200 turns per pole, how many series turns per pole are required to obtain zero voltage regulation. Assume a short-shunt connection and that the series winding has a resistance $R_{sr} = 0.01\ \Omega$.

Solution

$$V_t|_{NL} = 100\text{ V}$$

From Example 4.3(b), $R_f = 100\ \Omega$. Now, from Fig. 4.41*a*,

$$I_a = I_f + I_t$$

$$120 = I_f + I_t \tag{E.4.5a}$$

Also, from Fig. 4.41*a*

$$I_f R_f = I_t R_{sr} + V_t$$

$$I_f = \frac{V_t + I_t R_{sr}}{R_f}$$

$$I_f = \frac{100 + I_t \times 0.01}{100} \tag{E.4.5b}$$

From Eqs. E4.5a and E4.5b

$$I_f = 1.01 \text{ A}$$

$$I_t = 118.99 \text{ A}$$

From Fig. 4.41*a*,

$$\begin{aligned} E_a &= V_t + I_t R_{sr} + I_a R_a \\ &= 100 + 118.99 \times 0.01 + 120 \times 0.1 \\ &= 113.2 \text{ V} \end{aligned}$$

From the magnetization curve (Example 4.2 and Fig. 4.24) the shunt field current required to generate $E_a = 113.2$ V is 1.45 A ($= I_{f(eff)}$).

From Eq. 4.34

$$I_{f(eff)} = I_f + \frac{N_{sr}}{N_f} I_t - I_{f(AR)}$$

$$1.45 = 1.01 + \frac{N_{sr}}{1200} \times 118.99 - 0.06$$

$$N_{sr} = 5.04 \text{ turns per pole}$$

4.3.4 SERIES GENERATOR

The circuit diagram of a series generator is shown in Fig. 4.43. The series field winding provides the flux in the machine when the armature current flows through it. Note that the field circuit is not complete unless a load is connected to the machine. The equations governing the steady-state operation are

$$E_a = V_t + I_a(R_a + R_{sr}) \tag{4.35}$$

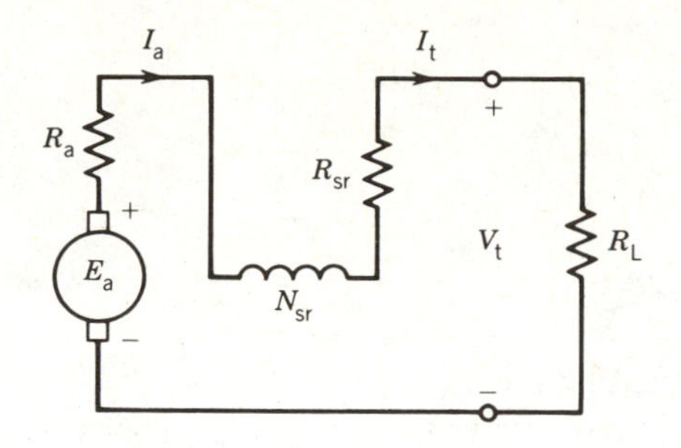

FIGURE 4.43
Equivalent circuit of a dc series generator.

$$I_t = I_a \tag{4.36}$$

The magnetization curve E_a versus I_a (Fig. 4.44) for the series machine can be obtained by separately exciting the series field. To obtain the terminal voltage–current characteristic (i.e., V_t versus I_t), draw a straight line (Fig. 4.44) having the slope $R_a + R_{sr}$. This straight line represents the voltage drop across R_a and R_{sr}. As shown in Fig. 4.44, the vertical distance between the magnetization curve and this straight line is the terminal voltage for a particular value of I_a. (If the effect of armature reaction is considered, the terminal voltage will be less, as shown in Fig. 4.44 by dashed lines, where *ab* represents armature reaction in equivalent armature current.) The terminal voltages for various values of the terminal current can thus be obtained from Fig. 4.44. These voltages are plotted in Fig. 4.45. If the load is a resistance of value R_L, the load characteristic, V_t $(= R_L I_t)$ versus I_t, is a straight line with slope R_L. The operating point for this load is the point of intersection (point *p* in Fig. 4.45) of the magnetization curve and the load characteristic. Note that if R_L is too large, the terminal voltage will be very small; that is, the series generator will not build up any appreciable voltage.

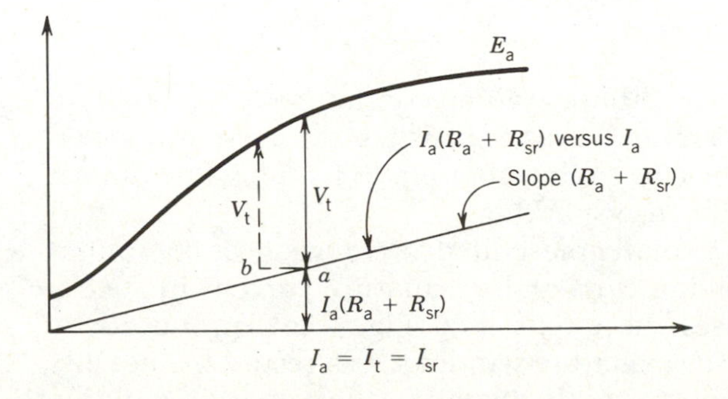

FIGURE 4.44
Magnetization curve (E_a versus I_a) and $I_a(R_a + R_{sr})$ versus I_a.

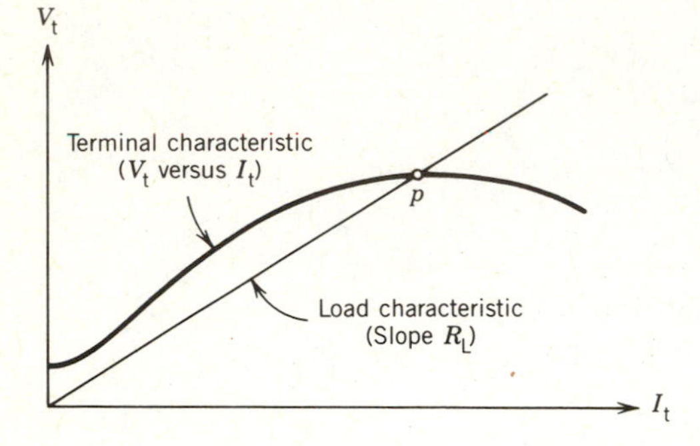

FIGURE 4.45
External characteristic (V_t versus I_t) of a series generator.

4.3.5 INTERPOLES OR COMMUTATOR POLES

The purpose of commutators and brushes in a dc machine is to reverse the current in a conductor when it goes from one pole to the next. This is illustrated in Fig. 4.46*a*. When the conductor *x* is under the north pole it carries a dot current, but after passing through the brush it comes under the south pole (conductor *y*) and thus carries the cross current. In the developed diagram shown in Fig. 4.46*b*, the position of a coil (or turn) undergoing commutation is shown. When the coil passes the brush its current changes direction. Figure 4.46*c* shows a linear change of current in the coil. This is an ideal situation, providing a smooth transfer of current. However, current commutation in a dc machine is not linear for two reasons.

Coil inductance. The coil (Fig. 4.46*b*) undergoing commutation has inductance, which will delay current change.

Reactance voltage. The coil undergoing commutation is in the interpolar region, as can be seen in Fig. 4.46*b*. The armature winding mmf acts along the q-axis and therefore produces flux in the interpolar region. Consequently, when the coil moves in this region, a voltage, called a reactance voltage, is induced in the coil. This reactance voltage delays current change in the coil.

The actual current through a coil undergoing commutation is shown in Fig. 4.46*d*. When the coil is about to leave the brushes, the current reversal is not complete. Therefore, the current has to jump to its full value almost instantaneously and this will cause sparking.

To improve commutation, a small pole, called an interpole or commutation pole, is created. Its winding carries the armature current in such a direction that its flux opposes the q-axis flux (Fig. 4.46*e*) produced by armature current flowing in the armature winding. As a result, the net flux in the interpolar region is almost zero. If current in the armature winding

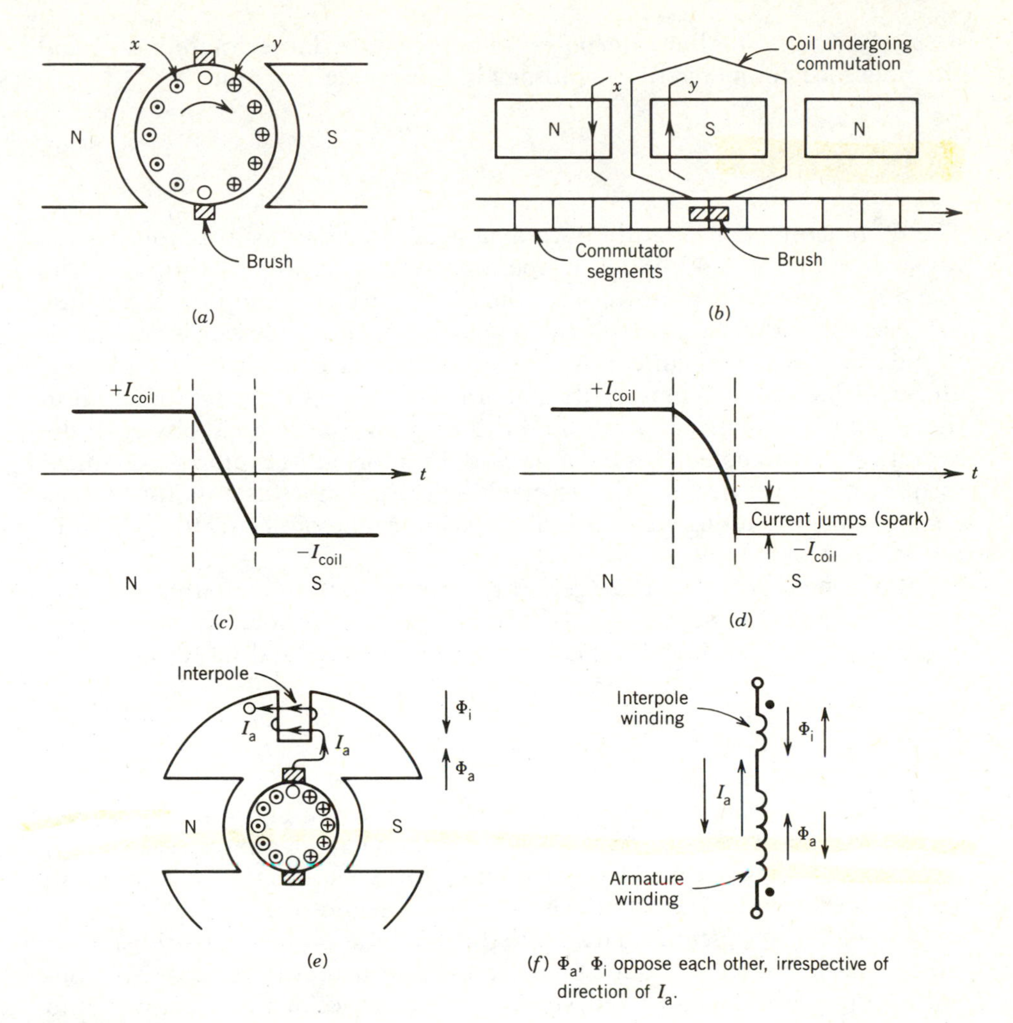

(*f*) Φ_a, Φ_i oppose each other, irrespective of direction of I_a.

FIGURE 4.46
Current communication in dc machine.

reverses, the current in the interpole also reverses and hence these fluxes always oppose, as shown in Fig. 4.46*f*.

Recall that the compensating winding on the pole face also provides flux in the q-axis. However, it cannot completely remove fluxes from the interpolar region. Similarly, interpoles cannot completely overcome the demagnetizing effects of armature reaction on the main poles. Consequently, both pole face compensating windings and interpoles are essential for improved performance of a dc machine. In almost all modern dc machines of large size, both interpoles and compensating windings are used. Figure

4.33*c* shows the smaller interpoles (in between the larger main poles) and the pole face compensating windings in a large dc machine.

4.4 DC MOTORS

The dc machine can operate both as a generator and as a motor. This is illustrated in Fig. 4.47. When it operates as a generator, the input to the machine is mechanical power and the output is electrical power. A prime mover rotates the armature of the dc machine, and dc power is generated in the machine. The prime mover can be a gas turbine, a diesel engine, or an electrical motor. When the dc machine operates as a motor, the input to the machine is electrical power and the output is mechanical power. If the armature is connected to a dc supply, the motor will develop mechanical torque and power. In fact, the dc machine is used more as a motor than as a generator. DC motors can provide a wide range of accurate speed and torque control.

In both modes of operation (generator and motor) the armature winding rotates in the magnetic field and carries current. Therefore, the same basic equations 4.9 and 4.17 hold good for both generator and motor action.

4.4.1 SHUNT MOTOR

A schematic diagram of a shunt dc motor is shown in Fig. 4.48. The armature circuit and the shunt field circuit are connected across a dc source of fixed voltage V_t. An external field rheostat (R_{fc}) is used in the field circuit to control the speed of the motor. The motor takes power from the dc source, and therefore the current I_t flows into the machine from the positive terminal of the dc source. As both field circuit and armature circuit are connected to a dc source of fixed voltage, the connections for separate and shunt excitation are the same. The behavior of the field circuit is independent of the armature circuit.

The governing equations for steady-state operation of the dc motor are as follows:

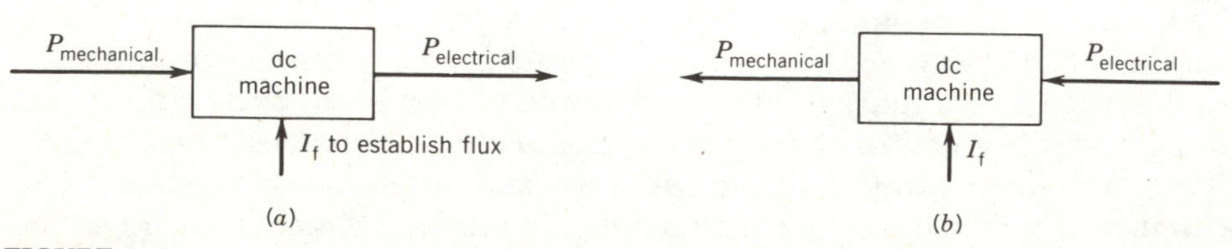

FIGURE 4.47
Reversibility of a dc machine. (*a*) Generator. (*b*) Motor.

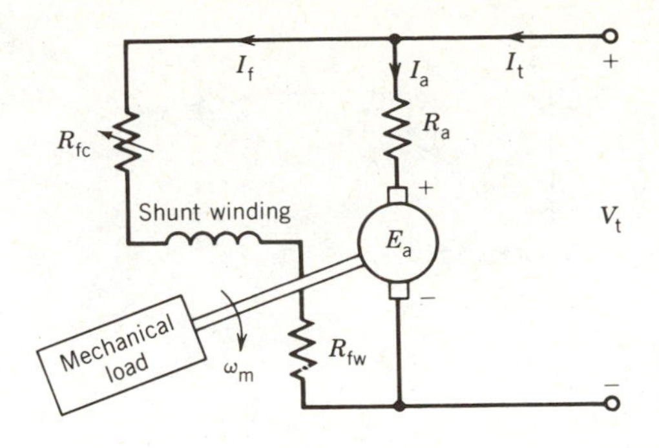

FIGURE 4.48
Shunt dc motor equivalent circuit.

$$V_t = I_a R_a + E_a \tag{4.35a}$$

$$I_t = I_a + I_f \tag{4.36a}$$

$$E_a = K_a \Phi \omega_m \tag{4.37}$$

$$= V_t - I_a R_a \tag{4.38}$$

The armature current I_a and the motor speed ω_m depend on the mechanical load connected to the motor shaft.

Power Flow and Efficiency

The power flow in a dc machine is shown in Fig. 4.49. The various losses in the machine are identified and their magnitudes as percentages of input power are shown. A short-shunt compound dc machine is considered as an example (Fig. 4.49*a*).

With the machine operating as a generator (Fig. 4.49*b*), the input power is the mechanical power derived from a prime mover. Part of this input power is lost as rotational losses required to rotate the machine against windage and friction (rotor core loss is also included in the rotational loss). The rest of the power is converted into electrical power $E_a I_a$. Part of this developed power is lost in R_a (which includes brush contact loss), part is lost in R_f (= $R_{fc} + R_{fw}$), and part is lost in R_{sr}. The remaining power is available as the output electrical power. Various powers and losses in a motoring operation are shown in Fig. 4.49*c*.

The percentage losses depend on the size of the dc machine. The range of percentage losses shown in Fig. 4.49 is for dc machines in the range 1 to 100 kW or 1 to 100 hp. Smaller machines have a larger percentage of losses, whereas larger machines have a smaller percentage of losses.

The efficiency of the machine is

$$\text{Eff} = \frac{P_{\text{output}}}{P_{\text{input}}}$$

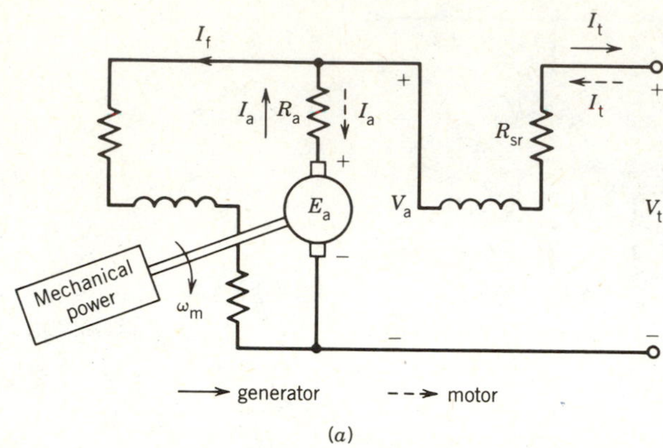

(a)

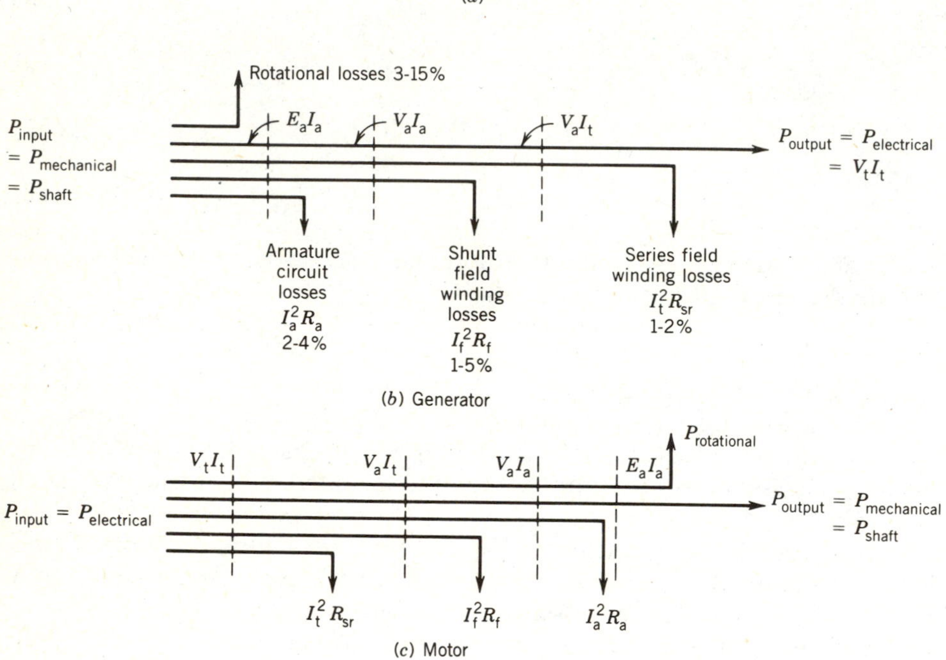

FIGURE 4.49
Power losses in a dc machine.

EXAMPLE 4.6

The dc machine (12 kW, 100 V, 1000 rpm) of Example 4.2 is connected to a 100 V dc supply and is operated as a dc shunt motor. At no-load condition, the motor runs at 1000 rpm and the armature takes 6 amperes.

(a) Find the value of the resistance of the shunt field control rheostat (R_{fc}).

(b) Find the rotational losses at 1000 rpm.

(c) Find the speed, electromagnetic torque, and efficiency of the motor when rated current flows in the armature.

(i) Consider that the air gap flux remains the same as that at no load.

(ii) Consider that the air gap flux is reduced by 5% when rated current flows in the armature because of armature reaction.

(d) Find the starting torque if the starting armature current is limited to 150% of its rated value.

(i) Neglect armature reaction.

(ii) Consider armature reaction, $I_{f(AR)} = 0.16$ A.

Solution

(a) No load, $I_a = 6$ A.

$$\begin{aligned} E_a &= V_t - I_a R_a \\ &= 100 - 6 \times 0.1 \\ &= 99.4 \text{ V} \end{aligned}$$

From the magnetization curve (Fig. 4.24), to generate $E_a = 99.4$ V at 1000 rpm requires $I_f = 0.99$ A.

$$R_f = R_{fc} + R_{fw} = \frac{V_t}{I_f} = \frac{100}{0.99} = 101 \ \Omega$$

$$\begin{aligned} R_{fc} &= 101 - R_{fw} \\ &= 101 - 80 \\ &= 21 \ \Omega \end{aligned}$$

(b) At no load the electromagnetic power developed is lost as rotational power.

$$P_{\text{rotational}} = E_a I_a = 99.6 \times 5 = 497.5 \text{ W}$$

(c) The motor is loaded and $I_a = I_a|_{\text{rated}} = 120$ A.

(i) No armature reaction, that is, $\Phi_{NL} = \Phi_{FL}$.

$$E_a|_{NL} = 99.4 \text{ V}$$

$$E_a|_{FL} = V_t - I_aR_a = 100 - 120 \times 0.1 = 88 \text{ V}$$

$$\frac{E_a|_{FL}}{E_a|_{NL}} = \frac{K_a\Phi_{FL}\omega_{FL}}{K_a\Phi_{NL}\omega_{NL}} = \frac{\omega_{FL}}{\omega_{NL}}$$

$$\omega_{FL} = \frac{E_a|_{FL}}{E_a|_{NL}}\,\omega_{NL} = \frac{88}{99.4} \times 1000 = 885.31 \text{ rpm}$$

$$\omega_m = \frac{885.31}{60} \times 2\pi = 92.71 \text{ rad/sec}$$

$$T = \frac{E_aI_a}{\omega_m} = \frac{88 \times 120}{92.71} = 113.9 \text{ N} \cdot \text{m}$$

$$P_{out} = E_aI_a - P_{rotational}$$

$$= 10{,}560 - 497.5$$

$$= 10{,}062.5 \text{ W}$$

$$P_{in} = V_tI_t = V_t(I_a + I_f)$$

$$= 100(120 + 0.99)$$

$$= 12{,}099 \text{ W}$$

$$\text{Eff} = \frac{P_{out}}{P_{in}} = \frac{10{,}062.5}{12{,}099} \times 100\% = 83.17\%$$

(ii) With armature reaction, $\Phi_{FL} = 0.95\Phi_{NL}$.

$$\frac{E_a|_{FL}}{E_a|_{NL}} = \frac{K_a\Phi_{FL}\omega_{FL}}{K_a\Phi_{NL}\omega_{NL}}$$

$$\frac{88}{99.4} = 0.95\,\frac{\omega_{FL}}{\omega_{NL}}$$

$$\omega_{FL} = \frac{88}{99.4} \times \frac{1}{0.95} \times 1000 = 931.91 \text{ rpm}$$

Note that the speed increases if flux decreases because of armature reaction.

$$\omega_m = \frac{931.91}{60} \times 2\pi = 97.59 \text{ rad/sec}$$

$$\text{T} = \frac{88 \times 120}{97.59} = 75.22 \text{ N} \cdot \text{m}$$

$\text{Eff} = \dfrac{10{,}062.5}{12{,}099} \times 100\% = 83.17\%$, assuming rotational losses do not change with speed.

(d) $T = K_a\Phi I_a$.

(i) If armature reaction is neglected, the flux condition under load can be obtained from the no-load condition.

$$E_a|_{NL} = 99.4 \text{ V} = K_a\Phi\omega_m = K_a\Phi \frac{1000}{60} \times 2\pi$$

$$K_a\Phi = 0.949 \text{ V/rad/sec}$$

$$I_a = 1.5 \times 120 = 180 \text{ A}$$

$$T_{start} = 0.949 \times 180 = 170.82 \text{ N} \cdot \text{m}$$

(ii) $I_f = 0.99$ A. When $I_a = 180$ A

$$I_{f(eff)} = I_f - I_{f(AR)} = 0.99 - 0.16 = 0.83 \text{ A}$$

From the magnetization curve (Fig. 4.24) the corresponding generated voltage is

$$E_a = 93.5 \text{ V } (= K_a\Phi\omega_m) \text{ at } 1000 \text{ rpm}$$

$$K_a\Phi = \frac{93.5}{\omega_m} = \frac{93.5}{1000 \times 2\pi/60} = 0.893 \text{ V/rad/sec}$$

$$T_{start} = 0.893 \times 180 = 160.71 \text{ N} \cdot \text{m}$$

EXAMPLE 4.7

The dc machine of Example 4.2 runs at 1000 rpm at no load ($I_a = 6$ A, Example 4.6) and at 932 rpm at full load ($I_a = 120$ A, Example 4.6) when operated as a shunt motor.

(a) Determine the armature reaction effect at full load in ampere-turns of the shunt field winding.

(b) How many series field turns per pole should be added to make this machine into a cumulatively compound motor (short-shunt) whose speed will be 800 rpm at full load? Neglect the resistance of the series field winding.

(c) If the series field winding is connected for differential compounding, determine the speed of the motor at full load.

Solution

(a) From Example 4.6,

$$I_f = 0.99 \text{ A}$$

At full load,

$$E_a = 100 - 120 \times 0.1 = 88 \text{ V at } 932 \text{ rpm}$$

The effective field current ($I_{f(eff)}$) at full load can be obtained from the magnetization curve (Fig. 4.24) of the machine, if we first find E_a at 1000 rpm.

$$E_a|_{1000} = \frac{1000}{932} \times 88 = 94.42 \text{ V}$$

From the magnetization curve, for $E_a = 94.42$ V at 1000 rpm,

$$I_{f(eff)} = 0.86 \text{ A} = I_f - I_{f(AR)}$$

$$I_{f(AR)} = I_f - 0.86 = 0.99 - 0.86$$

$$= 0.13 \text{ A}$$

$$\text{The corresponding ampere-turns} = N_f I_{f(AR)}$$

$$= 1200 \times 0.13$$

$$= 156 \text{ At/pole}$$

(b) $E_a = 88$ V at 800 rpm.

$$E_a|_{1000} = \frac{1000}{800} \times 88 = 110 \text{ V}$$

From the magnetization curve for $E_a = 110$ V at 1000 rpm,

$$I_{f(eff)} = 1.32 = I_f + \frac{N_{sv}}{N_f}(I_a + I_f) - I_{f(AR)}$$

$$1.32 = 0.99 + \frac{N_{sr}}{1200}(120 + 0.99) - 0.13$$

$$N_{sv} = 4.56 \text{ turns/pole}$$

(c) For differential compounding,

$$I_{f(eff)} = 0.99 - \frac{4.56 \times 120.99}{1200} - 0.13$$

$$= 0.99 - 0.46 - 0.13$$

$$= 0.4 \text{ A}$$

From the magnetization curve, at 1000 rpm and $I_f = 0.4$ A, $E_a = 65$ V. But $E_a = 88$ V at full load (parts a and b). If the operating speed is n rpm,

$$65 = K\Phi 1200$$

$$88 = K\Phi n$$

or

$$n = \frac{88}{65} \times 1000 = 1343.9 \text{ rpm}$$

Torque-Speed Characteristics

In many applications dc motors are used to drive mechanical loads. Some applications require that the speed remain constant as the mechanical load applied to the motor changes. On the other hand, some applications require that the speed be controlled over a wide range. An engineer who wishes to use a dc motor for a particular application must therefore know the relation between torque and speed of the machine. In this section the torque–speed characteristics of the various dc motors are discussed.

Consider the separately excited dc motor shown in Fig. 4.50. The voltage, current, speed, and torque are related as follows:

$$E_a = K_a\Phi\omega_m = V_t - I_aR_a \tag{4.39}$$

$$T = K_a\Phi I_a \tag{4.40}$$

From Eqs. 4.39, the speed is

$$\omega_m = \frac{V_t - I_aR_a}{K_a\Phi} \tag{4.41}$$

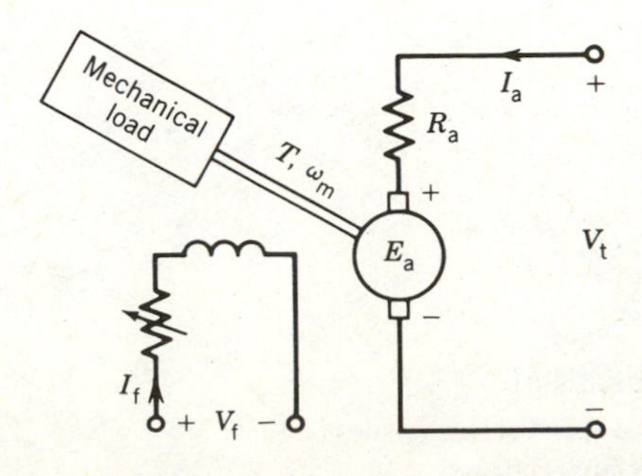

FIGURE 4.50
Separately excited dc motor.

From Eqs. 4.40 and 4.41

$$\omega_m = \frac{V_t}{K_a\Phi} - \frac{R_a}{(K_a\Phi)^2}T \tag{4.42}$$

If the terminal voltage V_t and machine flux Φ are kept constant, the torque–speed characteristic is as shown in Fig. 4.51. The drop in speed as the applied torque increases is small, providing a good speed regulation. In an actual machine, the flux Φ will decrease because of armature reaction as T or I_a increases, and as a result the speed drop will be less than that shown in Fig. 4.51. The armature reaction therefore improves the speed regulation in a dc motor.

Equation 4.42 suggests that speed control in a dc machine can be achieved by the following methods:

1. Armature voltage control (V_t).
2. Field control (Φ).
3. Armature resistance control (R_a).

In fact, speed in a dc machine increases as V_t increases and decreases as Φ or R_a increases. The characteristic features of these different methods of speed control of a dc machine will be discussed further.

Armature Voltage Control

In this method of speed control the armature circuit resistance (R_a) remains unchanged, the field current I_f is kept constant (normally at its rated value), and the armature terminal voltage (V_t) is varied to change the speed. If armature reaction is neglected, from Eq. 4.42,

$$\omega_m = K_1V_t - K_2T \tag{4.43}$$

where $K_1 = 1/K_a\Phi$

$K_2 = R_a/(K_a\Phi)^2$

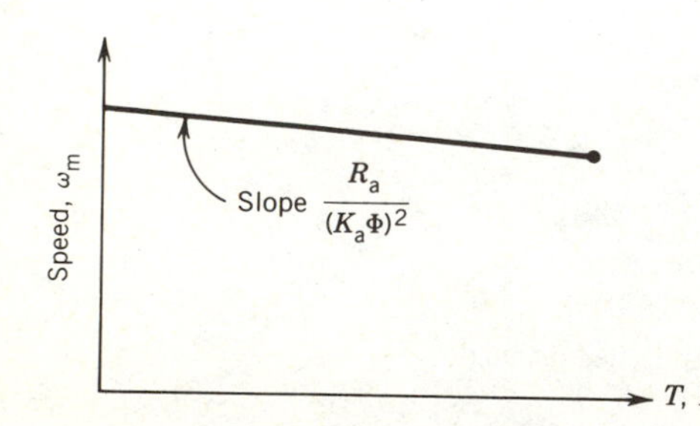

FIGURE 4.51
Torque-speed characteristics of a separately-excited dc motor.

For a constant load torque, such as applied by an elevator or hoist crane load, the speed will change linearly with V_t as shown in Fig. 4.52*a*. If the terminal voltage is kept constant and the load torque is varied, the speed can be adjusted by V_t as shown in Fig. 4.52*b*.

In an actual application, when speed is changed by changing the terminal voltage, the armature current is kept constant (needs a closed-loop operation). From Eq. 4.39, if I_a is constant,

$$E_a \propto V_t$$
$$\propto \omega_m$$

Therefore, as V_t increases, the speed increases linearly (Fig. 4.52*c*). From Eq. 4.40, if I_a remains constant, so does the torque (Fig. 4.52*c*). The input power from the source ($P = V_t I_a$) also changes linearly with speed (Fig. 4.52*c*). If R_a is neglected, the values of V_t, E_a, and P are zero at zero speed and change linearly with speed (Fig. 4.52*d*).

The armature voltage control scheme provides a smooth variation of speed control from zero to the base speed. The base speed is defined as the

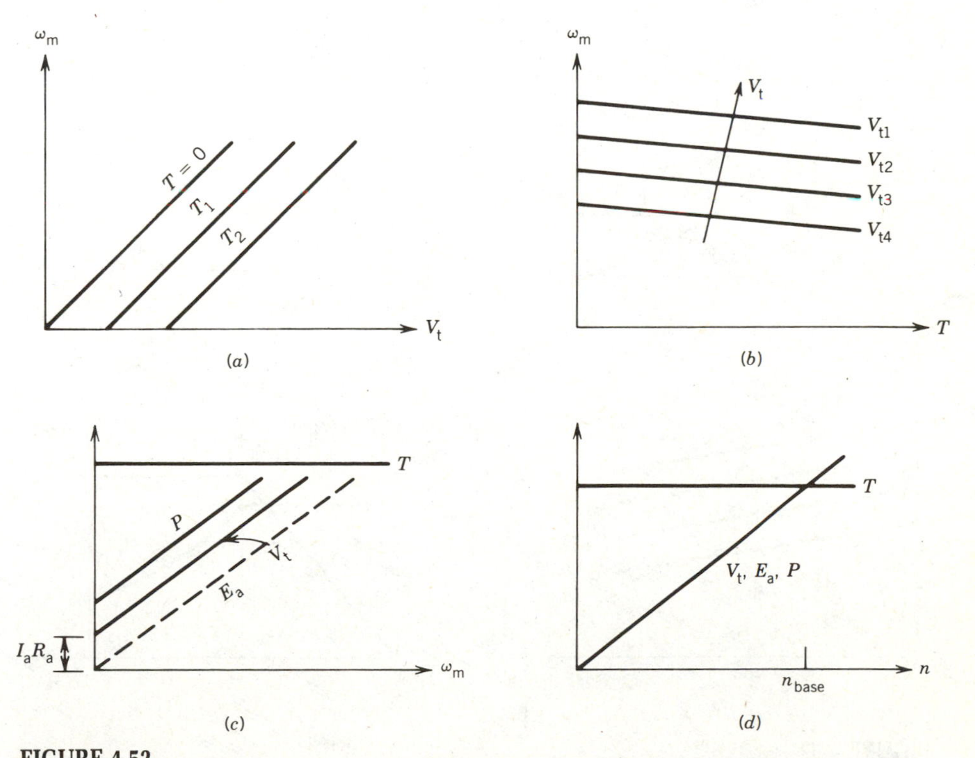

FIGURE 4.52
Armature voltage control of a dc motor. (*a*) Variable speed. (*b*) Adjustable speed. (*c*) Operation under constant torque. (*d*) Operation with $R_a = 0$.

speed obtained at rated terminal voltage. This method of speed control is, however, expensive because it requires a variable dc supply for the armature circuit.

Field Control

In this method the armature circuit resistance R_a and the terminal voltage V_t remain fixed and the speed is controlled by varying the current (I_f) of the field circuit. This is normally achieved by using a field circuit rheostat (R_{fc}) as shown in Fig. 4.53*a*.

If magnetic linearity is assumed, the flux in the machine (Φ) will be proportional to the field current (I_f). Therefore,

$$K_a\Phi = K_f I_f \tag{4.44}$$

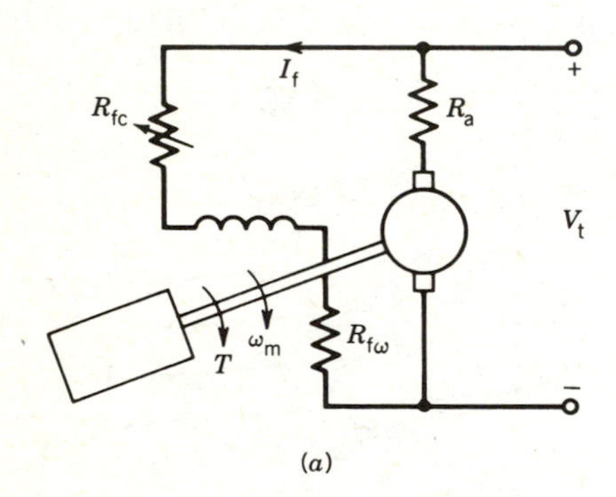

(*a*)

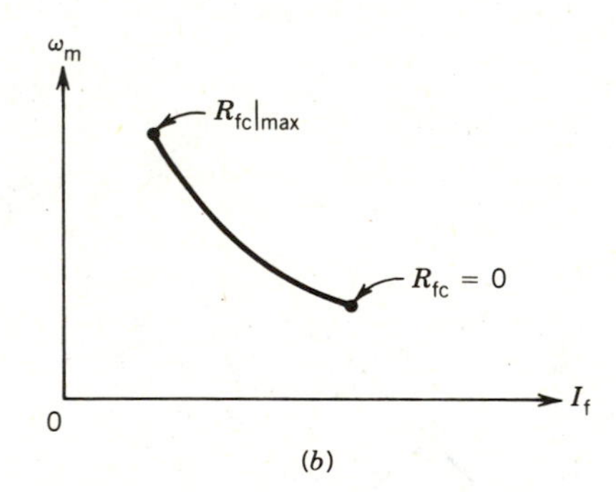

(*b*)

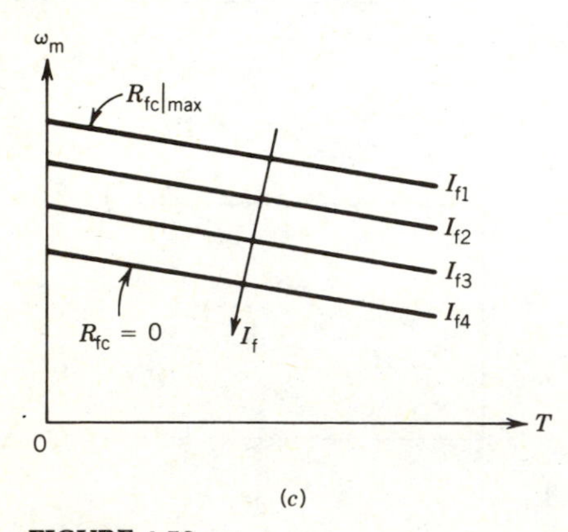

(*c*)

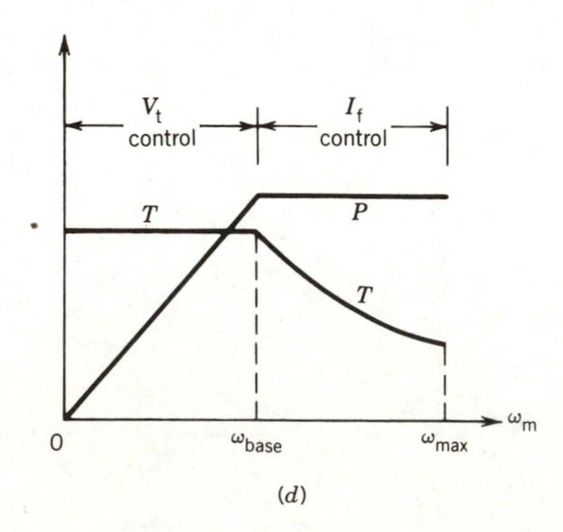

(*d*)

FIGURE 4.53
Field control.

From Eqs. 4.42 and 4.44,

$$\omega_m = \frac{V_t}{K_f I_f} - \frac{R_a}{(K_f I_f)^2} T \tag{4.45}$$

For the no-load condition, $T \approx 0$. From Eq. 4.45,

$$\omega_m \approx \frac{V_t}{K_f I_f}$$

The speed varies inversely with the field current as shown in Fig. 4.53*b*. Note that if the field circuit breaks (i.e., $I_f \to 0$), the speed can become dangerously high.

For a particular value of I_f, from Eq. 4.45,

$$\omega_m = K_3 - K_4 T \tag{4.46}$$

where $K_3 = \dfrac{V_t}{K_f I_f}$ represents no-load speed

$$K_4 = \frac{R_a}{(K_f I_f)^2}$$

At a particular value of I_f, the speed remains essentially constant at a particular level as the torque increases. The level of speed can be adjusted by I_f as shown in Fig. 4.53*c*. Thus, like armature voltage control, field control can also provide variable speed as well as adjustable speed operation.

Speed control from zero to a base speed is usually obtained by armature voltage control (V_t). Speed control beyond the base speed is obtained by decreasing the field current, called field weakening. At the base speed, the armature terminal voltage is at its rated value. If armature current is not to exceed its rated value (heating limit), speed control beyond the base speed is restricted to constant power, known as constant-power operation.

$$\begin{aligned} P &= V_t I_a, \text{ constant} \\ &\approx E_a I_a \\ T\omega_m &= E_a I_a \\ T &= \frac{E_a I_a}{\omega_m} \approx \frac{\text{constant}}{\omega_m} \end{aligned}$$

The torque, therefore, decreases with speed in the field weakening region. The features of armature voltage control (constant-torque operation) and field control (constant-power operation) are shown in Fig. 4.53*d*.

Field control is simple to implement and is less expensive, because the control is at the low power level of the field circuit. However, because of large inductance in the field circuit, change of field current will be slow, which will result in a sluggish response for the speed.

Armature Resistance Control

In this method, the armature terminal voltage V_t and the field current I_f (hence Φ) are kept constant at their rated values. The speed is controlled by changing resistance in the armature circuit. An armature circuit rheostat R_{ae}, as shown in Fig. 4.54*a*, is used for this purpose.

From Eq. 4.42,

$$\omega_m = \frac{V_t}{K_a\Phi} - \frac{R_a + R_{ae}}{(K_a\Phi)^2} T \tag{4.47}$$

If V_t and Φ remain unchanged,

$$\omega_m = K_5 - K_6 T \tag{4.48}$$

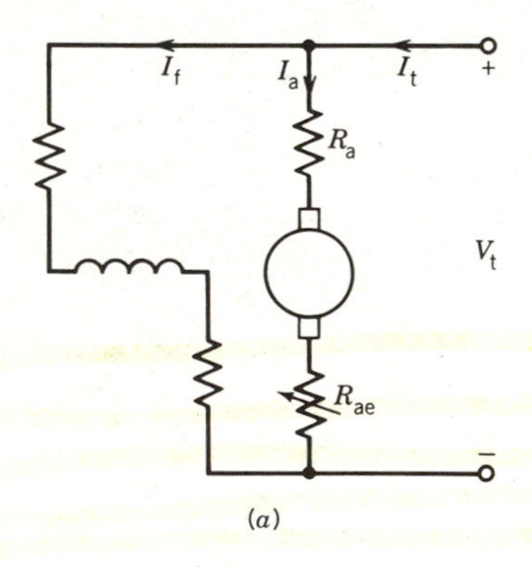

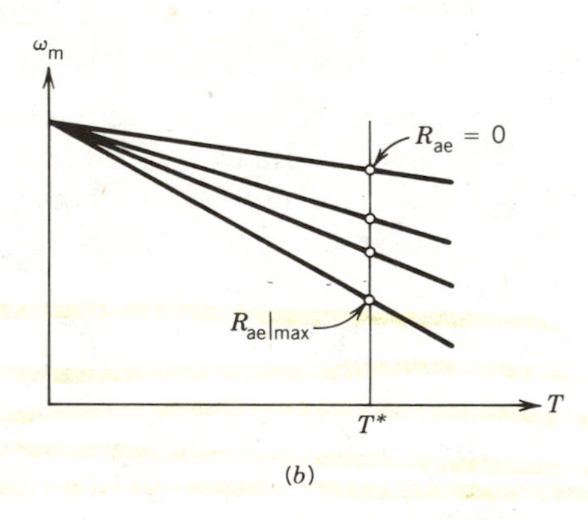

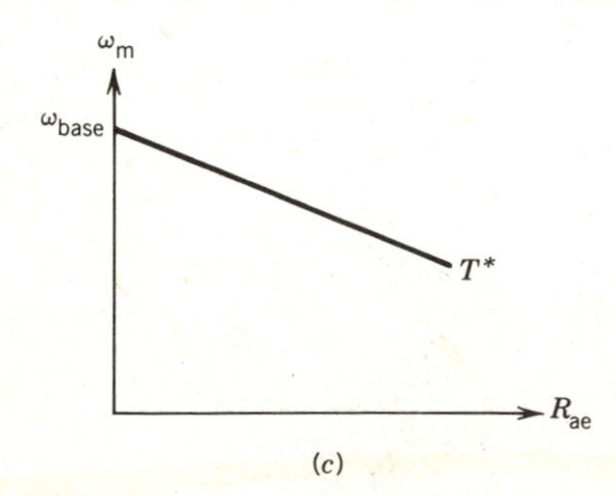

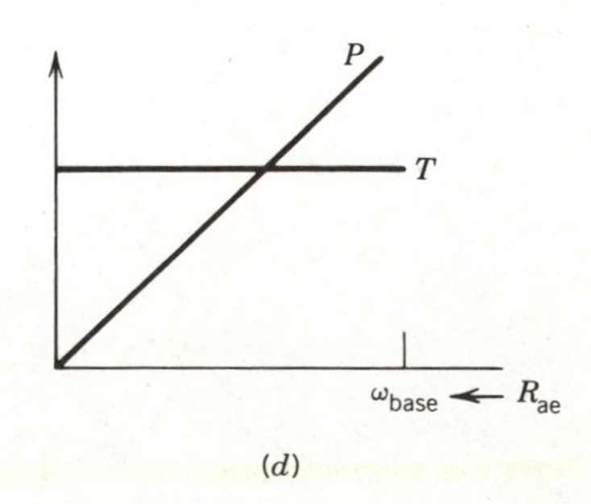

FIGURE 4.54
Armature resistance control.

where $K_5 = \dfrac{V_t}{K_a\Phi}$ represents no-load speed

$$K_6 = \frac{R_a + R_{ae}}{(K_a\Phi)^2}$$

The speed–torque characteristics for various values of the external armature circuit resistance are shown in Fig. 4.54*b*. The value of R_{ae} can be adjusted to obtain various speeds such that armature current I_a (hence torque $T = K_a\Phi I_a$) remains constant. Figure 4.54*b* shows the various values of R_{ae} required to operate at a particular value of torque, T^*. The speed resistance curve for a constant-torque operation is shown in Fig. 4.54*c*. The speed can be varied from zero to a base speed at constant torque, as shown in Fig. 4.54*d*, by changing the external resistance R_{ae}.

Armature resistance control is simple to implement. However, this method is less efficient because of losses in R_{ae}. Many transit system vehicles are still controlled by this method. The resistance R_{ae} should be designed to carry the armature current. It is therefore more expensive than the rheostat (R_{fc}) used in the field control method.

4.4.2 SERIES MOTOR

A schematic diagram of a series motor is shown in Fig. 4.55*a*. An external resistance R_{ae} is shown in series with the armature. This resistance can be used to control the speed of the series motor. The basic machine equations 4.9 and 4.17 hold good for series dc motors, where Φ is produced by the armature current flowing through the series field winding of turns N_{sr}.

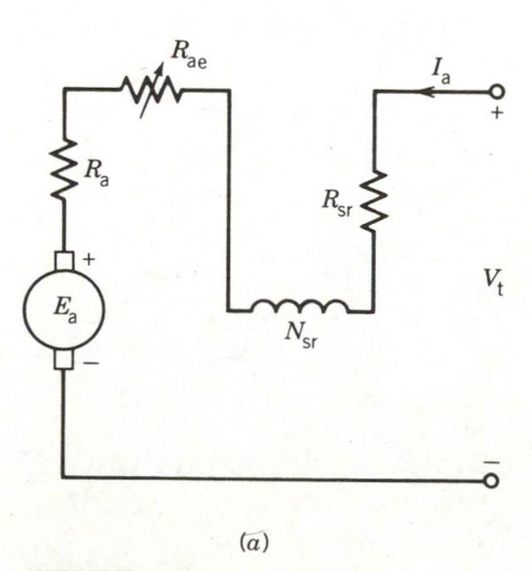

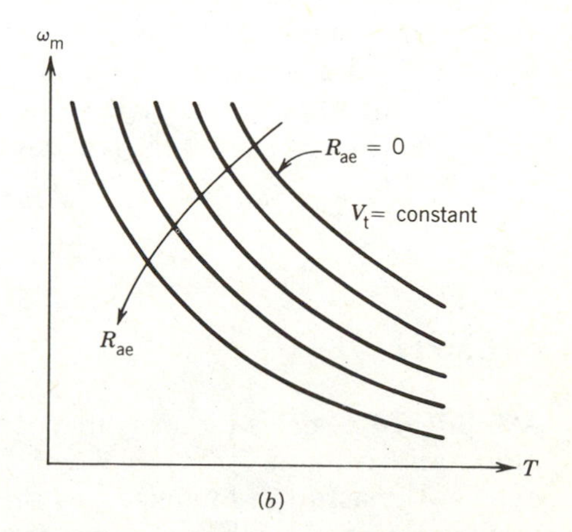

FIGURE 4.55
Series motor.

If magnetic linearity is assumed,

$$K_a\Phi = K_{sr}I_a \tag{4.49}$$

From Eqs. 4.9, 4.17, and 4.49,

$$E_a = K_{sr}I_a\omega_m \tag{4.50}$$

$$T = K_{sr}I_a^2 \tag{4.51}$$

Equation 4.51 shows that a series motor will develop unidirectional torque for both dc and ac currents. Also, from Fig. 4.55*a*,

$$E_a = V_t - I_a(R_a + R_{ae} + R_{sr}) \tag{4.52}$$

From Eqs. 4.50 and 4.52,

$$\omega_m = \frac{V_t}{K_{sr}I_a} - \frac{R_a + R_{sr} + R_{ae}}{K_{sr}} \tag{4.53}$$

From Eqs. 4.51 and 4.53,

$$\omega_m = \frac{V_t}{\sqrt{K_{sr}}\sqrt{T}} - \frac{R_a + R_{sr} + R_{ae}}{K_{sr}} \tag{4.54}$$

The torque–speed characteristics for various values of R_{ae} are shown in Fig. 4.55*b*. For a particular value of R_{ae}, the speed is almost inversely proportional to the square root of the torque. A high torque is obtained at low speed and a low torque is obtained at high speed—a characteristic known as the series motor characteristic. Series motors are therefore used where large starting torques are required, as in subway cars, automobile starters, hoists, cranes, and blenders.

The torque–speed characteristics of the various dc motors are shown in Fig. 4.56. The series motor provides a variable speed characteristic over a wide range.

EXAMPLE 4.8

A 220 V, 7 hp series motor is mechanically coupled to a fan and draws 25 amps and runs at 300 rpm when connected to a 220 V supply with no external resistance connected to the armature circuit (i.e., $R_{ae} = 0$). The torque required by the fan is proportional to the square of the speed. $R_a = 0.6\ \Omega$ and $R_{sr} = 0.4\ \Omega$. Neglect armature reaction and rotational loss.

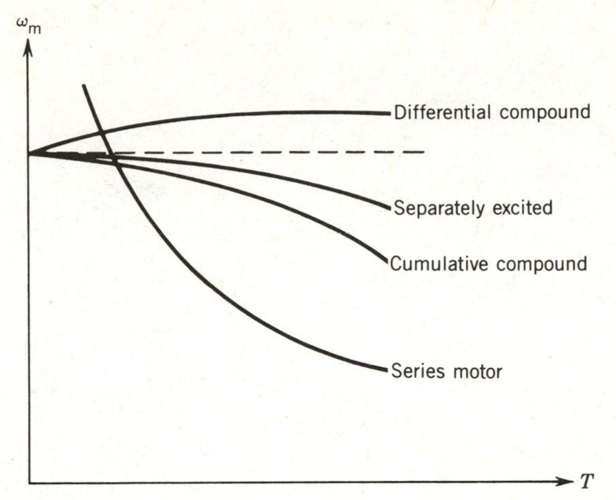

FIGURE 4.56
Torque–speed characteristics of different dc motors.

(a) Determine the power delivered to the fan and the torque developed by the machine.

(b) The speed is to be reduced to 200 rpm by inserting a resistance (R_{ae}) in the armature circuit. Determine the value of this resistance and the power delivered to the fan.

Solution

(a) From Fig. 4.55*a*

$$\begin{aligned} E_a &= V_t - I_a(R_a + R_{sr} + R_{ae}) \\ &= 220 - 25(0.6 + 0.4 + 0) \\ &= 195 \text{ V} \\ P &= E_a I_a \\ &= 195 \times 25 \\ &= 4880 \text{ W} \\ &= \frac{4880}{746} \text{ hp} = 6.54 \text{ hp} \\ T &= \frac{E_a I_a}{\omega_m} \\ &= \frac{4880}{300 \times 2\pi/60} \\ &= 155.2 \text{ N} \cdot \text{m} \end{aligned}$$

(b)

$$T = K_{sr} I_a^2$$

$$155.2 = K_{sr} 25^2$$

$$K_{sr} = 0.248$$

$$T|_{200\text{ rpm}} = \left(\frac{200}{300}\right)^2 \times 155.2$$

$$= 68.98\ \text{N}\cdot\text{m}$$

From Eq. 4.54

$$\frac{200}{60} \times 2\pi = \frac{200}{\sqrt{0.248}\,\sqrt{68.98}} - \frac{0.6 + 0.4 + R_{ae}}{0.248}$$

$$R_{ae} = 7\ \Omega$$

$$P = T\omega_m = 68.98 \times \frac{200}{60} \times 2\pi = 1444\ \text{W} \rightarrow 1.94\ \text{hp}$$

or

$$68.98 = 0.248 I_a^2$$

$$I_a = 16.68\ \text{amps}$$

$$E_a = K_{sr} I_a \omega_m$$

$$= 0.248 \times 16.68 \times \frac{200}{60} \times 2\pi$$

$$= 86.57\ \text{V}$$

$$E_a = V_t - I_a(R_a + R_{sr} + R_{ae})$$

$$86.57 = 220 - 16.68(0.6 + 0.4 + R_{ae})$$

$$R_{ae} = 7\ \Omega$$

$$P = E_a I_a = 86.57 \times 16.68$$

$$= 1444\ \text{W} \rightarrow 1.94\ \text{hp}$$

4.4.3 STARTER

If a dc motor is directly connected to a dc power supply, the starting current will be dangerously high. From Fig. 4.57*a*,

$$I_a = \frac{V_t - E_a}{R_a} \tag{4.55}$$

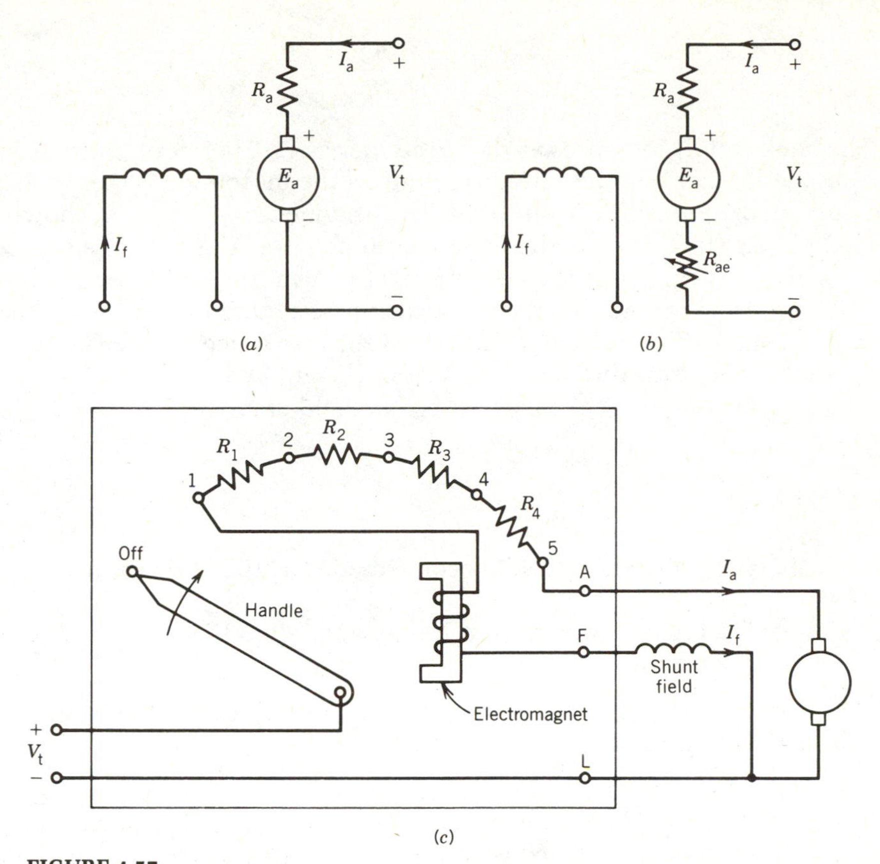

FIGURE 4.57
Development of a dc motor starter.

The back emf E_a ($= K_a\Phi\omega_m$) is zero at start. Therefore,

$$I_a\big|_{\text{start}} = \frac{V_t}{R_a} \tag{4.56}$$

Since R_a is small, the starting current is very large. The starting current can be limited to a safe value by the following methods:

1. Insert an external resistance, R_{ae} (Fig. 4.57*b*), at start.
2. Use a low dc terminal voltage (V_t) at start. This, of course, requires a variable-voltage supply.

With an external resistance in the armature circuit, the armature current as the motor speeds up is

$$I_a = \frac{V_t - E_a}{R_a + R_{ae}} \tag{4.57}$$

The back emf E_a increases as the speed increases. Therefore, the external resistance R_{ae} can be gradually taken out as the motor speeds up without the current exceeding a certain limit. This is done using a starter, shown in Fig. 4.57*c*. At start, the handle is moved to position 1. All the resistances, R_1, R_2, R_3, and R_4, appear in series with the armature and thereby limit the starting current. As the motor speeds up, the handle is moved to positions 2, 3, 4, and finally 5. At position 5 all the resistances in the starter are taken out of the armature circuit. The handle will be held in position 5 by the electromagnet, which is excited by the field current I_f.

EXAMPLE 4.9

The dc machine in Example 4.2 is connected to a 100 V dc supply.

(a) Determine the starting current if no starting resistance is used in the armature circuit.

(b) Determine the value of the starting resistance if the starting current is limited to twice the rated current.

(c) This dc machine is to be run as a motor, using a starter box. Determine the values of resistances required in the starter box such that the armature current I_a is constrained within 100 to 200% of its rated value (i.e., 1 to 2 pu) during start-up.

Solution

(a)

$$I_a|_{rated} = 100 \text{ A}$$

$$I_a|_{start} = \frac{V_t}{R_a} = \frac{100}{0.1} = 1000 \text{ A} = 10 I_a|_{rated} = 10 \text{ pu}$$

(b)

$$200 = \frac{100}{0.1 + R_{ae}}$$

$$R_{ae} = 0.4\ \Omega$$

(c) An arrangement of the resistances in the starter box is shown in Fig. E4.9*a*, where R_{ae1}, R_{ae2}, . . . represent total resistances of the box for positions 1, 2, . . . , respectively. The handle will be moved to a new position when I_a decreases to 100 A (rated armature current). The variation of current I_a and speed n with time is shown in Fig. E4.9*b*.

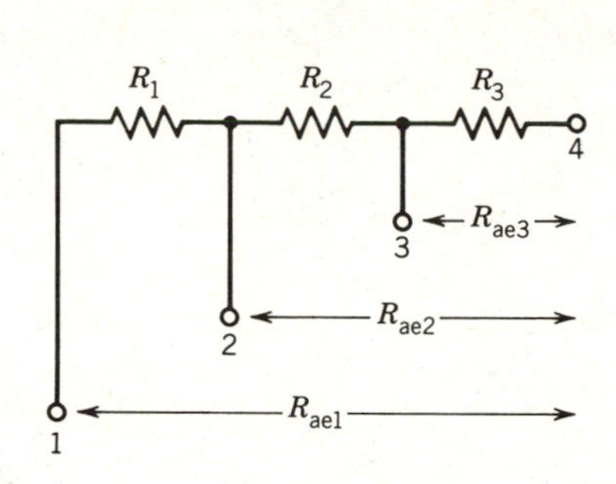

(a)

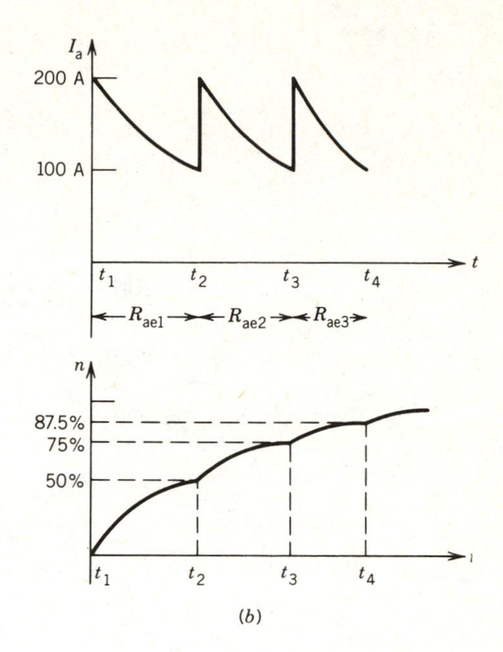

(b)

FIGURE E4.9

R_{ae1}. From part (b)

$$R_{ae1} = 0.4\ \Omega$$
$$= \text{total resistance in starter box}$$

R_{ae2}. At any speed,

$$\underset{\substack{\uparrow \\ \text{fixed}}}{V_t} = \underset{\substack{\uparrow \\ \text{increases} \\ \text{with} \\ \text{speed}}}{E_a} + \underset{\substack{\uparrow \\ \text{decreases} \\ \text{with} \\ \text{speed}}}{I_a(R_a + R_{ae})}$$

At $t = t_2^-$ (i.e., before the handle is moved to position 2),

$$I_a = 100\ \text{A}$$

and

$$\begin{aligned} E_{a2} &= V_t - I_a(R_a + R_{ae1}) \\ &= 100 - 100(0.1 + 0.4) \\ &= 50\ \text{V} \end{aligned}$$

At $t = t_2^+$ (i.e., after the handle is moved to position 2),

$$I_a = 200\ \text{A} = \frac{V_t - E_{a2}}{R_a + R_{ae2}}$$

or

$$200 = \frac{100 - 50}{0.1 + R_{ae2}}$$

$$R_{ae2} = 0.15\ \Omega$$

R_{ae3}. At $t = t_3^-$, $I_a = 100$ A.

$$\begin{aligned} E_{a3} &= 100 - 100(0.1 + 0.15) \\ &= 100 - 25 \\ &= 75\ \text{V} \end{aligned}$$

At $t = t_3^+$,

$$I_a = 200\ \text{A} = \frac{100 - 75}{0.1 + R_{ae3}}$$

$$R_{ae3} = 0.025\ \Omega$$

R_{ae4}. At $t = t_4^-$, $I_a = 100$ A.

$$\begin{aligned} E_{a4} &= 100 - 100(0.1 + 0.025) \\ &= 87.5\ \text{V} \end{aligned}$$

At $t = t_4^+$,

$$I_a = 200 = \frac{100 - 87.5}{0.1 + R_{ae4}}$$

$$R_{ae4} = -0.0375\ \Omega$$

The negative value of R_{ae4} indicates that it is not required, that is, $R_{ae4} = 0$. At $T = t_4^+$ (i.e., after the handle is moved to position 4), the armature current without any resistance in the box will not exceed 200 A. In fact, the value of I_a when the handle is moved to position 4 at $t = t_4$ is

$$\mathrm{I_a} = \frac{100 - 87.5}{0.1} = 125\ \text{A}$$

Therefore, three resistances in the starter box are required. Their values are

$$R_1 = R_{ae1} - R_{ae2} = 0.4 - 0.15 = 0.25\ \Omega$$
$$R_2 = R_{ae2} - R_{ae3} = 0.15 - 0.025 = 0.125\ \Omega$$
$$R_3 = R_{ae3} - R_{ae4} = 0.025 - 0 = 0.025\ \Omega$$

4.5 SPEED CONTROL

There are numerous applications where control of speed is required, as in rolling mills, cranes, hoists, elevators, machine tools, and transit system and locomotive drives. DC motors are extensively used in many of these applications. Control of the speed of dc motors below and above the base (or rated) speed can easily be achieved. Besides, the methods of control are simpler and less expensive than those applicable to ac motors. The technology of speed control of dc motors has evolved considerably over the past quarter-century. In the classical method a Ward–Leonard system with rotating machines is used for speed control of dc motors. Recently, solid-state converters have been used for this purpose. In this section, various methods of speed control of dc motors are discussed.

4.5.1 WARD–LEONARD SYSTEM

This system was introduced in the 1890s. The system, shown in Fig. 4.58*a*, uses a motor–generator (M–G) set to control the speed of the dc drive

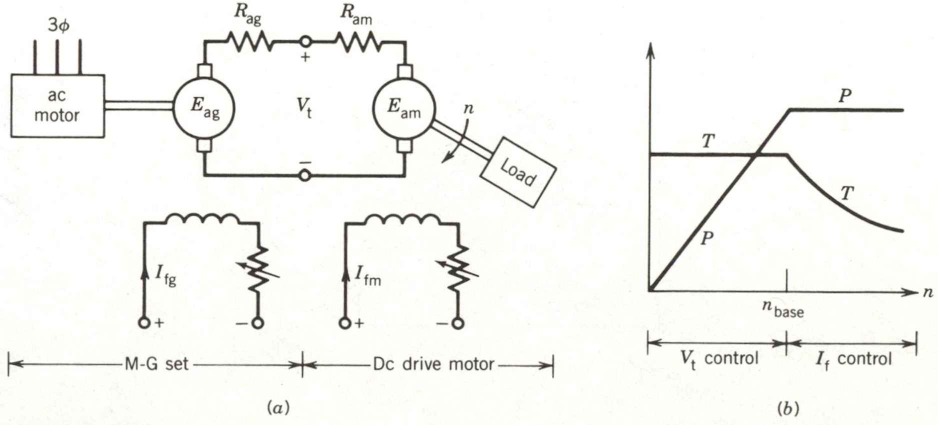

FIGURE 4.58
Ward–Leonard system.

motor. The motor of the M–G set (which is usually an ac motor) runs at a constant speed. By varying the generator field current I_{fg}, the generator voltage V_t is changed, which in turn changes the speed of the dc drive motor. The system is operated in two control modes.

V_t Control

In the armature voltage control mode, the motor current I_{fm} is kept constant at its rated value. The generator field current I_{fg} is changed such that V_t changes from zero to its rated value. The speed will change from zero to the base speed. The torque can be maintained constant during operation in this range of speed, as shown in Fig. 4.58*b*.

I_f Control

The field current control mode is used to obtain speed above the base speed. In this mode, the armature voltage V_t remains constant and the motor field current I_{fm} is decreased (field weakening) to obtain higher speeds. The armature current can be kept constant, thereby operating the motor in a constant-horsepower mode. The torque obviously decreases as speed increases, as shown in Fig. 4.58*b*.

4.5.2 SOLID-STATE CONTROL

In recent years, solid-state converters have been used as a replacement for rotating motor–generator sets to control the speed of dc motors. Figure 4.59 shows the block diagram of a solid-state converter system. The converters used are controlled rectifiers or choppers, which are discussed in Chapter 10.

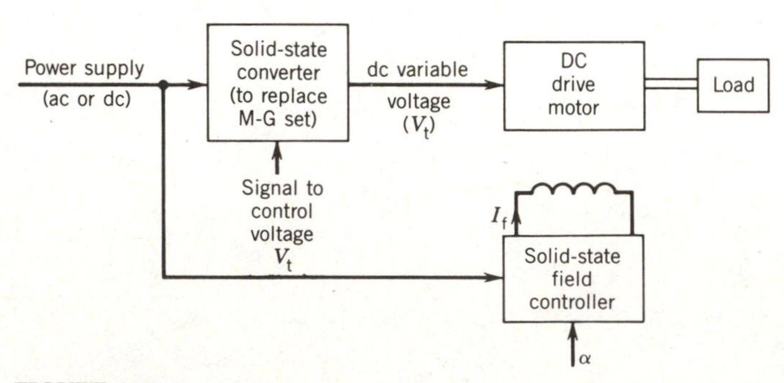

FIGURE 4.59
Block diagram of solid-state control of dc motors.

Controlled Rectifiers

If the supply is ac, controlled rectifiers can be used to convert a fixed ac supply voltage into a variable-voltage dc supply. The operation of the phase-controlled rectifiers is described in Chapter 10.

If all the switching devices in the converter are controlled devices, such as silicon-controlled rectifiers (SCRs), the converter is called a full converter. If some devices are SCRs and some are diodes, the converter is called a semiconverter. In Fig. 4.60, the firing angle α of the SCRs determines the average value (V_t) of the output voltage v_t. The control voltage V_c changes the firing angle α and therefore changes V_t. The relationship between the average output voltage V_t and the firing angle α is as follows.

Single-phase input. Assume that the dc current i_a is continuous. For a full-converter (from Eq. 10.3)

$$V_t = \frac{2\sqrt{2}V_p}{\pi} \cos \alpha \tag{4.58}$$

For a semiconverter (from Eq. 10.5)

$$V_t = \frac{\sqrt{2}V_p}{\pi}(1 + \cos \alpha) \tag{4.59}$$

Three-phase input. For a full converter (from Eq. 10.10)

$$V_t = \frac{3\sqrt{6}V_p}{\pi} \cos \alpha \tag{4.60}$$

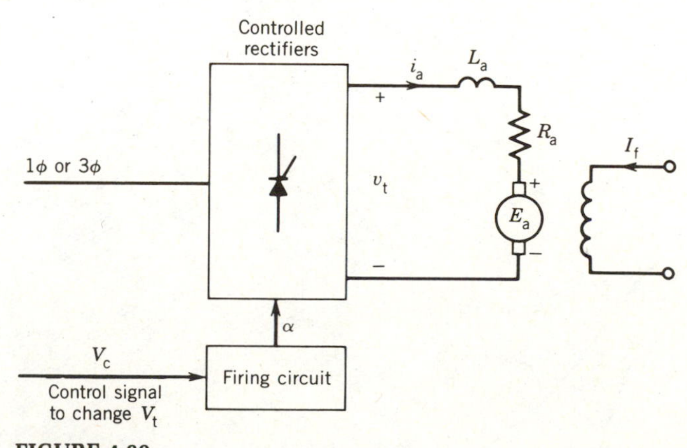

FIGURE 4.60
Speed control of dc motors by controlled rectifiers.

For a semiconverter (from Eq. 10.10a)

$$V_t = \frac{3\sqrt{6}V_p}{2\pi}(1 + \cos\alpha) \tag{4.61}$$

where V_p is the rms value of the ac supply phase voltage. The variation of the motor terminal voltage V_t as a function of the firing angle α is shown in Fig. 4.61 for both semiconverter and full-converter systems. If the I_aR_a drop is neglected ($V_t \simeq E_a$) the curves in Fig. 4.61 also show the variation of E_a (hence speed) with the firing angle.

Although instantaneous values of voltage v_t and current i_a are not constant but change with time, in terms of average values the basic dc machine equations still hold good.

$$V_t = E_a + I_aR_a \tag{4.61a}$$

$$E_a = K_a\Phi\omega_m \tag{4.61b}$$

$$T = K_a\Phi I_a \tag{4.61c}$$

EXAMPLE 4.10

The speed of a 10 hp, 220 V, 1200 rpm separately excited dc motor is controlled by a single-phase full converter as shown in Fig. 4.60 (or Fig. 10.21*a*). The rated armature current is 40 A. The armature resistance is $R_a = 0.25\ \Omega$ and armature inductance is $L_a = 10$ mH. The ac supply voltage is 265 V. The motor voltage constant is $K_a\Phi = 0.18$ V/rpm. Assume that motor current is continuous and ripple-free.

For a firing angle $\alpha = 30°$ and rated motor current, determine the

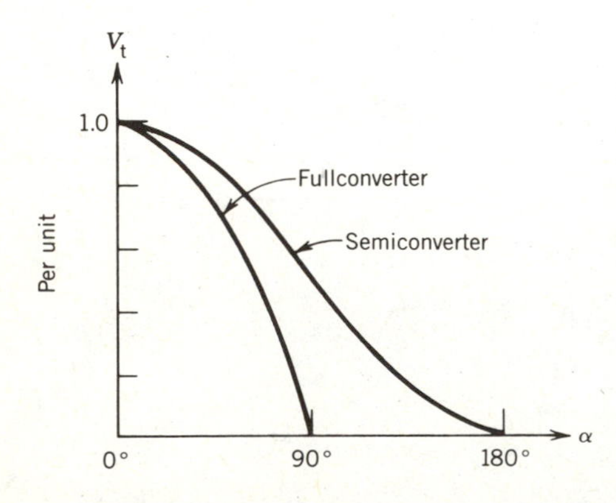

FIGURE 4.61
Controlled-rectifier characteristics.

(a) Speed of the motor.

(b) Motor torque.

(c) Power to the motor.

Solution

(a) From Eq. 4.58 the average terminal voltage is

$$V_t = \frac{2\sqrt{2} \times 265}{\pi} \cos 30^\circ$$

$$= 206.6 \text{ V}$$

The back emf is

$$E_a = V_t - I_a R_a$$

$$= 206.6 - 40 \times 0.25$$

$$= 196.6 \text{ V}$$

Hence the speed in rpm is

$$N = \frac{196.6}{0.18} = 1092.2 \text{ rpm}$$

(b)

$$K_a \Phi = 0.18 \text{ V/rpm}$$

$$= \frac{0.18 \times 60}{2\pi} \text{ V} \cdot \text{sec/rad}$$

$$= 1.72 \text{ V} \cdot \text{sec/rad}$$

$$T = 1.72 \times 40$$

$$= 68.75 \text{ N} \cdot \text{m}$$

(c) The power to the motor is

$$P = (i_a)^2_{rms} R_a + E_a I_a$$

Since i_a is ripple-free (i.e., constant),

$$(i_a)_{rms} = (i_a)_{average} = I_a$$

$$P = I_a^2 R_a + E_a I_a$$

$$= V_t I_a$$

$$= 206.6 \times 40$$
$$= 8264 \text{ W}$$

EXAMPLE 4.11

The speed of a 125 hp, 600 V, 1800 rpm, separately excited dc motor is controlled by a 3ϕ (three-phase) full converter as shown in Fig. 4.60 (or Fig. 10.27*a*). The converter is operated from a 3ϕ, 480 V, 60 Hz supply. The rated armature current of the motor is 165 A. The motor parameters are $R_a = 0.0874\ \Omega$, $L_a = 6.5$ mH, and $K_a\Phi = 0.33$ V/rpm. The converter and ac supply are considered to be ideal.

(a) Find no-load speeds at firing angles $\alpha = 0°$ and $\alpha = 30°$. Assume that, at no load, the armature current is 10% of the rated current and is continuous.

(b) Find the firing angle to obtain the rated speed of 1800 rpm at rated motor current.

(c) Compute the speed regulation for the firing angle obtained in part (b).

Solution

(a) *No-load condition.* The supply phase voltage is

$$V_p = \frac{480}{\sqrt{3}} = 277 \text{ V}$$

From Eq. 4.60 the motor terminal voltage is

$$V_t = \frac{3\sqrt{6} \times 277}{\pi} \cos\alpha = 648 \cos\alpha$$

For $\alpha = 0°$

$$V_t = 648 \text{ V}$$
$$E_a = V_t - I_a R_a$$
$$= 648 - (16.5 \times 0.0874)$$
$$= 646.6 \text{ V}$$

No-load speed is

$$N_0 = \frac{E_a}{K_a\Phi} = \frac{646.6}{0.33} = 1959 \text{ rpm}$$

For $\alpha = 30°$

$$V_t = 648 \cos 30° = 561.2 \text{ V}$$
$$E_a = 561.2 - (16.5 \times 0.0874) = 559.8 \text{ V}$$

The no-load speed is

$$N_0 = \frac{559.8}{0.33} = 1696 \text{ rpm}$$

(b) *Full-load condition.* The motor back emf E_a at 1800 rpm is

$$E_a = 0.33 \times 1800 = 594 \text{ V}$$

The motor terminal voltage at rated current is

$$V_t = 594 + (165 \times 0.0874)$$
$$= 608.4 \text{ V}$$

Therefore,

$$648 \cos \alpha = 608 \text{ V}$$
$$\cos \alpha = \frac{608.4}{648} = 0.94$$
$$\alpha = 20.1°$$

(c) *Speed regulation.* At full load the motor current is 165 A and the speed is 1800 rpm. If the load is thrown off, keeping the firing angle the same at $\alpha = 20.1°$, the motor current decreases to 16.5 A. Therefore

$$E_a = 608.4 - (16.5 \times 0.0874)$$
$$= 606.96 \text{ V}$$

and the no-load speed is

$$N_0 = \frac{606.96}{0.33} = 1839.3 \text{ rpm}$$

The speed regulation is

$$\frac{1839.3 - 1800}{1800} \times 100\% = 2.18\%$$

Choppers

A solid-state chopper converts a fixed-voltage dc supply into a variable-voltage dc supply. A schematic diagram of a chopper is shown in Fig. 4.62*a*. The chopper is a high-speed on–off switch as illustrated in Fig. 4.62*b*. The switch S can be a conventional thyristor (i.e., SCR), a gate turn-off (GTO) thyristor, or a power transistor. The operation of choppers is described in detail in Chapter 10.

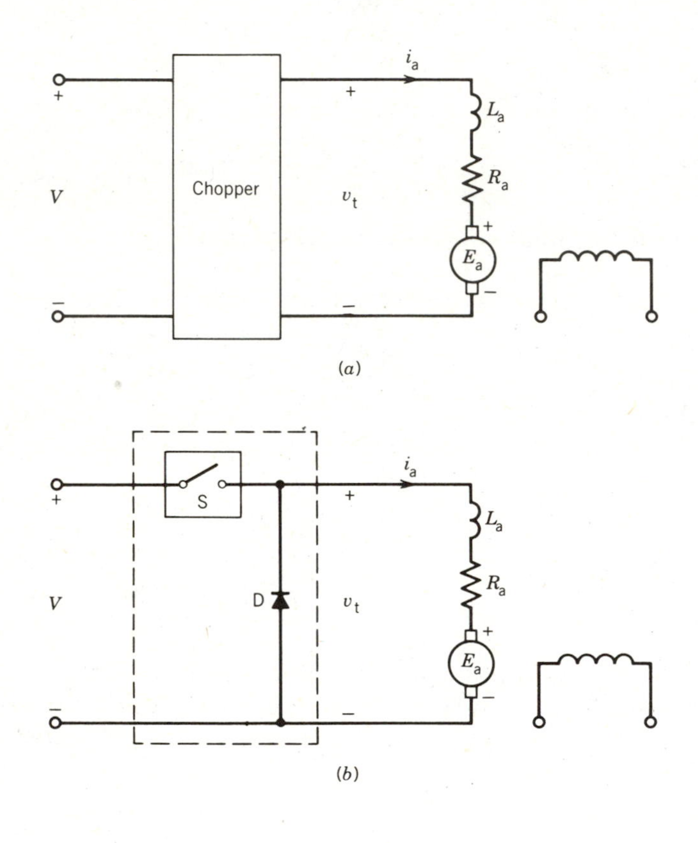

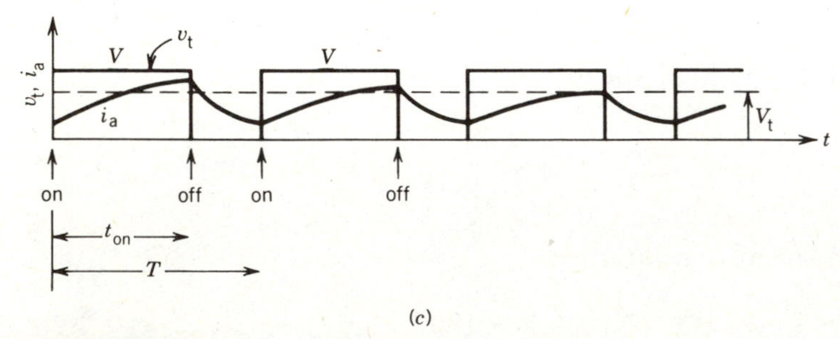

FIGURE 4.62
Chopper circuit and its operation.

When the switching device in the chopper (Fig. 4.62*b*) is on, $v_t = V$ (supply voltage) and motor current i_a increases. When it is off, motor current i_a decays through the diode (D), making $v_t = 0$. The waveforms of voltage v_t and current i_a are shown in Fig. 4.62*c*. The output voltage v_t is a chopped voltage derived from the input voltage V. The average output voltage V_t, which determines the speed of the dc motor, is

$$V_t = \frac{t_{on}}{T} V \tag{4.62}$$

$$= \alpha V \tag{4.63}$$

where t_{on} is the on time of the chopper

T is the chopping period

α is the duty ratio of the chopper

From Eq. 4.63 it is obvious that the motor terminal voltage varies linearly with the duty ratio of the chopper.

EXAMPLE 4.12

The speed of a separately excited dc motor is controlled by a chopper as shown in Fig. 4.62*a*. The dc supply voltage is 120 V, the armature circuit resistance $R_a = 0.5\ \Omega$, the armature circuit inductance $L_a = 20$ mH, and the motor constant is $K_a\Phi = 0.05$ V/rpm. The motor drives a constant-torque load requiring an average armature current of 20 A. Assume that motor current is continuous.

Determine the

(a) Range of speed control.

(b) Range of the duty cycle α.

Solution

Minimum speed is zero, at which $E_a = 0$. Therefore from Eq. 4.61a

$$V_t = I_a R_a = 20 \times 0.5 = 10 \text{ V}$$

From Eq. 4.63

$$10 = 120\alpha$$

$$\alpha = \frac{1}{12}$$

Maximum speed corresponds to $\alpha = 1$, at which $V_t = V = 120$ V. Therefore

$$E_a = V_t - I_a R_a$$
$$= 120 - (20 \times 0.5)$$
$$= 110 \text{ V}$$

From Eq. 4.61b

$$N = \frac{E_a}{K_a \Phi} = \frac{110}{0.05} = 2200 \text{ rpm}$$

(a) The range of speed is $0 < N < 2200$ rpm.

(b) The range of the duty cycle is $\frac{1}{12} < \alpha < 1$.

4.5.3 CLOSED-LOOP OPERATION

DC motors are extensively used in many drives where speed control is desired. In many applications where a constant speed is required, open-loop operation of dc motors may not be satisfactory. In open-loop operation, if load torque changes, the speed will change too. In a closed-loop system, the speed can be maintained constant by adjusting the motor terminal voltage as the load torque changes. The basic block diagram of a closed-loop speed control system is shown in Fig. 4.63. If an additional load torque is applied, the motor speed momentarily decreases and the speed error ε_N increases, which increases the control signal V_c. The control signal increases the converter output voltage (the control signal decreases the firing angle if the converter is a phase-controlled rectifier or increases the duty ratio if the converter is a chopper). An increase in the motor

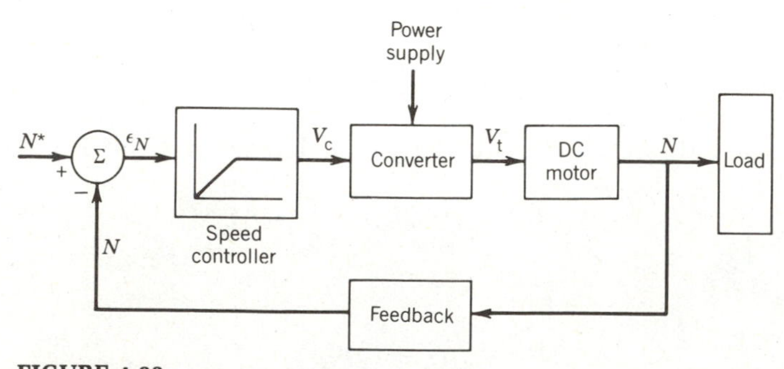

FIGURE 4.63
Closed-loop speed control system.

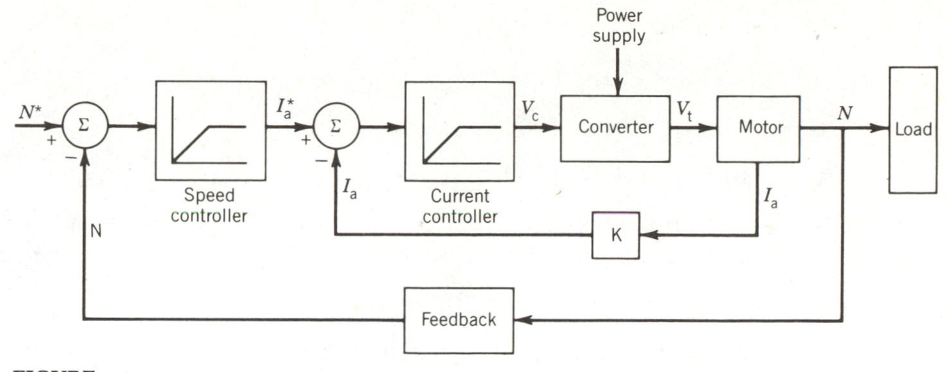

FIGURE 4.64
Closed-loop speed control with inner current loop.

armature voltage develops more torque to restore the speed of the motor. The system thus passes through a transient period until the developed torque matches the applied load torque.

There are other advantages of closed-loop operation, such as greater accuracy, improved dynamic response, and stability of operation. In a closed-loop system the drive characteristics can also be made to operate at constant torque or constant horsepower over a certain speed range, a requirement in traction systems. Circuit protection can also be provided in a closed-loop system. In fact, most industrial drive systems operate as closed-loop feedback systems.

In a dc motor, the armature resistance (R_a) and inductance (L_a) are small. The time constant ($\tau_1 = L_a/R_a$) of the armature circuit is also small. Consequently, a small change in the armature terminal voltage may result in a quick and large change in the armature current, which may damage the solid-state devices used in the converter. An inner current loop can be provided so that the motor current can be clamped to a specified value.[1] A block diagram of such a system is shown in Fig. 4.64. The output of the speed controller represents a torque command. Because torque is proportional to armature current, the output of the speed controller also represents the current command I_a^*, which is then compared with the actual current I_a. A limit on the output of the speed controller will therefore clamp the value of the motor current I_a.

The speed controller and current controller can have proportional (P) or proportional–integral (PI) control.[1] The selection depends on the requirement of drive performance.

[1] P. C. Sen, *Thyristor DC Drives*, Wiley-Interscience, New York, 1981.

PROBLEMS

4.1 Two dc machines of the following rating are required:

DC machine 1: 120 V, 1500 rpm, four poles

DC machine 2: 240 V, 1500 rpm, four poles

Coils are available which are rated at 4 volts and 5 amperes. For the same number of coils to be used for both machines, determine the

(a) Type of armature winding for each machine.

(b) Number of coils required for each machine.

(c) kW rating of each machine.

4.2 A dc machine (6 kW, 120 V, 1200 rpm) has the following magnetization characteristics at 1200 rpm.

I_f (A)	0.0	0.1	0.2	0.3	0.4	0.5	0.6	0.8	1.0	1.2
E_a (V)	5	20	40	60	79	93	102	114	120	125

The machine parameters are $R_a = 0.2\ \Omega$, $R_{fw} = 100\ \Omega$. The machine is driven at 1200 rpm and is separately excited. The field current is adjusted at $I_f = 0.8$ A. A load resistance $R_L = 2\ \Omega$ is connected to the armature terminals. Neglect armature reaction effect.

(a) Determine the quantity $K_a\Phi$ for the machine.

(b) Determine E_a and I_a.

(c) Determine torque T and load power P_L.

4.3 Repeat Problem 4.2 if the speed is 800 rpm.

4.4 The dc machine in Problem 4.2 has a field control resistance whose value can be changed from 0 to 150 Ω. The machine is driven at 1200 rpm. The machine is separately excited and the field winding is supplied from a 120 V supply.

(a) Determine the maximum and minimum values of the no-load terminal voltage.

(b) The field control resistance (R_{fc}) is adjusted to provide a no-load terminal voltage of 120 V. Determine the value of R_{fc}. Determine the terminal voltage at full load for no armature reaction and also if $I_{f(AR)} = 0.1$ A.

4.5 The dc machine in Problem 4.4 is self-excited.

(a) Determine the maximum and minimum values of the no-load terminal voltage.

(b) R_{fc} is adjusted to provide a no-load terminal voltage of 120 V. Determine the value of R_{fc}.

(i) Assume no armature reaction. Determine the terminal voltage at rated armature current. Determine the maximum current the armature can deliver. What is the terminal voltage for this situation?

(ii) Assume that $I_{f(AR)} = 0.1$ A at $I_a = 50$ A and consider armature reaction proportional to armature current. Repeat part (i).

4.6 A dc machine (10 kW, 250 V, 1000 rpm) has $R_a = 0.2\ \Omega$ and $R_{fw} = 133\ \Omega$. The machine is self-excited and is driven at 1000 rpm. The data for the magnetization curve are

I_f	(A)	0	0.1	0.2	0.3	0.4	0.5	0.75	1.0	1.5	2.0
E_a	(V)	10	40	80	120	150	170	200	220	245	263

(a) Determine the generated voltage with no field current.

(b) Determine the critical field circuit resistance.

(c) Determine the value of the field control resistance (R_{fc}) if the no-load terminal voltage is 250 V.

(d) Determine the value of the no-load generated voltage if the generator is driven at 800 rpm and $R_{fc} = 0$.

(e) Determine the speed at which the generator is to be driven such that no-load voltage is 200 V with $R_{fc} = 0$.

4.7 The self-excited dc machine in Problem 4.6 delivers rated load when driven at 1000 rpm. The rotational loss is 500 watts.

(a) Determine the generated voltage.

(b) Determine the developed torque.

(c) Determine current in the field circuit. Neglect the armature reaction effect.

(d) Determine the efficiency.

4.8 A dc shunt machine (24 kW, 240 V, 1000 rpm) has $R_a = 0.12\ \Omega$, $N_f =$ 600 turns/pole. The machine is operated as a separately excited dc generator and is driven at 1000 rpm. When $I_f = 1.8$ A, the no-load terminal voltage is 240 V. When the generator delivers full-load current, the terminal voltage drops to 225 V.

(a) Determine the generated voltage and developed torque when the generator delivers full load.

(b) Determine the voltage drop due to armature reaction.

(c) The full-load terminal voltage can be made the same as the no-load terminal voltage by increasing the field current to 2.2 A or by using series winding on each pole. Determine the number of turns per pole of the series winding required if I_f is kept at 1.8 A.

4.9 A dc shunt generator (20 kW, 200 V, 1800 rpm) has $R_a = 0.1\ \Omega$, $R_{fw} =$ 150 Ω. Assume that $E_a = V_t$ at no load. Data for the magnetization curve at 1800 rpm are

I_f (A)	0.0	0.125	0.25	0.5	0.625	0.75	0.875	1.0	1.25	1.5
E_a (V)	5	33.5	67	134	160	175	190	200	214	223

1. The machine is self-excited.

   (a) Determine the maximum generated voltage.

   (b) At full-load condition, $V_t = V_t$ (rated), $I_a = I_a$ (rated), $I_f =$ 1.25 A. Determine the value of the field control resistance (R_{fc}).

   (c) Determine the electromagnetic power and torque developed at full-load condition.

   (d) Determine the armature reaction effect in equivalent field amperes ($I_{f(AR)}$) at full load.

   (e) Determine the maximum value of the armature current assuming that $I_{f(AR)}$ is proportional to I_a.

2. The shunt generator is now connected as a *long* shunt compound generator.

   (a) Show the generator connection.

   (b) Determine the number of turns per pole of the series field winding required to make the no-load and full-load terminal voltage $V_t = 200$ V. $R_{sr} = 0.04\ \Omega$ and $N_f = 1200$ turns/pole.

4.10 A dc machine is connected across a 240-volt line. It rotates at 1200 rpm and is generating 230 volts. The armature current is 40 amps.

(a) Is the machine functioning as a generator or as a motor?

(b) Determine the resistance of the armature circuit.

(c) Determine power loss in the armature circuit resistance and the electromagnetic power.

(d) Determine the electromagnetic torque in newton-meters.

(e) If the load is thrown off, what will the generated voltage and the rpm of the machine be, assuming

(i) No armature reaction.

(ii) 10% reduction of flux due to armature reaction at 40 amps armature current.

4.11 A dc shunt motor (50 hp, 250 V) is connected to a 230 V supply and delivers power to a load drawing an armature current of 200 amperes and running at a speed of 1200 rpm. $R_a = 0.2\ \Omega$.

(a) Determine the value of the generated voltage at this load condition.

(b) Determine the value of the load torque. The rotational losses are 500 watts.

(c) Determine the efficiency of the motor if the field circuit resistance is 115 Ω.

4.12 A dc shunt machine (23 kW, 230 V, 1500 rpm) has $R_a = 0.1\ \Omega$.

1. The dc machine is connected to a 230 V supply. It runs at 1500 rpm at no-load and 1480 rpm at full-load armature current.

(a) Determine the generated voltage at full load.

(b) Determine the percentage reduction of flux in the machine due to armature reaction at full-load condition.

2. The dc machine now operates as a separately excited generator and the field current is kept the same as in part 1. It delivers full load at rated voltage.

(a) Determine the generated voltage at full-load.

(b) Determine the speed at which the machine is driven.

(c) Determine the terminal voltage if the load is thrown-off.

4.13 A 240 V, 2 hp, 1200 rpm dc shunt motor drives a load whose torque varies directly as the speed. The armature resistance of the motor is 0.75 Ω. With I_f = 1 A, the motor draws a line current of 7 A and rotates at 1200 rpm. Assume magnetic linearity and neglect armature reaction effect.

(a) The field current is now reduced to 0.7 A. Determine the operating speed of the motor.

(b) Determine the line current, mechanical power developed, and efficiency for the operating condition of part (a). Assume that rotational losses are 150 W.

4.14 A 125 V, 5 kW, 1800 rpm dc shunt motor requires only 5 volts to send full-load current through the armature when the armature is held stationary.

(a) Determine the armature current if full-line voltage is impressed across the armature at starting.

(b) Determine the value of the external resistance needed in series with the armature to limit the starting current to 1.5 times the full-load current.

(c) The motor is coupled to a mechanical load by a belt. Determine the generated voltage at full-load condition (V_t = 125 V, n = 1800 rpm, $I_a = I_{a(rated)}$). If the belt breaks, determine the speed of the motor. Neglect rotational losses and assume 10% reduction of flux due to armature reaction at full load.

4.15 A dc motor is mechanically connected to a constant-torque load. When the armature is connected to a 120 V dc supply, it draws an armature current of value 10 A and runs at 1800 rpm. The armature resistance is R_a = 0.1 Ω. Accidentally, the field circuit breaks and the flux drops to the residual flux, which is only 5% of the original flux.

(a) Determine the value of the armature current immediately after the field circuit breaks (i.e., before the speed has had time to change from 1800 rpm).

(b) Determine the theoretical final speed of the motor after the field circuit breaks.

4.16 At standstill, a dc series motor draws 5 amperes and develops 5 N · m torque when connected to a 5 V dc supply. The series motor is mechanically coupled to a load. It draws 10 amperes when connected to a 120 V dc supply and drives the load at 300 rpm. Assume magnetic linearity.

(a) Determine the torque developed by the motor.

(b) Determine the value of the external resistance required to be connected in series with the motor.

4.17 A dc series motor (230 V, 12 hp, 1200 rpm) is connected to a 230 V supply, draws a current of 40 amperes, and rotates at 1200 rpm. $R_a = 0.25\ \Omega$ and $R_{sr} = 0.1\ \Omega$. Assume magnetic linearity.

(a) Determine the power and torque developed by the motor.

(b) Determine the speed, torque, and power if the motor draws 20 amperes.

4.18 Repeat Example 4.9 if the armature current is constrained within 100 to 150 percent of its rated value during start-up.

4.19 The Ward–Leonard speed control system shown in Fig. 4.58*a* uses two identical dc machines of rating 250 V, 5 kW, 1200 rpm. The armature resistance of each machine is 0.5 ohms. The generator is driven at a constant speed of 1200 rpm. The magnetization characteristic of each machine at 1200 rpm is as follows:

I_f (A)	0.0	0.1	0.2	0.3	0.4	0.5	0.6	0.7	0.8	1.0	1.2	1.4
E_a (V)	5	60	120	160	190	212	230	242	250	262	270	273

Neglect the effect of armature reaction.

(a) If the motor field current I_{fm} is kept constant at 0.8 A, determine the maximum and minimum values of the generator field current, I_{fg}, required for the motor to operate in a speed range of 200 to 1200 rpm at full-load armature current.

(b) The generator field current is kept at 1.0 amps and the motor field current is reduced to 0.2 amp. Determine the speed of the motor at full-load armature current.

4.20 Repeat Example 4.10 if a single-phase semiconverter (Fig. 4.60 or Fig. 10.24*a*) is used to control the speed of the dc motor.

4.21 Repeat Example 4.11 if a three-phase semiconverter is used to control the speed of the dc motor.

4.22 A dc series motor drives an elevator load that requires a constant torque of 200 N · m. The dc supply voltage is 400 V and the combined resistance of the armature and series field winding is 0.75 Ω. Neglect rotational losses and armature reaction effect.

(a) The speed of the elevator is controlled by a solid-state chopper. At 50% duty cycle (i.e., $\alpha = 0.5$) of the chopper, the motor current is 40 amps. Determine the speed and the horsepower output of the motor and the efficiency of the system.

(b) The elevator is controlled by inserting resistance in series with the armature of the series motor. For the speed of part (a), determine the values of the series resistance, horsepower output of the motor, and efficiency of the system.

CHAPTER five

INDUCTION (ASYNCHRONOUS) MACHINES

The induction machine is the most rugged and the most widely used machine in industry. Like the dc machine discussed in the preceding chapter, the induction machine has a stator and a rotor mounted on bearings and separated from the stator by an air gap. However, in the induction machine both stator winding and rotor winding carry alternating currents. The alternating current (ac) is supplied to the stator winding directly and to the rotor winding by induction—hence the name induction machine.

The induction machine can operate both as a motor and as a generator. However, it is seldom used as a generator supplying electrical power to a load. The performance characteristics as a generator are not satisfactory for most applications. The induction machine is extensively used as a motor in many applications.

The induction motor is used in various sizes. Small single-phase induction motors (in fractional horsepower rating; see Chapter 7) are used in many household appliances, such as blenders, lawn mowers, juice mixers, washing machines, refrigerators, and stereo turntables.

Large three-phase induction motors (in tens or hundreds of horsepower) are used in pumps, fans, compressors, paper mills, textile mills, and so forth.

The linear version of the induction machine has been developed primarily for use in transportation systems.

The induction machine is undoubtedly a very useful electrical machine. Single-phase induction motors are discussed in Chapter 7. Two-phase induction motors are used primarily as servomotors in a control system. These motors are discussed in Chapter 8. Three-phase induction motors are the most important ones and are most widely used in industry. In this chapter the operation, characteristic features, and steady-state performance of the three-phase induction machine are studied in detail.

5.1 CONSTRUCTIONAL FEATURES

Unlike dc machines, induction machines have a uniform air gap. A pictorial view of the three-phase induction machine is shown in Fig. 5.1*a*. The stator is composed of laminations of high-grade sheet steel. A three-phase winding is put in slots cut on the inner surface of the stator frame as shown in Fig. 5.1*b*. The rotor also consists of laminated ferromagnetic material, with slots cut on the outer surface. The rotor winding may be either of two types, the *squirrel-cage type* or the *wound-rotor type*. The squirrel-cage winding consists of aluminum or copper bars embedded in the rotor slots and shorted at both ends by aluminum or copper end rings as shown in Fig. 5.2*a*. The wound-rotor winding has the same form as the stator winding. The terminals of the rotor winding are connected to three slip rings, as shown in Fig. 5.2*b*. Using stationary brushes pressing against the slip rings, the rotor terminals can be connected to an external circuit. In fact, an external three-phase resistor can thus be connected for the purpose of speed control of the induction motor, as shown in Fig. 5.39 and discussed in Section 5.13.6. It is obvious that the squirrel-cage induction machine is simpler, more economical, and more rugged than the wound-rotor induction machine.

The three-phase winding on the stator and on the rotor (in the wound-rotor type) is a distributed winding. Such windings make better use of iron and copper and also improve the mmf waveform and smooth out the torque developed by the machine. The winding of each phase is distributed over several slots. When current flows through a distributed winding it produces an essentially sinusoidal space distribution of mmf. The properties of a distributed winding are discussed in Appendix A.

(*a*)

(*b*)

FIGURE 5.1

Three-phase induction machine. (*a*) Induction machine with enclosure. (*b*) Stator with three-phase winding. (Courtesy of Westinghouse Canada Inc.)

(a) (b)

FIGURE 5.2
Rotor of an induction machine. (a) Squirrel-cage rotor. (b) Wound-rotor type. (Courtesy of Westinghouse Canada Inc.)

Figure 5.3*a* shows a cross-sectional view of a three-phase squirrel-cage induction machine. The three-phase stator winding, which in practice would be a distributed winding, is represented by three concentrated coils for simplicity. The axes of these coils are 120 electrical degrees apart. Coil aa′ represents all the distributed coils assigned to the phase-a winding for one pair of poles. Similarly, coil bb′ represents the phase-b distributed winding, and coil cc′ represents the phase-c distributed winding. The ends of these phase windings can be connected in a wye (Fig. 5.3*b*) or a delta (Fig. 5.3*c*) to form the three-phase connection. As shown in the next section, if balanced three-phase currents flow through these three-phase distributed windings, a rotating magnetic field of constant amplitude and speed will be produced in the air gap and will induce current in the rotor circuit to produce torque.

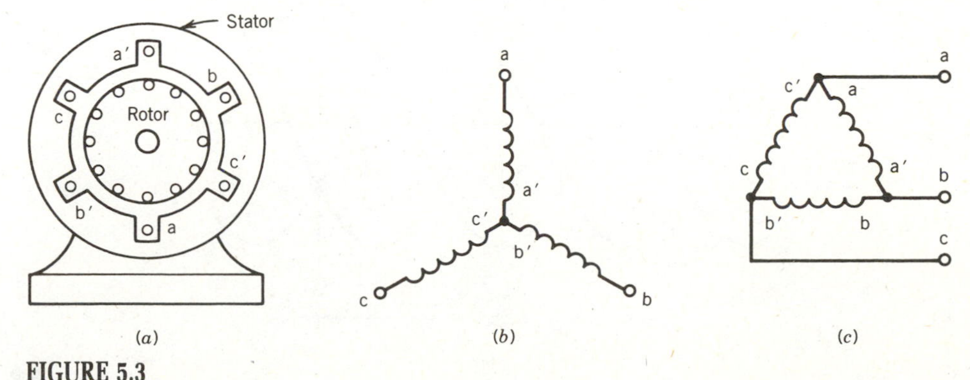

FIGURE 5.3
Three-phase squirrel-cage induction machine. (a) Cross-sectional view. (b) Y-connected stator winding. (c) Δ-connected stator winding.

5.2 ROTATING MAGNETIC FIELD

In this section we study the magnetic field produced by currents flowing in the polyphase windings of an ac machine. In Fig. 5.4*a* the three-phase windings, represented by aa′, bb′, and cc′, are displaced from each other by 120 electrical degrees in space around the inner circumference of the stator. A two-pole machine is considered. The concentrated coils represent the actual distributed windings. When a current flows through a phase coil, it produces a sinusoidally distributed mmf wave centered on the axis of the coil representing the phase winding. If an alternating current flows through the coil, it produces a pulsating mmf wave, whose amplitude and direction depend on the instantaneous value of the current flowing through the winding. Figure 5.4*b* illustrates the mmf distribution in space

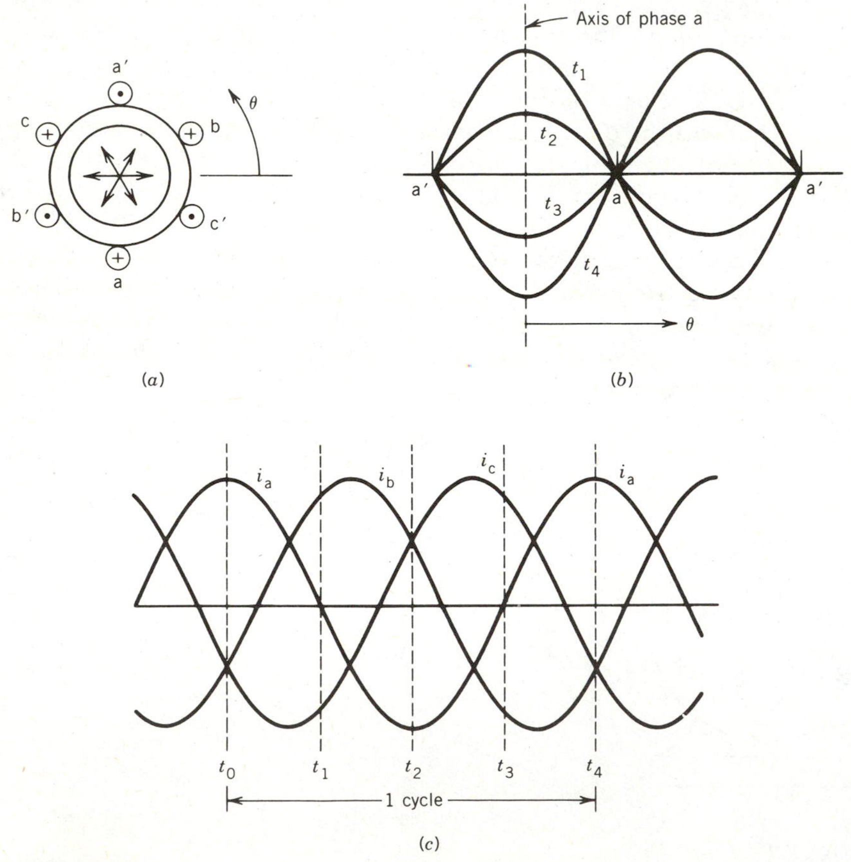

FIGURE 5.4
Pulsating mmf.

at various instants due to an alternating current flow in coil aa′. Each phase winding will produce similar sinusoidally distributed mmf waves, but displaced by 120 electrical degrees in space from each other.

Let us now consider a balanced three-phase current flowing through the three-phase windings. The currents are

$$i_a = I_m \cos \omega t \tag{5.1}$$

$$i_b = I_m \cos(\omega t - 120°) \tag{5.2}$$

$$i_c = I_m \cos(\omega t + 120°) \tag{5.3}$$

These instantaneous currents are shown in Fig. 5.4*c*. The reference directions, when positive-phase currents flow through the windings, are shown by dots and crosses in the coil sides in Fig. 5.4*a*. When these currents flow through the respective phase windings, each produces a sinusoidally distributed mmf wave in space, pulsating along its axis and having a peak located along the axis. Each mmf wave can be represented by a space vector along the axis of its phase with magnitude proportional to the instantaneous value of the current. The resultant mmf wave is the net effect of the three component mmf waves, which can be computed either graphically or analytically.

5.2.1 GRAPHICAL METHOD

Let us consider situations at several instants of time and find out the magnitude and direction of the resultant mmf wave. From Fig. 5.4*c*, at instant $t = t_0$, the currents in the phase windings are as follows:

$$i_a = I_m \quad \text{flowing in phase winding a} \tag{5.4}$$

$$i_b = -\frac{I_m}{2} \quad \text{flowing in phase winding b} \tag{5.5}$$

$$i_c = -\frac{I_m}{2} \quad \text{flowing in phase winding c} \tag{5.6}$$

The current directions in the representative coils are shown in Fig. 5.5*a* by dots and crosses. Because the current in the phase-a winding is at its maximum, its mmf has its maximum value and is represented by a vector $F_a = F_{max}$ along the axis of phase a, as shown in Fig. 5.5*a*. The mmfs of phases b and c are shown by vectors F_b and F_c, respectively, each having magnitude $F_{max}/2$ and shown in the negative direction along their respective axes. The resultant of the three vectors is a vector $F = \frac{3}{2}F_{max}$ acting in the positive direction along the phase-a axis. Therefore, at this instant, the resultant mmf wave is a sinusoidally distributed wave which is the same

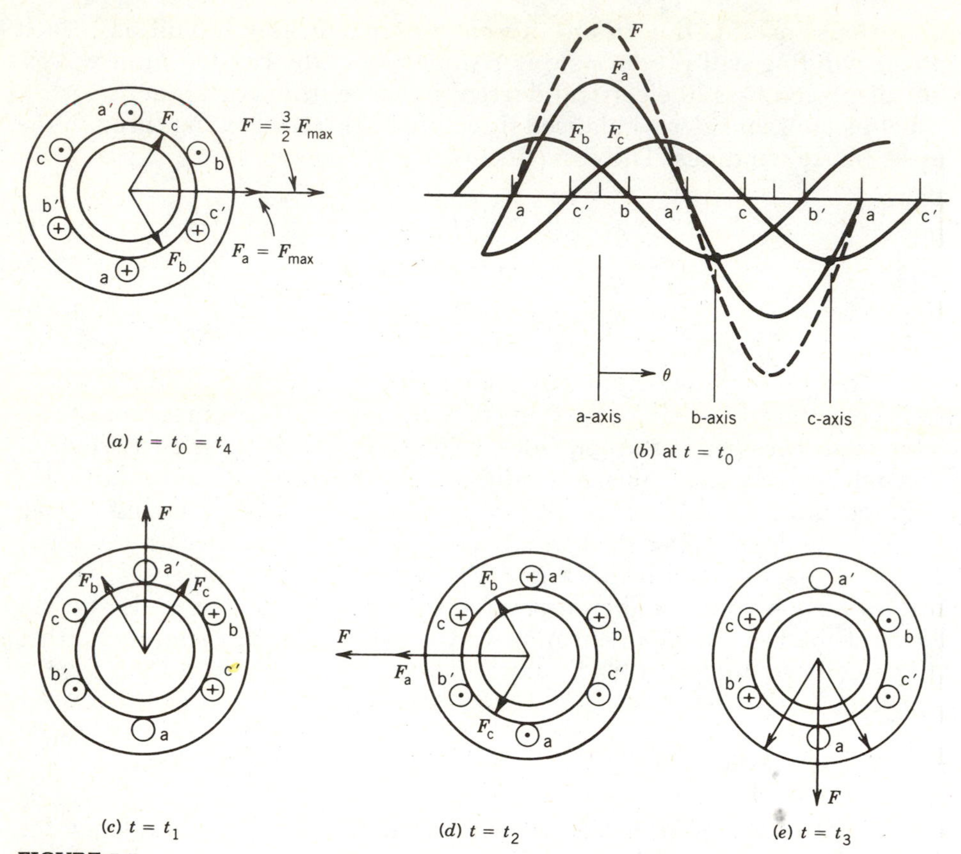

FIGURE 5.5
Rotating magnetic field by graphical method: mmf's at various instants.

as that due to phase-a mmf alone but with $\frac{3}{2}$ the amplitude of the phase-a mmf wave. The component mmf waves and the resultant mmf wave at this instant ($t = t_0$) are shown in Fig. 5.5*b*.

At a later instant of time t_1 (Fig. 5.4*c*), the currents and mmfs are as follows:

$$i_a = 0, \qquad F_a = 0 \tag{5.7}$$

$$i_b = \frac{\sqrt{3}}{2} I_m, \qquad F_b = \frac{\sqrt{3}}{2} F_{max} \tag{5.8}$$

$$i_c = -\frac{\sqrt{3}}{2} I_m, \qquad F_c = -\frac{\sqrt{3}}{2} F_{max} \tag{5.9}$$

The current directions, component mmf vectors, and resultant mmf vector are shown in Fig. 5.5*c*. Note that the resultant mmf vector F has the same amplitude $\frac{3}{2}F_{max}$ at $t = t_1$ as it had at $t = t_0$, but it has rotated counterclockwise by 90° (electrical degrees) in space.

Currents at other instants $t = t_2$ and $t = t_3$ are also considered, and their effects on the resultant mmf vector are shown in Figs. 5.5*d* and 5.5*e*, respectively. It is obvious that as time passes, the resultant mmf wave retains its sinusoidal distribution in space with constant amplitude but moves around the air gap. In one cycle of the current variation, the resultant mmf wave comes back to the position of Fig. 5.5*a*. Therefore, the resultant mmf wave makes one revolution per cycle of the current variation in a two-pole machine. In a p-pole machine, one cycle of variation of the current will make the mmf wave rotate by $2/p$ revolutions. The revolutions per minute n (rpm) of the traveling wave in a p-pole machine for a frequency f cycles per second for the currents are

$$n = \frac{2}{p} f60 = \frac{120f}{\mathrm{p}} \tag{5.10}$$

It can be shown that if i_a flows through the phase-a winding but i_b flows through the phase-c winding and i_c flows through the phase-b winding, the traveling mmf wave will rotate in the clockwise direction. Thus, a reversal of the phase sequence of the currents in the windings makes the rotating mmf rotate in the opposite direction.

5.2.2 ANALYTICAL METHOD

Again we shall consider a two-pole machine with three phase windings on the stator. An analytical expression will be obtained for the resultant mmf wave at any point in the air gap, defined by an angle θ. The origin of the angle θ can be chosen to be the axis of phase a, as shown in Fig. 5.6*a*. At any instant of time, all three phases contribute to the air gap mmf along the path defined by θ. The mmf along θ is

$$F(\theta) = F_a(\theta) + F_b(\theta) + F_c(\theta) \tag{5.11}$$

At any instant of time, each phase winding produces a sinusoidally distributed mmf wave with its peak along the axis of the phase winding and amplitude proportional to the instantaneous value of the phase current. The contribution from phase a along θ is

$$F_a(\theta) = Ni_a \cos \theta \tag{5.12}$$

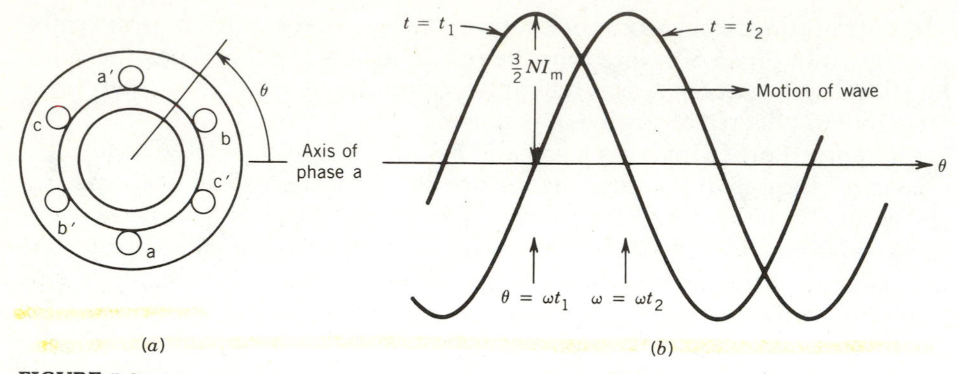

FIGURE 5.6
Motion of the resultant mmf.

where N is the effective number of turns in phase a

i_a is the current in phase a

Because the phase axes are shifted from each other by 120 electrical degrees, the contributions from phases b and c are, respectively,

$$F_b(\theta) = Ni_b \cos(\theta - 120°) \tag{5.13}$$

$$F_c(\theta) = Ni_c \cos(\theta + 120°) \tag{5.14}$$

The resultant mmf at point θ is

$$F(\theta) = Ni_a \cos\theta + Ni_b \cos(\theta - 120°) + Ni_c \cos(\theta + 120°) \tag{5.15}$$

The currents i_a, i_b, and i_c are functions of time and are defined by Eqs. 5.1, 5.2, and 5.3, and thus

$$\begin{aligned} F(\theta, t) = {} & NI_m \cos\omega t \cos\theta \\ & + NI_m \cos(\omega t - 120°) \cos(\theta - 120°) \\ & + NI_m \cos(\omega t + 120°) \cos(\theta + 120°) \end{aligned} \tag{5.16}$$

Using the trigonometric identity

$$\cos A \cos B = \tfrac{1}{2}\cos(A - B) + \tfrac{1}{2}\cos(A + B) \tag{5.17}$$

each term on the right-hand side of Eq. 5.16 can be expressed as the sum of two cosine functions, one involving the difference and the other the sum of the two angles. Therefore,

$$
\begin{aligned}
F(\theta, t) = {} & \tfrac{1}{2}NI_m \cos(\omega t - \theta) + \tfrac{1}{2}NI_m \cos(\omega t + \theta) \\
& + \tfrac{1}{2}NI_m \cos(\omega t - \theta) + \tfrac{1}{2}NI_m \cos(\omega t + \theta - 240^\circ) \\
& + \underbrace{\tfrac{1}{2}NI_m \cos(\omega t - \theta)}_{\text{Forward-rotating components}} + \underbrace{\tfrac{1}{2}NI_m \cos(\omega t + \theta + 240^\circ)}_{\text{Backward-rotating components}}
\end{aligned}
\tag{5.18}
$$

$$
= \tfrac{3}{2}NI_m \cos(\omega t - \theta) \tag{5.19}
$$

The expression of Eq. 5.19 represents the resultant mmf wave in the air gap. This represents an mmf rotating at the constant angular velocity ω ($= 2\pi f$). At any instant of time, say t_1, the wave is distributed sinusoidally around the air gap (Fig. 5.6*b*) with the positive peak acting along $\theta = \omega t_1$. At a later instant, say t_2, the positive peak of the sinusoidally distributed wave is along $\theta = \omega t_2$; that is, the wave has moved by $\omega(t_2 - t_1)$ around the air gap.

The angular velocity of the rotating mmf wave is $\omega = 2\pi f$ radians per second and its rpm for a *p*-pole machine is given by Eq. 5.10.

It can be shown in general that an *m*-phase distributed winding excited by balanced *m*-phase currents will produce a sinusoidally distributed rotating field of constant amplitude when the phase windings are wound $2\pi/m$ electrical degrees apart in space. Note that a rotating magnetic field is produced without physically rotating any magnet. All that is necessary is to pass a polyphase current (ac) through the polyphase windings of the machine.

5.3 INDUCED VOLTAGES

In the preceding section it was shown that when balanced polyphase currents flow through a polyphase distributed winding, a sinusoidally distributed magnetic field rotates in the air gap of the machine. This effect can be visualized as one produced by a pair of magnets, for a two-pole machine, rotating in the air gap, the magnetic field (i.e., flux density) being sinusoidally distributed with the peak along the center of the magnetic poles. The result is illustrated in Fig. 5.7. The rotating field will induce voltages in the phase coils aa′, bb′, and cc′. Expressions for the induced voltages can be obtained by using Faraday's laws of induction.

The flux density distribution in the air gap can be expressed as

$$
B(\theta) = B_{max} \cos \theta \tag{5.20}
$$

The air gap flux per pole, Φ_p, is

$$\Phi_p = \int_{-\pi/2}^{\pi/2} B(\theta) lr \, d\theta = 2B_{max} \, lr \tag{5.21}$$

where l is the axial length of the stator

r is the radius of the stator at the air gap

Let us consider that the phase coils are full-pitch coils of N turns (the coil sides of each phase are 180 electrical degrees apart as shown in Fig. 5.7). It is obvious that as the rotating field moves (or the magnetic poles rotate) the flux linkage of a coil will vary. The flux linkage for coil aa′ will be maximum ($= N\Phi_p$) at $\omega t = 0°$ (Fig. 5.7*a*) and zero at $\omega t = 90°$. The flux linkage $\lambda_a(\omega t)$ will vary as the cosine of the angle ωt. Hence

$$\lambda_a(\omega t) = N\Phi_p \cos \omega t \tag{5.22}$$

Therefore, the voltage induced in phase coil aa′ is obtained from Faraday's law as

$$e_a = -\frac{d\lambda_a}{dt} = \omega N \Phi_p \sin \omega t \tag{5.23}$$

$$= E_{max} \sin \omega t \tag{5.23a}$$

The voltages induced in the other phase coils are also sinusoidal, but phase-shifted from each other by 120 electrical degrees. Thus,

$$e_b = E_{max} \sin(\omega t - 120°) \tag{5.24}$$

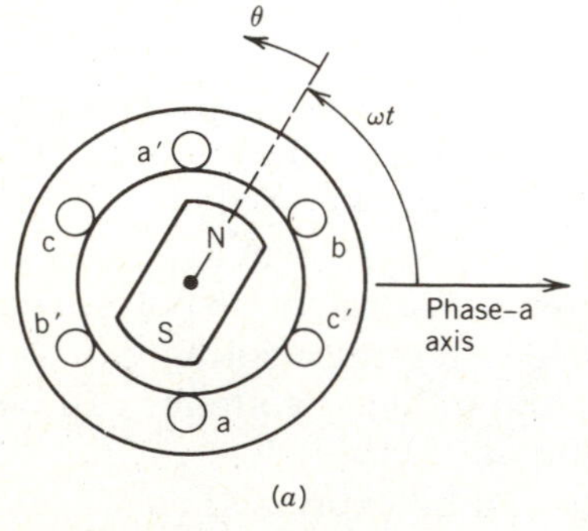

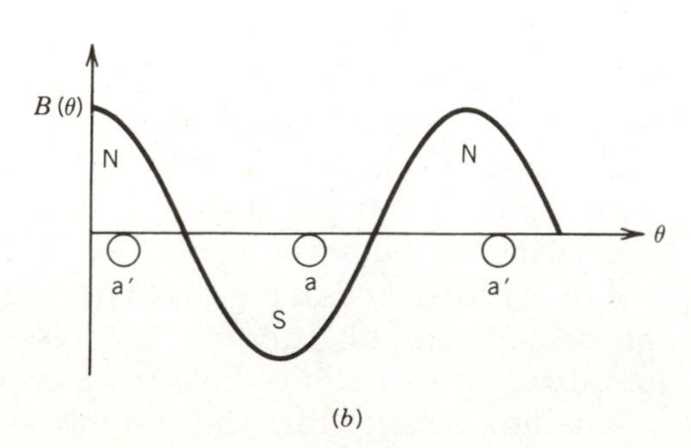

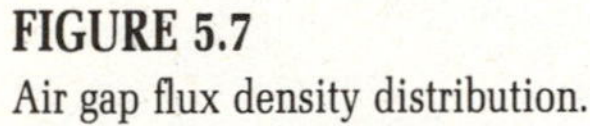

FIGURE 5.7
Air gap flux density distribution.

$$e_c = E_{max} \sin(\omega t + 120°) \tag{5.25}$$

From Eq. 5.23, the rms value of the induced voltage is

$$E_{rms} = \frac{\omega N \Phi_p}{\sqrt{2}} = \frac{2\pi f}{\sqrt{2}} N \Phi_p$$

$$= 4.44 f N \Phi_p \tag{5.26}$$

where f is the frequency in hertz. Equation 5.26 has the same form as that for the induced voltage in transformers (Eq. 1.40). However, Φ_p in Eq. 5.26 represents the flux per pole of the machine.

Equation 5.26 shows the rms voltage per phase. The N is the total number of series turns per phase with the turns forming a concentrated full-pitch winding. In an actual ac machine each phase winding is distributed in a number of slots (see Appendix A) for better use of the iron and copper and to improve the waveform. For such a distributed winding, the emf's induced in various coils placed in different slots are not in time phase, and therefore the phasor sum of the emf's is less than their numerical sum when they are connected in series for the phase winding. A reduction factor K_W, called the winding factor, must therefore be applied. For most three-phase machine windings K_W is about 0.85 to 0.95. Therefore, for a distributed phase winding, the rms voltage per phase is

$$E_{rms} = 4.44 f N_{ph} \Phi_p K_W \tag{5.27}$$

where N_{ph} is the number of turns in series per phase.

5.4 POLYPHASE INDUCTION MACHINE

We shall now consider the various modes of operation of a polyphase induction machine. We shall first consider the standstill behavior of the machine and then study its behavior in the running condition.

5.4.1 STANDSTILL OPERATION

Let us consider a three-phase wound-rotor induction machine with the rotor circuit left open-circuited. If the three-phase stator windings are connected to a three-phase supply, a rotating magnetic field will be produced in the air gap. The field rotates at synchronous speed n_s given by Eq. 5.10. This rotating field induces voltages in both stator and rotor windings at the same frequency f_1. The magnitudes of these voltages, from Eq. 5.27, are

$$E_1 = 4.44 f_1 N_1 \Phi_p K_{W1} \tag{5.28}$$

$$E_2 = 4.44 f_1 N_2 \Phi_p K_{W2} \tag{5.29}$$

Therefore,

$$\frac{E_1}{E_2} = \frac{N_1}{N_2} \frac{K_{W1}}{K_{W2}} \tag{5.30}$$

The winding factors K_{W1} and K_{W2} for the stator and rotor windings are normally the same. Thus,

$$\frac{E_1}{E_2} \simeq \frac{N_1}{N_2} = \text{turns ratio} \tag{5.31}$$

5.4.2 PHASE SHIFTER

Notice that the rotor can be held in such a position that the axes of the corresponding phase windings in the stator and the rotor make an angle β (Fig. 5.8*a*). In such a case, the induced voltage in the rotor winding will be phase-shifted from that of the stator winding by the angle β (Figs. 5.8*b* and 5.8*c*). Thus, a stationary wound-rotor induction machine can be used as a phase shifter. By turning the rotor mechanically, a phase shift range of 360° can be achieved.

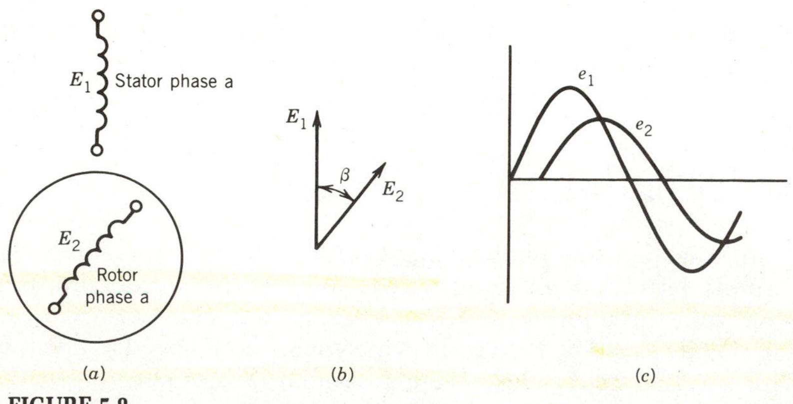

FIGURE 5.8
Induction machine as phase shifter.

5.4.3 INDUCTION REGULATOR

The stationary polyphase induction machine can also be used as a source of variable polyphase voltage if it is connected as an induction regulator as shown in Fig. 5.9*a*. The phasor diagram is shown in Fig. 5.9*b* to illustrate the principle. As the rotor is rotated through 360°, the output voltage V_o follows a circular locus of variable magnitude. If the induced voltages E_1 and E_2 are of the same magnitude (i.e., identical stator and rotor windings) the output voltage may be adjusted from zero to twice the supply voltage.

The induction regulator has the following advantages over a variable autotransformer:

- A continuous stepless variation of the output voltage is possible.
- No sliding electrical connections are necessary.

However, the induction regulator suffers from the disadvantages of higher leakage inductances, higher magnetizing current, and higher costs.

5.4.4 RUNNING OPERATION

If the stator windings are connected to a three-phase supply and the rotor circuit is closed, the induced voltages in the rotor windings produce rotor currents that interact with the air gap field to produce torque. The rotor, if free to do so, will then start rotating. According to Lenz's law, the rotor rotates in the direction of the rotating field such that the relative speed between the rotating field and the rotor winding decreases. The rotor will eventually reach a steady-state speed n that is less than the synchronous

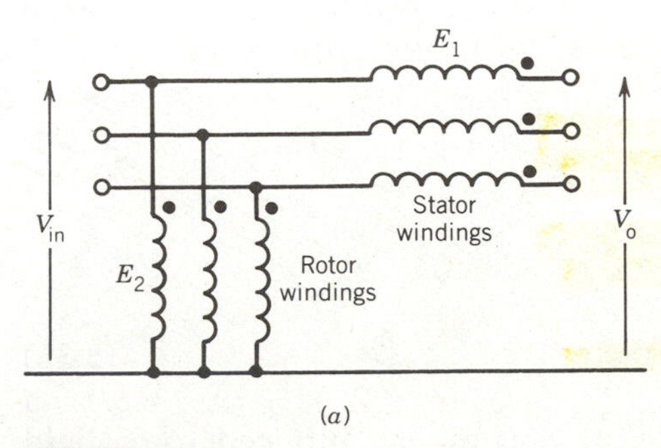

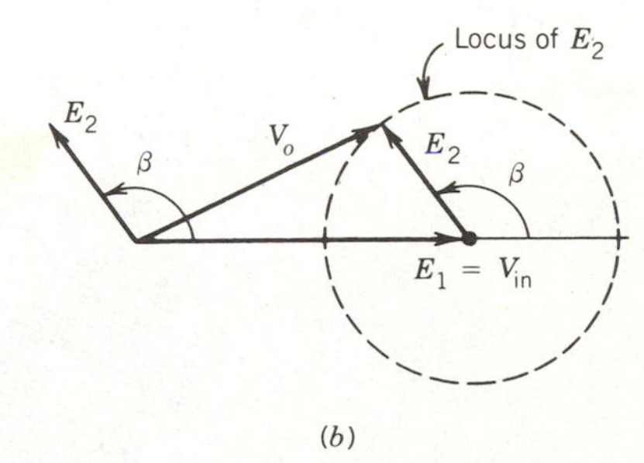

FIGURE 5.9
Induction regulator.

speed n_s at which the stator rotating field rotates in the air gap. It is obvious that at $n = n_s$ there will be no induced voltage and current in the rotor circuit and hence no torque.

The difference between the rotor speed n and the synchronous speed n_s of the rotating field is called the *slip* s and is defined as

$$s = \frac{n_s - n}{n_s} \tag{5.32}$$

If you were sitting on the rotor, you would find that the rotor was slipping behind the rotating field by the *slip rpm* $= n_s - n = sn_s$. The frequency f_2 of the induced voltage and current in the rotor circuit will correspond to this slip rpm, because this is the relative speed between the rotating field and the rotor winding. Thus, from Eq. 5.10

$$\begin{aligned} f_2 &= \frac{p}{120}(n_s - n) \\ &= \frac{p}{120} sn_s \\ &= sf_1 \end{aligned} \tag{5.33}$$

This rotor circuit frequency f_2 is also called *slip frequency*. The voltage induced in the rotor circuit at slip s is

$$\begin{aligned} E_{2s} &= 4.44 f_2 N_2 \Phi_p K_{W2} \\ &= 4.44 s f_1 N_2 \Phi_p K_{W2} \\ &= sE_2 \end{aligned} \tag{5.34}$$

where E_2 is the induced voltage in the rotor circuit at standstill, that is, at the stator frequency f_1.

The induced currents in the three-phase rotor windings also produce a rotating field. Its speed (rpm) n_2 with respect to the rotor is

$$\begin{aligned} n_2 &= \frac{120 f_2}{p} \\ &= \frac{120 s f_1}{p} \\ &= sn_s \end{aligned} \tag{5.35}$$

Because the rotor itself is rotating at n rpm, the induced rotor field rotates in the air gap at speed $n + n_2 = (1 - s)n_s + sn_s = n_s$ rpm. Therefore, both the stator field and the induced rotor field rotate in the air gap at the same synchronous speed n_s. The stator magnetic field and the rotor magnetic field are therefore stationary with respect to each other. The interaction between these two fields can be considered to produce the torque. As the magnetic fields tend to align, the stator magnetic field can be visualized as dragging the rotor magnetic field.

EXAMPLE 5.1

A 3ϕ, 460 V, 100 hp, 60 Hz, four-pole induction machine delivers rated output power at a slip of 0.05. Determine the

(a) Synchronous speed and motor speed.

(b) Speed of the rotating air gap field.

(c) Frequency of the rotor circuit.

(d) Slip rpm.

(e) Speed of the rotor field relative to the

- (i) rotor structure.
- (ii) stator structure.
- (iii) stator rotating field.

(f) Rotor induced voltage at the operating speed, if the stator-to-rotor turns ratio is 1 : 0.5.

Solution

(a)
$$n_s = \frac{120f}{p} = \frac{120 \times 60}{4} = 1800 \text{ rpm}$$
$$n = (1 - s)n_s = (1 - 0.05)1800 = 1710 \text{ rpm}$$

(b) 1800 rpm (same as synchronous speed)

(c) $f_2 = sf_1 = 0.05 \times 60 = 3$ Hz

(d) slip rpm $= sn_s = 0.05 \times 1800 = 90$ rpm

(e)
- (i) 90 rpm
- (ii) 1800 rpm
- (iii) 0 rpm

(f) Assume that the induced voltage in the stator winding is the same as the applied voltage. Now,

$$
\begin{aligned}
E_{2s} &= sE_2 \\
&= s\frac{N_2}{N_1}E_1 \\
&= 0.05 \times 0.5 \times \frac{460\text{ V}}{\sqrt{3}} \\
&= 6.64\text{ V/phase}
\end{aligned}
$$

5.5 THREE MODES OF OPERATION

The induction machine can be operated in three modes: motoring, generating, and plugging. To illustrate these three modes of operation consider an induction machine mechanically coupled to a dc machine, as shown in Fig. 5.10*a*.

5.5.1 MOTORING

If the stator terminals are connected to a three-phase supply, the rotor will rotate in the direction of the stator rotating magnetic field. This is the natural (or motoring) mode of operation of the induction machine. The steady-state speed n is less than the synchronous speed n_s as shown in Fig. 5.10*b*.

5.5.2 GENERATING

The dc motor can be adjusted so that the speed of the system is higher than the synchronous speed and the system rotates in the same direction as the stator rotating field as shown in Fig. 5.10*c*. The induction machine will produce a generating torque, that is, a torque acting opposite to the rotation of the rotor (or acting opposite to the stator rotating magnetic field). The generating mode of operation is utilized in some drive applications to provide regenerative braking. For example, suppose an induction machine is fed from a variable-frequency supply to control the speed of a drive system. To stop the drive system, the frequency of the supply is gradually reduced. In the process, the instantaneous speed of the drive system is higher than the instantaneous synchronous speed because of the inertia of the drive system. As a result, the generating action of the induction machine will cause the power flow to reverse and the kinetic energy of the

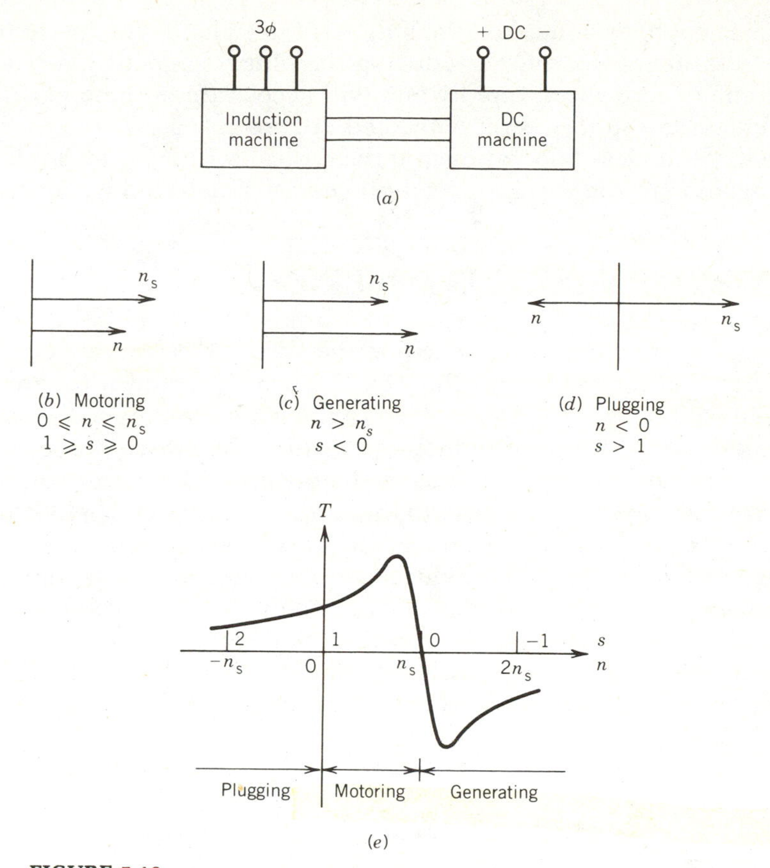

FIGURE 5.10
Three modes of operation of an induction machine.

drive system will be fed back to the supply. The process is known as regenerative braking.

5.5.3 PLUGGING

If the dc motor of Fig. 5.10*a* is adjusted so that the system rotates in a direction opposite to the stator rotating magnetic field (Fig. 5.10*d*), the torque will be in the direction of the rotating field but will oppose the motion of the rotor. This torque is a braking torque.

This mode of operation is sometimes utilized in drive applications where the drive system is required to stop very quickly. Suppose an induction motor is running at a steady-state speed. If its terminal phase sequence is changed suddenly, the stator rotating field will rotate opposite to the rotation of the rotor, producing the plugging operation. The motor will come to zero speed rapidly and will accelerate in the opposite direction, unless the supply is disconnected at zero speed.

The three modes of operation and the typical torque profile of the induction machine in the various speed ranges are illustrated in Fig. 5.10*e*.

5.6 INVERTED INDUCTION MACHINE

In a wound-rotor induction machine the three-phase supply can be connected to the rotor windings through the slip rings and the stator terminals can be shorted, as shown in Fig. 5.11. A rotor-fed induction machine is also known as an inverted induction machine. The three-phase rotor current will produce a rotating field in the air gap, which will rotate at the synchronous speed with respect to the rotor. If the rotor is held stationary, this rotating field will also rotate in the air gap at the synchronous speed. Voltage and current will be induced in the stator windings and a torque will be developed. If the rotor is allowed to move, it will rotate, according to Lenz's law, opposite to the rotation of the rotating field so that the induced voltage in the stator winding is decreased. At a particular speed, therefore, the frequency of the stator circuit will correspond to the slip rpm.

5.7 EQUIVALENT CIRCUIT MODEL

The preceding sections have provided an appreciation of the physical behavior of the induction machine. We now proceed to develop an equivalent circuit model that can be used to study and predict the performance of the induction machine with reasonable accuracy. In this section a steady-state per-phase equivalent circuit will be derived.

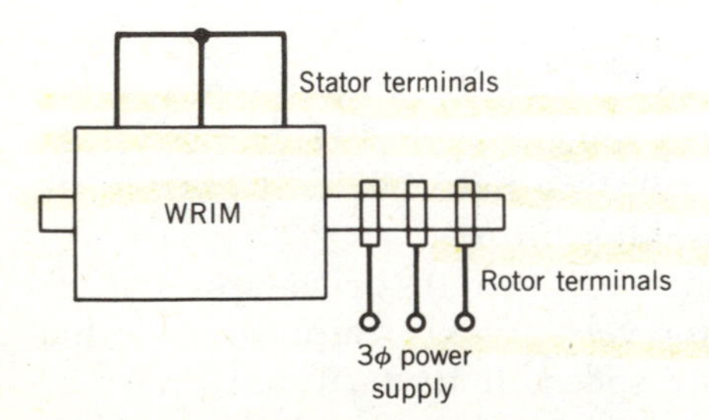

FIGURE 5.11
Inverted wound-rotor induction machine.

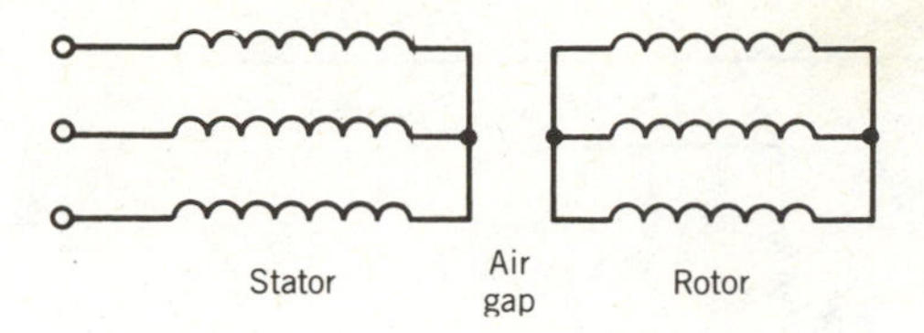

FIGURE 5.12
Three-phase wound-rotor induction motor.

For convenience, consider a three-phase wound-rotor induction machine as shown in Fig. 5.12. In the case of a squirrel-cage rotor, the rotor circuit can be represented by an equivalent three-phase rotor winding. If currents flow in both stator and rotor windings, rotating magnetic fields will be produced in the air gap. Because they rotate at the same speed in the air gap, they will produce a resultant air gap field rotating at the synchronous speed. This resultant air gap field will induce voltages in both stator windings (at the supply frequency f_1) and rotor windings (at slip frequency f_2). It appears that the equivalent circuit may assume a form identical to that of a transformer.

5.7.1 STATOR WINDING

The stator winding can be represented as shown in Fig. 5.13*a*, where

V_1 = per-phase terminal voltage

R_1 = per-phase stator winding resistance

L_1 = per-phase stator leakage inductance

E_1 = per-phase induced voltage in the stator winding

L_m = per-phase stator magnetizing inductance

R_c = per-phase stator core loss resistance

Note that there is no difference in form between this equivalent circuit and that of the transformer primary winding. The difference lies only in the magnitude of the parameters. For example, the excitation current I_ϕ is considerably larger in the induction machine because of the air gap. In induction machines it is as high as 30 to 50 percent of the rated current, depending on the motor size, whereas it is only 1 to 5 percent in transformers. Moreover, the leakage reactance X_1 is larger because of the air gap and also because the stator and rotor windings are distributed along the periphery of the air gap rather than concentrated on a core, as in the transformer.

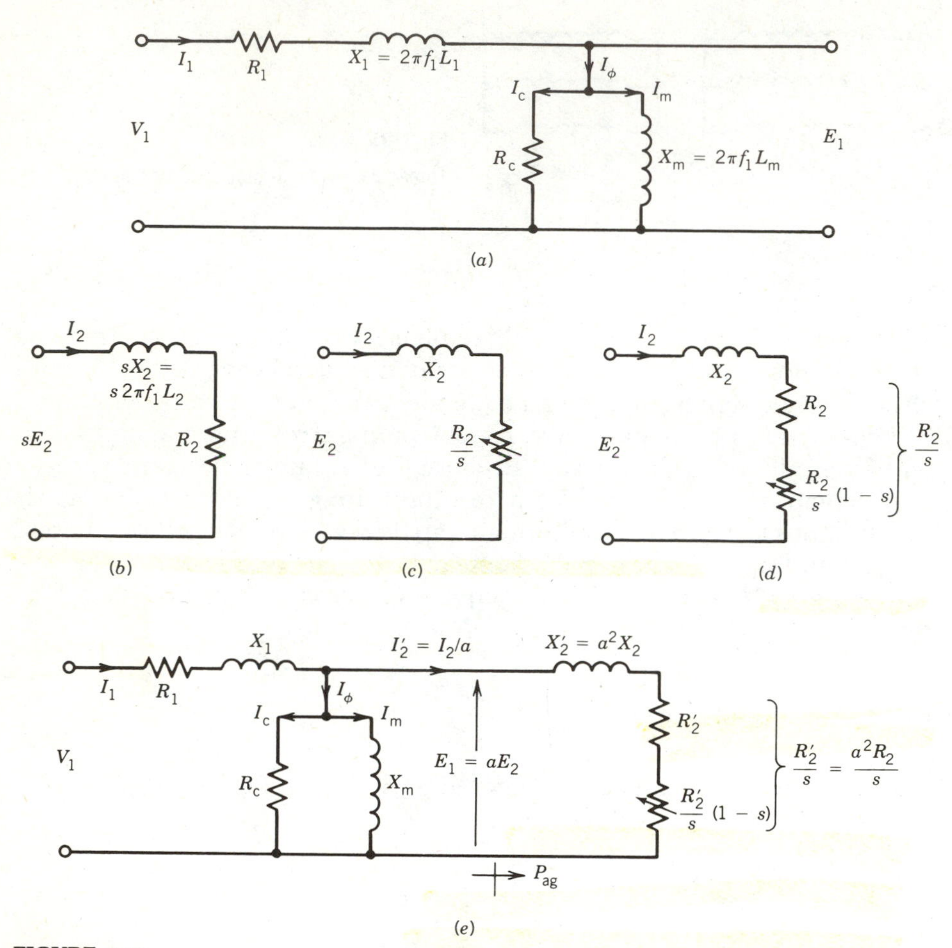

FIGURE 5.13
Development of the induction machine equivalent circuit.

5.7.2 ROTOR CIRCUIT

The rotor equivalent circuit at slip s is shown in Fig. 5.13b, where

E_2 = per-phase induced voltage in rotor at standstill (i.e., at stator frequency f_1)

R_2 = per-phase rotor circuit resistance

L_2 = per-phase rotor leakage inductance

Note that this circuit is at frequency f_2. The rotor current I_2 is

$$I_2 = \frac{sE_2}{R_2 + jsX_2} \tag{5.36}$$

The power involved in the rotor circuit is

$$P_2 = I_2^2 R_2 \tag{5.37}$$

which represents the rotor copper loss per phase.

Equation 5.36 can be rewritten as

$$I_2 = \frac{E_2}{(R_2/s) + jX_2} \tag{5.38}$$

Equation 5.38 suggests the rotor equivalent circuit of Fig. 5.13*c*. Although the magnitude and phase angle of I_2 are the same in Eqs. 5.36 and 5.38, there is a significant difference between these two equations and the circuits (Figs. 5.13*b* and 5.13*c*) they represent. The current I_2 in Eq. 5.36 is at slip frequency f_2, whereas I_2 in Eq. 5.38 is at line frequency f_1. In Eq. 5.36 the rotor leakage reactance sX_2 varies with speed but resistance R_2 remains fixed, whereas in Eq. 5.38 the resistance R_2/s varies with speed but the leakage reactance X_2 remains unaltered. The per-phase power associated with the equivalent circuit of Fig. 5.13*c* is

$$P = I_2^2 \frac{R_2}{s} = \frac{P_2}{s} \tag{5.39}$$

Because induction machines are operated at low slips (typical values of slip s are 0.01 to 0.05) the power associated with Fig. 5.13*c* is considerably larger. Note that the equivalent circuit of Fig. 5.13*c* is at the stator frequency and therefore this is the rotor equivalent circuit as seen from the stator. The power in Eq. 5.39 therefore represents the power that crosses the air gap and thus includes the rotor copper loss as well as the mechanical power developed. Equation 5.39 can be rewritten as

$$P = P_{ag} = I_2^2 \left[R_2 + \frac{R_2}{s}(1 - s) \right] \tag{5.40}$$

$$= I_2^2 \frac{R_2}{s} \tag{5.40a}$$

The corresponding equivalent circuit is shown in Fig. 5.13*d*. The speed-dependent resistance $R_2(1 - s)/s$ represents the mechanical power developed by the induction machine.

$$P_{\text{mech}} = I_2^2 \frac{R_2}{s}(1 - s)$$

$$= (1 - s)P_{\text{ag}} \tag{5.41}$$

$$= \frac{1 - s}{s} P_2 \tag{5.42}$$

and

$$P_2 = I_2^2 R_2 = sP_{\text{ag}} \tag{5.43}$$

Thus,

$$P_{\text{ag}} : P_2 : p_{\text{mech}} = 1 : s : 1 - s \tag{5.44}$$

Equation 5.44 indicates that, of the total power input to the rotor (i.e., power crossing the air gap, P_{ag}), a fraction s is dissipated in the resistance of the rotor circuit (known as rotor copper loss) and the fraction $1 - s$ is converted into mechanical power. Therefore, for efficient operation of the induction machine, it should operate at a low slip so that more of the air gap power is converted into mechanical power. Part of the mechanical power will be lost to overcome the windage and friction. The remainder of the mechanical power will be available as output shaft power.

5.7.3 COMPLETE EQUIVALENT CIRCUIT

The stator equivalent circuit, Fig. 5.13*a*, and the rotor equivalent circuit of Fig. 5.13*c* or 5.13*d* are at the same line frequency f_1 and therefore can be joined together. However, E_1 and E_2 may be different if the turns in the stator winding and the rotor winding are different. If the turns ratio ($a = N_1/N_2$) is considered, the equivalent circuit of the induction machine is that shown in Fig. 5.13*e*. Note that the form of the equivalent circuit is identical to that of a two-winding transformer, as expected.

EXAMPLE 5.2

A 3ϕ, 15 hp, 460 V, four-pole, 60 Hz, 1728 rpm induction motor delivers full output power to a load connected to its shaft. The windage and friction loss of the motor is 750 W. Determine the

(a) Mechanical power developed.

(b) Air gap power.

(c) Rotor copper loss.

Solution

(a) Full-load shaft power = $15 \times 746 = 11{,}190$ W

$$\text{Mechanical power developed} = \text{shaft power} + \text{windage and friction loss}$$
$$= 11{,}190 + 750 = 11{,}940 \text{ W}$$

(b) Synchronous speed $n_s = \dfrac{120 \times 60}{4} = 1800$ rpm

Slip $s = \dfrac{1800 - 1728}{1800} = 0.04$

Air gap power $P_{ag} = \dfrac{11{,}940}{1 - 0.04} = 12{,}437.5$ W

(c) Rotor copper loss $P_2 = 0.04 \times 12{,}437.5$
$= 497.5$ W

5.7.4 VARIOUS EQUIVALENT CIRCUIT CONFIGURATIONS

The equivalent circuit shown in Fig. 5.13*e* is not convenient to use for predicting the performance of the induction machine. As a result, several simplified versions have been proposed in various textbooks on electric machines. There is no general agreement on how to treat the shunt branch (i.e., R_c and X_m), particularly the resistance R_c representing the core loss in the machine. Some of the commonly used versions of the equivalent circuit are discussed here.

Approximate Equivalent Circuit

If the voltage drop across R_1 and X_1 is small and the terminal voltage V_1 does not appreciably differ from the induced voltage E_1, the magnetizing branch (i.e., R_c and X_m) can be moved to the machine terminals as shown in Fig. 5.14*a*. This approximation of the equivalent circuit will considerably simplify computation, because the excitation current (I_ϕ) and the load component (I_2') of the machine current can be directly computed from the terminal voltage V_1 by dividing it by the corresponding impedances.

Note that if the induction machine is connected to a supply of fixed voltage and frequency, the stator core loss is fixed. At no load, the machine will operate close to synchronous speed. Therefore, the rotor frequency f_2 is very small and hence rotor core loss is very small. At a lower speed f_2 increases and so does the rotor core loss. The total core losses thus increase as the speed falls. On the other hand, at no load, friction and windage

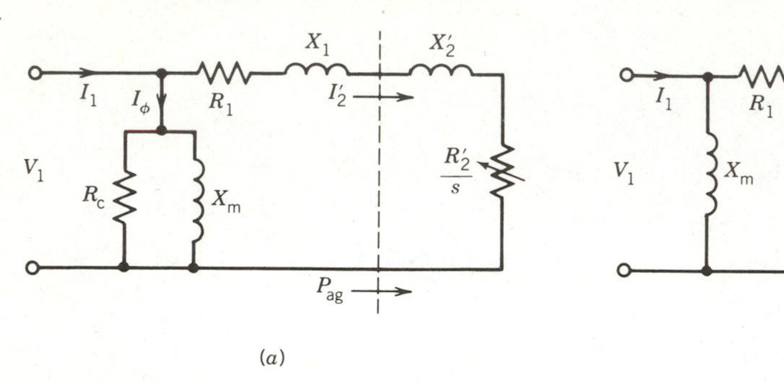

FIGURE 5.14
Approximate equivalent circuit.

losses are maximum and as speed falls these losses decrease. Therefore, if a machine operates from a constant-voltage and constant-frequency source, the sum of core losses and friction and windage losses remains essentially constant at all operating speeds. These losses can thus be lumped together and termed the constant *rotational losses* of the induction machine. If the core loss is lumped with the windage and friction loss, R_c can be removed from the equivalent circuit, as shown in Fig. 5.14*b*.

IEEE-Recommended Equivalent Circuit

In the induction machine, because of its air gap, the exciting current I_ϕ is high—of the order of 30 to 50 percent of the full-load current. The leakage reactance X_1 is also high. The IEEE recommends that, in such a situation, the magnetizing reactance X_m not be moved to the machine terminals (as is done in Fig. 5.14*b*) but be retained at its appropriate place, as shown in Fig. 5.15. The resistance R_c is, however, omitted, and the core loss is lumped with the windage and friction losses. This equivalent circuit (Fig. 5.15) is to be preferred for situations in which the induced voltage E_1 differs appreciably from the terminal voltage V_1.

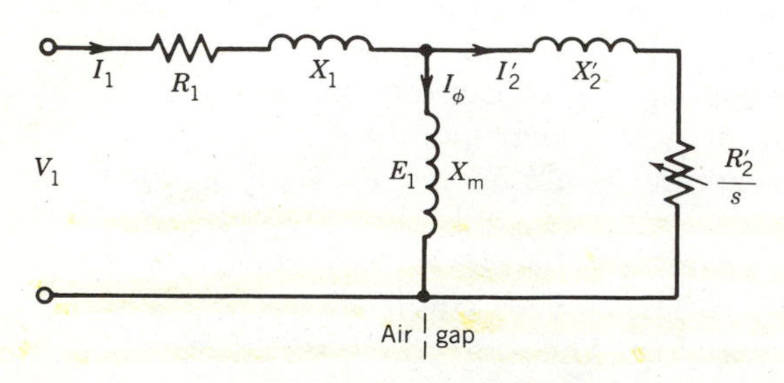

FIGURE 5.15
IEEE-recommended equivalent circuit.

5.7.5 THEVENIN EQUIVALENT CIRCUIT

In order to simplify computations, V_1, R_1, X_1, and X_m can be replaced by the Thevenin equivalent circuit values V_{th}, R_{th}, and X_{th}, as shown in Fig. 5.16, where

$$V_{th} = \frac{X_m}{[R_1^2 + (X_1 + X_m)^2]^{1/2}} V_1 \tag{5.45}$$

If $R_1^2 << (X_1 + X_m)^2$, as is usually the case,

$$V_{th} \approx \frac{X_m}{X_1 + X_m} V_1 \tag{5.45a}$$

$$= K_{th} V_1 \tag{5.45b}$$

The Thevenin impedance is

$$Z_{th} = \frac{jX_m(R_1 + jX_1)}{R_1 + j(X_1 + X_m)}$$

$$= R_{th} + jX_{th}$$

If $R_1^2 << (X_1 + X_m)^2$,

$$R_{th} \simeq \left(\frac{X_m}{X_1 + X_m}\right)^2 R_1 \tag{5.46}$$

$$= K_{th}^2 R_1 \tag{5.46a}$$

and since $X_1 << X_m$,

$$X_{th} \simeq X_1 \tag{5.47}$$

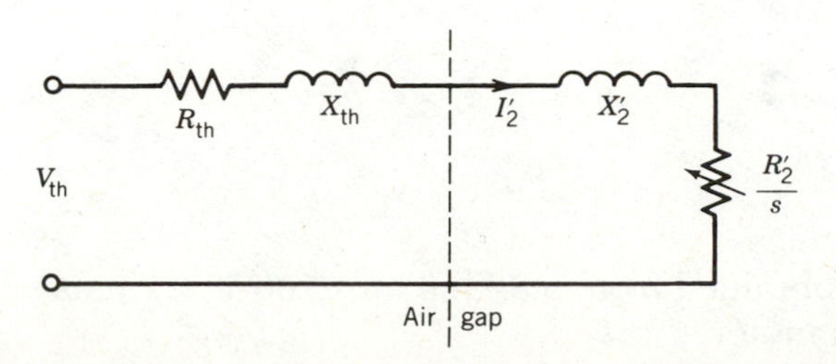

FIGURE 5.16
Thevenin equivalent circuit.

5.8 NO-LOAD TEST, BLOCKED-ROTOR TEST, AND EQUIVALENT CIRCUIT PARAMETERS

The parameters of the equivalent circuit, R_c, X_m, R_1, X_1, X_2, and R_2, can be determined from the results of a no-load test, a blocked-rotor test and from measurement of the dc resistance of the stator winding. The no-load test on an induction machine, like the open-circuit test on a transformer, gives information about exciting current and rotational losses. This test is performed by applying balanced polyphase voltages to the stator windings at the rated frequency. The rotor is kept uncoupled from any mechanical load. The small power loss in the machine at no load is due to the core loss and the friction and windage loss. The total rotational loss at the rated voltage and frequency under load is usually considered to be constant and equal to its value at no load.

The blocked-rotor test on an induction machine, like the short-circuit test on a transformer, gives information about leakage impedances. In this test the rotor is blocked so that the motor cannot rotate, and balanced polyphase voltages are applied to the stator terminals. The blocked-rotor test should be performed under the same conditions of rotor current and frequency that will prevail in the normal operating conditions. For example, if the performance characteristics in the normal running condition (i.e., low-slip region) are required, the blocked-rotor test should be performed at a reduced voltage and rated current. The frequency also should be reduced because the rotor effective resistance and leakage inductance at the reduced frequency (corresponding to lower values of slip) may differ appreciably from their values at the rated frequency. This will be particularly true for double-cage or deep-bar rotors, as discussed in Section 5.11, and also for high-power motors.

The IEEE recommends a frequency of 25 percent of the rated frequency for the blocked-rotor test. The leakage reactances at the rated frequency can then be obtained by considering that the reactance is proportional to frequency. However, for normal motors of less than 20-hp rating, the effects of frequency are negligible and the blocked-rotor test can be performed directly at the rated frequency.

The determination of the equivalent circuit parameters from the results of the no-load and blocked-rotor tests is illustrated by the following example.

EXAMPLE 5.3

The following test results are obtained from a 3ϕ, 60 hp, 2200 V, six-pole, 60 Hz squirrel-cage induction motor.

(1) No-load test:

$$\begin{aligned} \text{Supply frequency} &= 60 \text{ Hz} \\ \text{Line voltage} &= 2200 \text{ V} \\ \text{Line current} &= 4.5 \text{ A} \\ \text{Input power} &= 1600 \text{ W} \end{aligned}$$

(2) Blocked-rotor test:

$$\begin{aligned} \text{Frequency} &= 15 \text{ Hz} \\ \text{Line voltage} &= 270 \text{ V} \\ \text{Line current} &= 25 \text{ A} \\ \text{Input power} &= 9000 \text{ W} \end{aligned}$$

(3) Average dc resistance per stator phase:

$$R_1 = 2.8 \ \Omega$$

(a) Determine the no-load rotational loss.

(b) Determine the parameters of the IEEE-recommended equivalent circuit of Fig. 5.15.

(c) Determine the parameters (V_{th}, R_{th}, X_{th}) for the Thevenin equivalent circuit of Fig. 5.16.

Solution

(a) From the no-load test, the no-load power is

$$P_{NL} = 1600 \text{ W}$$

The no-load rotational loss is

$$\begin{aligned} P_{Rot} &= P_{NL} - 3I_1^2 R_1 \\ &= 1600 - 3 \times 4.5^2 \times 2.8 \\ &= 1429.9 \text{ W} \end{aligned}$$

(b) *IEEE-recommended equivalent circuit.* For the no-load condition, R_2'/s is very high. Therefore, in the equivalent circuit of Fig. 5.15, the magnetizing reactance X_m is shunted by a very high resistive branch representing the rotor circuit. The reactance of this parallel combination is almost the same as X_m. Therefore the total reactance X_{NL}, measured at no load at the stator terminals, is essentially $X_1 + X_m$. The equivalent circuit at no load is shown in Fig. E5.3*a*.

$$V_1 = \frac{2200}{\sqrt{3}} = 1270.2 \text{ V/phase}$$

The no-load impedance is

$$Z_{NL} = \frac{V_1}{I_1} = \frac{1270.2}{4.5} = 282.27 \ \Omega$$

The no-load resistance is

$$R_{NL} = \frac{P_{NL}}{3I_1^2} = \frac{1600}{3 \times 4.5^2} = 26.34 \ \Omega$$

The no-load reactance is

$$\begin{aligned} X_{NL} &= (Z_{NL}^2 - R_{NL}^2)^{1/2} \\ &= (282.27^2 - 26.34^2)^{1/2} \\ &= 281.0 \ \Omega \end{aligned}$$

Thus, $X_1 + X_m = X_{NL} = 281.0 \ \Omega$.

For the blocked-rotor test the slip is 1. In the equivalent circuit of Fig. 5.15, the magnetizing reactance X_m is shunted by the low-impedance branch $jX_2' + R_2'$. Because $|X_m| >> |R_2' + jX_2'|$, the impedance X_m can be neglected and the equivalent circuit for the blocked-rotor test reduces to the form shown in Fig. E5.3*b*. From the blocked-rotor test, the blocked-rotor resistance is

$$\begin{aligned} R_{BL} &= \frac{P_{BL}}{3I_1^2} \\ &= \frac{9000}{3 \times 25^2} \\ &= 4.8 \ \Omega \end{aligned}$$

Therefore, $R_2' = R_{BL} - R_1 = 4.8 - 2.8 = 2 \ \Omega$. The blocked-rotor impedance at 15 Hz is

$$Z_{BL} = \frac{V_1}{I_1} = \frac{270}{\sqrt{3} \times 25} = 6.24 \ \Omega$$

The blocked-rotor reactance at 15 Hz is

$$\begin{aligned} X_{BL} &= (6.24^2 - 4.8^2)^{1/2} \\ &= 3.98 \ \Omega \end{aligned}$$

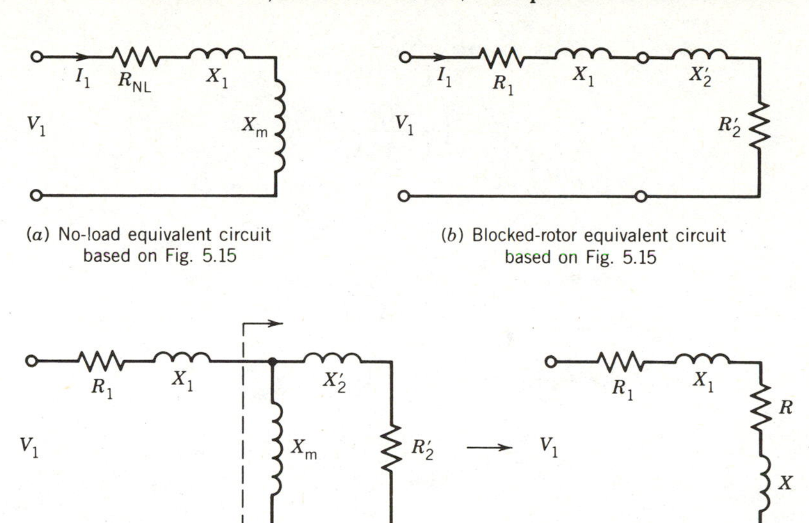

(c) Blocked-rotor equivalent circuit for improved value for R_2'

FIGURE E5.3

Its value at 60 Hz is

$$X_{BL} = 3.98 \times \frac{60}{15} = 15.92\ \Omega$$

$$X_{BL} \simeq X_1 + X_2'$$

Hence,

$$X_1 = X_2' = \frac{15.92}{2} = 7.96\ \Omega \quad \text{(at 60 Hz)}$$

The magnetizing reactance is therefore

$$X_m = 281.0 - 7.96 = 273.04\ \Omega$$

Comments: The rotor equivalent resistance R_2' plays an important role in the performance of the induction machine. A more accurate determination of R_2' is recommended by the IEEE as follows: The blocked resistance R_{BL} is the sum of R_1 and an equivalent resistance, say R, which is the resistance of $R_2' + jX_2'$ in parallel with X_m as shown in Fig. E5.3*c*; therefore,

$$R = \frac{X_m^2}{R_2'^2 + (X_2' + X_m)^2} R_2'$$

If $X_2' + X_m >> R_2'$, as is usually the case,

$$R \simeq \left(\frac{X_m}{X_2' + X_m}\right)^2 R_2'$$

or

$$R_2' = \left(\frac{X_2' + X_m}{X_m}\right)^2 R$$

Now $R = R_{BL} - R_1 = 4.8 - 2.8 = 2\ \Omega$. So,

$$R_2' = \left(\frac{7.96 + 273.04}{273.04}\right)^2 \times 2 = 2.12\ \Omega$$

(c) From Eq. 5.45

$$V_{th} \simeq \frac{273.04}{7.96 + 273.04} V_1$$
$$= 0.97 V_1$$

From Eq. 5.46a

$$R_{th} \simeq 0.97^2 R_1 = 0.97^2 \times 2.8 = 2.63\ \Omega$$

From Eq. 5.47

$$X_{th} \simeq X_1 = 7.96\ \Omega$$

5.9 PERFORMANCE CHARACTERISTICS

The equivalent circuits derived in the preceding section can be used to predict the performance characteristics of the induction machine. The important performance characteristics in the steady state are the efficiency, power factor, current, starting torque, maximum (or pull-out) torque, and so forth.

The mechanical torque developed T_{mech} per phase is given by

$$P_{mech} = T_{mech}\omega_{mech} = I_2^2 \frac{R_2}{s}(1 - s) \tag{5.48}$$

where

$$\omega_{mech} = \frac{2\pi n}{60} \tag{5.48a}$$

The mechanical speed ω_{mech} is related to the synchronous speed by

$$\omega_{mech} = (1 - s)\omega_{syn} \tag{5.49}$$

$$= \frac{n_{syn}}{60} 2\pi(1 - s)$$

and

$$\omega_{syn} = \frac{120f}{p60} \times 2\pi = \frac{4\pi f_1}{p} \tag{5.50}$$

From Eqs. 5.48, 5.49, and 5.40a

$$T_{mech}\omega_{syn} = I_2^2 \frac{R_2}{s} = P_{ag} \tag{5.51}$$

$$T_{mech} = \frac{1}{\omega_{syn}} P_{ag} \tag{5.52}$$

$$= \frac{1}{\omega_{syn}} I_2^2 \frac{R_2}{s} \tag{5.52a}$$

$$= \frac{1}{\omega_{syn}} I_2'^2 \frac{R_2'}{s} \tag{5.53}$$

From the equivalent circuit of Fig. 5.16 and Eq. 5.53

$$T_{mech} = \frac{1}{\omega_{syn}} \frac{V_{th}^2}{(R_{th} + R_2'/s)^2 + (X_{th} + X_2')^2} \frac{R_2'}{s} \tag{5.54}$$

Note that if the approximate equivalent circuits (Fig. 5.14) are used to determine I_2', in Eq. 5.54, V_{th}, R_{th}, and X_{th} should be replaced by V_1, R_1, and X_1, respectively. The prediction of performance based on the approximate equivalent circuit (Fig. 5.14) may differ by 5 percent from those based on the equivalent circuit of Fig. 5.15 or 5.16.

For a three-phase machine, Eq. 5.54 should be multiplied by three to obtain the total torque developed by the machine. The torque–speed characteristic is shown in Fig. 5.17. At low values of slip,

$$R_{th} + \frac{R_2'}{s} >> X_{th} + X_2' \quad \text{and} \quad \frac{R_2'}{s} >> R_{th}$$

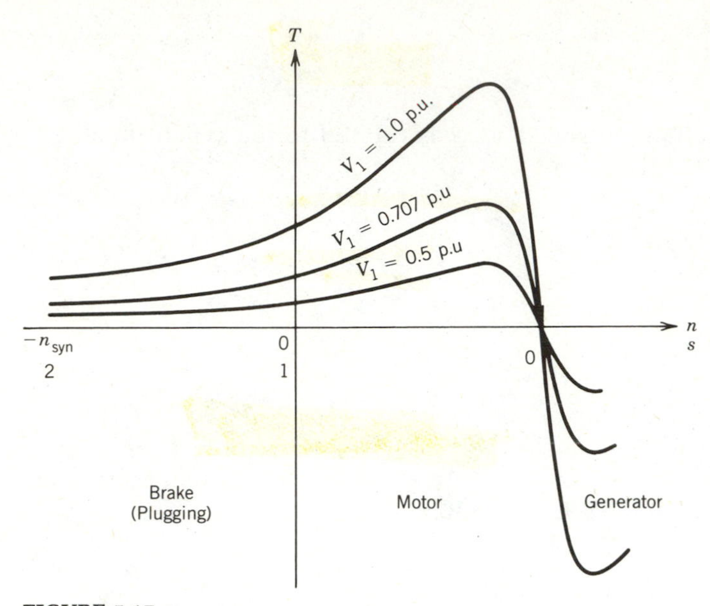

FIGURE 5.17
Torque–speed profile at different voltages.

and thus

$$T_{mech} \approx \frac{1}{\omega_{syn}} \frac{V_{th}^2}{R_2'} s \tag{5.55}$$

The linear torque–speed relationship is evident in Fig. 5.17 near the synchronous speed. At larger values of slip,

$$R_{th} + \frac{R_2'}{s} << X_{th} + X_2'$$

and

$$T_{mech} \approx \frac{1}{\omega_{syn}} \frac{V_{th}^2}{(X_{th} + X_2')^2} \frac{R_2'}{s} \tag{5.56}$$

The torque varies almost inversely with slip near $s = 1$, as seen from Fig. 5.17.

Equation 5.54 also indicates that at a particular speed (i.e., a fixed value of s) the torque varies as the square of the supply voltage V_{th} (hence V_1). Figure 5.17 shows the T–n profile at various supply voltages. This aspect will be discussed further in a later section on speed control of induction machines by changing the stator voltage.

An expression for maximum torque can be obtained by setting $dT/ds = 0$. Differentiating Eq. 5.54 with respect to slip s and equating the result to zero gives the following condition for maximum torque:

$$\frac{R_2'}{s_{T_{\max}}} = [R_{\text{th}}^2 + (X_{\text{th}} + X_2')^2]^{1/2} \tag{5.57}$$

This expression can also be derived from the fact that the condition for maximum torque corresponds to the condition for maximum air gap power (Eq. 5.52). This occurs, by the familiar impedance-matching principle in circuit theory, when the impedance of R_2'/s equals in magnitude the impedance between it and the supply voltage V_1 (Fig. 5.16) as shown in Eq. 5.57. The slip $s_{T_{\max}}$ at maximum torque $T_{\max}$ is

$$s_{T_{\max}} = \frac{R_2'}{[R_{\text{th}}^2 + (X_{\text{th}} + X_2')^2]^{1/2}} \tag{5.58}$$

The maximum torque per phase from Eqs. 5.54 and 5.58 is

$$T_{\max} = \frac{1}{2\omega_{\text{syn}}} \frac{V_{\text{th}}^2}{R_{\text{th}} + [R_{\text{th}}^2 + (X_{\text{th}} + X_2')^2]^{1/2}} \tag{5.59}$$

Equation 5.59 shows that the maximum torque developed by the induction machine is independent of the rotor circuit resistance. However, from Eq. 5.58 it is evident that the value of the rotor circuit resistance R_2 determines the speed at which this maximum torque will occur. The torque–speed characteristics for various values of R_2 are shown in Fig. 5.18. In a wound-rotor induction motor, external resistance is added to the rotor circuit to make the maximum torque occur at standstill so that high start-

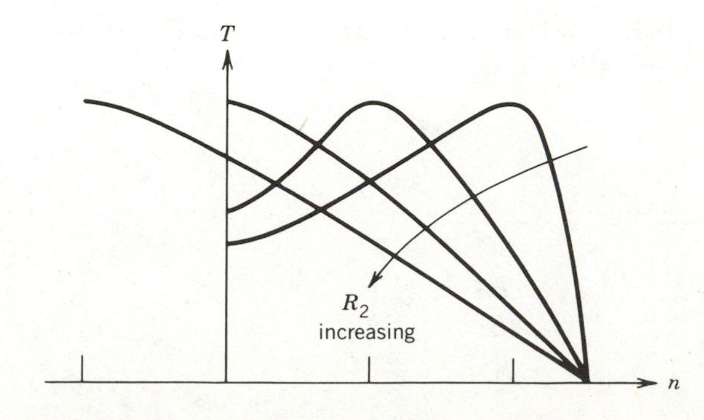

FIGURE 5.18
Torque–speed characteristics for varying R_2.

ing torque can be obtained. As the motor speeds up, the external resistance is gradually decreased and finally taken out completely. Some induction motors are, in fact, designed so that maximum torque is available at start, that is, at zero speed.

If the stator resistance R_1 is small (hence R_{th} is negligibly small), from Eqs. 5.58 and 5.59,

$$s_{T_{max}} \simeq \frac{R_2'}{X_{th} + X_2'} \tag{5.60}$$

$$T_{max} \simeq \frac{1}{2\omega_{syn}} \frac{V_{th}^2}{X_{th} + X_2'} \tag{5.61}$$

Equation 5.61 indicates that the maximum torque developed by an induction machine is inversely proportional to the sum of the leakage reactances.

From Eq. 5.54, the ratio of the maximum developed torque to the torque developed at any speed is

$$\frac{T_{max}}{T} = \frac{(R_{th} + R_2'/s)^2 + (X_{th} + X_2')^2}{(R_{th} + R_2'/s_{T_{max}})^2 + (X_{th} + X_2')^2} \frac{s}{s_{T_{max}}} \tag{5.62}$$

If R_1 (hence R_{th}) is negligibly small,

$$\frac{T_{max}}{T} \simeq \frac{(R_2'/s)^2 + (X_{th} + X_2')^2}{(R_2'/s_{T_{max}})^2 + (X_{th} + X_2')^2} \frac{s}{s_{T_{max}}} \tag{5.63}$$

From Eqs. 5.60 and 5.63

$$\begin{aligned} \frac{T_{max}}{T} &= \frac{(R_2'/s)^2 + (R_2'/s_{T_{max}})^2}{2(R_2'/s_{T_{max}})^2} \times \frac{s}{s_{T_{max}}} \\ &= \frac{s_{T_{max}}^2 + s^2}{2s_{T_{max}}s} \end{aligned} \tag{5.64}$$

Equation 5.64 shows the relationship between torque at any speed and the maximum torque in terms of their slip values.

Stator Current

From Fig. 5.15, the input impedance is

$$\begin{aligned} Z_1 &= R_1 + jX_1 + X_m \,//\, \left(\frac{R_2'}{s} + jX_2'\right) \\ &= R_1 + jX_1 + X_m \,//\, Z_2' \end{aligned} \tag{5.65}$$

$$= R_1 + jX_1 + \frac{jX_m(R'_2/s + jX'_2)}{R'_2/s + j(X_m + X'_2)} \tag{5.65a}$$

$$= |Z_1| \underline{/\theta_1} \tag{5.65b}$$

The stator current is

$$I_1 = \frac{V_1}{Z_1} = I_\phi + I'_2 \tag{5.65c}$$

At synchronous speed (i.e., $s = 0$), R'_2/s is infinite and so $I'_2 = 0$. The stator current I_1 is the exciting current I_ϕ. At larger values of slip Z'_2 ($= R'_2/s + jX'_2$) is low and therefore I'_2 (and hence I_1) is large. In fact, the typical starting current (i.e., at $s = 1$) is five to eight times the rated current. The typical stator current variation with speed is shown in Fig. 5.19.

Input Power Factor

The supply power factor is given by

$$\text{PF} = \cos \theta_1$$

where θ_1 is the phase angle of the stator current I_1. This phase angle θ_1 is the same as the impedance angle of the equivalent circuit of Fig. 5.15. The typical power factor variation with speed is shown in Fig. 5.20.

Efficiency

In order to determine the efficiency of the induction machine as a power converter, the various losses in the machine are first identified. These losses are illustrated in the power flow diagram of Fig. 5.21. For a 3ϕ machine the power input to the stator is

$$P_{in} = 3V_1I_1 \cos \theta_1 \tag{5.66}$$

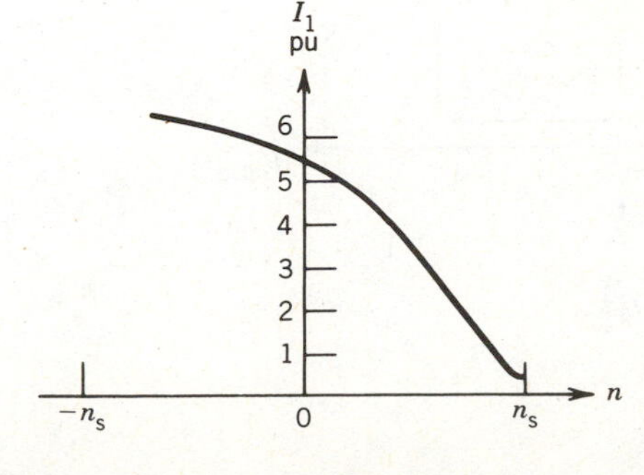

FIGURE 5.19
Stator current as a function of speed.

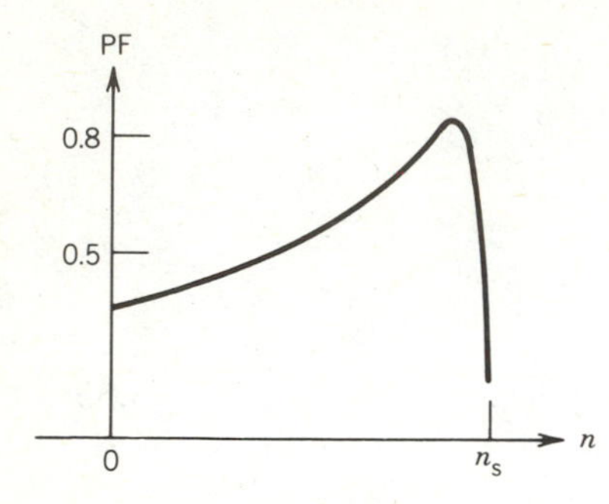

FIGURE 5.20
Power factor as a function of speed.

The power loss in the stator windings is

$$P_1 = 3I_1^2R_1 \tag{5.67}$$

where R_1 is the ac resistance (including skin effect) of each phase winding at the operating temperature and frequency.

Power is also lost as hysteresis and eddy current loss in the magnetic material of the stator core.

The remaining power, P_{ag}, crosses the air gap. Part of it is lost in the resistance of the rotor circuit.

$$P_2 = 3I_2^2R_2 \tag{5.68}$$

where R_2 is the ac resistance of the rotor winding. If it is a wound-rotor machine, R_2 also includes any external resistance connected to the rotor circuit through slip rings.

Power is also lost in the rotor core. Because the core losses are dependent on the frequency f_2 of the rotor, these may be negligible at normal operating speeds, where f_2 is very low.

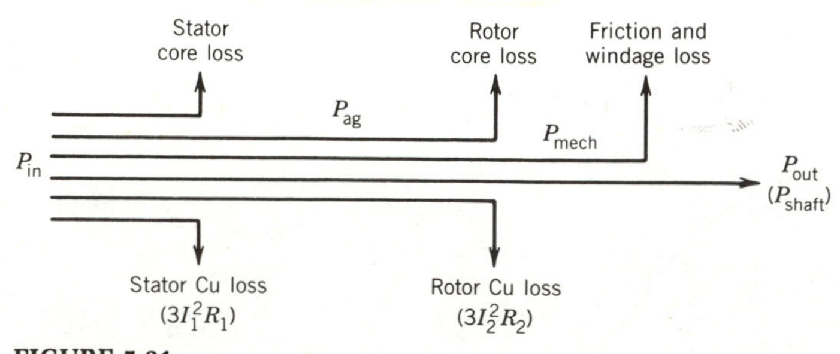

FIGURE 5.21
Power flow in an induction motor.

The remaining power is converted into mechanical form. Part of this is lost as windage and friction losses, which are dependent on speed. The rest is the mechanical output power P_{out}, which is the useful power output from the machine.

The efficiency of the induction motor is

$$\text{Eff} = \frac{P_{out}}{P_{in}} \tag{5.69}$$

The efficiency is highly dependent on slip. If all losses are neglected except those in the resistance of the rotor circuit,

$$P_{ag} = P_{in}$$

$$P_2 = sP_{ag}$$

$$P_{out} = P_{mech} = P_{ag}(1 - s)$$

and the ideal efficiency is

$$\text{Eff}_{(ideal)} = \frac{P_{out}}{P_{in}} = 1 - s \tag{5.70}$$

Sometimes $\text{Eff}_{(ideal)}$ is also called the *internal efficiency* as it represents the ratio of the power output to the air gap power. The ideal efficiency as a function of speed is shown in Fig. 5.22. It indicates that an induction machine must operate near its synchronous speed if high efficiency is desired. This is why the slip is very low for normal operation of the induction machine.

If other losses are included, the actual efficiency is lower than the ideal efficiency of Eq. 5.70 as shown in Fig. 5.22. The full-load efficiency of a large induction motor may be as high as 95 percent.

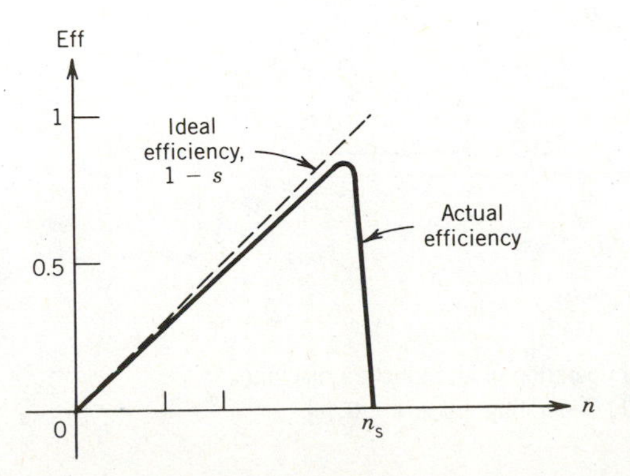

FIGURE 5.22
Efficiency as a function of speed.

5.10 POWER FLOW IN THREE MODES OF OPERATION

It was pointed out in Section 5.5 that the induction machine can be operated in three modes: motoring, generating, and plugging. The power flow in the machine will depend on the mode of operation. However, the equations derived in Section 5.9 for various power relationships hold good for all modes of operation. If the appropriate sign of the slip s is used in these expressions, the sign of the power will indicate the actual power flow. For example, in the generating mode, the slip is negative. Therefore, from Eq. 5.40a the air gap power P_{ag} is negative (note that the copper loss P_2 in the rotor circuit is always positive). This implies that the actual power flow across the air gap in the generating mode is from rotor to stator.

The power flow diagram in the three modes of operation is shown in Fig. 5.23. The core losses and the friction and windage losses are all lumped together as a constant rotational loss.

In the motoring mode, slip s is positive. The air gap power P_{ag} (Eq. 5.40a) and the developed mechanical power P_{mech} (Eq. 5.42) are positive, as shown in Fig. 5.23*a*.

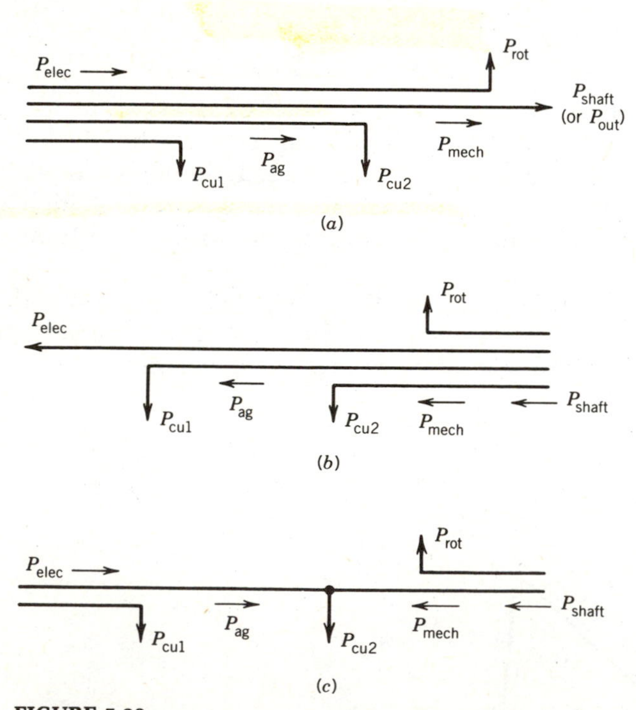

FIGURE 5.23
Power flow for various modes of operation of an induction machine. (*a*) Motoring mode, $0 < s < 1$. (*b*) Generating mode, $s < 0$. (*c*) Plugging mode, $s > 1$.

In the generating mode s is negative and therefore both P_{ag} and P_{mech} are negative, as shown in Fig. 5.23*b*. In terms of the equivalent circuit of Fig. 5.13*e* the resistance $[(1 - s)/s]R_2$ is negative, which indicates that this resistance represents a source of energy.

In the plugging mode, s is greater than one and therefore P_{ag} is positive but P_{mech} is negative as shown in Fig. 5.23*c*. In this mode the rotor rotates opposite to the rotating field and therefore mechanical energy must be put into the system. Power therefore flows from both sides, and as a result the loss in the rotor circuit, P_2, is enormously increased. In terms of the equivalent circuit of Fig. 5.13*e*, the resistance $[(1 - s)/s]R_2$ is negative and represents a source of energy.

Example 5.7 will further illustrate the power flow in the three modes of operation of an induction machine.

EXAMPLE 5.4

A three-phase, 460 V, 1740 rpm, 60 Hz, four-pole wound-rotor induction motor has the following parameters per phase:

$$R_1 = 0.25 \text{ ohms}, \qquad R_2' = 0.2 \text{ ohms}$$

$$X_1 = X_2' = 0.5 \text{ ohms}, \qquad X_m = 30 \text{ ohms}$$

The rotational losses are 1700 watts. With the rotor terminals short-circuited, find

(a) (i) Starting current when started direct on full voltage.

(ii) Starting torque.

(b) (i) Full-load slip.

(ii) Full-load current.

(iii) Ratio of starting current to full-load current.

(iv) Full-load power factor.

(v) Full-load torque.

(vi) Internal efficiency and motor efficiency at full load.

(c) (i) Slip at which maximum torque is developed.

(ii) Maximum torque developed.

(d) How much external resistance per phase should be connected in the rotor circuit so that maximum torque occurs at start?

Solution

(a)

$$V_1 = \frac{460}{\sqrt{3}} = 265.6 \text{ volts/phase}$$

(i) At start $s = 1$. The input impedance is

$$Z_1 = 0.25 + j0.5 + \frac{j30(0.2 + j0.5)}{0.2 + j30.5}$$

$$= 1.08\underline{/66^\circ}\ \Omega$$

$$I_{st} = \frac{265.6}{1.08\underline{/66^\circ}} = 245.9\underline{/-66^\circ}\ \text{A}$$

(ii)

$$\omega_{syn} = \frac{1800}{60} \times 2\pi = 188.5 \text{ rad/sec}$$

$$V_{th} = \frac{265.6(j30.0)}{(0.25 + j30.5)} \approx 261.3 \text{ V}$$

$$Z_{th} = \frac{j30(0.25 + j0.5)}{0.25 + j30.5} = 0.55\underline{/63.9^\circ}$$

$$= 0.24 + j0.49$$

$$R_{th} = 0.24\ \Omega$$

$$X_{th} = 0.49 \simeq X_1$$

$$T_{st} = \frac{P_{ag}}{\omega_{syn}} = \frac{I_2'^2 R_2'/s}{\omega_{syn}}$$

$$= \frac{3}{188.5} \frac{261.3^2}{(0.24 + 0.2)^2 + (0.49 + 0.5)^2} \times \frac{0.2}{1}$$

$$= \frac{3}{188.5} \times (241.2)^2 \times \frac{0.2}{1}$$

$$= 185.2 \text{ N} \cdot \text{m}$$

(b) (i)

$$s = \frac{1800 - 1740}{1800} = 0.0333$$

(ii)

$$\frac{R_2'}{s} = \frac{0.2}{0.0333} = 6.01\ \Omega$$

$$Z_1 = (0.2 + j0.5) + \frac{(j30)(6.01 + j0.5)}{6.01 + j30.5}$$

$$= 6.15\underline{/19.8^\circ}\ \Omega$$

$$I_{FL} = \frac{265.6}{6.15\angle 19.8^\circ}$$

$$= 43.19\angle -19.8^\circ \text{ A}$$

(iii)
$$\frac{I_{st}}{I_{FL}} = \frac{245.9}{43.19} = 5.69$$

(iv)
$$\text{PF} = \cos(19.8^\circ) = 0.94 \quad \text{(lagging)}$$

(v)
$$\text{T} = \frac{3}{188.5} \frac{(261.3)^2}{(0.24 + 6.01)^2 + (0.49 + 0.5)^2} \times 6.01$$

$$= \frac{3}{188.5} \times 41.29^2 \times 6.01$$

$$= 163.11 \text{ N} \cdot \text{m}$$

(vi) Air gap power:

$$P_{ag} = T\omega_{syn} = 163.11 \times 188.5 = 30{,}746.2 \text{ W}$$

Rotor copper loss:

$$P_2 = sP_{ag} = 0.0333 \times 30{,}746.2 = 1023.9 \text{ W}$$

$$P_{mech} = (1 - 0.0333)30{,}746.2 = 29{,}722.3 \text{ W}$$

$$P_{out} = P_{mech} - P_{rot} = 29{,}722.3 - 1700 = 28{,}022.3 \text{ W}$$

$$P_{input} = 3V_1I_1 \cos \theta_1$$

$$= 3 \times 265.6 \times 43.19 \times 0.94 = 32349 \text{ W}$$

$$\text{Eff}_{motor} = \frac{28{,}022.3}{32349} \times 100 = 86.6\%$$

$$\text{Eff}_{internal} = (1 - s) = 1 - 0.0333 = 0.967 \rightarrow 96.7\%$$

(c) (i) From Eq. 5.58

$$s_{T_{max}} = \frac{0.2}{[0.24^2 + (0.49 + 0.05)^2]^{1/2}} = \frac{0.2}{1.0186} = 0.1963$$

(ii) From Eq. 5.59

$$T_{max} = \frac{3}{2 \times 188.5} \frac{261.3^2}{0.24 + [0.24^2 + (0.49 + 0.5)^2]^{1/2}}$$

$$= 431.68 \text{ N} \cdot \text{m}$$

$$\frac{T_{max}}{T_{FL}} = \frac{431.68}{163.11} = 2.65$$

(d)

$$s_{T\max} = 1 = \frac{R_2' + R_{ext}'}{[0.24^2 + (0.49 + 0.5)^2]^{1/2}} = \frac{R_2' + R_{ext}'}{1.0186}$$

$$R_{ext}' = 1.0186 - 0.2 = 0.8186\ \Omega/\text{phase}$$

EXAMPLE 5.5

A three-phase, 460 V, 60 Hz, six-pole wound-rotor induction motor drives a constant load of 100 N · m at a speed of 1140 rpm when the rotor terminals are short-circuited. It is required to reduce the speed of the motor to 1000 rpm by inserting resistances in the rotor circuit. Determine the value of the resistance if the rotor winding resistance per phase is 0.2 ohms. Neglect rotational losses. The stator-to-rotor turns ratio is unity.

Solution

The synchronous speed is

$$n_s = \frac{120 \times 60}{6} = 1200 \text{ rpm}$$

Slip at 1140 rpm:

$$s_1 = \frac{1200 - 1140}{1200} = 0.05$$

Slip at 1000 rpm:

$$s_2 = \frac{1200 - 1000}{1200} = 0.167$$

From the equivalent circuits, it is obvious that if the value of R_2'/s remains the same, the rotor current I_2 and the stator current I_1 will remain the same, and the machine will develop the same torque (Eq. 5.54). Also, if the rotational losses are neglected, the developed torque is the same as the load torque. Therefore, for unity turns ratio,

$$\frac{R_2}{s_1} = \frac{R_2 + R_{ext}}{s_2}$$

$$\frac{0.2}{0.05} = \frac{0.2 + R_{ext}}{0.167}$$

$$R_{ext} = 0.468\ \Omega/\text{phase}$$

EXAMPLE 5.6

The rotor current at start of a three-phase, 460 volt, 1710 rpm, 60 Hz, four-pole, squirrel-cage induction motor is six times the rotor current at full load.

(a) Determine the starting torque as percent of full load torque.

(b) Determine the slip and speed at which the motor develops maximum torque.

(c) Determine the maximum torque developed by the motor as percent of full load torque.

Solution

Note that the equivalent circuit parameters are not given. Therefore, the equivalent circuit parameters cannot be used directly for computation.

(a) The synchronous speed is

$$n_s = \frac{120 \times 60}{4} = 1800 \text{ rpm}$$

The full-load slip is

$$s_{FL} = \frac{1800 - 1710}{1800} = 0.05$$

From Eq. 5.52a

$$T = \frac{I_2^2 R_2}{\omega_{syn} s} \propto \frac{I_2^2 R_2}{s}$$

Thus,

$$\frac{T_{st}}{T_{FL}} = \left| \frac{I_{2(st)}}{I_{2(FL)}} \right|^2 s_{FL}$$

$$T_{st} = 6^2 \times 0.05 \times T_{FL} = 1.8 T_{FL}$$

$$= 180\% \ T_{FL}$$

(b) From Eq. 5.64,

$$\frac{T_{st}}{T_{max}} = \frac{2 s_{T_{max}}}{1 + s_{T_{max}}^2}$$

$$\frac{T_{FL}}{T_{max}} = \frac{2s_{T_{max}}s_{FL}}{s_{T_{max}}^2 + s_{FL}^2}$$

From these two expressions,

$$\frac{T_{st}}{T_{FL}} = \frac{s_{T_{max}}^2 + s_{FL}^2}{s_{FL} + s_{FL} \times s_{T_{max}}^2}$$

$$1.8 = \frac{s_{T_{max}}^2 + 0.0025}{0.05 + 0.05 \times s_{T_{max}}^2}$$

$$s_{T_{max}}^2 + 0.0025 = 0.09 + 0.09s_{T_{max}}^2$$

$$s_{T_{max}} = \left(\frac{0.0875}{0.91}\right)^{1/2} = 0.31$$

$$\text{Speed at maximum torque} = (1 - 0.31) \times 1800$$
$$= 1242 \text{ rpm}$$

(c) From Eq. 5.64,

$$T_{max} = \left|\frac{1 + s_{T_{max}}^2}{2s_{T_{max}}}\right| T_{st}$$
$$= \frac{1 + 0.31^2}{2 \times 0.31} \times 1.8T_{FL}$$
$$= 3.18T_{FL} = 318\% \, T_{FL}$$

EXAMPLE 5.7

A three-phase source of variable frequency is required for an experiment. The frequency-changer system is as shown in Fig. E5.7. The induction machine is a three-phase, six-pole, wound-rotor type whose stator terminals are connected to a three-phase, 460 volt, 60 Hz supply. The variable-frequency output is obtained from the rotor terminals. The frequency is to be controlled over the range 15–120 Hz.

(a) Determine the speed in rpm of the system to give 15 Hz and 120 Hz.

(b) If the open-circuit rotor voltage is 240 volts when the rotor is at standstill, determine the rotor voltage available on open circuit with 15 Hz and 120 Hz.

(c) If all the losses in the machine are neglected, what fraction of the output power is supplied by the ac supply and what fraction is supplied by the dc machine at 15 Hz and 120 Hz?

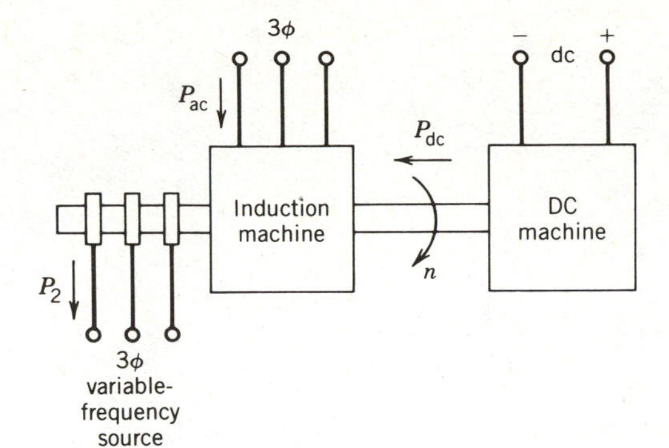

FIGURE E5.7

Solution

(a) For $f_2 = 15$ Hz, the slip is

$$s = \pm\frac{f_2}{f_1} = \pm\frac{15}{60} = \pm\frac{1}{4}$$

The synchronous speed is

$$n_s = \frac{120 \times 60}{6} = 1200 \text{ rpm}$$

The speed of the system for $f_2 = 15$ Hz is

$$n = (1 \pm s)n_s = (1 \pm \tfrac{1}{4}) \times 1200$$
$$= 900 \text{ and } 1500 \text{ rpm}$$

For $f_2 = 120$ Hz,

$$s = \pm\frac{120}{60} = \pm 2.0$$
$$n = (1 \pm 2.0)1200$$
$$= -1200 \text{ and } 3600 \text{ rpm}$$

(b)

$$sE_2 = s \times 240$$

For $f_2 = 15$ Hz, $\quad sE_2 = 60$ V

For $f_2 = 120$ Hz, $\quad sE_2 = 480$ V

(c) Power from the supply:

$$P_{ac} = P_{ag} = \frac{P_2}{s}$$

Power from the shaft:

$$P_{dc} = -(1 - s)P_{ag} = -\frac{(1 - s)}{s} \times P_2$$

For $f_2 = 15$ Hz,

$$P_{ac} = \frac{P_2}{+(1/4)}, \frac{P_2}{-(1/4)} = +4P_2, -4P_2$$

$$P_{dc} = \frac{-[1 - (1/4)]}{+(1/4)} P_2, \frac{-[1 + (1/4)]}{-(1/4)} P_2$$

$$= -3P_2, +5P_2$$

For $f_2 = 120$ Hz,

$$P_{ac} = \frac{P_2}{2.0}, \frac{P_2}{-2.0} = 0.5P_2, -0.5P_2$$

$$P_{dc} = \frac{-(1 - 2.0)}{+2.0} P_2, \frac{-(1 + 2.0)}{-2.0} P_2$$

$$= 0.5P_2, 1.5P_2$$

The results are summarized in the following table:

f_2 (Hz)	rpm	Mode of Operation of Induction Machine	Slip	sE_2 (V)	Stator Input, P_{ac}	Shaft Input, P_{dc}	Rotor Output
15	900	Motor	+(1/4)	60	$4P_2$	$-3P_2$	P_2
	1500	Generator	−(1/4)	60	$-4P_2$	$5P_2$	P_2
120	−1200	Plugging	+2.0	480	$0.5P_2$	$0.5P_2$	P_2
	3600	Generator	−2.0	480	$-0.5P_2$	$1.5P_2$	P_2

For practical reasons, high-speed operation should be avoided. The speed range is therefore 900 to −1200 rpm for the varying output frequency in the range 15 to 120 Hz. The speed 900 rpm implies that the dc machine rotates in the same direction as the rotating field, whereas −1200 rpm implies that the dc machine rotates opposite to the rotating field.

5.11 EFFECTS OF ROTOR RESISTANCE

In a conventional squirrel-cage motor at full load, the slip and the current are low but the power factor and the efficiency are high. However, at start, the torque and power factor are low but the current is high. If the load requires a high starting torque (Fig. 5.24), the motor will accelerate slowly. This will make a large current flow for a longer time, thereby creating a heating problem.

The resistances in the rotor circuit greatly influence the performance of an induction motor. A low rotor resistance is required for normal operation, when running, so that the slip is low and the efficiency high. However, a higher rotor resistance is required for starting so that the starting torque and power factor are high and the starting current is low. An induction motor with a fixed rotor circuit resistance therefore requires a compromise design of the rotor for starting and running conditions. Various types of induction motors are available in which the rotor circuit resistance is changed or can change with speed to suit the particular application.

5.11.1 WOUND-ROTOR MOTORS

In wound-rotor induction motors external resistances can be connected to the rotor winding through slip rings (Fig. 5.39*a*). Equation 5.58 shows that the slip at which maximum torque occurs is directly proportional to the rotor circuit resistance.

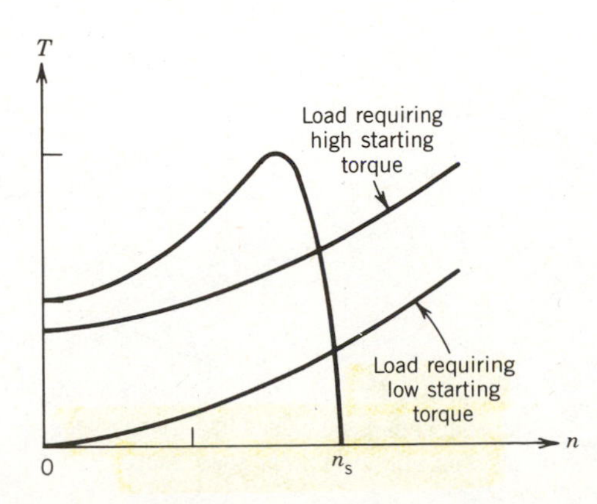

FIGURE 5.24
Loads with different torque requirements.

$$s_{T_{max}} \propto (R_{w2} + R_{ext}) \tag{5.71}$$

where R_{w2} is the per-phase rotor winding resistance

R_{ext} is the per-phase resistance connected externally to the rotor winding

The external resistance R_{ext} can be chosen to make the maximum torque occur at standstill ($s_{T_{max}} = 1$) if high starting torque is desired. This external resistance can be decreased as the motor speeds up, making maximum torque available over the whole accelerating range, as shown in Fig. 5.25. Equation 5.59 indicates that the maximum torque remains the same, as it is independent of the rotor circuit resistance.

Note that most of the rotor I^2R loss is dissipated in the external resistances. Thus the rotor heating is lower during the starting and acceleration period than it would be if the resistances were incorporated in the rotor windings. The external resistance is eventually cut out so that under running conditions the rotor resistance is only the rotor winding resistance, which is designed to be low to make the rotor operate at high efficiency and low full-load slip.

Apart from high starting torque requirements, the external resistance can also be used for varying the running speed. This will be discussed in Section 5.13.6.

The disadvantage of the wound-rotor induction machine is its higher cost than the squirrel-cage motor.

5.11.2 DEEP-BAR SQUIRREL-CAGE MOTORS

The rotor frequency changes with speed. At standstill, the rotor frequency equals the supply frequency. As the motor speeds up, the rotor frequency

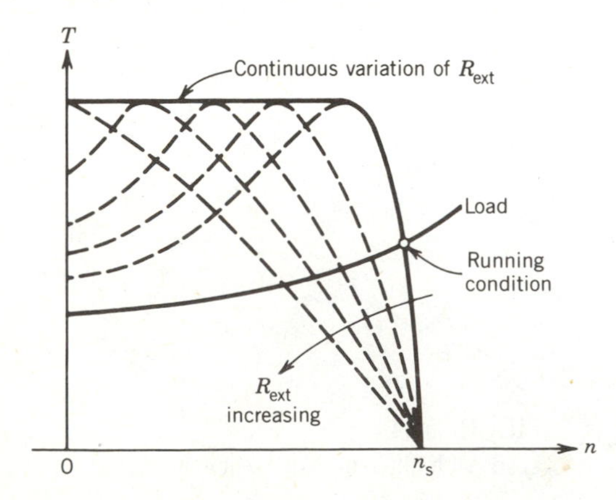

FIGURE 5.25 Maximum torque obtained by varying rotor resistance throughout the speed range.

decreases to a low value. At full-load running condition the rotor frequency is in the range of 1 to 3 Hz with a 60 Hz supply connected to the stator terminals. This fact can be utilized to change the rotor resistance automatically. The rotor bars can be properly shaped and arranged so that their effective resistance at supply frequency (say 60 Hz) is several times the resistance at full-load rotor frequency (1 to 3 Hz). This change in the resistance of the rotor bars is due to what is commonly known as the skin effect.

Figure 5.26*a* shows a squirrel-cage rotor with deep and narrow bars. The slot leakage fluxes produced by the current in the bar are also shown in the figure. All the leakage flux lines will close paths below the slot. It is obvious that the leakage inductance of the bottom layer is higher than that of the top layer because the bottom layer is linked with more leakage flux. The current in the high-reactance bottom layer will be less than the current in the low-reactance upper layer; that is, the rotor current will be

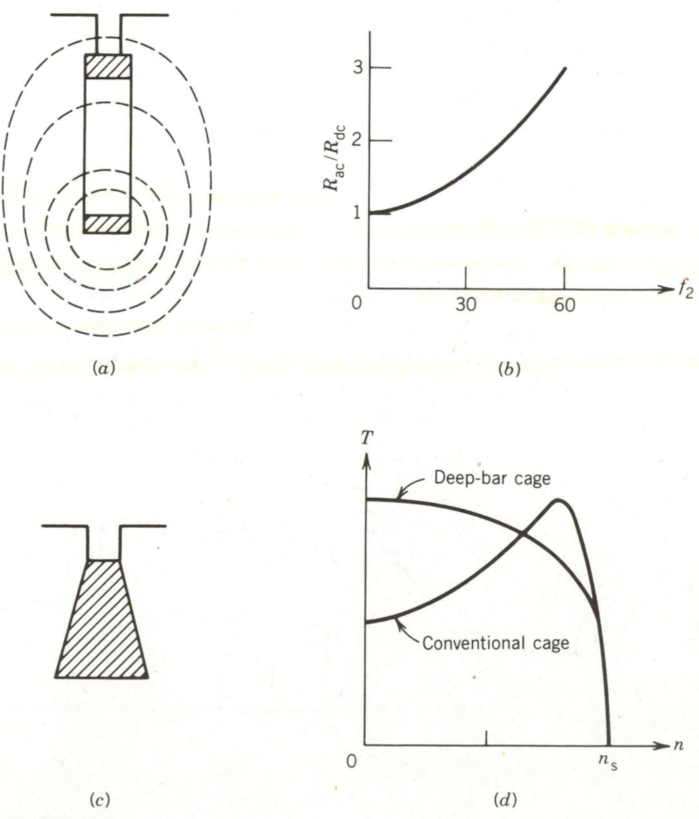

FIGURE 5.26
Deep-bar rotor and characteristics.

pushed to the top of the bars. The result will be an increase in the effective resistance of the bar. Because this nonuniform current distribution depends on the reactance, it is more pronounced at high frequency (i.e., when the motor is at standstill) than at low frequency (i.e., when the motor is running at full speed). The deep rotor bars may be designed so that the effective rotor resistance at standstill is several times its effective resistance at rated speed. At full speed, the rotor frequency is very low (1–3 Hz), the rotor current is almost uniformly distributed over the cross section of the rotor bar, and the rotor ac effective resistance, R_{ac}, is almost the same as the dc resistance, R_{dc}. A typical variation of the ratio R_{ac}/R_{dc} with rotor frequency is shown in Fig. 5.26*b*.

The rotor bars can take other shapes, such as broader at the base than at the top of the slot, as shown in Fig. 5.26*c*. This shape will further increase the ratio R_{ac}/R_{dc} with frequency.

With proper design of the rotor bars, a flat-topped torque–speed characteristic, as shown in Fig. 5.26*d*, can be obtained.

5.11.3 DOUBLE-CAGE ROTORS

Low starting current and high starting torque may also be obtained by building a double-cage rotor. In this type of rotor construction, the squirrel-cage windings consist of two layers of bars (as shown in Fig. 5.27*a*), each layer short-circuited by end rings. The outer cage bars have a smaller cross-sectional area than the inner bars and are made of relatively higher-resistivity material. Thus the resistance of the outer cage is greater than the resistance of the inner cage. The leakage inductance of the inner cage is increased by narrowing the slots above it. At standstill most of the rotor current flows through the upper cage, thereby increasing the effective

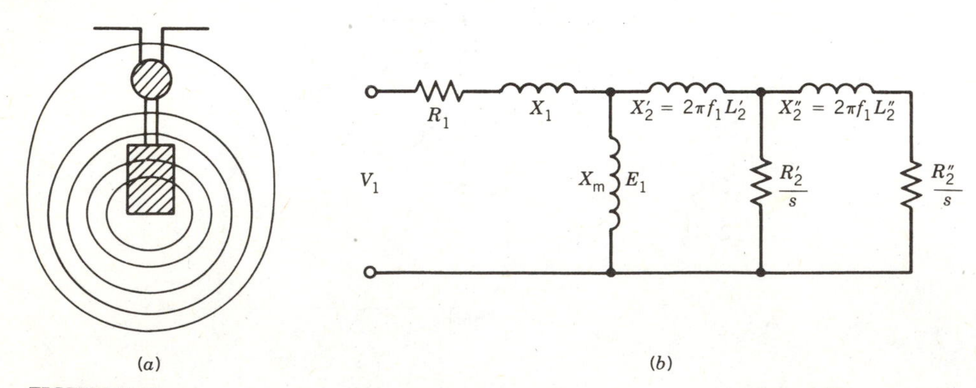

FIGURE 5.27
Double-cage rotor bars and the equivalent circuit.

resistance of the rotor circuit. At low rotor frequencies, corresponding to low slips (such as at full-load running condition), reactance is negligible, the lower-cage bars share the current with the outer bars, and the rotor resistance approaches that of two cages in parallel. Because both cages carry current under normal running conditions, the rating of the motor is somewhat increased.

In both double-cage and deep-bar rotors, the effective resistance and leakage inductance vary with the rotor frequency. The equivalent circuit of a double-cage rotor can be represented by the circuit shown in Fig. 5.27*b*, where

$L_2' =$ per-phase leakage inductance of the upper-cage bars

$R_2' =$ per-phase resistance of the upper-cage bars

$L_2'' =$ per-phase leakage inductance of the lower-cage bars

$R_2'' =$ per-phase resistance of the lower-cage bars

It should be recognized that the values of these parameters depend on the rotor frequency.

Both double-cage and deep-bar cage rotors can be designed to provide the good starting characteristics resulting from higher rotor resistance and, concurrently, the good running characteristics resulting from low rotor resistance. The rotor design in these cases is based on a compromise between starting and running performance. However, these motors are not as flexible as the wound-rotor machine with external rotor resistance. In fact, when starting requirements are very severe, the wound-rotor motor should be used.

5.12 CLASSES OF SQUIRREL-CAGE MOTORS

Industrial needs are diverse. To meet the various starting and running requirements of a variety of industrial applications, several standard designs of squirrel-cage motors are available from manufacturers' stock. The torque–speed characteristics of the most common designs, readily available and standardized in accordance with the criteria established by the National Electrical Manufacturers' Association (NEMA), are shown in Fig. 5.28. The most significant design variable in these motors is the effective resistance of the rotor cage circuits.

Class A Motors

Class A motors are characterized by normal starting torque, high starting current, and low operating slip. The motors have low rotor circuit resis-

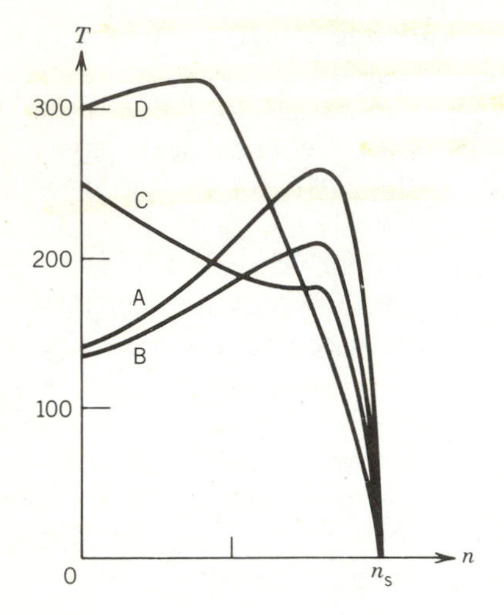

FIGURE 5.28
Torque–speed characteristics for different classes of induction motors.

tance and therefore operate efficiently with a low slip ($0.005 < s < 0.015$) at full load. These machines are suitable for applications where the load torque is low at start (such as fan or pump loads) so that full speed is achieved rapidly, thereby eliminating the problem of overheating during starting. In larger machines, low-voltage starting is required to limit the starting current.

Class B Motors

Class B motors are characterized by normal starting torque, low starting current, and low operating slip. The starting torque is almost the same as that in class A motors, but the starting current is about 75 percent of that for class A. The starting current is reduced by designing for relatively high leakage reactance by using either deep-bar rotors or double-cage rotors. The high leakage reactance lowers the maximum torque. The full-load slip and efficiency are as good as those of class A motors.

Motors of this class are good general-purpose motors and have a wide variety of industrial applications. They are particularly suitable for constant-speed drives, where the demand for starting torque is not severe. Examples are drives for fans, pumps, blowers, and motor–generator sets.

Class C Motors

Class C motors are characterized by high starting torque and low starting current. A double-cage rotor is used with higher rotor resistance than is found in class B motors. The full-load slip is somewhat higher and the efficiency lower than for class A and class B motors. Class C motors are suitable for driving compressors, conveyors, crushers, and so forth.

Class D Motors

Class D motors are characterized by high starting torque, low starting current, and high operating slip. The rotor cage bars are made of high-resistance material such as brass instead of copper. The torque–speed characteristic is similar to that of a wound-rotor motor with some external resistance connected to the rotor circuit. The maximum torque occurs at a slip of 0.5 or higher. The full-load operating slip is high (8 to 15 percent) and therefore the running efficiency is low. The high losses in the rotor circuit require that the machine be large (and hence expensive) for a given power. These motors are suitable for driving intermittent loads requiring rapid acceleration and high-impact loads such as punch presses or shears. In the case of impact loads, a flywheel is fitted to the system. As the motor speed falls appreciably with load impact, the flywheel delivers some of its kinetic energy during the impact.

5.13 SPEED CONTROL

An induction motor is essentially a constant-speed motor when connected to a constant-voltage and constant-frequency power supply. The operating speed is very close to the synchronous speed. If the load torque increases, the speed drops by a very small amount. It is therefore suitable for use in substantially constant-speed drive systems. Many industrial applications, however, require several speeds or a continuously adjustable range of speeds. Traditionally, dc motors have been used in such adjustable-speed drive systems. However, dc motors are expensive, require frequent maintenance of commutators and brushes, and are prohibitive in hazardous atmospheres. Squirrel-cage induction motors, on the other hand, are cheap, rugged, have no commutators, and are suitable for high-speed applications. The availability of solid-state controllers, although more complex than those used for dc motors, has made it possible to use induction motors in variable-speed drive systems.

In this section various methods for controlling the speed of an induction motor are discussed.

5.13.1 POLE CHANGING

Because the operating speed is close to the synchronous speed, the speed of an induction motor can be changed by changing the number of poles of the machine (Eq. 5.10). This can be done by changing the coil connections of the stator winding. Normally, poles are changed in the ratio 2 to 1. This method provides two synchronous speeds. If two independent sets of polyphase windings are used, each arranged for pole changing, four synchronous speeds can be obtained for the induction motor.

Squirrel-cage motors are invariably used in this scheme because the rotor can operate with any number of stator poles. It is obvious, however, that the speed can be changed only in discrete steps and that the elaborate stator winding makes the motor expensive.

5.13.2 LINE VOLTAGE CONTROL

Recall that the torque developed in an induction motor is proportional to the square of the terminal voltage. A set of T–n characteristics with various terminal voltages is shown in Fig. 5.29. If the rotor drives a fan load the speed can be varied over the range n_1 to n_2 by changing the line voltage. Note that the class D motor will allow speed variation over a wider speed range.

The terminal voltage V_1 can be varied by using a three-phase autotransformer or a solid-state voltage controller as shown in Fig. 5.30. The autotransformer provides a sinusoidal voltage for the induction motor, whereas the motor terminal voltage with a solid-state controller is nonsinusoidal. Speed control with a solid-state controller is commonly used with small squirrel-cage motors driving fan loads. In large power applications an input filter is required; otherwise, large harmonic currents will flow in the supply line.

The thyristor voltage controller shown in Fig. 5.30*b* is simple to understand but complicated to analyze. The operation of the voltage controller is discussed in Chapter 10. The command signal for a particular set speed fires the thyristors at a particular firing angle (α) to provide a particular

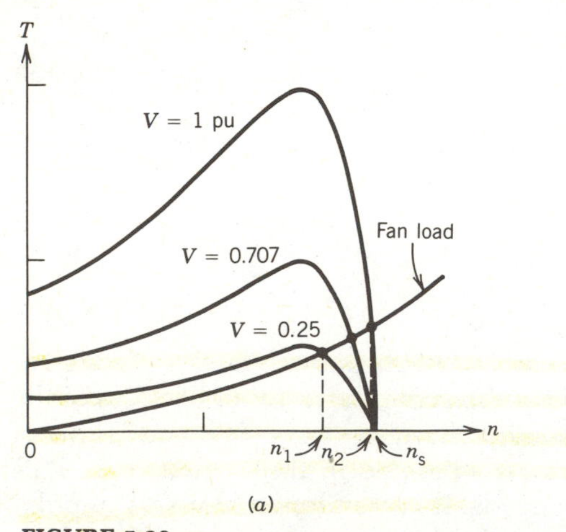

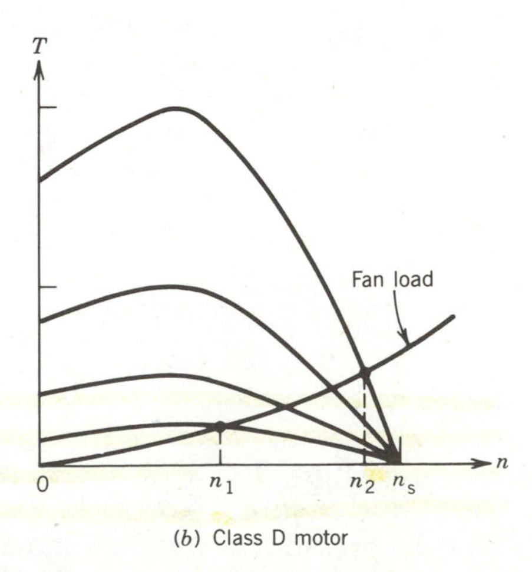

FIGURE 5.29
Torque–speed characteristics for various terminal voltages.

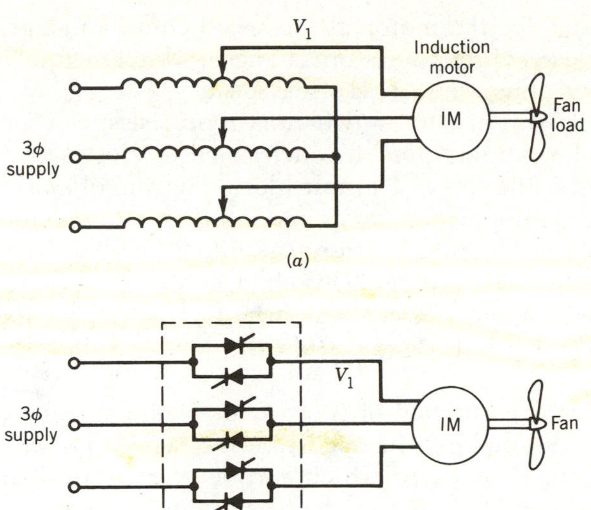

(a)

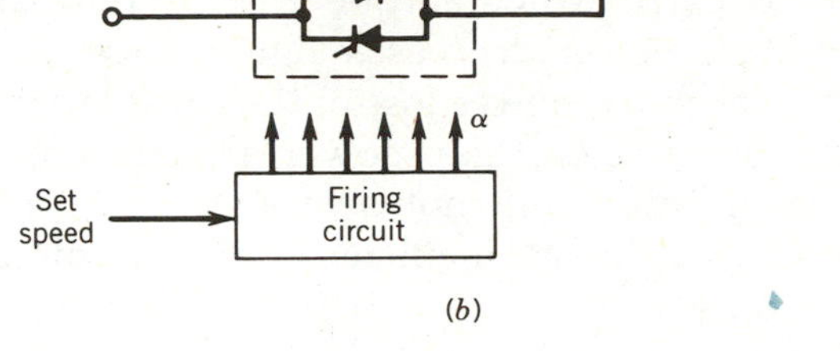

(b)

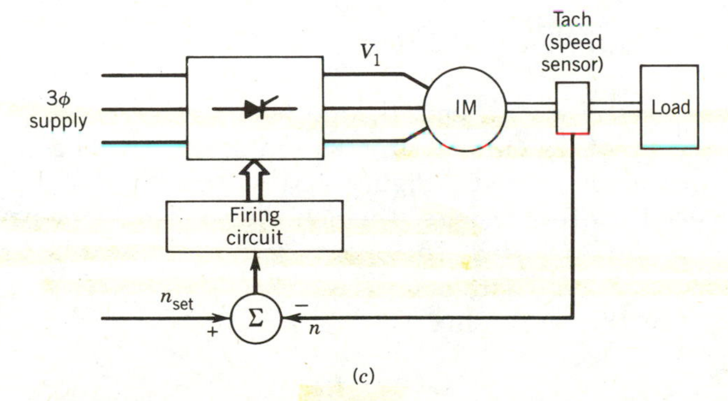

(c)

FIGURE 5.30

Starting and speed control. (*a*) Autotransformer voltage controller. (*b*) Solid-state voltage controller. (*c*) Closed-loop operation of voltage controller.

terminal voltage for the motor. If the speed command signal is changed, the firing angle (α) of the thyristors changes, which results in a new terminal voltage and thus a new operating speed.

Open-loop operation is not satisfactory if precise speed control is desired for a particular application. In such a case, closed-loop operation is needed. Figure 5.30*c* shows a simple block diagram of a drive system with closed-loop operation. If the motor speed falls because of any disturbance, such as supply voltage fluctuation, the difference between the set speed and the motor speed increases. This changes the firing angle of the thyristors to increase the terminal voltage, which in turn develops more torque. The increased torque tends to restore the speed to the value prior to the disturbance.

Note that for this method of speed control the slip increases at lower speeds (Fig. 5.29), making the operation inefficient. However, for fans, or similar centrifugal loads in which torque varies approximately as the square of the speed, the power decreases significantly with decrease in speed. Therefore, although the power lost in the rotor circuit ($= sP_{ag}$) may be a significant portion of the air gap power, the air gap power itself is small and therefore the rotor will not overheat. The voltage controller circuits are simple and, although inefficient, are suitable for fan, pump, and similar centrifugal drives.

5.13.3 LINE FREQUENCY CONTROL

The synchronous speed and hence the motor speed can be varied by changing the frequency of the supply. Application of this speed control method requires a frequency changer. Figure 5.31 shows a block diagram of an open-loop speed control system in which the supply frequency of the induction motor can be varied. The operation of the controlled rectifier (ac to dc) and the inverter (dc to ac) is described in Chapter 10.

From Eq. 5.27 the motor flux is

$$\Phi_p \propto \frac{E}{f} \tag{5.72}$$

If the voltage drop across R_1 and X_1 (Fig. 5.15) is small compared to the terminal voltage V_1, that is, $V_1 \simeq E_1$, then

$$\Phi_p \propto \frac{V}{f} \tag{5.73}$$

To avoid high saturation in the magnetic system, the terminal voltage of the motor must be varied in proportion to the frequency. This type of control is known as *constant volts per hertz*. At low frequencies, the voltage

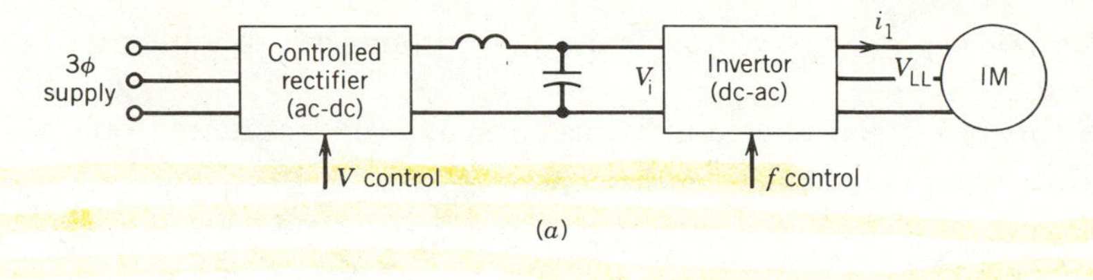

(a)

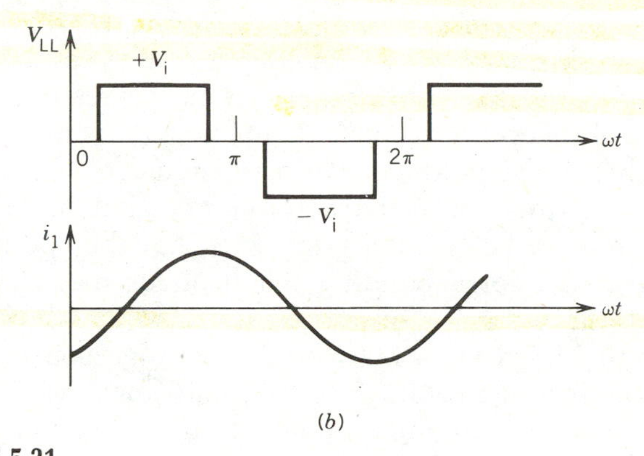

(b)

FIGURE 5.31
Open-loop speed control of an induction motor by input voltage and frequency control.

drop across R_1 and X_1 (Fig. 5.15) is comparable to the terminal voltage V_1, and therefore Eq. 5.73 is no longer valid. To maintain the same air gap flux density, the ratio V/f is increased for lower frequencies. The required variation of the supply voltage with frequency is shown in Fig. 5.32. In Fig. 5.31 the machine voltage will change if the input voltage to the inverter V_i is changed; V_i can be changed by changing the firing angle of the controlled rectifier. If the output voltage of the inverter can be changed in the inverter itself (as in pulse-width-modulated inverters), the controlled rec-

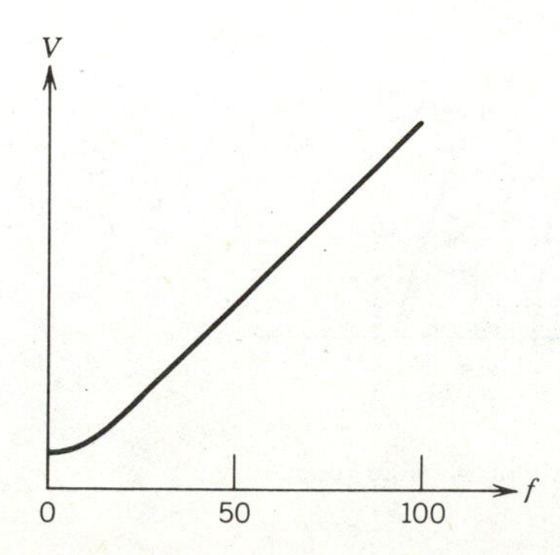

FIGURE 5.32
Required variation in voltage with change in frequency to maintain constant air gap flux density.

tifier can be replaced by a simple diode rectifier circuit, which will make V_i constant.

The torque–speed characteristics for variable-frequency operation are shown in Fig. 5.33. At the base frequency f_{base} the machine terminal voltage is the maximum that can be obtained from the inverter. Below this frequency, the air gap flux is maintained constant by changing V_1 with f_1; hence the same maximum torques are available. Beyond f_{base}, since V_1 cannot be further increased with frequency, the air gap flux decreases and so does the maximum available torque. This corresponds to the field-weakening control scheme used with dc motors. Constant-horsepower operation is possible in the field-weakening region.

In Fig. 5.33 the torque–speed characteristic of a load is superimposed on the motor torque–speed characteristic. Note that the operating speeds $n_1 \cdots n_8$ are close to the corresponding synchronous speeds. In this method of speed control, therefore, the operating slip is low and efficiency is high.

The inverter in Fig. 5.31 is known as a voltage source inverter. The motor line-to-line terminal voltage is a quasi-square wave of 120° width. However, because of motor inductance the motor current is essentially sinusoidal. A current source inverter (see Chapter 10) can be used to control the speed of an induction motor. The open-loop block diagram of a drive system using a current source inverter is shown in Fig. 5.34. The magnitude of the current is controlled by the rectifier. The filter inductor in the dc link smooths out the current. The motor current waveform is a quasi-square wave having 120° pulse width. The motor terminal voltage is

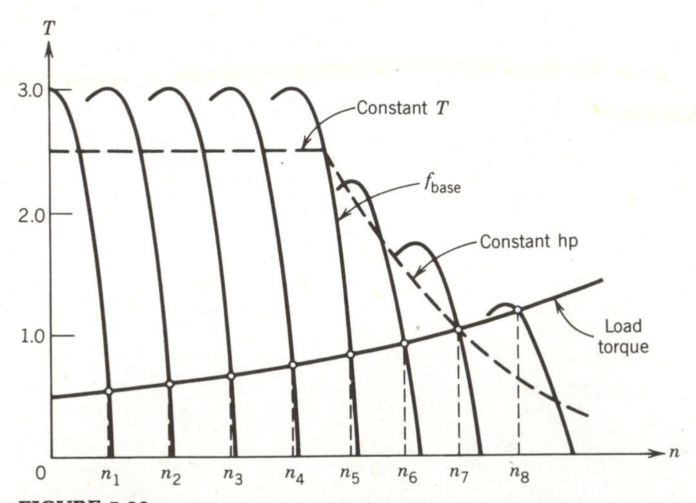

FIGURE 5.33

Torque–speed characteristics of an induction motor with variable-voltage, variable-frequency control.

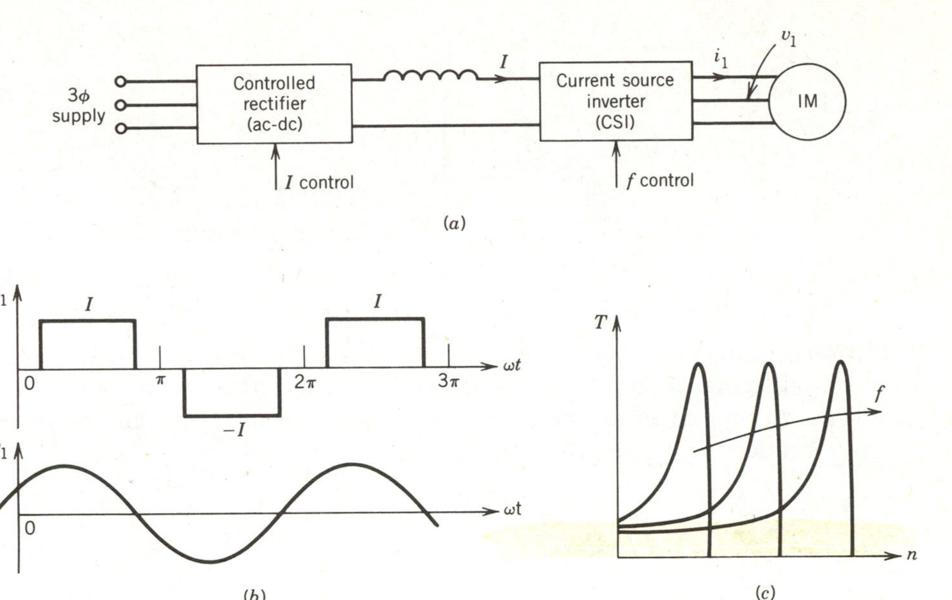

FIGURE 5.34
Induction motor drive system with a current source inverter and the corresponding characteristics.

essentially sinusoidal. The torque–speed characteristics of an induction motor fed from a current source inverter are also shown in Fig. 5.34. These characteristics have a very steep slope near synchronous speed. Although a current source inverter is rugged and desirable from the standpoint of protection of solid-state devices, the drive system should be properly operated, otherwise the system will not be stable.

5.13.4 CONSTANT-SLIP FREQUENCY OPERATION

For efficient operation of an induction machine, it is desirable to operate it at a fixed or controlled slip frequency (which is also the rotor circuit frequency). High efficiency and high power factor are obtained if the slip frequency f_2 is maintained below the breakdown frequency f_{2b}, which is the rotor circuit frequency at which the maximum torque is developed.

Consider the block diagram of Fig. 5.35. The signal f_n represents a frequency corresponding to the speed of the motor. To this a signal f_2 representing the slip (or rotor circuit) frequency is added or subtracted. The resultant f_1 represents the stator frequency:

$$f_1 = f_n \pm f_2 \tag{5.74}$$

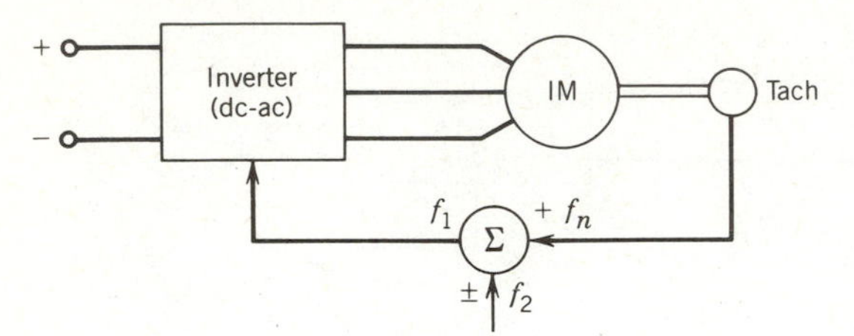

FIGURE 5.35
Constant slip frequency (f_2) operation.

Addition of f_2 to f_n will correspond to motoring action and subtraction of f_2 from f_n will correspond to regenerative braking action of the induction machine. At any speed of the motor the signal f_2 will represent the rotor circuit frequency, that is, the slip frequency.

5.13.5 CLOSED-LOOP CONTROL

Most industrial drive systems operate as closed-loop feedback systems. Figure 5.36 shows a block diagram of a speed control system employing slip frequency regulation and constant-volt/hertz operation. At the first summer junction the difference between the set speed n^* and the actual speed n represents the slip speed n_{sl} and hence the slip frequency. If the slip frequency nears the breakdown frequency, its value is clamped, thereby restricting the operation below the breakdown frequency. At the second summer, the slip frequency is added to the frequency f_n (representing the motor speed) to generate the stator frequency f_1. The function

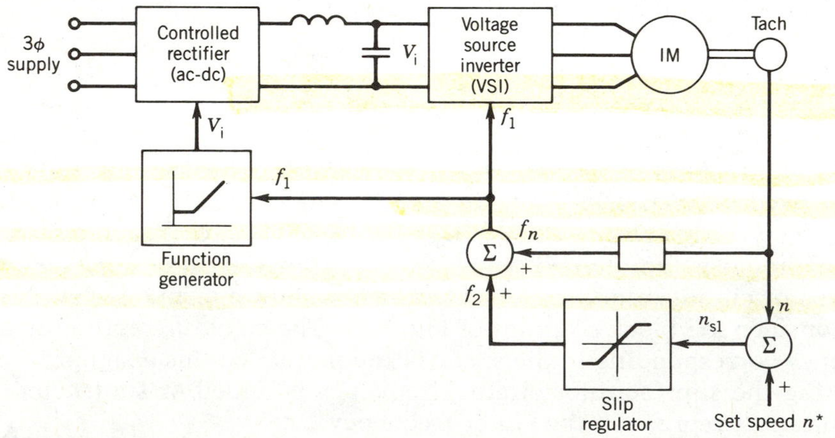

FIGURE 5.36
Speed control system employing slip frequency regulation and constant V/f operation.

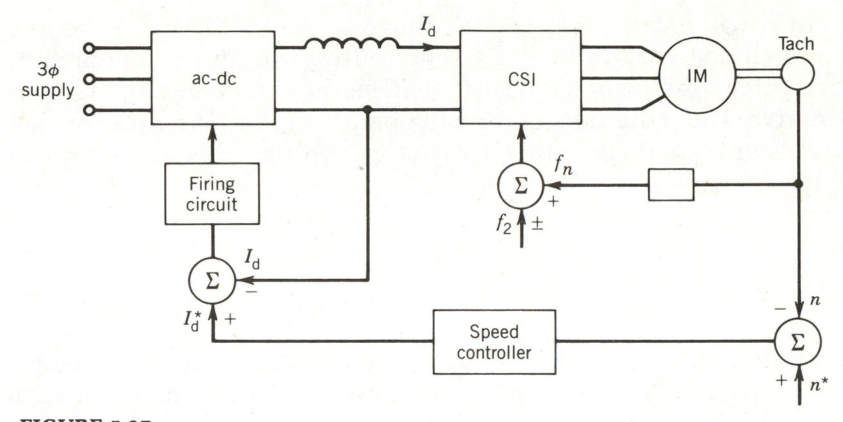

FIGURE 5.37
Speed control system with constant slip frequency (using a current source inverter).

generator provides a signal such that a voltage is obtained from the controlled rectifier for *constant-volt/hertz* operation.

A simple speed control system using a current source inverter is shown in Fig. 5.37. The slip frequency is kept constant and the speed is controlled by controlling the dc link current I_d and hence the magnitude of the motor current.

In traction applications, such as subway cars or other transit vehicles, the torque is controlled directly. A typical control scheme for transit drive systems is shown in Fig. 5.38. As the voltage available for a transit system is a fixed-voltage dc supply, a pulse-width-modulated (PWM) voltage source inverter is considered in which the output voltage can be varied. It can be shown (Example 5.8) that if the slip frequency is kept constant, the torque varies as the square of the stator current. The torque command is fed through a square root function generator to produce the current reference I^*. The signal representing the difference between I^* and the actual

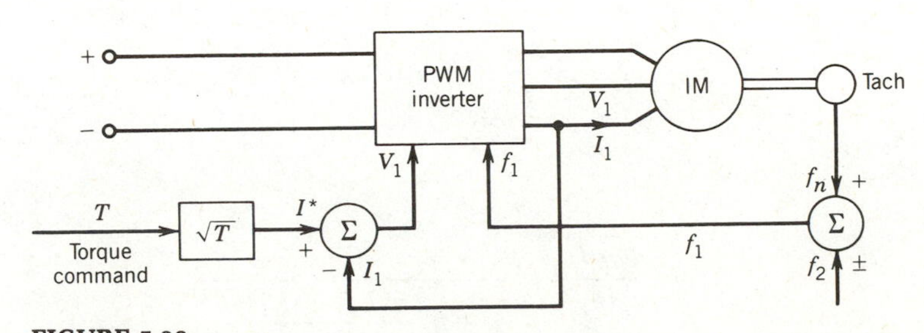

FIGURE 5.38
Typical speed control scheme for transit drive systems.

current I_1 will change the output voltage of the PWM inverter to make I_1 close to the desired value of I^* representing the torque command. For regenerative braking of the transit vehicle, the sign of the slip frequency f_2 is negative. The induction motor will operate in the generating mode ($f_n > f_1$) and feed back the kinetic energy stored in the drive system to the dc supply.

EXAMPLE 5.8

Show that if the slip frequency is kept constant, the torque developed by an induction machine is proportional to the square of the input current.

Solution

For variable-frequency operation, the terminal voltage V_1 is changed with frequency f_1 to maintain machine flux at a desired level. In the low-frequency region, V_1 is low. The voltage drop across R_1 and X_1 may be comparable to V_1. Therefore V_1 cannot be assumed to be equal to E_1. In the equivalent circuit, the shunt branch X_m should not be moved to the machine terminals. Therefore, for variable-frequency operation, the equivalent circuit of Fig. 5.15 is more appropriate to use for prediction of performance. From Fig. 5.15,

$$I_2' = \frac{jX_m}{(R_2'/s) + j(X_2' + X_m)} I_1 \tag{5.75}$$

$$(I_2')^2 = \frac{(X_m)^2}{(R_2'/s)^2 + (X_2' + X_m)^2} I_1^2 \tag{5.76}$$

$$T = \frac{1}{\omega_{syn}} I_2'^2 \frac{R_2}{s} \tag{5.77}$$

where

$$\omega_{syn} = \frac{4\pi}{p} f_1 \quad \text{and} \quad s = \frac{f_2}{f_1}$$

From Eqs. 5.76 and 5.77,

$$T = \frac{4\pi p L_m}{R_2'} I_1^2 \frac{f_2}{1 + \left|\frac{2\pi(L_m + L_2')f_2}{R_2'}\right|^2} \tag{5.78}$$

Equation 5.78 shows that if f_2 remains constant,

$$T \propto I_1^2 \tag{5.79}$$

EXAMPLE 5.9

Show that if the rotor frequency f_2 is kept constant, the torque developed by an induction machine is proportional to the square of the flux in the air gap.

Solution

From the equivalent circuit of Fig. 5.15,

$$I_2' = \frac{E_1}{|(R_2' f_1/f_2)^2 + (2\pi f_1 L_2')^2|^{1/2}} \tag{5.80}$$

From Eqs. 5.77 and 5.80,

$$T = \frac{p}{4\pi R_2'}\left(\frac{E_1}{f_1}\right)^2 \frac{f_2}{1 + (2\pi f_2 L_2'/R_2')^2} \tag{5.81}$$

If f_2 remains constant, from Eqs. 5.81 and 5.72,

$$T \propto (E_1/f_1)^2 \propto \Phi_p^2 \tag{5.82}$$

5.13.6 ROTOR RESISTANCE CONTROL

In Section 5.11.1 it was pointed out that the speed of a wound-rotor induction machine can be controlled by connecting external resistance in the rotor circuit through slip rings, as shown in Fig. 5.39*a*. The torque–speed characteristics for four external resistances are shown in Fig. 5.39*b*. The load T–n characteristic is also shown by the dashed line. By varying the external resistance $0 < R_{ex} < R_{ex4}$, the speed of the load can be controlled in the range $n_1 < n < n_5$. Note that by proper adjustment of the external resistance ($R_{ex} = R_{ex2}$), maximum starting torque can be obtained for the load.

The scheme shown in Fig. 5.39*a* requires a three-phase resistance bank, and for balanced operation all three resistances should be equal for any setting of the resistances. Manual adjustment of the resistances may not be

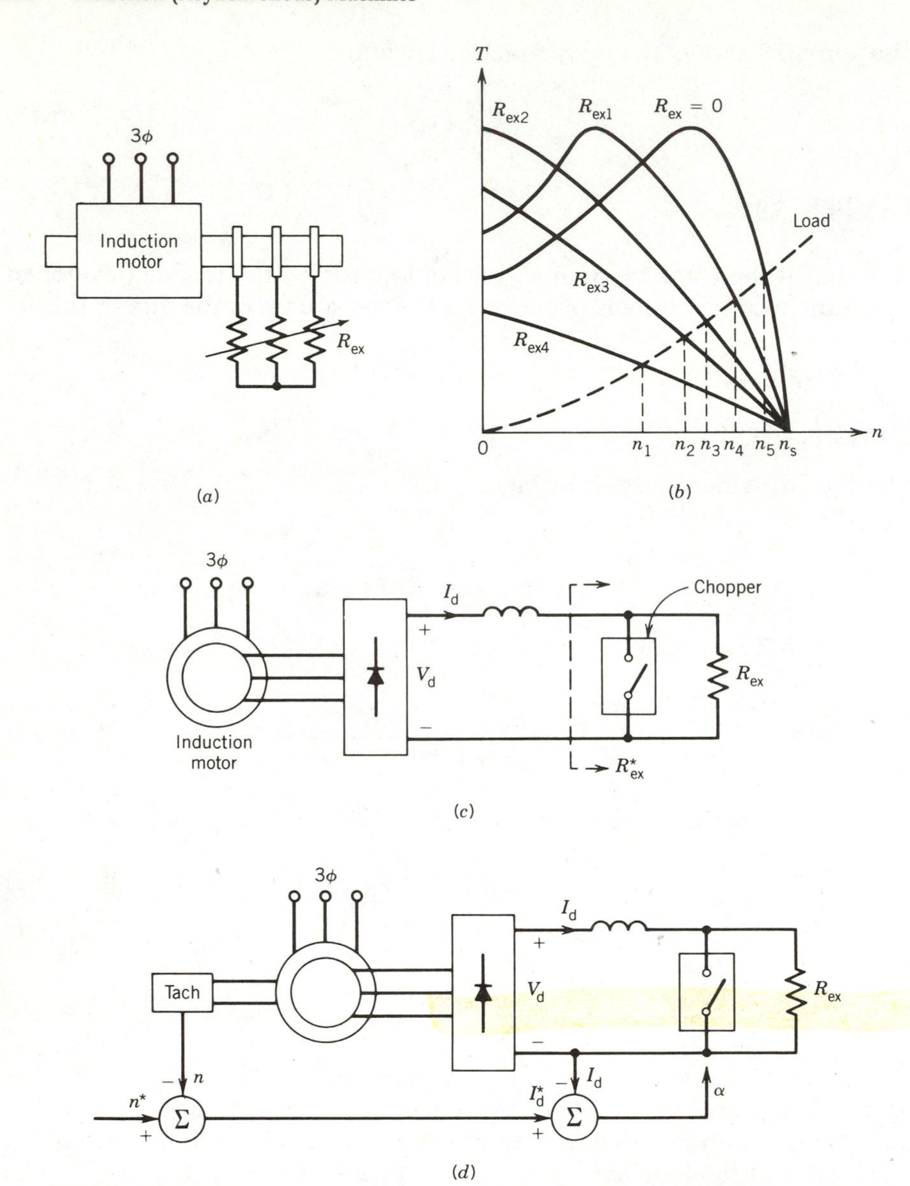

FIGURE 5.39
Speed control with rotor resistance control: open-loop and closed-loop schemes.

satisfactory for some applications, particularly for a closed-loop feedback control system. Solid-state control of the external resistance may provide smoother operation. A block diagram of a solid-state control scheme with open-loop operation is shown in Fig. 5.39*c*. The three-phase rotor power is rectified by a diode bridge. The effective value R_{ex}^* of the external resistance R_{ex} can be changed by varying the on-time (also called the duty ratio α, Chapter 10) of the chopper connected across R_{ex}. It can be shown that[1]

$$R_{ex}^* = (1 - \alpha)R_{ex} \tag{5.83}$$

When $\alpha = 0.0$, that is, when the chopper is off all the time, $R_{ex}^* = R_{ex}$. When $\alpha = 1.0$, that is, the chopper is on all the time, R_{ex} is short-circuited by the chopper and so $R_{ex}^* = 0$. In this case, the rotor circuit resistance consists of the rotor winding resistance only. Therefore, by varying α in the range $1 > \alpha > 0$, the effective resistance is varied in the range $0 < R_{ex}^* < R_{ex}$, and torque–speed characteristics similar to those shown in Fig. 5.39*b* are obtained.

The rectified voltage V_d (Fig. 5.39*c*) depends on the speed and hence the slip of the machine. At standstill, let the induced voltage in the rotor winding be E_2 (Eq. 5.29). From Eq. 5.34 and Eq. 10.9 (for a 3ϕ full-wave diode rectifier with six diodes) the rectified voltage V at slip s is

$$V_d = s|V_d|_{s=1} = s\,\frac{3\sqrt{6}}{\pi}\,E_2 \tag{5.84}$$

The electrical power in the rotor circuit is

$$P_2 = sP_{ag}$$

If the power lost in the rotor winding is neglected, the power P_2 is the dc output power of the rectifier. Hence,

$$sP_{ag} \simeq V_d I_d \tag{5.85}$$

From Eqs. 5.84 and 5.85,

$$sT\omega_{syn} = s\,\frac{3\sqrt{6}}{\pi}\,E_2 I_d$$

$$T \propto I_d \tag{5.86}$$

---

[1] P. C. Sen and K. H. J. Ma, Rotor Chopper Control for Induction Motor Drives: TRC Strategy, *IA-IEEE Transactions*, vol. IA-11, no. 1, pp. 43–49, 1975.

This linear relationship between the developed power and the rectified current is an advantage from the standpoint of closed-loop control of this type of speed drive system. A block diagram for closed-loop operation of the solid-state rotor resistance control system is shown in Fig. 5.39*d*. The actual speed n is compared with the set speed n^*, and the error signal represents the torque command or the current reference I_d^*. This current demand I_d^* is compared with the actual current I_d, and the error signal changes the duty ratio α of the chopper to make current I_d close to the value I_d^*.

The major disadvantage of the rotor resistance control method is that the efficiency is low at reduced speed because of higher slips (Eq. 5.70). However, this control method is often employed because of its simplicity. In applications where low-speed operation is only a small proportion of the work, the low efficiency is acceptable. A typical application of the rotor resistance control method is the hoist drive of a shop crane. This method also can be used in fan or pump drives, where speed variation over a small range near the top speed is required.

5.13.7 ROTOR SLIP ENERGY RECOVERY

In the scheme just discussed, if the slip power lost in the resistance could be returned to the ac source, the overall efficiency of the drive system would be very much increased. A method for recovering the slip power is shown in Fig. 5.40*a*. The rotor power is rectified by the diode bridge. The rectified current is smoothed out by the smoothing choke. The output of the rectifier is then connected to the dc terminals of the inverter, which inverts this dc power to ac power and feeds it back to the ac source. The inverter is a controlled rectifier operated in the inversion mode (see Chapter 10).

At no load the torque required is very small and from Eq. 5.86 $I_d \simeq 0$. From Fig. 5.40*a*,

$$V_d = V_i$$

If the no-load slip is s_0, then from Eqs. 5.84 and 10.10 (Chapter 10),

$$s_0 \frac{3\sqrt{6}}{\pi} E_2 = \frac{-3\sqrt{6}}{\pi} V_1 \cos \alpha$$

or

$$s_0 = -(V_1/E_2) \cos \alpha \tag{5.87}$$

The firing angle α of the inverter will therefore set the no-load speed. If the load is applied, the speed will decrease. The torque–speed characteristics

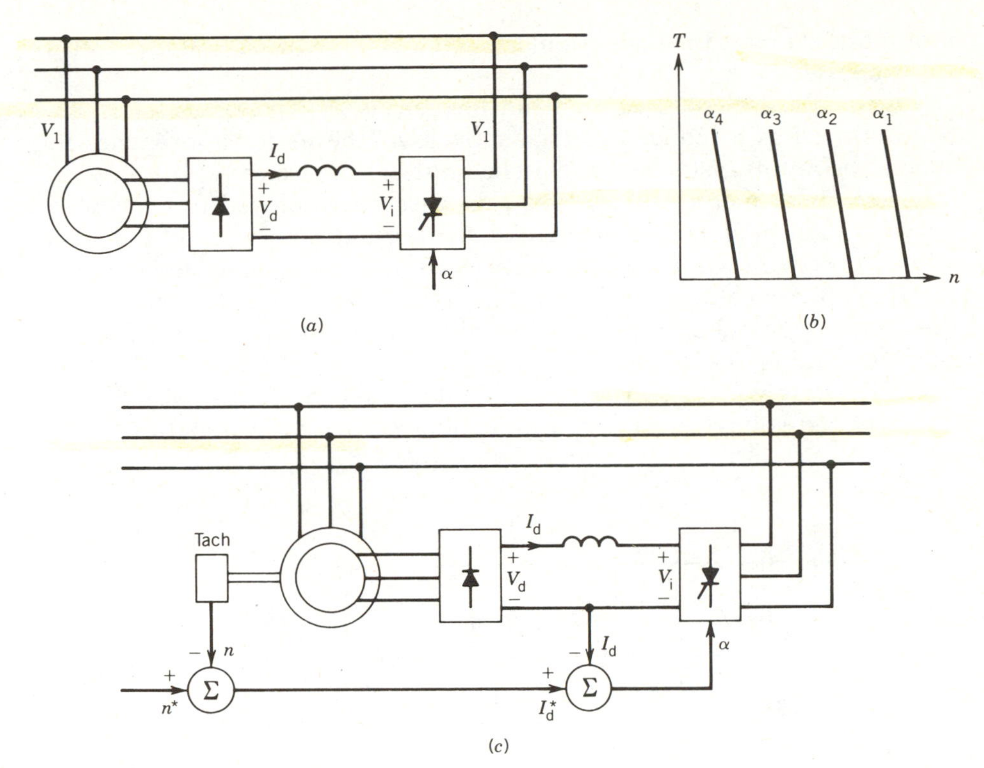

FIGURE 5.40

Slip power recovery. (*a*) Open-loop operation. (*b*) Torque–speed characteristics for different firing angles. (*c*) Closed-loop speed control.

at various firing angles are shown in Fig. 5.40*b*. These characteristics are similar to those of a dc separately excited motor at various armature voltages.

As shown earlier, the torque developed by the machine is proportional to the dc link current I_d (Eq. 5.86). A closed-loop speed control system using the slip power recovery technique is shown in Fig. 5.40*c*.

This method of speed control is useful in large power applications where variation of speed over a wide range involves a large amount of slip power.

5.14 STARTING OF INDUCTION MOTORS

Squirrel-cage induction motors are frequently started by connecting them directly across the supply line. A large starting current of the order of 500 to 800 percent of full-load current may flow in the line. If this causes appreciable voltage drop in the line it may affect other drives connected to the line. Also, if a large current flows for a long time it may overheat the motor and damage the insulation. In such a case, reduced-voltage starting must be used.

A three-phase step-down autotransformer, as shown in Fig. 5.41*a*, may be employed as a reduced-voltage starter. As the motor approaches full speed, the autotransformer is switched out of the circuit.

A star–delta method of starting may also be employed to provide reduced voltage at start. In this method, the normal connection of the stator windings is delta while running. If these windings are connected in star at start, the phase voltage is reduced, resulting in less current at starting. As the motor approaches full speed, the windings will be connected in delta, as shown in Fig. 5.41*b*.

A solid-state voltage controller, as shown in Fig. 5.41*c*, can also be used as a reduced-voltage starter. The controller can provide smooth starting.

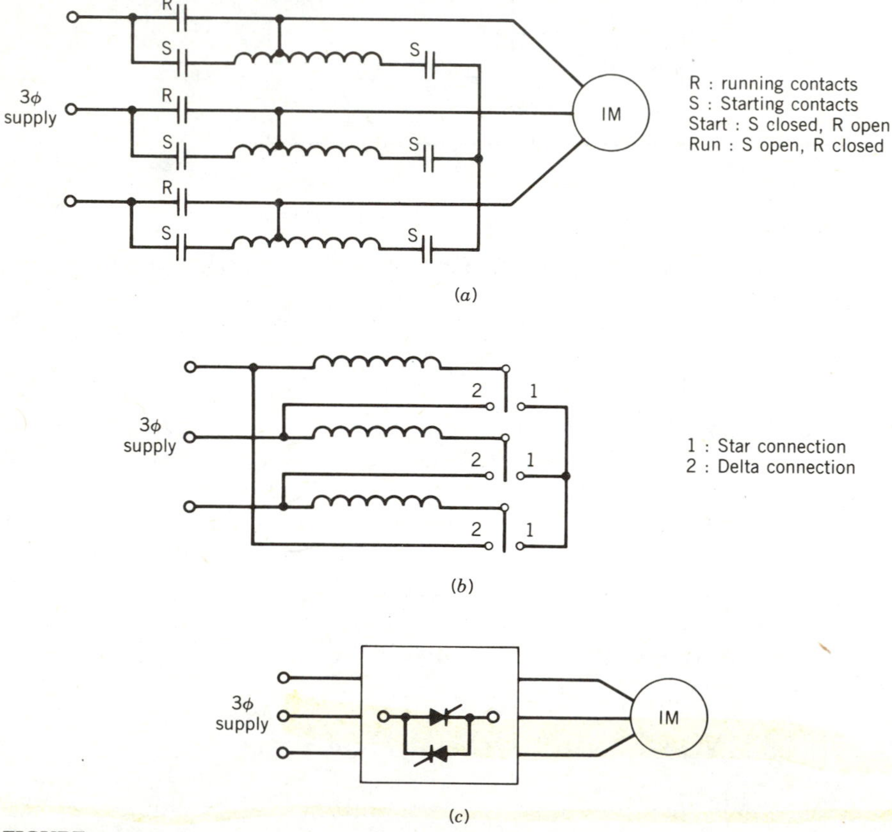

FIGURE 5.41

Starting methods for squirrel-cage induction motors. (a) One-step starting autotransformer. (b) Star–delta starting. (c) Solid-state voltage controller for starting.

This arrangement can also be used to control the speed of the induction motor, as discussed earlier in Section 5.13.

Note that although reduced-voltage starting reduces the starting current, it also results in a decrease in the starting torque, because the torque developed is proportional to the square of the terminal voltage (see Eq. 5.54).

5.15 TIME AND SPACE HARMONICS

Induction machines are often controlled by voltage source inverters or current source inverters, as discussed in Section 5.13.3. The machine currents are therefore nonsinusoidal. They contain fundamental and harmonic components of current. The harmonic currents produce rotating fields in the air gap that rotate at higher speeds than the rotating field produced by the fundamental current. The time harmonic currents and their rotating fields produce parasitic torques in the machine.

The winding of a phase of an induction machine is distributed over a finite number of slots in the machine. As a result, when current flows through the winding the mmf produced is nonsinusoidally distributed in the air gap (see Appendix A). The air gap flux, therefore, consists of fundamental and harmonic components of fluxes. These harmonic fluxes produced by a distributed winding are known as space harmonics. The space harmonics also produce parasitic torques in the machine.

5.15.1 TIME HARMONICS

While considering the effects of time harmonics we shall assume that when current flows through a phase winding it produces sinusoidally distributed mmf in the air gap. In other words, we shall assume a sinusoidally distributed winding (see Appendix A) and no space harmonics.

Let phase currents in the three-phase induction machine be as follows:

$$i_{\mathrm{a}} = \sum_{h=1}^{\infty} I_{h(\max)} \cos h\omega t \tag{5.88}$$

$$i_{\mathrm{b}} = \sum_{h=1}^{\infty} I_{h(\max)} \cos h(\omega t - 120°) \tag{5.89}$$

$$i_{\mathrm{c}} = \sum_{h=1}^{\infty} I_{h(\max)} \cos h(\omega t + 120°) \tag{5.90}$$

where h is the order of harmonics

$I_{h(\max)}$ is the amplitude of the hth-order harmonic current

Assume that turns of a phase winding are sinusoidally distributed. From Fig. 5.6a the mmf along θ due to current in phase a is

$$F_a = Ni_a \cos\theta \tag{5.91}$$

where N is the turns per phase. From Eqs. 5.88 and 5.91

$$F_a(\theta, t) = \sum_{h=1}^{\infty} NI_{h(\max)} \cos h\omega t \cos\theta \tag{5.92}$$

$$= \sum_{h=1}^{\infty} F_{h(\max)} \cos h\omega t \cos\theta \tag{5.93}$$

where

$$F_{h(\max)} = NI_{h(\max)} \tag{5.94}$$

Similarly, contributions from phases b and c are, respectively,

$$F_b(\theta, t) = \sum_{h=1}^{\infty} F_{h(\max)} \cos h(\omega t - 120°) \cos(\theta - 120°) \tag{5.95}$$

$$F_c(\theta, t) = \sum_{h=1}^{\infty} F_{h(\max)} \cos h(\omega t + 120°) \cos(\theta + 120°) \tag{5.96}$$

The resultant mmf along θ is

$$F(\theta, t) = F_a(\theta, t) + F_b(\theta, t) + F_c(\theta, t) \tag{5.97}$$

$$= \sum_{h=1}^{\infty} F_{h(\max)} [\cos(h\,\omega t) \cos\theta + \cos h(\omega t - 120°) \cos(\theta - 120°)$$

$$+ \cos h(\omega t + 120°) \cos(\theta + 120°) \tag{5.98}$$

Fundamental mmf

From Eq. 5.98

$$F_1(\theta, t) = F_{1(\max)}[\cos\omega t \cos\theta + \cos(\omega t - 120°) \cos(\theta - 120°)$$

$$+ \cos(\omega t + 120°) \cos(\theta + 120°)]$$

$$= \tfrac{3}{2} F_{1(\max)} \cos(\theta - \omega t) \tag{5.99}$$

The fundamental mmf is therefore a rotating mmf that rotates in the forward direction (i.e., in the direction of θ) at an angular speed of ω radians per second.

Third Harmonic mmf

From Eq. 5.98

$$\begin{aligned} F_3(\theta, t) &= F_{3(\max)}\{\cos 3\omega t[\cos\theta + \cos(\theta - 120°) + \cos(\theta + 120°)]\} \\ &= F_{3(\max)} \times 0 \\ &= 0 \end{aligned} \tag{5.100}$$

Note that in a three-phase, three-wire system third harmonic current is absent. Therefore, $F_{3(\max)}$ is also zero.

Fifth Harmonic mmf

From Eq. 5.98, it can be shown that

$$F_5(\theta, t) = \tfrac{3}{2} F_{5(\max)} \cos(\theta + 5\omega t) \tag{5.101}$$

The fifth harmonic mmf wave is also a rotating wave that rotates in the opposite direction (with respect to the rotation of the fundamental wave) and at five times the speed of the fundamental wave.

Seventh Harmonic mmf

From Eq. 5.98, it can be shown that

$$F_7(\theta, t) = \tfrac{3}{2} F_{7(\max)} \cos(\theta - 7\omega t) \tag{5.102}$$

The seventh harmonic mmf wave therefore rotates in the same direction as the fundamental wave but at seven times the speed of the fundamental wave.

Other Harmonic mmf Waves

In general, all odd harmonic mmf waves of order $h = 6m \pm 1$, where m is an integer, are present. These are represented by

$$F_h(\theta, t) = \tfrac{3}{2} F_{h(\max)} \cos(\theta \pm h\omega t) \tag{5.103}$$

The hth-order mmf wave rotates at a speed $h\omega$ radians per second. It rotates in the same direction as the fundamental wave if $h = 6m + 1$ and in the opposite direction if $h = 6m - 1$.

TABLE 5.1
Synchronous Speeds for Different Time Harmonic mmf Waves in a 3ϕ, Four-Pole, 60 Hz Induction Machine

Current Harmonic	Synchronous Speed (rpm)
1	1800
3	0
5	$-5 \times 1800 = -9000$
7	$7 \times 1800 = 12{,}600$

Effects on T–n Characteristic

For a four-pole, three-phase, 60 Hz induction machine the speeds of the rotating mmf waves corresponding to various time harmonic currents are shown in Table 5.1. The torque–speed characteristics corresponding to fundamental, fifth, and seventh harmonic currents are shown in Fig. 5.42. In the normal region of operation of the induction motor, the magnitudes of the parasitic torques are very small. Therefore, time harmonics produce no significant effects on the operation of the induction motor. This is further illustrated in Example 5.10.

EXAMPLE 5.10

The speed of a 3ϕ, 5 hp, 208 V, 1740 rpm, 60 Hz, four-pole induction motor is controlled by a current source inverter (Fig. 5.34*a*). The phase current is

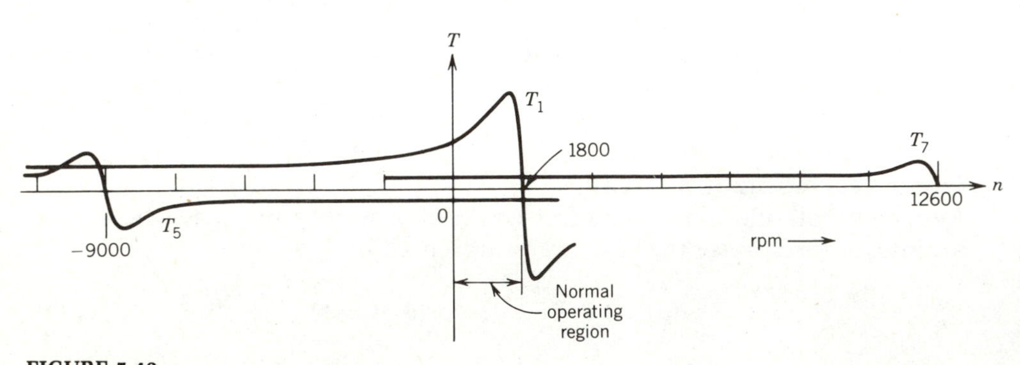

FIGURE 5.42
T–n characteristics for different time harmonic currents.

a quasi-square wave of 120° pulse width as shown in Fig. 5.34*b*. The phase current can be expressed in a Fourier series as follows:

$$i = 1.1I \sin \omega t - 0.22I \sin 5\omega t - 0.16I \sin 7\omega t + \cdots$$

The parameters of the single-phase equivalent circuit of the induction machine at fundamental frequency (60 Hz) are

$$R_1 = 0.5\ \Omega, \qquad X_1 = X_2' = 1.0\ \Omega$$

$$R_2' = 0.5\ \Omega, \qquad X_m = 35\ \Omega$$

At full load the induction machine draws a peak current of 10 amps ($=I$ in Fig. 5.34*b*).

(a) Draw the equivalent circuit for the *h*th harmonic current.

(b) Determine the torques produced by the fundamental current.

(c) Determine the parasitic torques produced by the fifth and seventh harmonic currents.

Solution

(a) The equivalent circuit for the *h*th harmonic current is shown in Fig. E5.10.

(b) $h = 1$:

$$n_s = \frac{120f}{P} = \frac{120 \times 60}{4} = 1800 \text{ rpm}$$

$$s_1 = \frac{1800 - 1740}{1800} = 0.0333$$

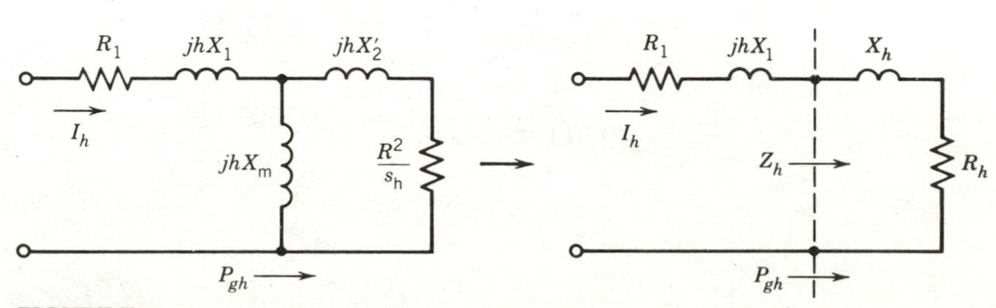

FIGURE E5.10

$$\frac{R_2'}{s_1} = \frac{0.5}{0.0333} = 15\ \Omega$$

$$X_m = 35\ \Omega$$

$$X_2' = 1\ \Omega$$

$$Z_1 = \frac{j35(15 + j1)}{15 + j1 + j35}$$

$$= 12.08 + j6.0$$

$$= R_1 + jX_1$$

$$P_{g1} = 3 \times I_1^2 R_1 = 3 \times \left(\frac{1.1 \times 10}{\sqrt{2}}\right)^2 \times 12.08$$

$$= 2192.5\ \text{W}$$

$$T_1 = \frac{P_{g1}}{\omega_{syn}} = \frac{2192.5}{(1800/60) \times 2\pi} = 11.63\ \text{N} \cdot \text{m}$$

(c) $h = 5$:

$$n_s = -\frac{120 \times 5 \times 60}{4} = -9000\ \text{rpm}$$

$$s_5 = \frac{-9000 - 1740}{-9000} = 1.19$$

$$\frac{R_2'}{s_5} = \frac{0.5}{1.19} = 0.42\ \Omega$$

$$hX_m = 5 \times 35 = 175\ \Omega$$

$$hX_2' = 5 \times 1 = 5\ \Omega$$

$$Z_5 = \frac{j175(0.42 + j5)}{0.42 + j5 + j175}$$

$$= 0.4 + j4.86\ \Omega$$

$$P_{g5} = 3 \times \left(\frac{0.22 \times 10}{\sqrt{2}}\right)^2 \times 0.4 = 2.9\ \text{W}$$

$$T_5 = \frac{2.9}{-(9000/60) \times 2\pi} = -0.00307\ \text{N} \cdot \text{m}$$

$h = 7$

$$n_s = \frac{120 \times 7 \times 60}{4} = 12{,}600\ \text{rpm}$$

$$s_7 = \frac{12{,}600 - 1740}{12{,}600} = 0.862$$

$$\frac{R_2'}{s_7} = \frac{0.5}{0.862} = 0.58\ \Omega$$

$$hX_m = 7 \times 35 = 245\ \Omega$$

$$hX_2' = 7 \times 1 = 7\ \Omega$$

$$Z_h = \frac{j245(0.58 + j7)}{0.58 + j7 + j245}$$

$$= 0.549 + j6.807\ \Omega$$

$$P_{g7} = 3 \times \left(\frac{0.16 \times 10}{\sqrt{2}}\right)^2 \times 0.549$$

$$= 2.107\ \text{W}$$

$$T_7 = \frac{2.107}{(12{,}600/60) \times 2\pi} = 0.0016\ \text{N} \cdot \text{m}$$

Note that the parasitic torques produced by time harmonics are insignificant compared to the fundamental torque.

5.15.2 SPACE HARMONICS

An ideal sinusoidal distribution of mmf is possible only if the machine has an infinitely large number of slots and the turns of a winding are sinusoidally distributed in the slots. This is not practically possible to attain. In a practical machine the winding is distributed in a finite number of slots. As a result, when current flows through a winding the mmf distribution in space has a stairlike waveform as shown in Fig. 5.43 (see also Appendix A).

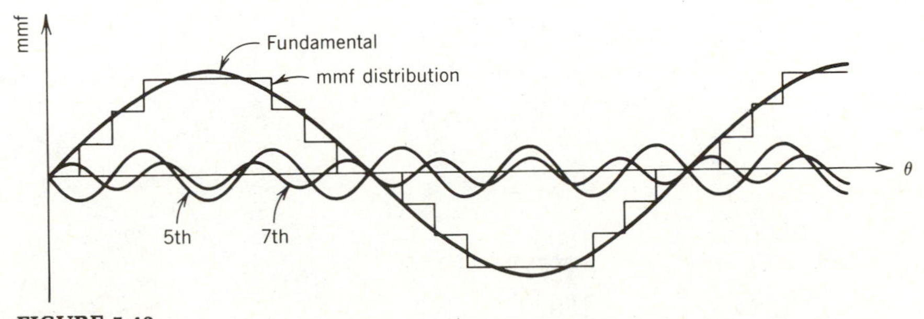

FIGURE 5.43
MMF distribution in the air gap.

The mmf distribution contains a fundamental and a family of space harmonics of order $h = 6m \pm 1$, where m is a positive number. The fundamental and the fifth harmonic component of mmf are also shown in Fig. 5.43.

In a three-phase machine when sinusoidally varying currents flow through the windings the space harmonic waves rotate at $(1/h)$ times the speed of the fundamental wave. The space harmonic waves rotate in the same direction as the fundamental wave if $h = 6m + 1$ and in the opposite direction if $h = 6m - 1$.

A space harmonic wave of order h is equivalent to a machine with hp number of poles. Therefore, the synchronous speed of the hth space harmonic wave is

$$n_{s(h)} = \frac{n_s}{h} = \frac{120f}{hp} \tag{5.104}$$

Effects on *T–n* Characteristics

For a three-phase, four-pole, 60 Hz machine, the synchronous speeds of the space harmonics are shown in Table 5.2. The torque speed characteristics for the fundamental flux and fifth and seventh space harmonic fluxes are shown in Fig. 5.44. The effects of space harmonics are significant. If the effect of seventh harmonic torque is appreciable, the motor may settle to a lower speed—such as the operating point A instead of the desired operating point B. The motor therefore crawls. To reduce the crawling effect, the fifth and seventh space harmonics should be reduced, and this can be done by using a chorded (or short-pitched) winding, as discussed in Appendix A.

TABLE 5.2
Synchronous Speeds for Space Harmonics of a 3ϕ, Four-Pole, 60 Hz Induction Machine

Space Harmonic	Synchronous Speed (rpm)
1	1800
5	$-\dfrac{1800}{5} = -360$
7	$\dfrac{1800}{7} = 257.1$
11	$-\dfrac{1800}{11} = -163.6$
13	$\dfrac{1800}{13} = 138.5$

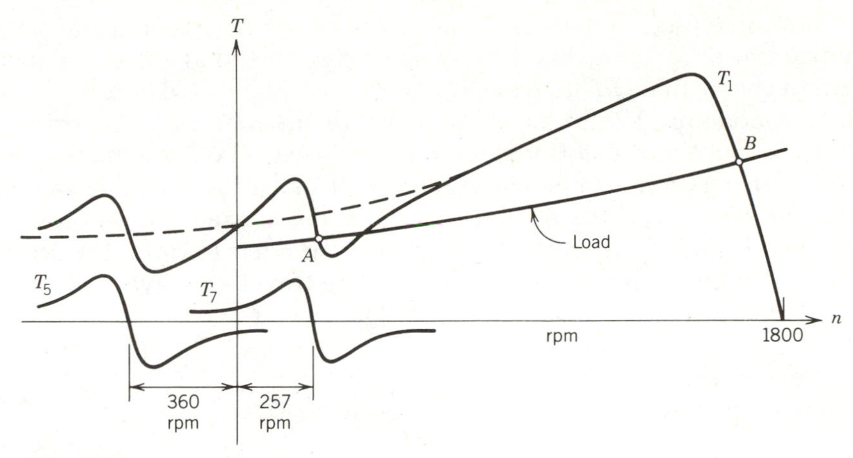

FIGURE 5.44
Parasitic torques due to space harmonics.

5.16 LINEAR INDUCTION MOTOR (LIM)

A linear version of the induction machine can produce linear or translational motion. Consider the cross-sectional view of the rotary induction machine shown in Fig. 5.45*a*. Instead of a squirrel-cage rotor, a cylinder of conductor (usually made of aluminum) enclosing the rotor's ferromagnetic core is considered. If the rotary machine of Fig. 5.45*a* is cut along the line *xy* and unrolled, a linear induction machine, shown in Fig. 5.45*b*, is obtained. Instead of the terms stator and rotor, it is more appropriate to call them primary and secondary members, respectively, of the linear induction machine.

If a three-phase supply is connected to the stator of a rotary induction machine, a flux density wave rotates in the air gap of the machine. Simi-

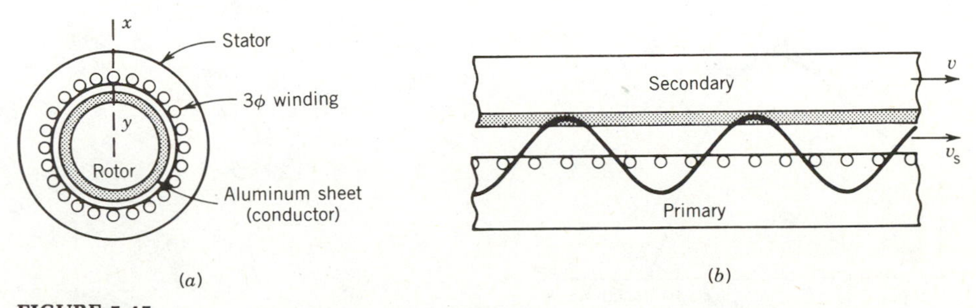

FIGURE 5.45
Induction motors. (*a*) Rotary induction motor. (*b*) Linear induction motor (LIM).

larly, if a three-phase supply is connected to the primary of a linear induction machine a traveling flux density wave is created that travels along the length of the primary. This traveling wave will induce current in the secondary conductor. The induced current will interact with the traveling wave to produce a translational force F (or thrust). If one member is fixed and the other is free to move, the force will make the movable member move. For example, if the primary in Fig. 5.45*b* is fixed, the secondary is free to move, and the traveling wave moves from left to right, the secondary will also move to the right, following the traveling wave.

LIM Performance

The synchronous velocity of the traveling wave is

$$V_s = 2T_p f \text{ m/sec} \tag{5.105}$$

where T_p is the pole pitch and f is the frequency of the supply. Note that the synchronous velocity does not depend on the number of poles. If the velocity of the moving member is V, then the slip is

$$s = \frac{V_s - V}{V_s} \tag{5.106}$$

The per-phase equivalent circuit of the linear induction motor has the same form as that of the rotary induction motor as shown in Fig. 5.15. The thrust–velocity characteristic of the linear induction motor also has the same form as the torque–speed characteristic of a rotary induction motor, as shown in Fig. 5.46. The thrust is given by

$$F = \frac{\text{air gap power, } P_g}{\text{synchronous velocity, } V_s}$$

$$= \frac{3I_2'^2 R_2'/s}{V_s} \text{ newtons} \tag{5.107}$$

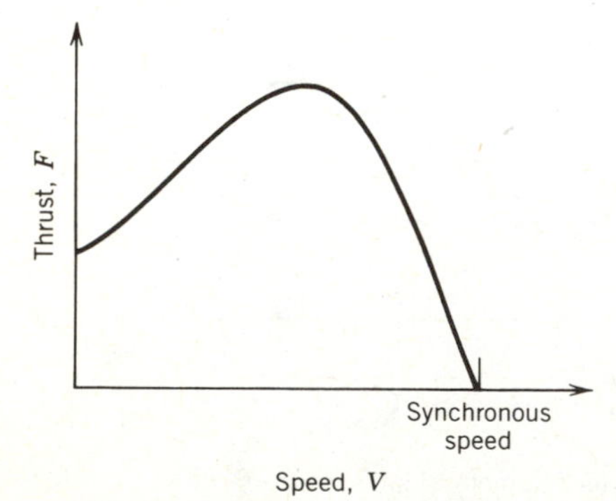

FIGURE 5.46
Thrust–speed characteristic of a LIM.

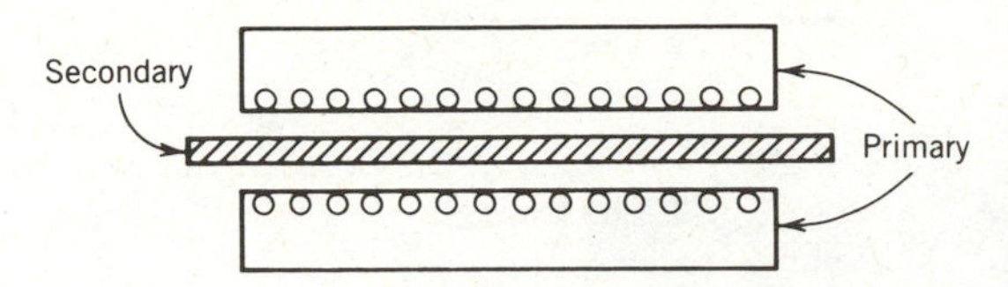

FIGURE 5.47
Double-sided LIM (DLIM).

A linear induction motor requires a large air gap, typically 15–30 mm, whereas the air gap for a rotary induction motor is small, typically 1–1.5 mm. The magnetizing reactance X_m is therefore quite low for the linear induction motor. Consequently, the excitation current is large and the power factor is low. The LIM also operates at a larger slip. The loss in the secondary is therefore high, making the efficiency low.

The LIM shown in Fig. 5.45*b* is called a *single-sided LIM or SLIM*. Another version is used in which primary is on both sides of the secondary, as shown in Fig. 5.47. This is known as a *double-sided LIM or DLIM*.

Applications

An important application of a LIM is in transportation. Usually a short primary is on the vehicle and a long secondary is on the track, as shown in Fig. 5.48. A transportation test vehicle using such a LIM is shown in Fig. 5.49.

A LIM can also be used in other applications, such as materials handling, pumping of liquid metal, sliding-door closers, and curtain pullers.

End Effect

Note that the LIM primary has an entry edge at which a new secondary conductor continuously comes under the influence of the magnetic field. The secondary current at the entry edge will tend to prevent the buildup of air gap flux. As a result, the flux density at the entry edge will be significantly less than the flux density at the center of the LIM. The LIM primary also has an exit edge at which the secondary conductor continuously leaves. A current will persist in the secondary conductor after it has left the exit edge in order to maintain the flux. This current produces extra

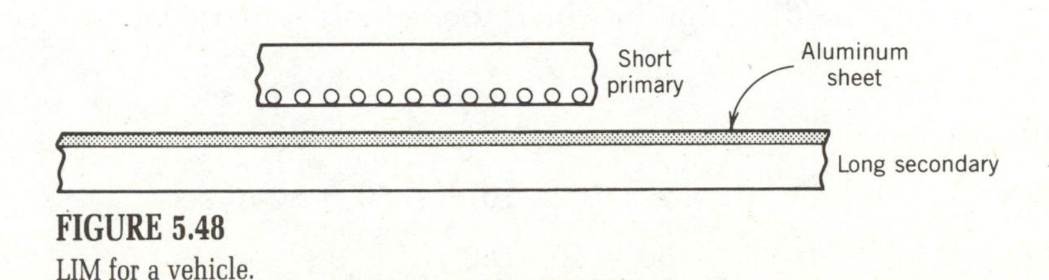

FIGURE 5.48
LIM for a vehicle.

FIGURE 5.49
Transportation test vehicle using LIM. (Courtesy of Urban Transportation Development Corporation, Kingston, Canada.)

resistive loss. These phenomena at the entry edge and the exit edge are known as *end effects* in a linear machine. The end effect reduces the maximum thrust that the motor can produce. Naturally, the end effect is more pronounced at high speed.

EXAMPLE 5.11

The linear induction motor shown in Fig. 5.48 has 98 poles and a pole pitch of 50 cm.

(a) Determine the synchronous speed and the vehicle speed in km/hr if frequency is 50 Hz and slip is 0.25.

(b) If the traveling wave moves left to right with respect to the vehicle, determine the direction in which the vehicle will move.

Solution

(a)
$$V_s = 2 \times 50 \times 10^{-2} \times 50 = 50 \text{ m/sec}$$
$$= \frac{50 \times 60 \times 60}{1000} \text{ km/hr}$$

$$= 180 \text{ km/hr}$$

$$V = (1 - 0.25)\, 180 = 135 \text{ km/hr}$$

(b) Right to left.

PROBLEMS

5.1 A three-phase, 5 hp, 208 V, 60 Hz induction motor runs at 1746 rpm when it delivers rated output power.

(a) Determine the number of poles of the machine.

(b) Determine the slip at full load.

(c) Determine the frequency of the rotor current.

(d) Determine the speed of the rotor field with respect to the

i. Stator.

ii. Stator rotating field.

5.2 A 3ϕ, 10 hp, 208 V, six-pole, 60 Hz, wound-rotor induction machine has a stator-to-rotor turns ratio of 1 : 0.5 and both stator and rotor windings are connected in star.

(a) The stator of the induction machine is connected to a 3ϕ, 208 V, 60 Hz supply and the motor runs at 1140 rpm.

i. Determine the operating slip

ii. Determine the voltage induced in the rotor per phase and frequency of the induced voltage.

iii. Determine the rpm of the rotor field with respect to the rotor and with respect to the stator.

(b) If the stator terminals are shorted and the rotor terminals are connected to a 3ϕ, 208 V, 60 Hz supply and the motor runs at 1164 rpm,

i. Determine the direction of rotation of the motor with respect to that of the rotating field.

ii. Determine the voltage induced in the stator per phase and its frequency.

5.3 The following test results are obtained from a 3ϕ, 100 hp, 460 V, eight-pole, star-connected squirrel-cage induction machine.

No-load test: 460 V, 60 Hz, 40 A, 4.2 kW

Blocked-rotor test: 100 V, 60 Hz, 140 A, 8.0 kW

(a) Determine the parameters of the equivalent circuit.

(b) The motor is connected to a 3ϕ, 460 V, 60 Hz supply and runs at 873 rpm. Determine the input current, input power, air gap power, rotor copper loss, mechanical power developed, output power, and efficiency of the motor.

5.4 A 3ϕ, 100 kVA, 460 V, 60 Hz, eight-pole induction machine has the following equivalent circuit parameters:

$$R_1 = 0.07\ \Omega, \qquad X_1 = 0.2\ \Omega$$
$$R_2' = 0.05\ \Omega, \qquad X_2' = 0.2\ \Omega$$
$$X_m = 6.5\ \Omega$$

(a) Derive the Thevenin equivalent circuit for the induction machine.

(b) If the machine is connected to a 3ϕ, 460 V, 60 Hz supply, determine the starting torque, the maximum torque the machine can develop, and the speed at which the maximum torque is developed.

(c) If the maximum torque is to occur at start, determine the external resistance required in each rotor phase. Assume a turns ratio (stator to rotor) of 1.2.

5.5 A 3ϕ, 25 hp, 460 V, 60 Hz, 1760 rpm, wound-rotor induction motor has the following equivalent circuit parameters:

$$R_1 = 0.25\ \Omega, \qquad X_1 = 1.2\ \Omega$$
$$R_2' = 0.2\ \Omega, \qquad X_2' = 1.1\ \Omega$$
$$X_m = 35\ \Omega$$

The motor is connected to a 3ϕ, 460 V, 60 Hz, supply.

(a) Determine the number of poles of the machine.

(b) Determine the starting torque.

(c) Determine the value of the external resistance required in each phase of the rotor circuit such that the maximum torque occurs at starting.

5.6 A three-phase, 460 V, 60 Hz induction machine produces 100 hp at the shaft at 1746 rpm. Determine the efficiency of the motor if rotational losses are 3500 W and stator copper losses are 3000 W.

5.7 A 440 V, 60 Hz, six-pole, 3ϕ induction motor is taking 50 kVA at 0.8 power factor and is running at a slip of 2.5 percent. The stator copper losses are 0.5 kW and rotational losses are 2.5 kW. Compute

(a) The rotor copper losses.

(b) The shaft hp.

(c) The efficiency.

(d) The shaft torque.

5.8 A 3ϕ wound-rotor induction machine is mechanically coupled to a 3ϕ synchronous machine as shown in Fig. P5.8. The synchronous machine has four poles and the induction machine has six poles. The stators of the two machines are connected to a 3ϕ, 60 Hz power supply. The rotor of the induction machine is connected to a 3ϕ resistive load. Neglect rotational losses and stator resistance losses. The load power is 1 pu. The synchronous machine rotates at the synchronous speed.

(a) The rotor rotates in the direction of the stator rotating field of the induction machine. Determine the speed, frequency of the current in the resistive load, and power taken by the synchronous machine and by the induction machine from the source.

(b) Repeat (a) if the phase sequence of the stator of the induction machine is reversed.

5.9 A 3ϕ induction machine is mechanically coupled to a dc shunt machine. The rating and parameters of the machines are as follows:

Induction machine:

3ϕ, 5 kVA, 208 V, 60 Hz, four-pole, 1746 rpm

$R_1 = 0.25\ \Omega$, $X_1 = 0.55\ \Omega$, $R_2' = 0.35\ \Omega$, $X_2' = 1.1\ \Omega$, $X_m = 38\ \Omega$

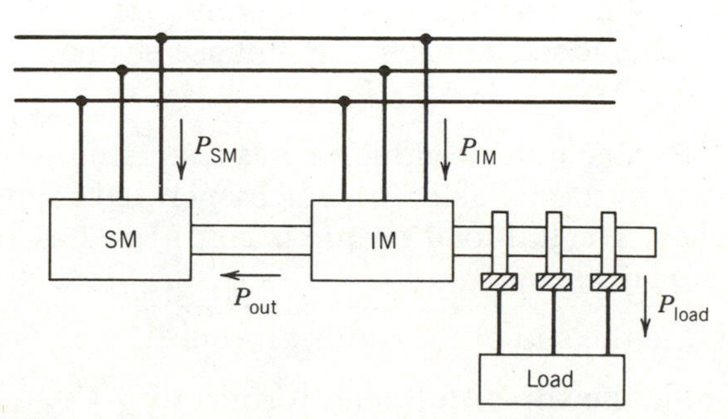

FIGURE P5.8

DC machine:

220 V, 5 kW, 1750 rpm

$R_a = 0.4\ \Omega$, $R_{fw} = 100\ \Omega$, $R_{fc} = 100\ \Omega$

The induction machine is connected to a 3ϕ, 208 V, 60 Hz supply and the dc machine is connected to a 220 V dc supply. The rotational loss of each machine of the M–G set may be considered constant at 225 W.

The system rotates at 1710 rpm in the direction of the rotating field of the induction machine.

(a) Determine the mode of operation of the induction machine.

(b) Determine the current taken by the induction machine.

(c) Determine the real and reactive power at the terminals of the induction machine and indicate their directions.

(d) Determine the copper loss in the rotor circuit.

(e) Determine the armature current (and its direction) of the dc machine.

5.10 The field current of the dc machine in the M–G set of Problem 5.9 is decreased so that the speed of the set increases to 1890 rpm. Repeat parts (a) to (e) of Problem 5.9.

5.11 The M–G set in Problem 5.9 is rotating at 1710 rpm in the direction of the rotating field. The phase sequence of the supply connected to the induction machine is suddenly reversed. Repeat parts (a) to (e) of Problem 5.9.

5.12 A 3ϕ, 460 V, 250 hp, eight-pole wound-rotor induction motor controls the speed of a fan. The torque required for the fan varies as the square of the speed. At full load (250 hp) the motor slip is 0.03 with the slip rings short-circuited. The slip–torque relationship of the motor can be assumed to be linear from no load to full load. The resistance of each rotor phase is 0.02 ohms. Determine the value of resistance to be added to each rotor phase so that the fan runs at 600 rpm.

5.13 A 3ϕ, squirrel-cage induction motor has a starting torque of 1.75 pu and a maximum torque of 2.5 pu when operated from rated voltage and frequency. The full-load torque is considered as 1 pu of torque. Neglect stator resistance.

(a) Determine the slip at maximum torque.

(b) Determine the slip at full-load torque.

(c) Determine the rotor current at starting in per unit—consider the full-load rotor current as 1 pu.

(d) Determine the rotor current at maximum torque in per unit of full-load rotor current.

5.14 A 3ϕ, 460 V, 60 Hz, four-pole wound-rotor induction motor develops full-load torque at a slip of 0.04 when the slip rings are short-circuited. The maximum torque it can develop is 2.5 pu. The stator leakage impedance is negligible. The rotor resistance measured between two slip rings is 0.5 Ω.

(a) Determine the speed of the motor at maximum torque.

(b) Determine the starting torque in per unit. (Full-load torque is one per-unit torque.)

(c) Determine the value of resistance to be added to each phase of the rotor circuit so that maximum torque is developed at the starting condition.

(d) Determine the speed at full-load torque with the added rotor resistance of part (c).

5.15 The approximate per-phase equivalent circuit for a 3ϕ, 60 Hz, 1710 rpm double-cage rotor induction machine is shown in Fig. P5.15. The standstill rotor impedances referred to the stator are as follows:

Outer cage: $4.0 + j1.5\ \Omega$

Inner cage: $0.5 + j4.5\ \Omega$

If stator impedance is neglected,

(a) Determine the ratio of currents in the outer and inner cages for standstill and full-load conditions.

(b) Determine the starting torque of the motor as percent of the full-load torque.

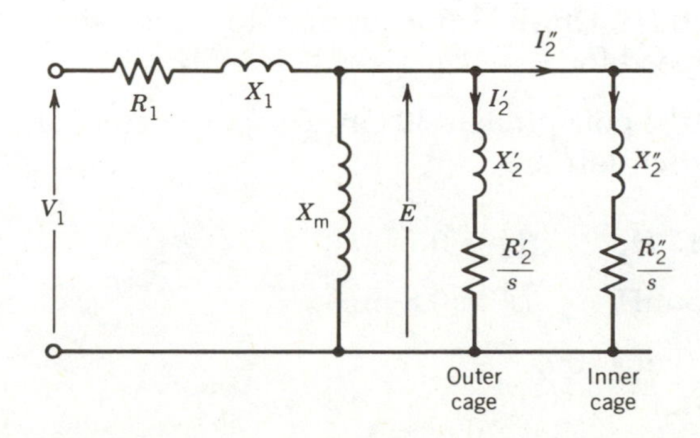

FIGURE P5.15

(c) Determine the ratio of torques due to the outer and inner cages for standstill and full-load conditions.

5.16 A 3ϕ, 460 V, 60 Hz, 1755 rpm, 100 hp, four-pole squirrel-cage induction motor has negligible stator resistance and leakage inductance. The motor is to be operated from a 50 Hz supply.

(a) Determine the supply voltage if the air gap flux is to remain at the same value if it were operated from a 3ϕ, 460 V, 60 Hz supply.

(b) Determine the speed at full-load torque if the motor operates from the 50 Hz supply of part (a).

5.17 A 3ϕ, 460 V, 60 Hz, 50 hp, 1180 rpm induction motor has the following parameters.

$$R_1 = 0.191\ \Omega$$

$$R_2' = 0.0707\ \Omega$$

$$L_1 = 2\text{ mH} \quad \text{(stator leakage inductance)}$$

$$L_2' = 2\text{ mH} \quad \text{(rotor leakage inductance, referred to stator)}$$

$$L_m = 44.8\text{ mH}$$

(a) Determine the values of the rated current and rated torque (use the equivalent circuit of Fig. 5.15).

(b) Use

$$I_{rated} = 1\text{ pu of current}$$

$$T_{rated} = 1\text{ pu of torque}$$

$$V_{rated} = 1\text{ pu of voltage}$$

$$n_{syn} = 1\text{ pu of speed}$$

Plot in per-unit values torque versus speed for $V_1 = 1$ pu and stator frequency $f_1 = 60$ Hz. On the same graph, plot in per-unit values torque versus speed for $I_1 = 1$ pu and $f_1 = 60$ Hz.

5.18 A LIM has seven poles and the pole pitch is 30 cm. The parameters of the single-phase equivalent circuit are

$$R_1 = 0.15\ \Omega, \qquad R_2' = 0.25\ \Omega$$

$$L_1 = 0.5\text{ mH}, \qquad L_2' = 0.8\text{ mH}$$

$$L_m = 5.0\text{ mH}$$

The LIM is connected to a 3ϕ variable-voltage, variable-frequency supply. At 300 V, 50 Hz, the speed of the LIM Is 75 km/hr.

(a) Determine the slip.

(b) Determine the input current, input power, power factor, air gap power, mechanical power developed, power loss in the secondary, and thrust produced.

5.19 A 3ϕ, 460 V, 60 Hz, 1025 rpm squirrel-cage induction motor has the following equivalent circuit parameters:

$$R_1 = 0.06\ \Omega, \qquad X_1 = 0.25\ \Omega$$

$$R_2' = 0.3\ \Omega, \qquad X_2' = 0.35\ \Omega$$

$$X_m = 7.8\ \Omega$$

Neglect the core losses and windage and friction losses. Use the equivalent circuit of Fig. 5.15 for computation. Write a computer program to study the performance characteristics of this machine operating as a motor over the speed range zero to synchronous speed. The program should yield

(a) A computer printout in tabular form showing the variation of torque, input current, input power factor, and efficiency with speed.

(b) A plot of the performance characteristics.

(c) Input current, torque, and output horsepower at the rated speed of 1025 rpm.

CHAPTER SIX

SYNCHRONOUS MACHINES

A synchronous machine rotates at a constant speed in the steady state. Unlike induction machines, the rotating air gap field and the rotor in the synchronous machine rotate at the same speed, called the synchronous speed. Synchronous machines are used primarily as generators of electrical power. In this case they are called *synchronous generators* or *alternators*. They are usually large machines generating electrical power at hydro, nuclear, or thermal power stations. Synchronous generators with power ratings of several hundred MVA (mega-volt-amperes) are quite common in generating stations. It is anticipated that machines of several thousand MVA ratings will be used before the end of the twentieth century. Synchronous generators are the primary energy conversion devices of the world's electric power systems today. In spite of continuing research for more direct energy conversion techniques, it is conceded that synchronous generators will continue to be used well into the next century.

Like most rotating machines, a synchronous machine can also operate as both a generator and a motor. In large sizes (several hundred or thousand kilowatts) synchronous motors are used for pumps in generating stations, and in small sizes (fractional horsepower) they are used in electric clocks, timers, record turntables, and so forth where constant speed is desired. Most industrial drives run at variable speeds. In industry, synchronous motors are used mainly where a constant speed is desired. In industrial drives, therefore, synchronous motors are not as widely used as induction or dc motors. A linear version of the synchronous motor (LSM) is being considered for high-speed transportation systems of the future.

An important feature of a synchronous motor is that it can draw either lagging or leading reactive current from the ac supply system. A synchronous machine is a doubly excited machine. Its rotor poles are excited by a dc current and its stator windings are connected to the ac supply (Fig. 6.1). The air gap flux is therefore the resultant of the fluxes due to both rotor current and stator current. In induction

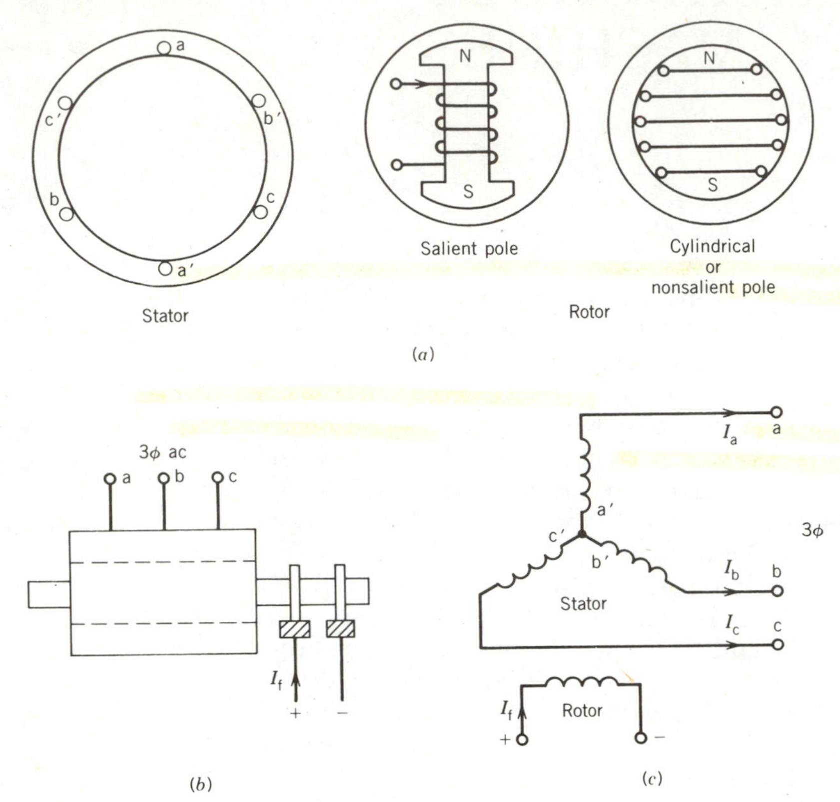

FIGURE 6.1
Basic structure of the three-phase synchronous machine.

machines, the only source of excitation is the stator current, because rotor currents are induced currents. Therefore, induction motors always operate at a lagging power factor, because lagging reactive current is required to establish flux in the machine. On the other hand, in a synchronous motor, if the rotor field winding provides just the necessary excitation, the stator will draw no reactive current; that is, the motor will operate at a unity power factor. If the rotor excitation current is decreased, lagging reactive current will be drawn from the ac source to aid magnetization by the rotor field current and the machine will operate at a lagging power factor. If the rotor field current is increased, leading reactive current will be drawn from the ac source to oppose magnetization by the rotor field current and the machine will operate at a leading power factor. Thus, by changing the field current, the power factor of the synchronous motor can be controlled. If the motor is not loaded but is simply floating on the ac supply system, it will thus behave as a variable inductor or capacitor as its rotor field current is changed. A synchronous machine with no load is called a *synchronous condenser*. It may be used in power transmission systems to regulate line voltage. In industry, synchronous motors are sometimes used with other induction motors and operated in an overexcited mode so that they draw leading current to compensate the lagging current drawn by the induction motors, thereby improving the overall plant power factor. Example 6.1 illustrates the use of synchronous motors for power factor improvement. The power factor characteristics of synchronous motors will be further discussed in a later section.

EXAMPLE 6.1

In a factory a 3ϕ, 4 kV, 400 kVA synchronous machine is installed along with other induction motors. The following are the loads on the machines:

Induction motors: 500 kVA at 0.8 PF lagging.

Synchronous motor: 300 kVA at 1.0 PF.

(a) Compute the overall power factor of the factory loads.

(b) To improve the factory power factor, the synchronous machine is overexcited (to draw leading current) without any change in its load. Without overloading the motor, to what extent can the factory power factor be improved? Find the current and power factor of the synchronous motor for this condition.

Solution

(a) Induction motors:

$$\text{Power} = 500 \times 0.8 = 400 \text{ kW}$$

$$\text{Reactive power} = 500 \times 0.6 = 300 \text{ kVAR}$$

Synchronous motor:

$$\text{Power} = 300 \text{ kW}$$

$$\text{Reactive power} = 0.0$$

Factory:

$$\text{Power} = 700 \text{ kW}$$

$$\text{Reactive power} = 300 \text{ kVAR}$$

$$\text{Complex power} = \sqrt{700^2 + 300^2} = 762 \text{ kVA}$$

$$\text{Power factor} = \frac{700}{762} = 0.92 \text{ lagging}$$

(b) The maximum leading kVAR that the synchronous motor can draw without exceeding its rating is

$$\sqrt{400^2 - 300^2} = 264.58 \text{ kVAR}$$

$$\text{Factory kVAR} = j300 - j264.48$$

$$= j35.42 \quad \text{(i.e., lagging)}$$

$$\text{New factory kVA} = \sqrt{700^2 + 35.42^2}$$

$$= 700.9 \text{ kVA}$$

$$\text{Improved factory power factor} = \frac{700}{700.9}$$

$$= 0.996$$

Synchronous motor current:

$$I_{SM} = \frac{400 \text{ kVA}}{\sqrt{3} \times 4 \text{ kV}} = 57.74 \text{ A}$$

Synchronous motor power factor:

$$PF_{SM} = \frac{300 \text{ kW}}{400 \text{ kVA}} = 0.75 \text{ lead}$$

6.1 CONSTRUCTION OF THREE-PHASE SYNCHRONOUS MACHINES

The stator of the three-phase synchronous machine has a three-phase distributed winding similar to that of the three-phase induction machine. Unlike the dc machine, the stator winding, which is connected to the ac supply system, is sometimes called the *armature* winding. It is designed for high voltage and current.

The rotor has a winding called the *field* winding, which carries direct current. The field winding on the rotating structure is normally fed from an external dc source through slip rings and brushes. The basic structure of the synchronous machine is illustrated in Fig. 6.1.

Synchronous machines can be broadly divided into two groups as follows:

1. High-speed machines with cylindrical (or non-salient pole) rotors.
2. Low-speed machines with salient pole rotors.

The cylindrical or non-salient pole rotor has one distributed winding and an essentially uniform air gap. These motors are used in large generators (several hundred megawatts) with two or sometimes four poles and are usually driven by steam turbines. The rotors are long and have a small diameter, as shown in Fig. 6.2. On the other hand, salient pole rotors have concentrated windings on the poles and a nonuniform air gap. Salient pole generators have a large number of poles, sometimes as many as 50, and operate at lower speeds. The synchronous generators in hydroelectric power stations are of the salient pole type and are driven by water turbines. These generators are rated for tens or hundreds of megawatts. The rotors are shorter but have a large diameter as shown in Fig. 6.3. Smaller salient pole synchronous machines in the range of 50 kW to 5 MW are also used. Such synchronous generators are used independently as emergency power supplies. Salient pole synchronous motors are used to drive pumps, cement mixers, and some other industrial drives.

In the following sections the steady-state performance of the cylindrical rotor synchronous machine will be studied first. Then the effects of saliency in the rotor poles will be considered.

FIGURE 6.2
High-speed cylindrical-rotor synchronous generator. (Courtesy of General Electric Canada Inc.)

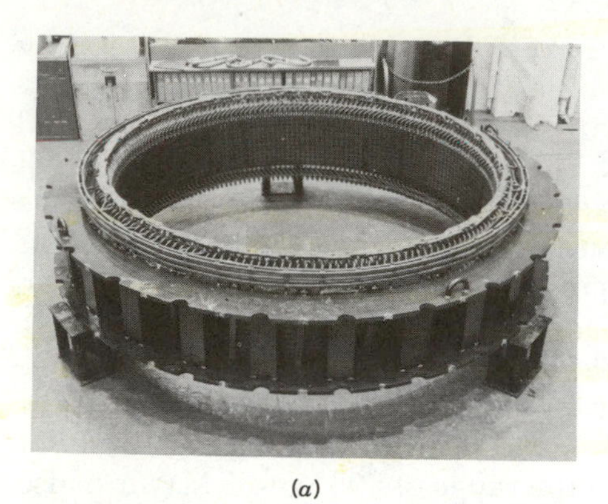

(a) (b)

FIGURE 6.3
Low-speed salient pole synchronous generator. (*a*) Stator. (*b*) Rotor. (Courtesy of General Electric Canada Inc.)

6.2 SYNCHRONOUS GENERATORS

Refer to Fig. 6.4*a* and assume that when the field current I_f flows through the rotor field winding, it establishes a sinusoidally distributed flux in the air gap. If the rotor is now rotated by the prime mover (which can be a turbine or diesel engine or dc motor or induction motor), a revolving field is produced in the air gap. This field is called the excitation field, because it is produced by the excitation current I_f. The rotating flux so produced will change the flux linkage of the armature windings aa′, bb′, and cc′ and will induce voltages in these stator windings. These induced voltages, shown in Fig. 6.4*b*, have the same magnitudes but are phase-shifted by 120

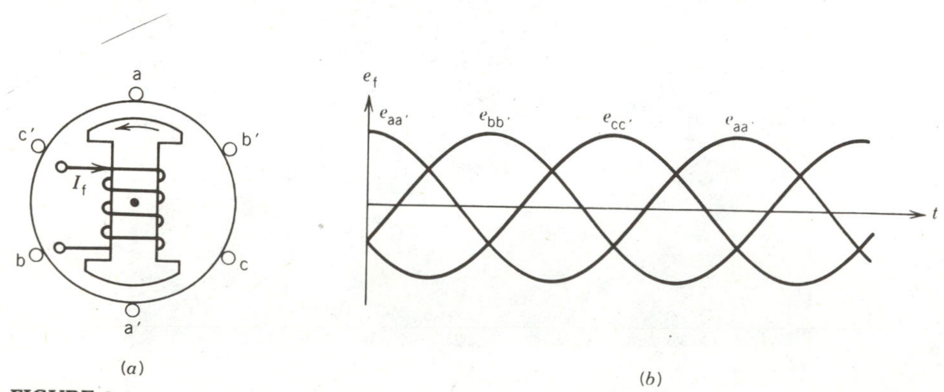

FIGURE 6.4
Excitation voltage in synchronous machines.

electrical degrees. They are called *excitation voltages* E_f. The rotor speed and frequency of the induced voltage are related by

$$n = \frac{120f}{p} \tag{6.1}$$

or

$$f = \frac{np}{120} \tag{6.2}$$

where n is the rotor speed in rpm
p is the number of poles

The excitation voltage in rms from Eq. 5.27 is

$$E_f = 4.44 f \Phi_f N K_w \tag{6.3}$$

where Φ_f is the flux per pole due to the excitation current I_f
N is the number of turns in each phase
K_w is the winding factor

From Eqs. 6.2 and 6.3

$$E_f \propto n\Phi_f \tag{6.4}$$

The excitation voltage is proportional to the machine speed and excitation flux, and the latter in turn depends on the excitation current I_f. The variation of the excitation voltage with the field current is shown in Fig. 6.5. The induced voltage at $I_f = 0$ is due to the residual magnetism. Initially the voltage rises linearly with the field current, but as the field

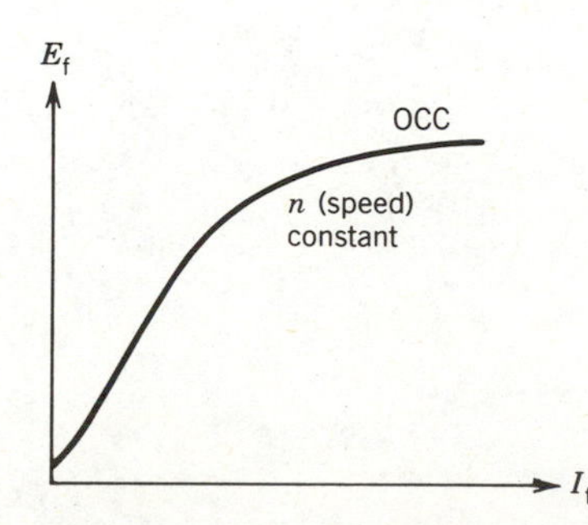

FIGURE 6.5
Open circuit characteristic (OCC) or magnetization characteristic of a synchronous machine.

current is further increased, the flux Φ_f does not increase linearly with I_f because of saturation of the magnetic circuit, and therefore E_f levels off. If the machine terminals are kept open, the excitation voltage is the same as the terminal voltage and can be measured using a voltmeter. The curve shown in Fig. 6.5 is known as the *open-circuit characteristic* (*OCC*) or *magnetization characteristic* of the synchronous machine.

If the stator terminals of the machine (Fig. 6.1*c*) are connected to a 3ϕ load, stator current I_a will flow. The frequency of I_a will be the same as that of the excitation voltage E_f. The stator currents flowing in the 3ϕ windings will also establish a rotating field in the air gap. The net air gap flux is the resultant of the fluxes produced by rotor current I_f and stator current I_a.

Let Φ_f be the flux due to I_f and Φ_a be the flux due to I_a, known as the *armature reaction* flux. Then,

$$\Phi_r = \Phi_f + \Phi_a = \text{resultant air gap flux, assuming no saturation}$$

It may be noted that the resultant and the component fluxes rotate in the air gap at the same speed, governed by Eq. 6.1. The space phasor diagram for these fluxes is shown in Fig. 6.6. The rotor field mmf F_f (due to I_f) and the flux Φ_f produced by the mmf F_f are represented along the same line. The induced voltage E_f lags the flux Φ_f by 90°. Assume that the stator current I_a lags E_f by an angle θ. The mmf F_a (due to the current I_a) and the flux Φ_a produced by the mmf F_a are along the same axis as the current I_a. The resultant mmf F_r is the vector sum of the mmf's F_f and F_a. Assuming no saturation, the resultant flux Φ_r is also the vector sum of the fluxes Φ_f and Φ_a. The space phasor relationship of mmf's and fluxes will be discussed further in later sections.

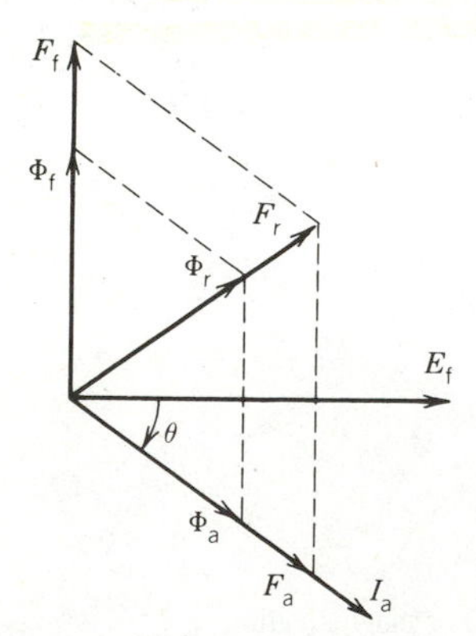

FIGURE 6.6
Space phasor diagram.

6.2.1 THE INFINITE BUS

Synchronous generators are rarely used to supply individual loads. These generators, in general, are connected to a power supply system known as an *infinite bus* or *grid*. Because a large number of synchronous generators of large sizes are connected together, the voltage and frequency of the infinite bus hardly change. Loads are tapped from the infinite bus at various load centers. A typical infinite bus or grid system is shown in Fig. 6.7. Transmission of power is normally at higher voltage levels (in hundreds of kilovolts) to achieve higher efficiency of power transmission. However, generation of electrical energy by the synchronous generators or alternators is at relatively lower voltage levels (20–30 kV). A transformer is used to step up the alternator voltage to the infinite bus voltage. At the load centers, the infinite bus (or grid) voltage is stepped down through several stages to bring the voltage down to the domestic voltage level (115/230 V) or industrial voltage levels such as 4.16 kV, 600 V, or 480 V.

In a power plant the synchronous generators are connected to or disconnected from the infinite bus, depending on the power demand on the grid system. The operation of connecting a synchronous generator to the infi-

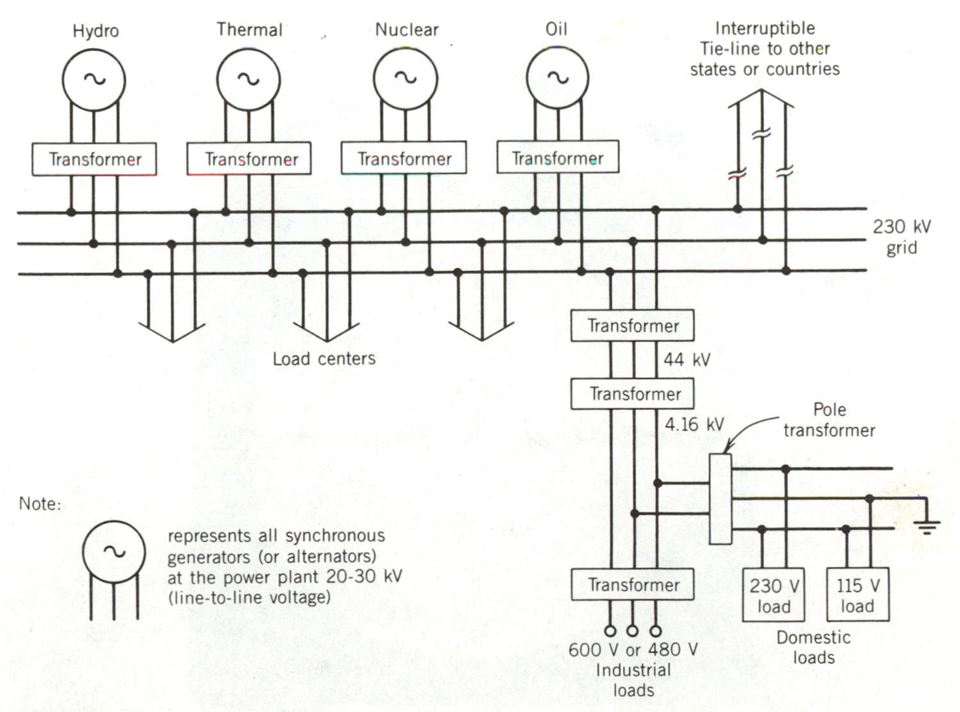

FIGURE 6.7
Infinite bus (or grid) system.

nite bus is known as *paralleling with the infinite bus*. Before the alternator can be connected to the infinite bus, the incoming alternator and the infinite bus must have the same

1. Voltage
2. Frequency
3. Phase sequence
4. Phase

In the power plant the satisfaction of these conditions is checked by an instrument known as a *synchroscope*, shown in Fig. 6.8. The position of the indicator indicates the phase difference between the voltages of the incoming machine and the infinite bus. The direction of motion of the indicator shows whether the incoming machine is running too fast or too slow, that is, whether the frequency of the incoming machine is higher or lower than that of the infinite bus. The phase sequence is predetermined because if phase sequence is not correct it will produce a disastrous situation. When the indicator moves very slowly (i.e., frequencies almost the same) and

FIGURE 6.8
Synchroscope.

passes through the zero phase point (vertical up position) the circuit breaker is closed and the alternator is connected to the infinite bus.

A set of *synchronizing lamps* can be used to check that the conditions for paralleling the incoming machine with the infinite bus are satisfied. In a laboratory such a set of lamps can be used to demonstrate what happens if the conditions are not satisfied. Figure 6.9 shows the schematic of the laboratory setup for this purpose. The prime mover can be a dc motor or an induction motor. It can be adjusted to a speed such that the frequency of the synchronous machine is the same as that of the infinite bus. For example, if the synchronous machine has four poles, the prime mover can be adjusted for 1800 rpm so that the frequency is 60 cycles—the same as that of the infinite bus. The field current I_f can then be adjusted so that the two voltmeters (V_1 and V_2) read the same. If the phase sequence is correct all the lamps will have the same brightness, and if the frequencies are not exactly the same the lamps will brighten and darken in step.

Let us examine what we expect to observe in the lamps if the conditions are not satisfied. The phenomena can be explained by drawing phasor diagrams for the voltages of the incoming machine and the infinite bus. Let

E_A, E_B, E_C	represent the phasor voltages of the infinite bus.
E_a, E_b, E_c	represent the phasor voltages of the incoming machine.
E_{Aa}, E_{Bb}, E_{Cc}	represent the phasor voltages of the synchronizing lamps. The magnitude of these will represent the brightness of the corresponding lamps.

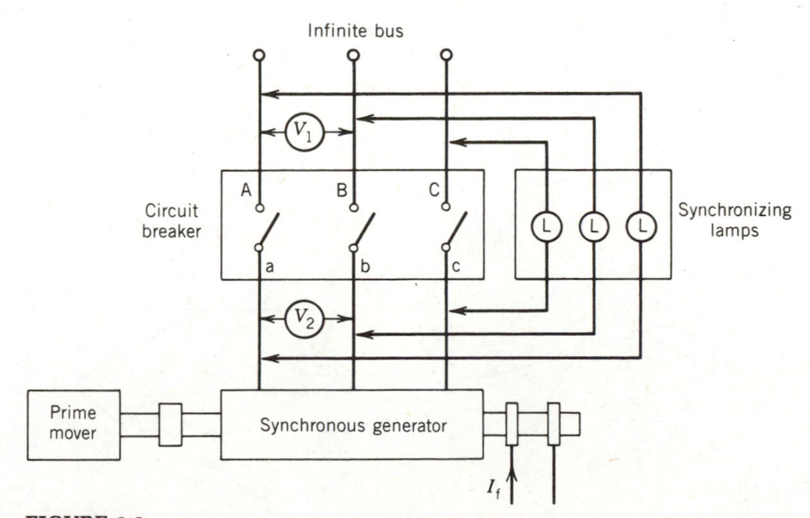

FIGURE 6.9
Schematic diagram for paralleling a synchronous generator with the infinite bus using synchronizing lamps.

1. *Voltages are not the same, but frequency and phase sequence are the same.*

Referring to Fig. 6.10*a*, one sees that the two sets of phasor voltages (E_A, E_B, E_C and E_a, E_b, E_c) rotate at the same speed. The lamp voltages E_{Aa}, E_{Bb}, and E_{Cc} have equal magnitudes and therefore all the three lamps will glow with the same intensity. To make the voltages equal, the field current I_f must be adjusted.

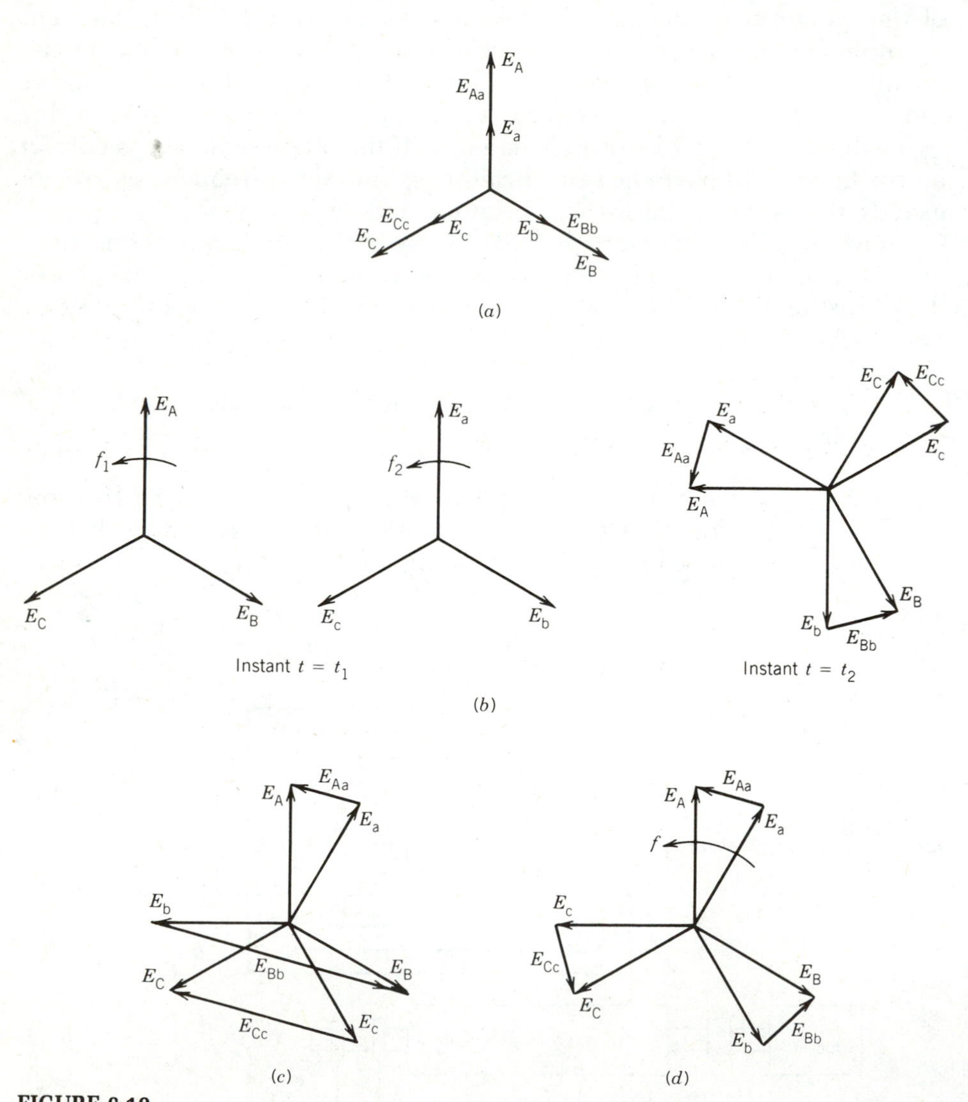

FIGURE 6.10
Phasor voltages of the incoming machine and infinite bus.

2. *Frequencies are not the same, but voltages and phase sequences are the same.*

The two sets of phasor voltages rotate at different speeds, depending on the frequencies. Assume that the phase voltages are in phase at an instant $t = t_1$ (Fig. 6.10*b*). At this instant, the voltages across the lamps are zero and therefore they are all dark. If $f_1 > f_2$ at a later instant $t = t_2$, phasors E_A, E_B, and E_C will move ahead of phasors E_a, E_b, and E_c. Equal voltages will appear across the three lamps and they will glow with the same intensity. It is therefore evident that if the frequencies are different, the lamps will darken and brighten in step.

To make the frequencies the same, the speed has to be adjusted until the lamps brighten and darken very slowly in step. It may be noted that as the speed of the incoming machine is adjusted, its voltages will change. Therefore, simultaneous adjustment of the field current I_f will also be necessary to keep the voltages the same.

3. *Phase sequences are not the same, but voltages and frequencies are the same.*

Let the phase sequence of the voltages of the infinite bus be E_A, E_B, E_C and of the incoming bus be E_a, E_c, E_b as shown in Fig. 6.10*c*. The voltages across the lamps are of different magnitudes and therefore the lamps will glow with different intensities. If the frequencies are slightly different, one set of phasor voltages will pass the other set of phasor voltages and the lamps will darken and brighten out of step.

To make the phase sequence the same, interchange connections to two terminals; for instance, connect a to B and b to A (Fig. 6.9).

4. *Phase is not the same, but voltage, frequency, and phase sequence are the same.*

The two sets of phasor voltages will maintain a steady phase difference (as shown in Fig. 6.10*d*) and the lamps will glow with the same intensity. To make the phase the same or the phase difference zero, the frequency of the incoming machine is slightly altered. At zero phase difference all the lamps will be dark, and if the circuit breaker is closed the incoming machine will be connected to the infinite bus. Once the synchronous machine is connected to the infinite bus, its speed cannot be changed further. However, the real power transfer from the machine to the infinite bus can be controlled by adjusting the prime mover power. The reactive power (and hence the machine power factor) can be controlled by adjusting the field current. Real and reactive power control will be discussed in detail in later sections.

6.3 SYNCHRONOUS MOTORS

When a synchronous machine is used as a motor, one should be able to connect it directly to the power supply like other motors, such as dc motors or induction motors. However, a synchronous motor is not self-starting. If the rotor field poles are excited by the field current and the stator terminals are connected to the ac supply, the motor will not start; instead, it vibrates. This can be explained as follows.

Let us consider a two-pole synchronous machine. If it is connected to a 3ϕ, 60 Hz ac supply, stator currents will produce a rotating field that will rotate at 3600 rpm in the air gap. Let us represent this rotating field by two stator poles rotating at 3600 rpm, as shown in Fig. 6.11*a*. At start ($t = 0$), let the rotor poles be at the position shown in Fig. 6.11*a*. The rotor will therefore experience a clockwise torque, making it rotate in the direction of the stator rotating poles. At $t = t_1$, let the stator poles move by half a revolution, shown in Fig. 6.11*b*. The rotor poles have hardly moved, because of the high inertia of the rotor. Therefore, at this instant the rotor experiences a counterclockwise torque tending to make it rotate in the direction opposite to that of the stator poles. The net torque on the rotor in one revolution will be zero, and therefore the motor will not develop any starting torque. The stator field is rotating so fast that the rotor poles cannot catch up or lock onto it. The motor will not speed up but will vibrate.

Two methods are normally used to start a synchronous motor: (a) use a variable-frequency supply or (b) start the machine as an induction motor. These methods will now be described.

Start with Variable-Frequency Supply

By using a frequency converter, a synchronous motor can be brought from standstill to its desired speed. The arrangement is shown schematically in

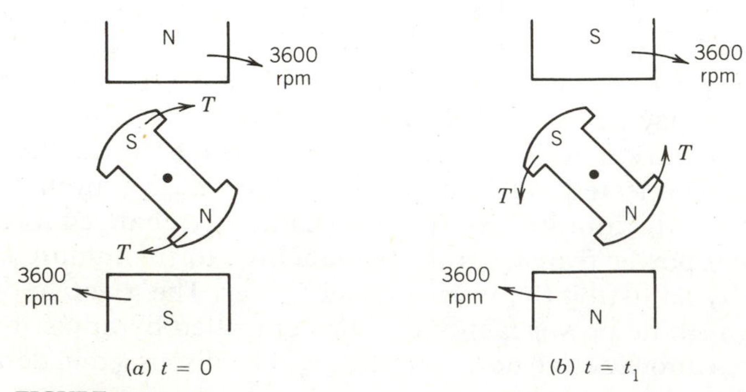

FIGURE 6.11
Torque on rotor at start.

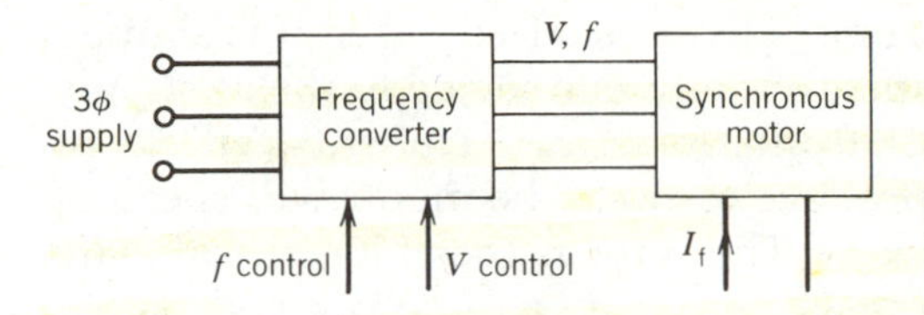

FIGURE 6.12
Starting of a synchronous motor using a variable-frequency supply.

Fig. 6.12. The motor is started with a low-frequency supply. This will make the stator field rotate slowly so that the rotor poles can follow the stator poles. Afterward, the frequency is gradually increased and the motor brought to its desired speed.

The frequency converter is a costly power conditioning unit and therefore this method is expensive. However, if the synchronous motor has to run at variable speeds, this method may be used.

Start as an Induction Motor

If the frequency converter is not available, or if the synchronous motor does not have to run at various speeds, it can be started as an induction motor. For this purpose an additional winding, which resembles the cage of an induction motor, is mounted on the rotor. This cage-type winding is known as a *damper* or *amortisseur winding* and is shown in Fig. 6.13.

FIGURE 6.13
Cage-type damper (or amortisseur) winding in a synchronous machine. (Courtesy of General Electric Canada Inc.)

To start the motor the field winding is left unexcited; often it is shunted by a resistance. If the motor terminals are now connected to the ac supply, the motor will start as an induction motor because currents will be induced in the damper winding to produce torque. The motor will speed up and will approach synchronous speed. The rotor is then closely following the stator field poles, which are rotating at the synchronous speed. Now if the rotor poles are excited by a field current from a dc source, the rotor poles, closely following the stator poles, will be locked to them. The rotor will then run at synchronous speed.

If the machine runs at synchronous speed, no current will be induced in the damper winding. The damper winding is therefore operative for starting. Note that if the rotor speed is different from the synchronous speed because of sudden load change or other transients, currents will be induced in the damper winding to produce a torque to restore the synchronous speed. The presence of this restorative torque is the reason for the name "damper" winding. Also note that a damper winding is not required to start a synchronous generator and parallel it with the infinite bus. However, both synchronous generators and motors have damper windings to damp out transient oscillations.

6.4 EQUIVALENT CIRCUIT MODEL

In the preceding sections the qualitative behavior of the synchronous machine as both a generator and a motor has been discussed to provide a "feel" for the machine behavior. We can now develop an equivalent circuit model that can be used to study the performance characteristics with sufficient accuracy. Since the steady-state behavior will be studied, the circuit time constants of the field and damper windings need not be considered. The equivalent circuit will be derived on a per-phase basis.

The current I_f in the field winding produces a flux Φ_f in the air gap. The current I_a in the stator winding produces flux Φ_a. Part of it, Φ_{al}, known as the *leakage flux,* links with the stator winding only and does not link with the field winding. A major part, Φ_{ar}, known as the *armature reaction flux,* is established in the air gap and links with the field winding. The resultant air gap flux Φ_r is therefore due to the two component fluxes, Φ_f and Φ_{ar}. Each component flux induces a component voltage in the stator winding. In Fig. 6.14*a*, E_f is induced by Φ_f, E_{ar} by Φ_{ar}, and the resultant voltage E_r by the resultant flux Φ_r. The excitation voltage E_f can be found from the open-circuit curve of Fig. 6.5. However, the voltage E_{ar}, known as the *armature reaction voltage,* depends on Φ_{ar} (and hence on I_a). From Fig. 6.14*a*,

$$E_r = E_{ar} + E_f \tag{6.5}$$

or

$$E_f = -E_{ar} + E_r \tag{6.6}$$

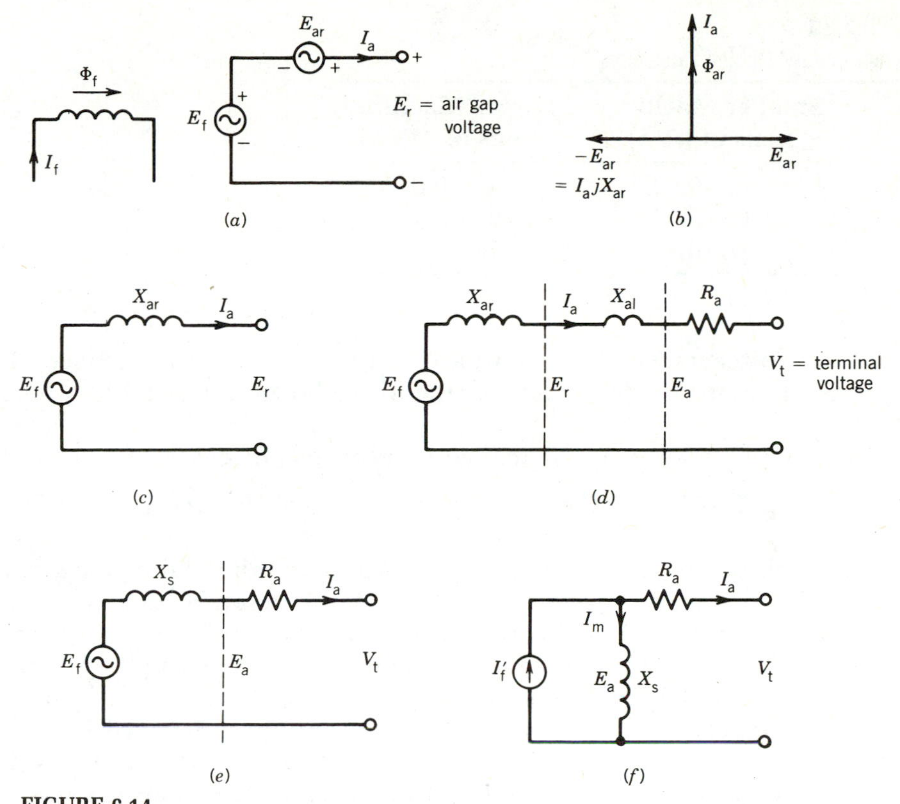

FIGURE 6.14
Equivalent circuit of a synchronous machine.

From the phasor diagram of Fig. 6.14*b*, the voltage E_{ar} lags Φ_{ar} (or I_a) by 90°. Therefore, I_a lags the phasor $-E_{ar}$ by 90°. In Eq. 6.6, the voltage $-E_{ar}$ can thus be represented as a voltage drop across a reactance X_{ar} due to the current I_a. Equation 6.6 can be written as

$$E_f = I_a jX_{ar} + E_r \tag{6.7}$$

This reactance X_{ar} is known as the *reactance of armature reaction* or the *magnetizing reactance* and is shown in Fig. 6.14*c*. If the stator winding resistance R_a and the leakage reactance X_{al} (which accounts for the leakage flux Φ_{al}) are included, the per-phase equivalent circuit is represented by the circuit of Fig. 6.14*d*. The resistance R_a is the *effective resistance* and is approximately 1.6 times the dc resistance of the stator winding. The effective resistance includes the effects of the operating temperature and the skin effect caused by the alternating current flowing through the armature winding.

TABLE 6.1
Synchronous Machine Parameters

	Smaller Machines (tens of kVA)	Larger Machines (tens of MVA)
R_a	0.05–0.02	0.01–0.005
X_{al}	0.05–0.08	0.1–0.15
X_s	0.5–0.8	1.0–1.5

If the two reactances X_{ar} and X_{al} are combined into one reactance, the equivalent circuit model reduces to the form shown in Fig. 6.14*e*, where

$$X_s = X_{ar} + X_{al} \qquad \text{(called } \textit{synchronous reactance}\text{)}$$

$$Z_s = R_a + jX_s \qquad \text{(called } \textit{synchronous impedance}\text{)}$$

The synchronous reactance X_s takes into account all the flux, magnetizing as well as leakage, produced by the armature (stator) current.

The values of these machine parameters depend on the size of the machine. Table 6.1 shows their order of magnitude. The per-unit system is described in Chapter 2. A 0.1 pu impedance means that if the rated current flows, the impedance will produce a voltage drop of 0.1 (or 10%) of the rated value. In general, as the machine size increases, the per-unit resistance decreases but the per-unit synchronous reactance increases.

In an alternative form of the equivalent circuit the excitation voltage E_f and the synchronous reactance X_s can be replaced by a Norton equivalent circuit, as shown in Fig. 6.14*f*, where

$$I_f' = \frac{E_f}{X_s} \tag{6.7a}$$

It can be shown that[1]

$$|I_f'| = \frac{X_{ar}}{X_s} n I_f \tag{6.7b}$$

where $$n = \frac{\sqrt{2}}{3}\frac{N_{re}}{N_{se}} \tag{6.7c}$$

N_{re} is the effective field winding turns

N_{se} is the effective stator phase winding turns

[1] G. R. Slemon and A. Straughen, *Electric Machines*, Addison–Wesley, Reading, Mass., 1980.

The equivalent circuit of Fig. 6.14*f* is useful for determining the actual field current I_f and also for assessing the performance of a synchronous motor if it is fed from a current source power supply.

6.4.1 DETERMINATION OF THE SYNCHRONOUS REACTANCE X_s

The synchronous reactance is an important parameter in the equivalent circuit of the synchronous machine. This reactance can be determined by performing two tests, an open-circuit test and a short-circuit test.

Open-Circuit Test

The synchronous machine is driven at the synchronous speed, and the open-circuit terminal voltage V_t (= E_f) is measured as the field current I_f is varied (see Fig. 6.15*a*). The curve showing the variation of E_f with I_f is known as the *open-circuit characteristic* (OCC, shown in Fig. 6.15*c*). Because the terminals are open, this curve shows the variation of the excitation voltage E_f with the field current I_f. Note that as the field current is increased, the magnetic circuit shows saturation effects. The line passing through the linear part of the OCC is called the *air gap line*. The excitation voltage would have changed along this line if there were no magnetic saturation effects in the machine.

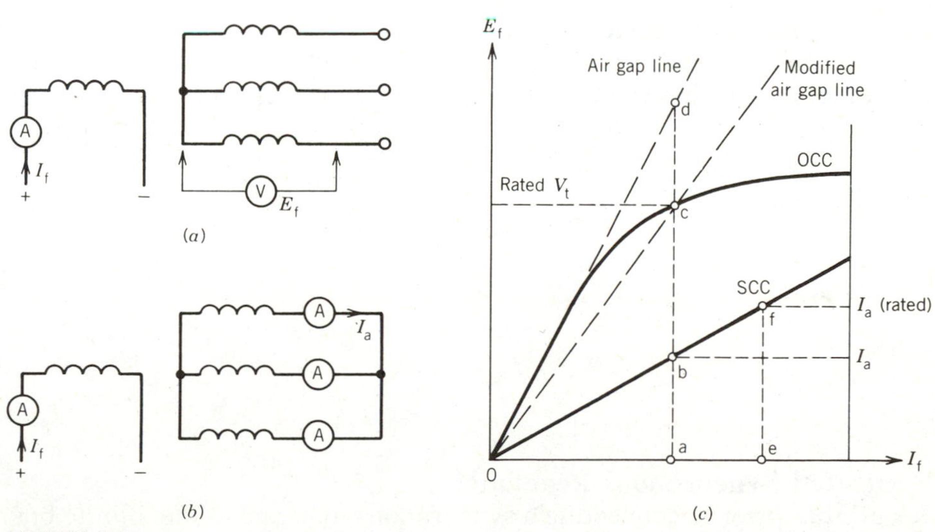

FIGURE 6.15
Open-circuit and short-circuit characteristics. (*a*) Circuit for open-circuit test. (*b*) Circuit for short-circuit test. (*c*) Characteristics.

Short-Circuit Test

The circuit arrangement for this test is shown in Fig. 6.15*b*. Ammeters are connected to each phase and the terminals are then shorted. The synchronous machine is driven at synchronous speed. The field current I_f is now varied and the average of the three armature currents is measured. The variation of the armature current with the field current is shown in Fig. 6.15*c* and is known as the *short-circuit characteristic* (SCC). Note that the SCC is a straight line. This is due to the fact that under short-circuit conditions, the magnetic circuit does not saturate because the air gap flux remains at a low level. This fact can be explained as follows.

The equivalent circuit under short-circuit conditions is shown in Fig. 6.16*a*. Because $R_a << X_s$ (see Table 6.1), the armature current I_a lags the excitation voltage E_f by almost 90°. The armature reaction mmf F_a therefore opposes the field mmf F_f and the resultant mmf F_r is very small, as can be seen from Fig. 6.16*b*. The magnetic circuit therefore remains unsaturated even if both I_f and I_a are large.

Also note from the equivalent circuit of Fig. 6.16a that the air gap voltage is $E_r = I_a(R_a + jX_{al})$. Because both R_a and X_{al} are small (see Table 6.1) at rated current, the air gap voltage will be less than 20 percent of the rated voltage (signifying unsaturated magnetic conditions at short-circuited operation). If the machine stays unsaturated, the excitation voltage E_f will increase linearly with the excitation current I_f along the air gap line and therefore the armature current will increase linearly with the field current.

Unsaturated Synchronous Reactance

This can be obtained from the air gap line voltage and the short-circuit current of the machine for a particular value of the field current. From Fig. 6.15*c*,

$$Z_{s(unsat)} = \frac{E_{da}}{I_{ba}} = R_a + jX_{s(unsat)} \tag{6.8}$$

If R_a is neglected

$$X_{s(unsat)} \simeq \frac{E_{da}}{I_{ba}} \tag{6.9}$$

Saturated Synchronous Reactance

Recall that prior to connecting a synchronous machine to the infinite bus, its excitation voltage is raised to the rated value. From Fig. 6.15*c*, this voltage is E_{ca} (= rated V_t) and the machine operates at some saturation level. If the machine is connected to the infinite bus, its terminal voltage

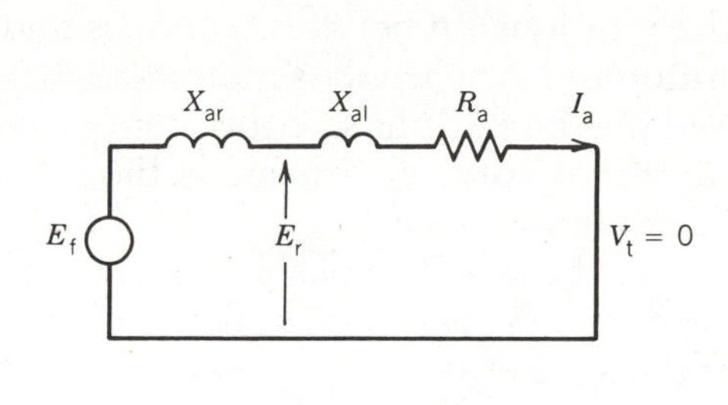

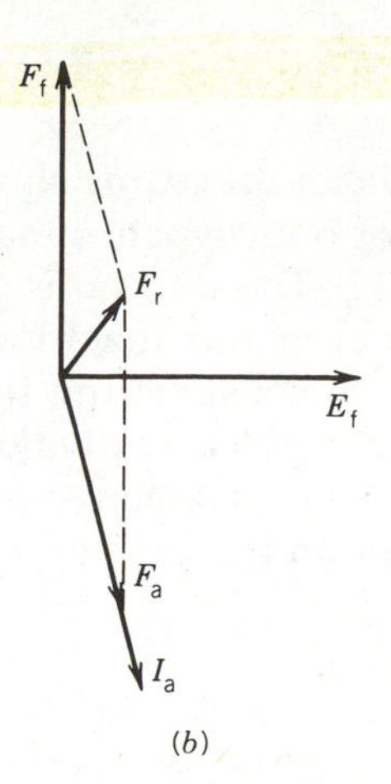

FIGURE 6.16
Short-circuit operation of a synchronous generator.

remains the same at the bus value. If the field current is now changed, the excitation voltage will change, but not along the OCC line. The excitation voltage E_f will change along the line 0c, known as the *modified air gap line.* This line represents the same magnetic saturation level as that corresponding to the operating point c. This can be explained as follows.

From the equivalent circuit of Fig. 6.14*d*,

$$E_r = V_t + I_a(R_a + jX_{al}) \tag{6.10}$$

If the drop across R_a and X_{al} is neglected,

$$E_r \simeq V_t \tag{6.10a}$$

Because V_t is constant, the air gap voltage remains essentially the same as the field current is changed. This implies that the air gap flux level (i.e., magnetic saturation level) remains practically unchanged and hence as I_f is changed, E_f will change linearly along the line 0c of Fig. 6.15*c*.

The saturated synchronous reactance at the rated voltage is obtained as follows:

$$Z_{s(sat)} = \frac{E_{ca}}{I_{ba}} = R_a + jX_{s(sat)} \tag{6.11}$$

If R_a is neglected

$$X_{s(sat)} \simeq \frac{E_{ca}}{I_{ba}} \tag{6.12}$$

6.4.2 PHASOR DIAGRAM

The phasor diagrams showing the relationship between voltages and currents for both synchronous generator and synchronous motor are shown in Fig. 6.17. The diagrams are based on the per-phase equivalent circuit of the synchronous machine. The terminal voltage is taken as the reference phasor in constructing the phasor diagram.

The per-phase equivalent circuit of the synchronous generator is shown in Fig. 6.17*a*. For convenience, the current I_a is shown as flowing out of the machine in the case of a synchronous generator.

$$E_f = V_t + I_a R_a + I_a jX_s = |E_f| \angle\delta \tag{6.13}$$

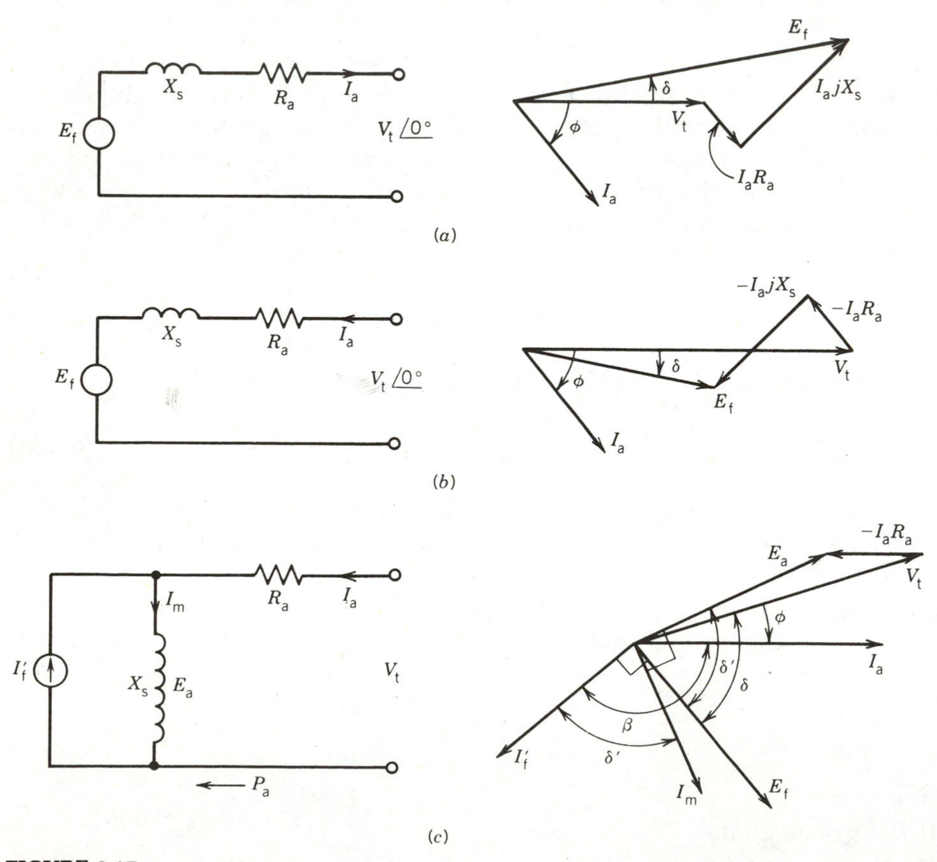

FIGURE 6.17
Phasor diagram for synchronous machines. Assume V_t, I_a, and ϕ known. (*a*) Synchronous generator. (*b*) Synchronous motor. (*c*) Synchronous motor for current source equivalent circuit.

The phasor for the excitation voltage E_f is obtained by adding the voltage drops I_aR_a and I_ajX_s to the terminal voltage V_t. The synchronous generator is considered to deliver a lagging current to the load or infinite bus represented by V_t.

In the case of a synchronous motor the current is shown (Fig. 6.17*b*) as flowing into the motor.

$$V_t = E_f + I_aR_a + I_ajX_s \tag{6.14}$$

$$E_f = V_t \angle 0^\circ - I_aR_a - I_ajX_s = |E_f| \angle -\delta \tag{6.15}$$

The phasor E_f is constructed by subtracting the voltage drops from the terminal voltage. Here also, the synchronous motor is considered to draw a lagging current from the infinite bus.

It is important to note that the angle δ between V_t and E_f is positive for the generating action and negative for the motoring action. This angle δ (known as the power angle) plays an important role in power transfer and in the stability of synchronous machine operation and will be discussed further in the following sections.

The phasor diagram based on the equivalent current source model of Fig. 6.14*f* is shown in Fig. 6.17*c*. In this case the stator current I_a is taken as the reference for convenience. Note that the angle between E_a and E_f is same as that between I_m and I_f'. If R_a is neglected, this angle is the power angle δ (angle between the phasors V_t and E_f). The angle between I_a and I_f' is called β.

EXAMPLE 6.2

The following data are obtained for a 3ϕ, 10 MVA, 14 kV, star-connected synchronous machine.

I_f (A)	Open-Circuit Voltage (kV) (line-to-line)	Air Gap Line Voltage (kV) (line-to-line)	Short-Circuit Current (A)
100	9.0		
150	12.0		
200	14.0	18	490
250	15.3		
300	15.9		
350	16.4		

The armature resistance is 0.07 Ω/phase.

(a) Find the unsaturated and saturated values of the synchronous reactance in ohms and also in pu.

(b) Find the field current required if the synchronous generator is connected to an infinite bus and delivers rated MVA at 0.8 lagging power factor.

(c) If the generator, operating as in part (b), is disconnected from the infinite bus without changing the field current, find the terminal voltage.

Solution

The data are plotted in Fig. E6.2.

$$\text{Base voltage} \qquad V_b = \frac{14{,}000}{\sqrt{3}} = 8083 \text{ V/phase}$$

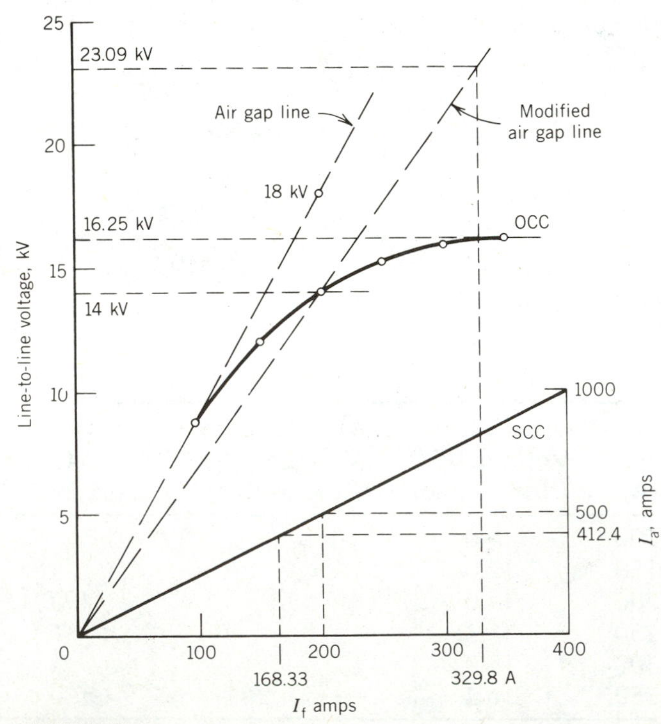

FIGURE E6.2

$$\text{Base current} \qquad I_b = \frac{10 \times 10^6}{\sqrt{3} \times 14{,}000} = 412.41 \text{ A}$$

$$\text{Base impedance} \qquad Z_b = \frac{8083}{412.41} = 19.6\ \Omega$$

(a) From Eq. (6.8) and Fig. E6.2,

$$Z_s|_{\text{unsat}} = \frac{18{,}000/\sqrt{3}}{490} = 21.21\ \Omega$$

$$X_s|_{\text{unsat}} = \sqrt{21.21^2 - 0.07^2} = 21.2\ \Omega$$

$$= \frac{21.2}{19.6} \text{ pu} = 1.08 \text{ pu}$$

From Eq. (6.11) and Fig. E6.2, and because R_a is very small,

$$Z_{s(\text{sat})} = \frac{14{,}000/\sqrt{3}}{490} = 16.5\ \Omega = (X^2_{s(\text{sat})} + 0.07^2)^{1/2}$$

$$X_{s(\text{sat})} \simeq 16.5\ \Omega = \frac{16.5}{19.6} = 0.84 \text{ pu}$$

(b) It will be convenient to carry out the calculation in pu values rather than in actual values.

$$V_t = 1\underline{/0^\circ} \text{ pu}$$

$$\text{PF} = 0.8 = \cos 36.9^\circ$$

$$I_a = 1\underline{/-36.9} \text{ pu}$$

$$Z_s = 0.84 \underline{\Big/ \tan^{-1}\frac{16.5}{0.07}} = 0.84\underline{/89.8^\circ} \text{ pu}$$

Now

$$E_f = V_t + I_a R_a + I_a jX_s$$

$$= V_t + I_a Z_s$$

$$= 1\underline{/0^\circ} + 1\underline{/-36.9^\circ} \cdot 0.85\underline{/89.8^\circ}$$

$$= 1\underline{/0^\circ} + 0.84\underline{/52.9^\circ}$$

$$= 1.5067 + j0.67$$

$$= 1.649\underline{/24^\circ} \text{ pu}$$

$$= 1.649 \times 14\underline{/24^\circ} \text{ kV}$$

$$= 23.09\underline{/24^\circ} \text{ kV}$$

Note that $\delta = 24°$ and is positive, as it should be for generator operation.

The required field current from the modified air gap line (Fig. E6.2) is

$$I_f = 1.649 \times 200 = 329.8 \text{ A}$$

(c) From the open-circuit data (Fig. E6.2) at $I_f = 329.8$ A the terminal voltage is

$$V_t = 16.25 \text{ kV} \quad \text{(line-to-line)}$$

6.5 POWER AND TORQUE CHARACTERISTICS

A synchronous machine is normally connected to a fixed-voltage bus and operates at a constant speed. There is a limit on the power a synchronous generator can deliver to the infinite bus and on the torque that can be applied to the synchronous motor without losing synchronism. Analytical expressions for the steady-state power transfer between the machine and the constant-voltage bus or the torque developed by the machine are derived in this section in terms of bus voltage, machine voltage, and machine parameters.

The per-phase equivalent circuit is shown again in Fig. 6.18 for convenience, where V_t is the constant bus voltage per phase and is considered as the reference phasor. Let

$$V_t = |V_t|\angle 0° \tag{6.16}$$

$$E_f = |E_f|\angle \delta \tag{6.17}$$

$$Z_s = R_a + jX_s = |Z_s|\angle \theta_s \tag{6.18}$$

where the quantities inside the vertical bars represent the magnitudes of the phasors.

The per-phase complex power S at the terminals is

$$S = V_t I_a^* \tag{6.19}$$

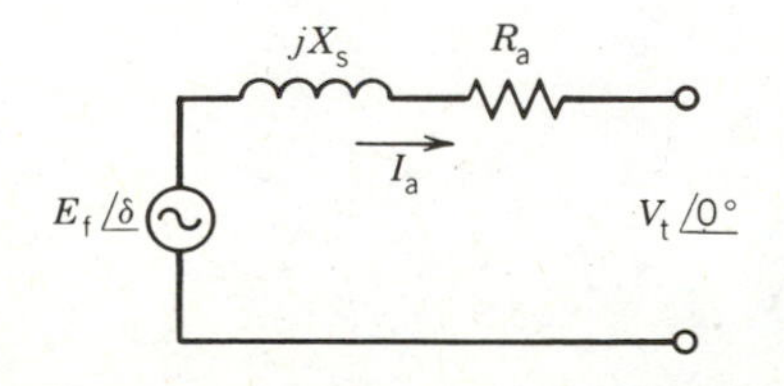

FIGURE 6.18
Per-phase equivalent circuit.

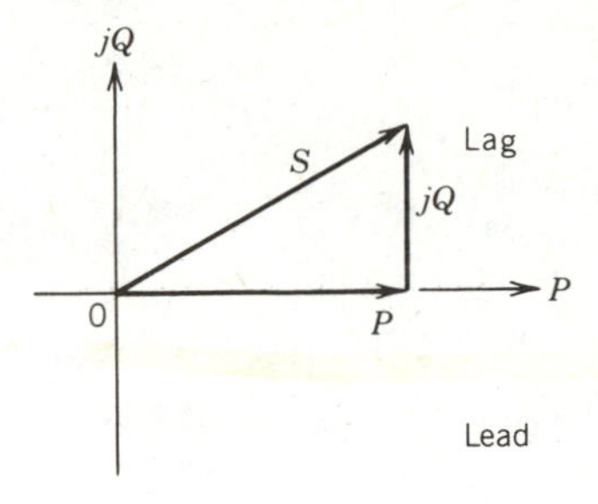

FIGURE 6.19
Complex power phasor.

The conjugate of the current phasor I_a is used to conform with the convention that lagging reactive power is considered as positive and leading reactive power as negative, as shown in Fig. 6.19.

From Fig. 6.18

$$I_a^* = \left|\frac{E_f - V_t}{Z_s}\right|^* = \frac{E_f^*}{Z_s^*} - \frac{V_t^*}{Z_s^*}$$

$$= \frac{|E_f|\underline{/-\delta}}{|Z_s|\underline{/-\theta_s}} - \frac{|V_t|\underline{/0}}{|Z_s|\underline{/-\theta_s}}$$

$$= \frac{|E_f|}{|Z_s|}\underline{/\theta_s - \delta} - \frac{|V_t|}{|Z_s|}\underline{/\theta_s} \tag{6.20}$$

From Eqs. 6.19 and 6.20

$$S = \frac{|V_t||E_f|}{|Z_s|}\underline{/\theta_s - \delta} - \frac{|V_t|^2}{|Z_s|}\underline{/\theta_s} \text{ VA/phase} \tag{6.21}$$

The real power P and the reactive power Q per phase are

$$P = \frac{|V_t||E_f|}{|Z_s|}\cos(\theta_s - \delta) - \frac{|V_t|^2}{|Z_s|}\cos\theta_s \text{ watt/phase} \tag{6.22}$$

$$Q = \frac{|V_t||E_f|}{|Z_s|}\sin(\theta_s - \delta) - \frac{|V_t|^2}{|Z_s|}\sin\theta_s \text{ VAR/phase} \tag{6.23}$$

If R_a is neglected, then $Z_s = X_s$ and $\theta_s = 90°$. From Eqs. 6.22 and 6.23 for a 3ϕ machine,

$$P_{3\phi} = \frac{3|V_t||E_f|}{|X_s|}\sin\delta \tag{6.24}$$

$$= P_{max}\sin\delta \text{ watts} \tag{6.25}$$

$$\text{where} \quad P_{max} = \frac{3|V_t||E_f|}{|X_s|} \tag{6.25a}$$

$$Q_{3\phi} = \frac{3|V_t||E_f|}{|X_s|}\cos\delta - \frac{3|V_t|^2}{|X_s|}\text{ VAR} \tag{6.26}$$

Because the stator losses are neglected in this analysis, the power developed at the terminals is also the air gap power. The developed torque of the machine is

$$T = \frac{P_{3\phi}}{\omega_{syn}} \tag{6.27}$$

$$= \frac{3}{\omega_{syn}}\frac{|V_t||E_f|}{X_s}\sin\delta \tag{6.28}$$

$$= T_{max}\sin\delta \text{ N}\cdot\text{m} \tag{6.29}$$

$$\text{where} \quad T_{max} = \frac{3}{\omega_{syn}}\frac{|V_t||E_f|}{X_s} = \frac{P_{max}}{\omega_{syn}} \tag{6.29a}$$

$$\omega_{syn} = \frac{n_{syn}}{60}2\pi$$

n_{syn} is the synchronous speed in rpm

Both power and torque vary sinusoidally with the angle δ (as shown in Fig. 6.20), which is called the *power angle* or *torque angle.* The machine can be loaded *gradually* up to the limit of P_{max} or T_{max}, which are known as *static stability limits.* The machine will lose synchronism if δ becomes greater than 90°. The maximum torque T_{max} is also known as the *pull-out* torque. Note that since V_t is constant, the pull-out torque can be increased by increasing the excitation voltage E_f. If a synchronous motor tends to pull

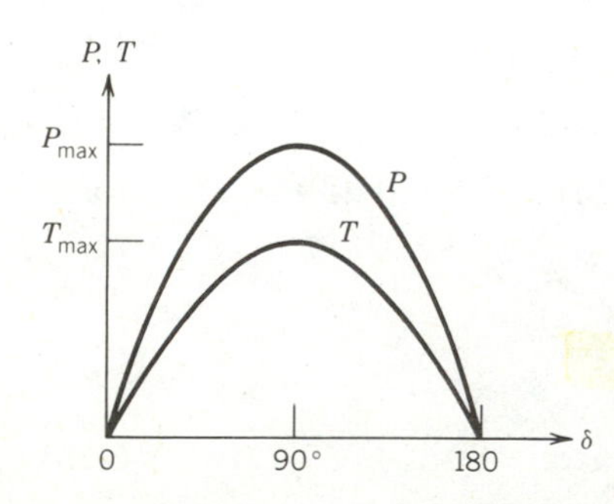

FIGURE 6.20
Power and torque–angle characteristics.

out of synchronism because of excessive load torque, the field current can be increased to develop high torque to prevent loss of synchronism. Similarly, in a synchronous generator, if the prime mover tends to drive the machine to supersynchronous speed by excessive driving torque, the field current can be increased to produce more counter torque to oppose such a tendency.

As the speed remains constant in a synchronous machine, the speed–torque characteristic is a straight line, parallel to the torque axis, as shown in Fig. 6.21.

Power and torque can also be expressed in terms of currents I_a, I'_f and the angle β between their phasors. Consider the equivalent circuit and the phasor diagram of Fig. 6.17*c*. The complex power S_a across the air gap is

$$S_a = I_a E_a^* \tag{6.29b}$$

Let

$$I_a = |I_a|\angle 0^\circ \tag{6.29c}$$

Now

$$\begin{aligned} E_a &= jX_s I_m \\ &= jX_s(I_a + I'_f) \\ &= |X_s||I_a|\angle 90^\circ + |X_s||I'_f|\angle 90 + \beta \end{aligned} \tag{6.29d}$$

$$E_a^* = |X_s \mid I_a|\angle -90^\circ + |X_s \mid I'_f|\angle -90 - \beta \tag{6.29e}$$

From Eqs. 6.29b, 6.29c, and 6.29e,

$$S_a = |X_s \mid I_a^2|\angle -90^\circ + |X_s \mid I_a \mid I'_f|\angle -90 - \beta \tag{6.29f}$$

From Eq. 6.29f, the real power transferred across the air gap is

$$\begin{aligned} P_a &= Re[S_a] \\ &= X_s I_a^2 \cos(-90^\circ) + |X_s \mid I_a \mid I'_f| \cos(-90 - \beta) \\ &= -\omega L_s I_a I'_f \sin \beta \end{aligned} \tag{6.29g}$$

where $\omega = 2\pi f$.

The torque developed is

$$T = \frac{3P_a}{\omega_{syn}} \tag{6.29h}$$

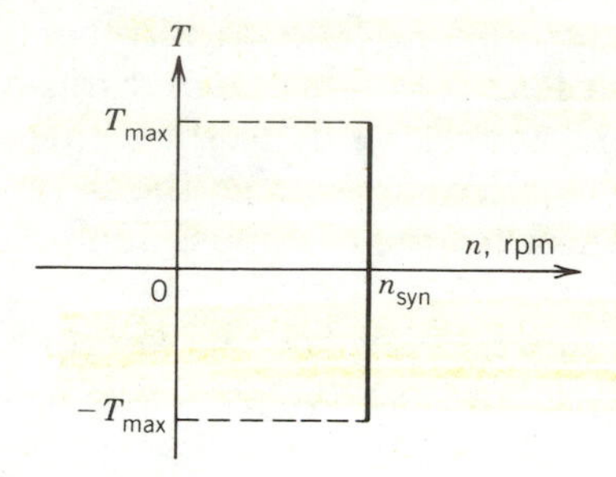

FIGURE 6.21
Torque–speed characteristics.

Now,

$$\omega_{syn} = \frac{n_{syn}2\pi}{60} = \frac{120f}{p}\frac{2\pi}{60} = \frac{4\pi f}{p} = \frac{2}{p}\,\omega \tag{6.29i}$$

From Eqs. 6.29g–i, the torque is

$$T = -\frac{3p}{2}L_s I_a I_f' \sin\beta \tag{6.29j}$$

Both power and torque vary sinusoidally with the angle β.

EXAMPLE 6.3

A 3ϕ, 5 kVA, 208 V, four-pole, 60 Hz, star-connected synchronous machine has negligible stator winding resistance and a synchronous reactance of 8 ohms per phase at rated terminal voltage.

The machine is first operated as a generator in parallel with a 3ϕ, 208 V, 60 Hz power supply.

(a) Determine the excitation voltage and the power angle when the machine is delivering rated kVA at 0.8 PF lagging. Draw the phasor diagram for this condition.

(b) If the field excitation current is now increased by 20 percent (without changing the prime mover power), find the stator current, power factor, and reactive kVA supplied by the machine.

(c) With the field current as in (a) the prime mover power is slowly increased. What is the steady-state (or static) stability limit? What are the corresponding values of the stator (or armature) current, power factor, and reactive power at this maximum power transfer condition?

Solution

The per-phase equivalent circuit for the synchronous generator is shown in Fig. E6.3*a*.

(a) $V_t = \dfrac{208}{\sqrt{3}} = 120$ V/phase

Stator current at rated kVA;

$$I_a = \frac{5000}{\sqrt{3} \times 208} = 13.9 \text{ A}$$

$$\phi = -36.9^\circ \quad \text{for lagging pf of 0.8}$$

From Fig. E6.3*a*

$$\begin{aligned} E_f &= V_t\underline{/0^\circ} + I_a jX_s \\ &= 120\underline{/0^\circ} + 13.9\underline{/-36.9^\circ} \cdot 8\underline{/90^\circ} \\ &= 206.9\underline{/25.5^\circ} \end{aligned}$$

$$\text{Excitation voltage} \qquad E_f = 206.9 \text{ V/phase}$$

$$\text{Power angle} \qquad \delta = +25.5^\circ$$

Note that because of generator action the power angle is positive.
The phasor diagram is shown in Fig. E6.3*b*.

(b) The new excitation voltage $E_f' = 1.2 \times 206.9 = 248.28$ V. Because power transfer remains same,

$$\frac{V_t E_f}{X_s} \sin\delta = \frac{V_t E_f'}{X_s} \sin\delta'$$

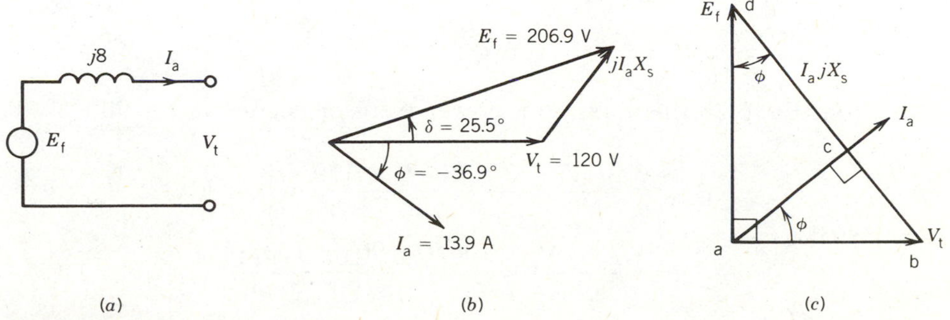

FIGURE E6.3

or

$$E_f \sin\delta = E_f' \sin\delta'$$

or

$$\sin\delta' = \frac{E_f}{E_f'}\sin\delta = \frac{\sin 25.5}{1.2}$$

$$\delta' = 21^\circ$$

The stator current is

$$I_a = \frac{E_f - V_t}{jX_s}$$

$$= \frac{248.28\underline{/21^\circ} - 120\underline{/0^\circ}}{8\underline{/90^\circ}}$$

$$= \frac{142.87\underline{/38.52^\circ}}{8\underline{/90^\circ}}$$

$$= 17.86\underline{/-51.5^\circ}\ \text{A}$$

$$\text{Power factor} = \cos 51.5 = 0.62 \quad \text{lag}$$

$$\text{Reactive kVA} = 3|V_t|\,|I_a| \sin 51.5$$

$$= 3 \times 120 \times 17.86 \times 0.78 \times 10^{-3}$$

$$= 5.03$$

or from Eq. 6.26

$$Q = 3\left(\frac{120 \times 248.28}{8}\cos 21^\circ - \frac{120^2}{8}\right) \times 10^{-3}$$

$$= 3(3476.86 - 1800)$$

$$= 5.03$$

(c) From Eq. 6.25 the maximum power transfer occurs at $\delta = 90^\circ$.

$$P_{max} = \frac{3E_fV_t}{X_s} = \frac{3 \times 206.9 \times 120}{8} = 9.32\ \text{kW}$$

$$I_a = \frac{E_f - V_t}{jX_s} = \frac{206.9\underline{/+90^\circ} - 120\underline{/0^\circ}}{8\underline{/90^\circ}}$$

$$= 29.9\underline{/30.1^\circ}\ \text{A}$$

$$\text{Stator current} \qquad I_a = 29.9 \text{ A}$$

$$\text{Power factor} = \cos 30.1° = 0.865 \quad \text{leading}$$

The stator current and power factor can also be obtained by drawing the phasor diagram for the maximum power transfer condition. The phasor diagram is shown in Fig. E6.3*c*.

Because $\delta = +90°$, E_f leads V_t by 90°. The distance *bd* between phasors V_t and E_f is the voltage drop I_aX_s and the current phasor I_a is in quadrature with I_aX_s.

From the phasor diagram,

$$|I_aX_s|^2 = |E_f|^2 + |V_t|^2$$

$$I_a = \left(\frac{206.9^2 + 120^2}{8^2}\right)^{1/2} = 29.9 \text{ A}$$

From the two triangles *abc* and *abd*,

$$\underline{/bac} = \underline{/adb} = \phi$$

$$\tan \phi = \frac{ab}{ad} = \frac{120}{206.9} = 0.58$$

$$\phi = 30.1°$$

$$\text{PF} = \cos 30.1° = 0.865 \quad \text{lead}$$

EXAMPLE 6.4

The synchronous machine in Example 6.3 is operated as a synchronous motor from the 3ϕ, 208 V, 60 Hz power supply. The field excitation is adjusted so that the power factor is unity when the machine draws 3 kW from the supply.

(a) Find the excitation voltage and the power angle. Draw the phasor diagram for this condition.

(b) If the field excitation is held constant and the shaft load is slowly increased, determine the maximum torque (i.e., pull-out torque) that the motor can deliver.

Solution

The per-phase equivalent circuit for motoring operation is shown in Fig. E6.4*a*.

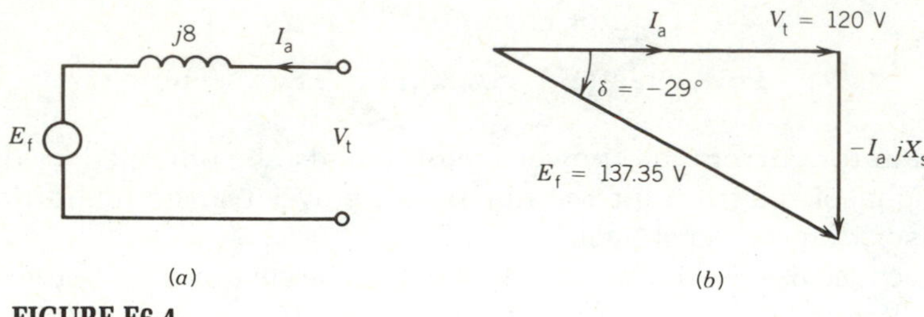

FIGURE E6.4

(a) $3V_tI_a \cos\phi = 3 \text{ kW} = 3V_tI_a$ for $\cos\phi = 1$.

$$I_a = \frac{3000}{3 \times 120} = 8.33 \text{ A}$$

$$\begin{aligned} E_f &= V_t - I_a jX_s \\ &= 120\underline{/0^\circ} - 8.33\underline{/0^\circ} \cdot 8\underline{/90^\circ} \\ &= 137.35\underline{/-29^\circ} \end{aligned}$$

$$\text{Excitation voltage} \qquad E_f = 137.35 \text{ V/phase}$$

$$\text{Power angle} \qquad \delta = -29^\circ$$

Note that because of motor action the power angle is negative.

The phasor diagram is shown in Fig. E.6.4*b*. E_f and δ can also be calculated from the phasor diagram.

$$\begin{aligned} E_f &= \sqrt{|V_t|^2 + |I_aX_s|^2} = \sqrt{120^2 + (8.33 \times 8)^2} \\ &= 137.35 \text{ V/phase} \end{aligned}$$

$$\tan\delta = \frac{|I_aX_s|}{|V_t|} = \frac{8.33 \times 8}{120} = 0.555$$

$$|\delta| = 29^\circ$$

$$\delta = -29^\circ$$

(b) Maximum torque will be developed at $\delta = 90^\circ$. From Eq. 6.25a,

$$P_{max} = \frac{3 \times 137.35 \times 120}{8} = 6185 \text{ W}$$

$$T_{max} = \frac{P_{max}}{\omega_{syn}} = \frac{6185}{(1800/60) \times 2\pi} = 32.8 \text{ N} \cdot \text{m}$$

EXAMPLE 6.5

A 3ϕ, 460 V, 60 Hz, 1200 rpm, 125 hp synchronous motor has the following equivalent circuit parameters:

$$R_a = 0.078\ \Omega, \qquad X_{al} = 0.15\ \Omega, \qquad X_{ar} = 1.85\ \Omega$$

$$N_{re}/N_{se} = 28.2$$

For rated conditions the field current is adjusted to make the motor power factor unity. Neglect all rotational losses and power lost in the field winding.

(a) For rated operating conditions, determine the motor current I_a, field current I_f, and power angle δ.

(b) Draw the phasor diagram.

Solution

(a) For rated conditions, from Fig. 6.17*c*,

$$P_{in} = \sqrt{3} \times 460 \times I_a = 3 \times 0.078 I_a^2 + 125 \times 746$$

$$I_a = 121.4\ \text{A}$$

Let

$$V_t = \frac{460}{\sqrt{3}}\underline{/0^\circ} = 265.6\underline{/0^\circ}$$

$$I_a = 121.4\underline{/0^\circ}$$

$$E_a = V_t - I_a R_a$$

$$= 265.6\underline{/0^\circ} - 121.4\underline{/0^\circ} \times 0.078$$

$$= 256.13\underline{/0^\circ}\ \text{V}$$

$$X_s = 0.15 + 1.85 = 1.9\ \Omega$$

$$I_m = \frac{256.13\underline{/0^\circ}}{1.9\underline{/90^\circ}} = 134.74\underline{/-90^\circ}\ \text{A}$$

$$I_f' = I_m - I_a$$

$$= 134.74\underline{/-90^\circ} - 121.4\underline{/0^\circ}$$

$$= 181.4\underline{/-132^\circ}$$

$$\beta = -132^\circ$$

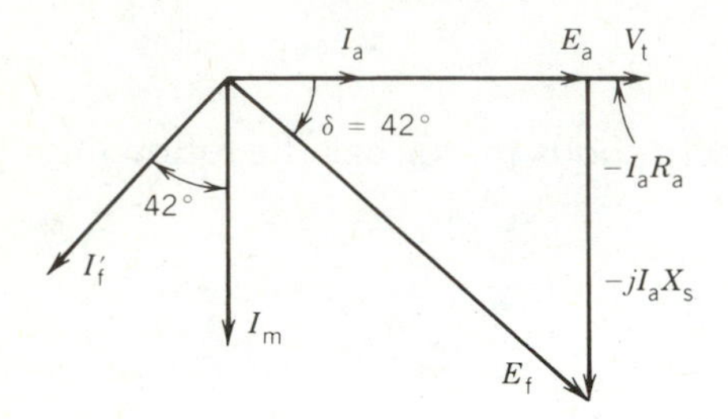

FIGURE E6.5

From Eq. 6.7c,

$$n = \frac{\sqrt{2}}{3} \times 28.2 = 13.29$$

From Eq. 6.7b,

$$I_f = \frac{181.4}{13.29} \times \frac{1.9}{1.85} = 14.02 \text{ A}$$

$$\delta = -132° + 90° = -42°$$

(b) The phasor diagram is drawn in Fig. E6.5.

Complex Power Locus

If the real power and reactive power given by Eqs. 6.24 and 6.26 are plotted in the per-phase complex S-plane, the locus will be a circle of radius $|V_t||E_f|/|X_s|$ with center at $0, -|V_t|^2/|X_s|$ as shown in Fig. 6.22. For an operating point x, the power angle δ and the power factor angle ϕ are indicated. The various circles shown in the figure correspond to various excitation voltages. The locus of the maximum power representing the steady-state limit is a horizontal line (passing through the center) for which $\delta = 90°$.

6.6 CAPABILITY CURVES

A synchronous machine cannot be operated at all points inside the region bounded by the circle shown in Fig. 6.22 without exceeding the machine rating. The region of operation is restricted by the following considerations.

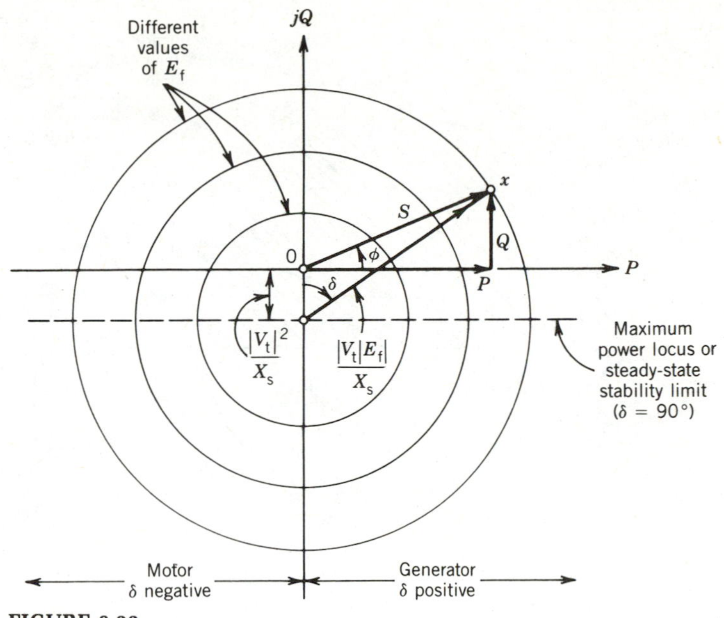

FIGURE 6.22
Complex power locus per phase.

1. Armature heating, determined by the armature current.
2. Field heating, determined by the field current.
3. Steady-state stability limit.

The capability curves that define the limiting region for each consideration can be drawn on the complex power plane for constant terminal voltage V_t, and the region of operation can be identified so that none of these limits is exceeded.

In Fig. 6.23, the circle with center at the origin 0 and radius S $(= V_t \cdot I_a)$ defines the region of operation for which armature heating will not exceed a specified limit. The circle with center at $Y(0, -|V_t|^2/|X_s|)$ and radius $|V_t|\,E_f|/|X_s|$ defines the region of operation for which field heating will not exceed a specified limit. The horizontal line XYZ specifies the steady-state stability limit. The shaded area bounded by these three capability curves defines the restricted region of operation for the synchronous machine. The intersecting points M (for generator) and N (for motor) of the armature heating and field heating curves determine the optimum operating points, because operation at these points makes the maximum utilization of the armature and field circuits.

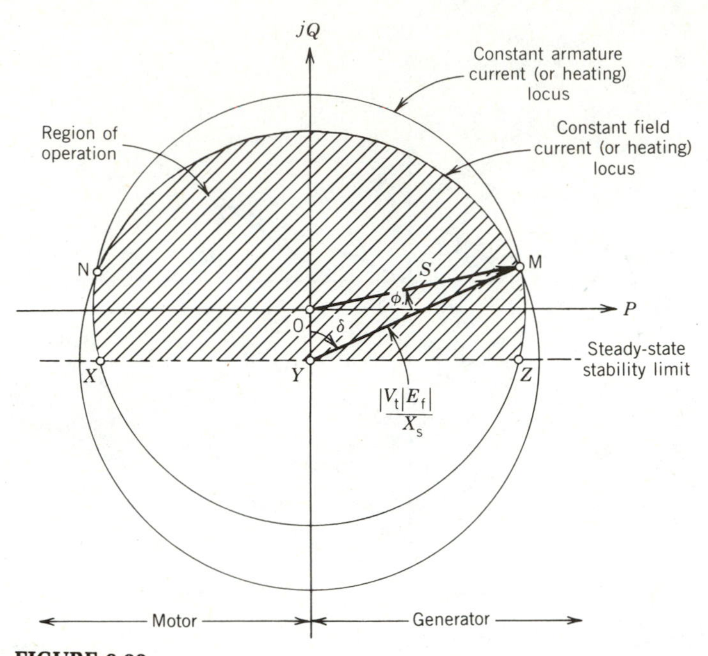

FIGURE 6.23
Capability curves of a synchronous machine.

6.7 POWER FACTOR CONTROL

An outstanding feature of the synchronous machine is that the power factor of the machine can be controlled by the field current. The field current can be adjusted to make the stator (or line) current lagging or leading as desired. This power factor characteristic can be explained by drawing phasor diagrams of machine voltages and currents.

Assume constant-power operation of a synchronous motor connected to an infinite bus. The equivalent circuit, neglecting the stator resistance, and the phasor diagram are shown in Figs. 6.24*a* and 6.24*b*, respectively. For a three-phase machine the power transfer is

$$P = 3V_t I_a \cos\phi \tag{6.30}$$

Because V_t is constant, for constant-power operation $|I_a \cos\phi|$ is constant; that is, the in-phase component of the stator current on the axis of the phasor V_t is constant. The locus of the stator current is therefore the vertical line passing through the current phasor for unity power factor. In Fig. 6.24*b*, phasor diagrams are drawn for three stator currents:

$$I_a = I_{a1}, \quad \text{lagging } V_t$$
$$= I_{a2}, \quad \text{in phase with } V_t$$
$$= I_{a3}, \quad \text{leading } V_t$$

For these stator currents the excitation voltages E_{f1}, E_{f2}, and E_{f3} (representing the field currents I_{f1}, I_{f2}, and I_{f3}, respectively) are drawn to satisfy the phasor relationship

$$E_f = V_t - jI_aX_s \tag{6.31}$$

The power can also be expressed as

$$P = 3\frac{V_tE_f}{X_s}\sin\delta \tag{6.32}$$

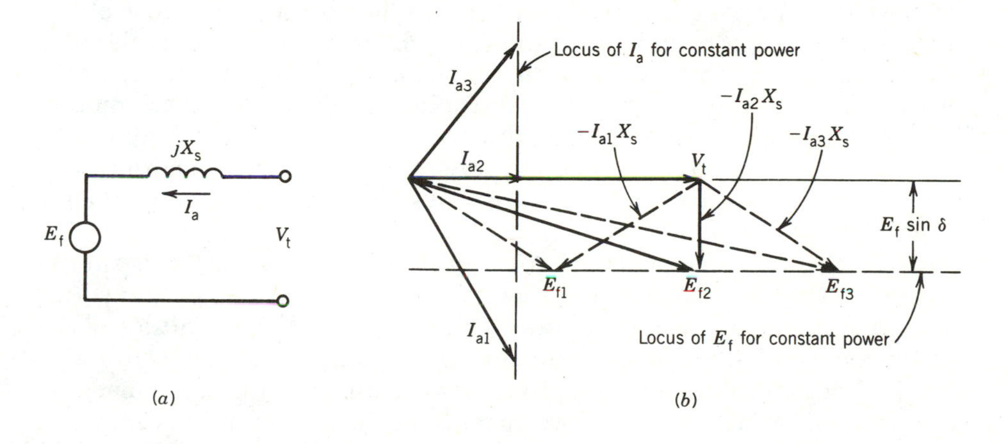

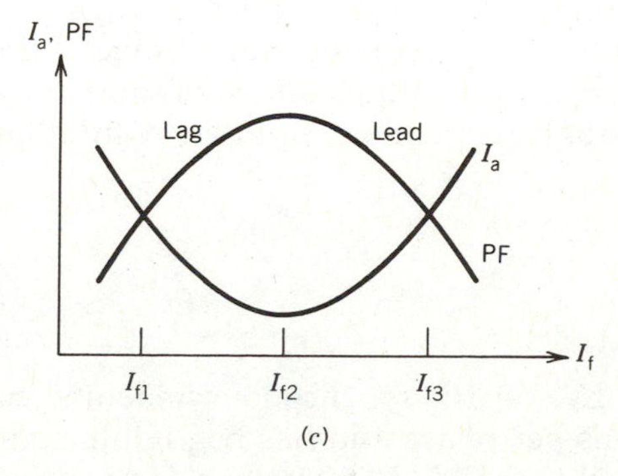

FIGURE 6.24
Power factor characteristics. (*a*) Equivalent circuit. (*b*) Phasor diagram. (*c*) Variation of I_a and PF with I_f.

Again, for constant-power operation, $|E_f \sin \delta|$ is constant. Thus, the locus of E_f (or I_f) is also a straight line parallel to the phasor V_t (see Fig. 6.24*b*) such that the vertical difference between the locus of E_f and the phasor V_t is constant and equals $|E_f \sin \delta|$.

The excitation voltage E_f changes linearly with the field current I_f. Therefore, as I_f is changed, E_f will change along the locus of E_f and I_a will change along the locus of I_a, signifying a change in the power factor angle ϕ of the stator current. For low field current I_{f1}, *underexcitation* ($E_f = E_{f1}$), the stator current ($I_a = I_{a1}$) is large and lagging. The stator current is minimum ($I_a = I_{a2}$) and at unity power factor for the field current I_{f2} ($E_f = E_{f2}$), which is called *normal excitation.* For larger field current I_{f2}, *overexcitation* ($E_f = E_{f3}$), the stator current ($I_a = I_{a3}$) is large again and leading. The variation of the stator current with the field current for constant-power operation is shown in Fig. 6.24*c*. This is known as the *V-curve* because of the characteristic shape. The variation of the power factor with the field current is the *inverted V-curve,* also shown in Fig. 6.24*c*.

This unique feature of power factor control by the field current can be utilized to improve the power factor of a plant. In a plant most of the motors are normally induction motors, which draw power at lagging power factors. Synchronous motors can be installed for some drives in the plant and made to operate in an overexcited mode so that these motors operate at leading power factors, thus compensating the lagging power factor of the induction motors and thereby improving the overall power factor of the plant. Example 6.1 illustrates this method of power factor improvement in a plant.

If the synchronous machine is not transferring any power but is simply *floating* on the infinite bus, the power factor is zero; that is, the stator current either leads or lags the stator voltage by 90°. The magnitude of the stator current changes as the field current is changed, but the stator current is always reactive. Looking from the machine terminals, the machine behaves as a variable inductor or capacitor as the field current is changed. An unloaded synchronous machine is called a *synchronous condenser* and may be used to regulate the receiving-end voltage of a long power transmission line. At present, solid-state control of inductors and capacitors is being used increasingly to achieve voltage regulation of transmission lines.

EXAMPLE 6.6

A 3 ϕ, 5 MVA, 11 kV, 60 Hz synchronous machine has a synchronous reactance of 10 ohms per phase and has negligible stator resistance. The machine is connected to the 11 kV, 60 Hz bus and is operated as a synchronous condenser.

1. Neglect rotational losses.
   (a) For normal excitation, find the stator current. Draw the phasor diagram.
   (b) If the excitation is increased to 150 percent of the normal excitation, find the stator current and power factor. Draw the phasor diagram.
   (c) If the excitation is decreased to 50 percent of the normal excitation, find the stator current and power factor. Draw the phasor diagram.
2. If the rotational losses are 80 kW, find the stator current and excitation voltage for normal excitation. Draw the phasor diagram.

Solution

1. (a) Power = $3V_t I_a \cos\phi$. For normal excitation, power factor = $\cos\phi$ = 1. Hence V_t and I_a are in phase. Since power is zero, I_a is zero. From Eq. 6.24, for no power transfer, δ is zero. If I_a is zero, both V_t and E_f are also the same in magnitude.

$$E_f = V_t = \frac{11}{\sqrt{3}} \text{ kV/phase} = 6.35 \text{ kV/phase}$$

The phasor diagram is shown in Fig. E6.6*a*.

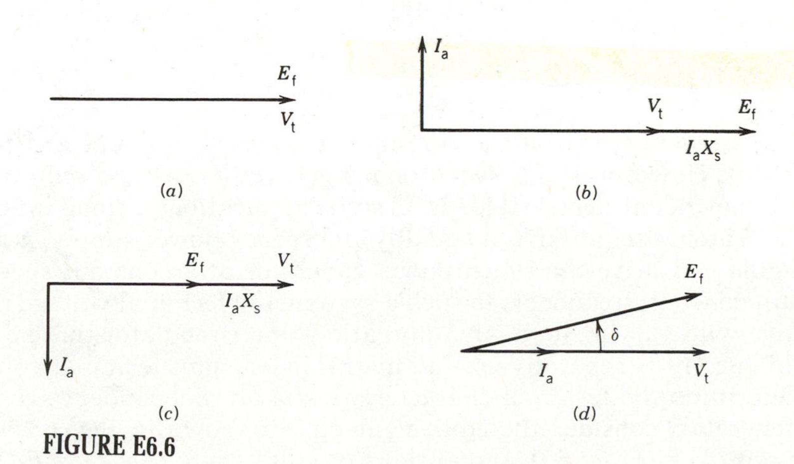

FIGURE E6.6

(b) Because power transfer is zero, δ is zero. Hence,

$$I_a = \frac{V_t\angle 0^\circ - E_f\angle 0^\circ}{jX_s} = \frac{6351 - 1.5 \times 6351}{10\angle 90^\circ}$$

$$= 317.55\angle 90^\circ \text{ A}$$

$$\text{PF} = \cos 90^\circ = 0 \quad \text{leading}$$

The phasor diagram is shown in Fig. E6.6*b*.

(c)

$$I_a = \frac{6351 - 0.5 \times 6351}{10\angle 90^\circ} = 317.55\angle -90^\circ \text{ A}$$

$$\text{PF} = \cos 90^\circ = 0 \quad \text{lagging}$$

The phasor diagram is shown in Fig. E6.6*c*.

2. For normal excitation, power factor = $\cos\theta = 1$.

$$\text{Power} = 3V_tI_a\cos\theta = 80{,}000 \text{ W}$$

$$I_a = \frac{80{,}000}{3 \times 6351 \times 1} = 4.2 \text{ A}$$

$$E_f = V_t - I_a jX_s$$

$$= 6351 - 4.2\angle 0^\circ \cdot 10\angle 90^\circ$$

$$= 6351 - j42$$

$$\simeq 6351\angle 0.4^\circ \text{ V/phase}$$

The phasor diagram is shown in Fig. E6.6*d*.

6.8 INDEPENDENT GENERATORS

As stated earlier, synchronous machines are normally connected to an infinite bus. However, small synchronous generators may be required to supply independent electrical loads. In some applications a small synchronous generator is required as a standby emergency power supply. A gasoline engine can drive the synchronous generator at a constant speed to maintain constant frequency. In such a system, the terminal voltage tends to change with varying load. An automatic voltage regulator that adjusts the field current is generally used to maintain constant terminal voltage.

To determine the terminal characteristics of an independent synchronous generator, consider the equivalent circuit shown in Fig. 6.25*a*. At open circuit $V_t = E_f$, $I_a = 0$, and at short circuit $V_t = 0$, $I_a = I_{sc} = E_f/X_s$. If the load current is changed from 0 to E_f/X_s, the terminal voltage will

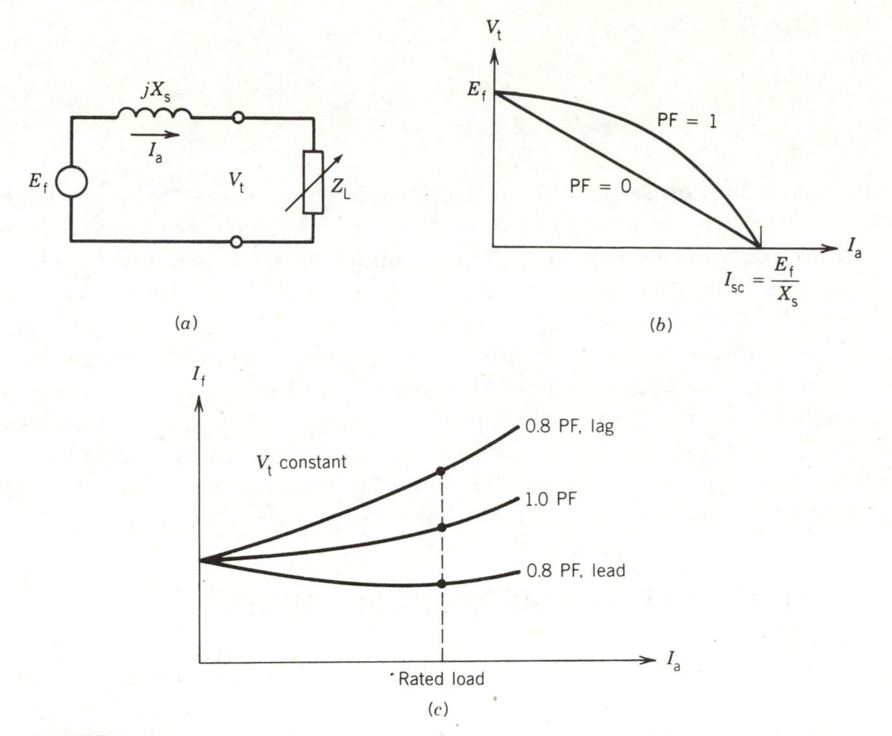

FIGURE 6.25
Performance characteristics of an independent synchronous generator. (*a*) Equivalent circuit. (*b*) Terminal voltage versus load current at constant field current. (*c*) I_f versus I_a for constant terminal voltage.

change from E_f to zero. Therefore, if the field current is held constant, the terminal voltage varies over a wide range (poor voltage regulation) as the load current is changed.

For a purely inductive load $Z_L = X_L$

$$\begin{aligned} V_t &= E_f - I_a X_s \\ &= I_{sc} X_s - I_a X_s \\ &= X_s (I_{sc} - I_a) \end{aligned} \tag{6.33}$$

For a purely resistive load $Z_L = R_L$

$$I_a = \frac{E_f}{\sqrt{R_L^2 + X_s^2}} = \frac{X_s I_{sc}}{\sqrt{R_L^2 + X_s^2}} \tag{6.34}$$

$$V_t = I_a R_L \tag{6.35}$$

From Eqs. 6.34 and 6.35

$$\frac{V_t^2}{(X_s I_{sc})^2} + \frac{I_a^2}{I_{sc}^2} = 1 \tag{6.36}$$

Equation 6.33 represents a linear decrease in the terminal voltage with the stator current I_a, whereas Eq. 6.36 represents a quarter-ellipse with rectangular coordinates V_t and I_a. These equations are plotted in Fig. 6.25*b*. The curves show that as the current is increased, the terminal voltage will fall rapidly if the load power factor falls, resulting in poor voltage regulation. To maintain constant terminal voltage with varying load current the field current is changed, which in turn changes the excitation voltage E_f. Figure 6.25*c* shows the variation in the field current required to maintain constant terminal voltage as the load current is increased. An automatic voltage regulator that senses the terminal voltage can adjust the field current to maintain constant terminal voltage with varying load.

6.9 SALIENT POLE SYNCHRONOUS MACHINES

Low-speed multipolar synchronous machines have salient poles and nonuniform air gaps. The magnetic reluctance is low along the poles and high between poles. Therefore, a particular armature reaction mmf will produce more flux if it is acting along the pole axis, called the *d-axis,* and less flux if it is acting along the interpolar axis, called the *q-axis.* In the cylindrical-rotor synchronous machine discussed in the preceding sections the same armature reaction mmf produces essentially the same flux irrespective of the rotor position because of the uniform air gap. It is therefore obvious that the magnetizing reactance X_{ar}, which represents the armature reaction flux in the cylindrical machine, can no longer be used to represent armature reaction flux in a salient pole machine.

Consider Fig. 6.26*a*, in which the stator current I_a is shown in phase with the excitation voltage E_f. The field mmf F_f and flux Φ_f are along the d-axis and the armature mmf F_a and flux Φ_{ar} are along the q-axis. Only the fundamental components of the fluxes are considered here. In Fig. 6.26*b* the stator current I_a is considered to lag the excitation voltage E_f by 90°. The armature reaction mmf F_a and flux Φ_{ar} act along the d-axis, directly opposing the field mmf F_f and flux Φ_f. Note that the same magnitude of the mmf F_a now acting along the d-axis (axis of high permeance) produces more armature reaction flux than that when I_a was in phase with E_f and its mmf was acting along the q-axis. The magnetizing reactance is more if I_a lags E_f than if I_a is in phase with E_f. Therefore, the magnetizing or armature reaction reactance is not unique in a salient pole machine but depends on the power factor of the stator current.

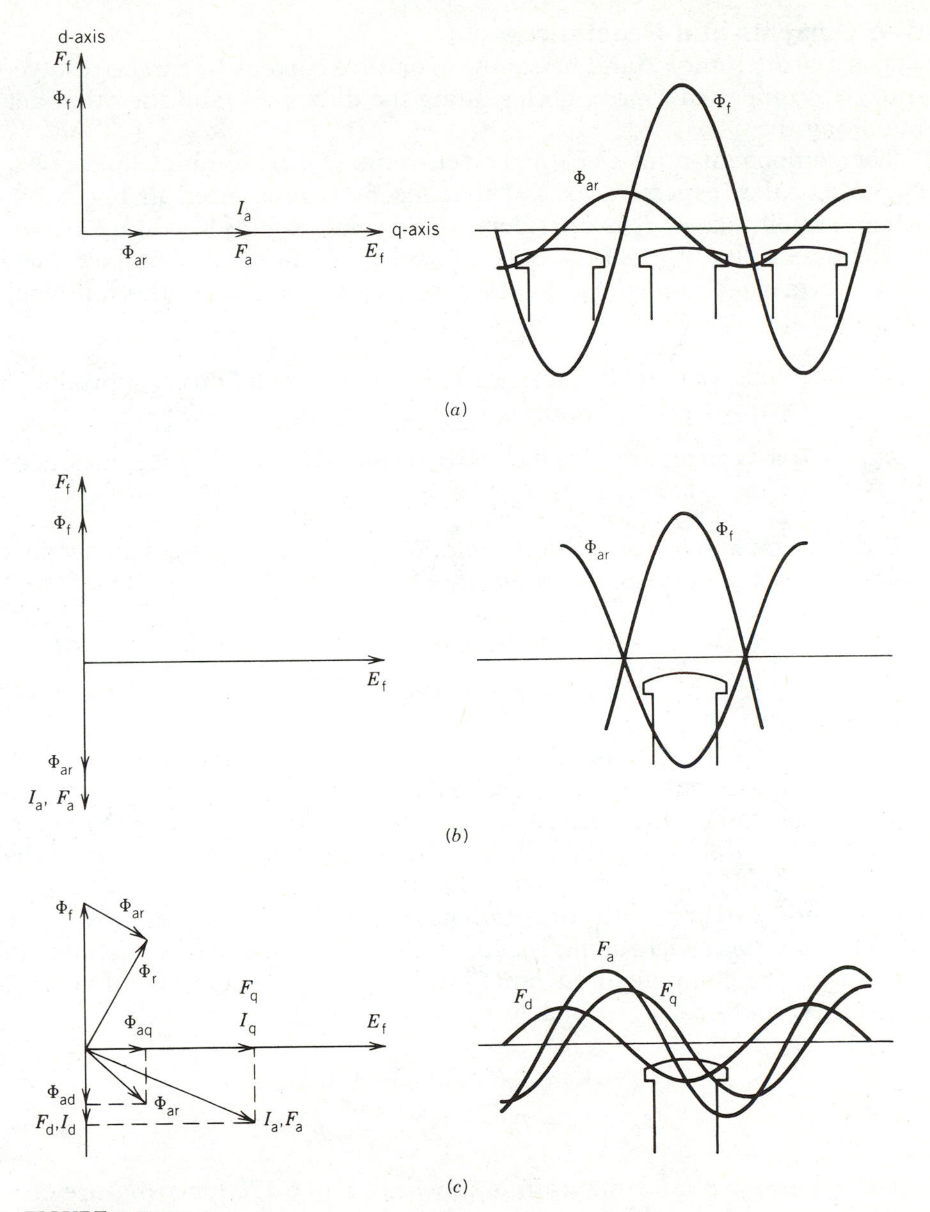

FIGURE 6.26
MMF and flux in salient pole synchronous machine.

d–q Currents and Reactances

The armature mmf F_a (and hence the armature current I_a) can be resolved into two components—one acting along the d-axis, F_d, and the other acting along the q-axis, F_q.

The component mmf's (F_d, F_q) or currents (I_d, I_q) produce fluxes (Φ_{ad}, Φ_{aq}) along the respective axes. This concept is illustrated in Fig. 6.26*c*, where stator current I_a is considered to lag the excitation voltage E_f.

The d-axis flux Φ_{ad} and the q-axis flux Φ_{aq} are along axes of fixed magnetic permeance, and these fluxes can be represented by the following reactances:

X_{ad} d-axis armature reactance to account for the flux Φ_{ad} produced by the d-axis current I_d.

X_{aq} q-axis armature reactance to account for the flux Φ_{aq} produced by the q-axis current I_q.

If the leakage inductance X_{al} is included to account for the leakage flux produced by the armature current, then

$$X_d = X_{ad} + X_{al}, \quad \textit{d-axis synchronous reactance} \tag{6.37}$$

$$X_q = X_{aq} + X_{al}, \quad \textit{q-axis synchronous reactance} \tag{6.38}$$

The armature leakage reactance X_{al} is assumed to be the same for both d-axis and q-axis currents, because leakage fluxes are primarily confined to the stator frame. Obviously, $X_d > X_q$ because reluctance along the q-axis is higher than that along the d-axis owing to the larger air gap along the q-axis. Normally, X_q is between 0.5 and 0.8 of X_d.

In the equivalent circuit for a salient pole synchronous machine, these d-axis and q-axis synchronous reactances must be considered, as shown in Fig. 6.27*a*. The component currents I_d and I_q produce component voltage drops jI_dX_d and jI_qX_q. The phasor relations are

$$E_f = V_t + I_aR_a + I_d jX_d + I_q jX_q \tag{6.39}$$

$$I_a = I_d + I_q \tag{6.40}$$

The generator phasor diagram is shown in Fig. 6.27*b* for armature current I_a lagging the excitation voltage E_f by an angle ψ (called the internal power factor angle). If the angle between E_f and I_a is known, the component currents I_q and I_d, respectively, are obtained by resolving the current I_a along E_f (which is along the q-axis) and perpendicular to it. However, normally the angle between I_a and V_t (which is the terminal power factor angle ϕ) is known and therefore the angle δ between V_t and E_f must be known to obtain the component currents I_d and I_q. The phasor diagram is drawn again in Fig. 6.27*c*, neglecting the armature resistance R_a.

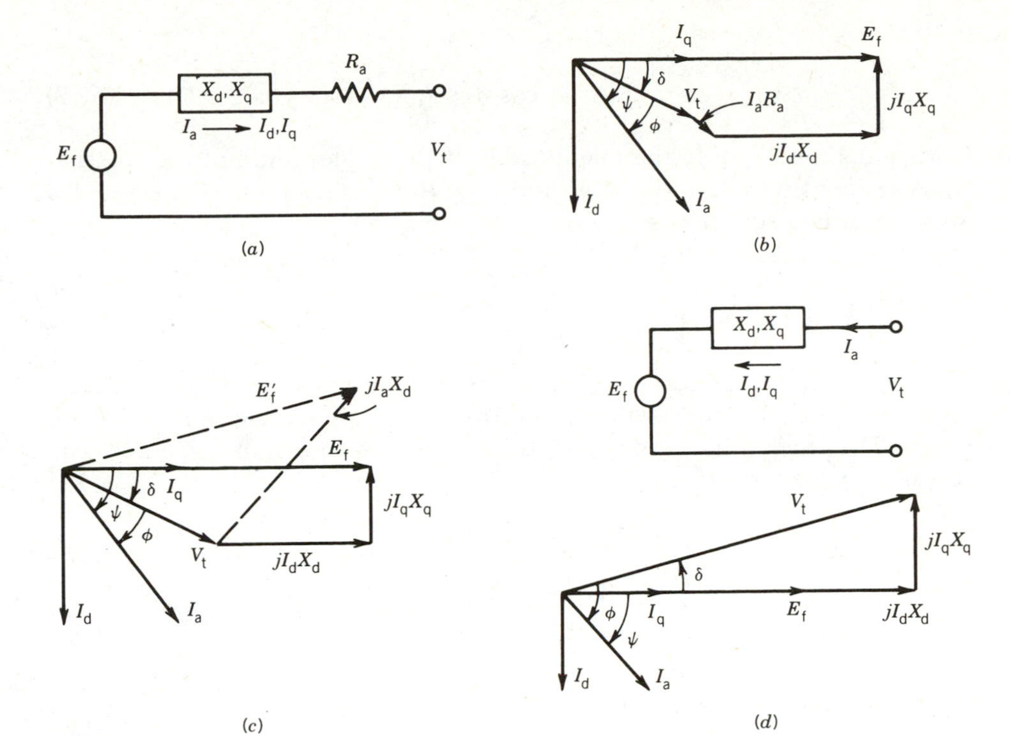

FIGURE 6.27
Equivalent circuit and phasor diagrams for the salient pole synchronous machine. (*a*),(*b*),(*c*) → Generator action. (*d*) → Motor action.

The equivalent circuit and the corresponding phasor diagram for motoring operation are shown in Fig. 6.27*d*. The phasor relation is

$$V_t = E_f + I_d jX_d + I_q jX_q \tag{6.41}$$

From the phasor diagram shown in Figs. 6.27*c* and 6.27*d*,

$$\psi = \phi \pm \delta \tag{6.42}$$

$$I_d = I_a \sin \psi = I_a \sin(\phi \pm \delta) \tag{6.43}$$

$$I_q = I_a \cos \psi = I_a \cos(\phi \pm \delta) \tag{6.44}$$

$$V_t \sin \delta = I_q X_q = I_a X_q \cos(\phi \pm \delta) \tag{6.45}$$

or

$$\tan \delta = \frac{I_a X_q \cos \phi}{V_t \pm I_a X_q \sin \phi} \tag{6.46}$$

and

$$E_f = V_t \cos \delta \pm I_d X_d \tag{6.47}$$

In Eqs. 6.42–6.47, only the magnitudes of the angles and not their actual signs are considered. In Eqs. 6.46 and 6.47, the plus sign is used if $\psi = \phi + \delta$ and the minus sign if $\psi = \phi - \delta$.

6.9.1 POWER TRANSFER

To simplify the derivation of expressions for the power and torque developed by a salient pole synchronous machine, neglect R_a and the core losses. The phasor diagram with E_f as reference is shown in Fig. 6.27*c*. The complex power per phase is

$$\begin{aligned} S &= V_t I_a^* \\ &= |V_t| \underline{/-\delta}\, (|I_q| - j|I_d|)^* \\ &= |V_t| \underline{/-\delta}\, (|I_q| + j|I_d|) \end{aligned} \tag{6.48}$$

From the phasor diagram (Fig. 6.27*c*)

$$|I_d| = \frac{|E_f| - |V_t| \cos \delta}{X_d} \tag{6.49}$$

$$|I_q| = \frac{|V_t| \sin \delta}{X_q} \tag{6.50}$$

If these values of I_d and I_d are substituted in Eq. 6.48,

$$S = \frac{|V_t|^2}{X_q} \sin \delta \underline{/-\delta} + \frac{|V_t||E_f|}{X_d} \underline{/90^\circ - \delta} - \frac{|V_t|^2}{X_d} \cos \delta \underline{/90^\circ - \delta} \tag{6.51}$$

$$= P + jQ$$

where P is the real power per phase and from Eq. 6.51

$$P = \frac{|V_t||E_f|}{X_d} \sin \delta + \frac{|V_t|^2 (X_d - X_q)}{2 X_d X_q} \sin 2\delta \tag{6.52}$$

$$= P_f + P_r \tag{6.52a}$$

and Q is the reactive power per phase and from Eq. 6.51

$$Q = \frac{|V_t||E_f|}{X_d} \cos \delta - |V_t|^2 \left| \frac{\sin^2 \delta}{X_q} + \frac{\cos^2 \delta}{X_d} \right| \tag{6.53}$$

If $X_d = X_q$ (i.e., no saliency), then from Eqs. 6.52 and 6.53

$$P = \frac{|V_t||E_f|}{X_d} \sin \delta \tag{6.54}$$

$$Q = \frac{|V_t||E_f|}{X_d} \cos \delta - \frac{|V_t|^2}{X_d} \tag{6.55}$$

These expressions for real and reactive power are the same as those of Eqs. 6.24 and 6.26 derived earlier for a cylindrical-rotor synchronous machine for which $X_d = X_q = X_s$.

Note carefully Eq. 6.52 for the real power in a salient pole synchronous machine. The first term, say P_f, represents power due to the excitation voltage E_f (the same as that obtained for cylindrical rotor machine). The second term, say P_r, represents the effects of salient poles and produces the *reluctance torque*. Note that the reluctance torque is independent of field excitation and vanishes if $X_d = X_q$.

The power angle characteristic is shown in Fig. 6.28, in which the excitation component P_f and the reluctance component P_r of the power are also indicated. The maximum resultant power is higher than that of a cylindrical-rotor machine for the same excitation voltage, and it occurs at δ less than 90°, making the curve steeper in the region of positive slope. This makes the machine respond quickly to changes in shaft torque.

A family of power angle characteristics at different values of excitation and constant terminal voltage is shown in Fig. 6.29. If the field excitation is reduced to zero, the machine can still develop power (or torque) because of the saliency of the rotor structure. This capability may extend the range of current operation if the machine is used as a synchronous condenser as

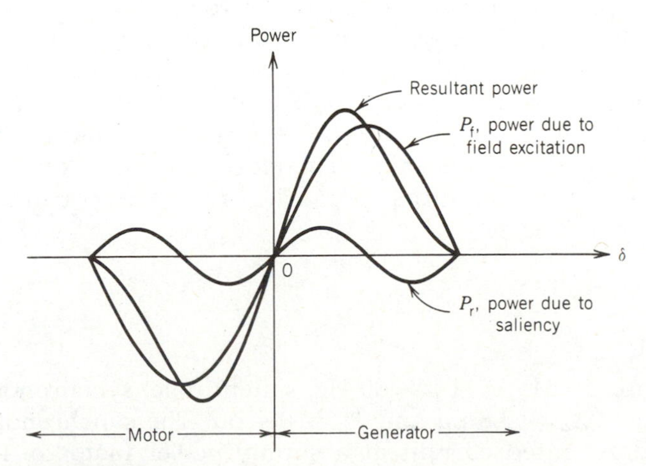

FIGURE 6.28
Power–angle characteristic of a salient pole synchronous machine.

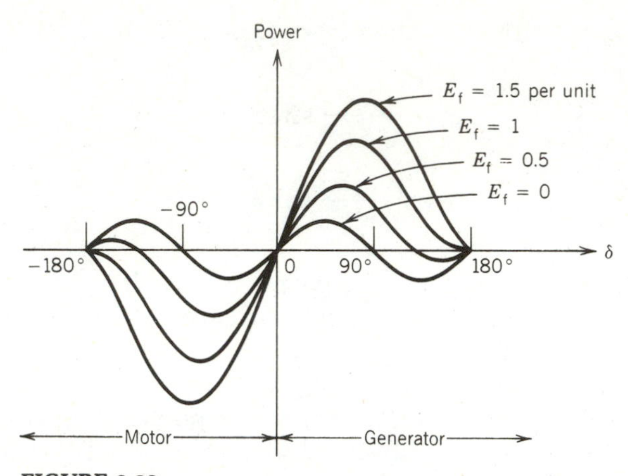

FIGURE 6.29
Power–angle characteristics for various field currents.

discussed in Section 6.7. In some power systems, there are situations where a synchronous condenser may have to be operated to draw large lagging current for proper voltage regulation. This result is achieved by under excitation of the synchronous machine. If the machine is a cylindrical-rotor machine, reducing the field current too far may make the developed torque less than the rotational torque required, and the machine will fall out of synchronism. If a salient pole machine is used the field current may be reduced to zero, even reversed, for the machine to draw more lagging current, and the reluctance torque can keep the machine in synchronism.

In a salient pole synchronous machine, if the excitation is varied over the normal operating range, the effects of the saliency on the power or torque developed are not significant. Only at low excitation does the power or torque due to saliency become important. In Fig. 6.27*c*, if saliency is neglected, the excitation voltage would be the one shown dashed and marked E_f', obtained from $E_f' = V_t + I_a jX_d$. Note that the magnitudes of E_f and E_f' are almost the same. Therefore, except at low excitation or when high accuracy is required, cylindrical-rotor theory can be used for a salient pole machine.

EXAMPLE 6.7

A three-phase, 50 MVA, 11 kV, 60 Hz, salient pole, synchronous machine has reactances $X_d = 0.8$ pu and $X_q = 0.4$ pu. The synchronous motor is loaded to draw rated current at a supply power factor of 0.8 lagging. Rotational losses are 0.15 pu. Neglect armature resistance losses.

(a) Determine the excitation voltage E_f in pu.

(b) Determine the power due to field excitation and that due to saliency of the machine.

(c) If the field current is reduced to zero, will the machine stay in synchronism?

(d) If the shaft load is removed before the field current is reduced to zero, determine the resultant supply current in pu and the supply power factor. Draw the phasor diagram for the machine in this condition.

Solution

(a) $V_t = 1\angle 0°$ pu, $I_a = 1\angle -36.9°$ pu.
From the phasor diagram (Fig. E6.7) $\psi = \phi - \delta$
From Eq. 6.46

$$\tan \delta = \frac{1 \times 0.4 \times 0.8}{1 - 1 \times 0.4 \times 0.6} = 0.42$$

$$\delta = 22.8°$$

From Fig. 6.27*d*

$$\psi = 36.9 - 22.8 = 14.1°$$

From Eqs. 6.43 and 6.44

$$I_d = 1 \sin 14.1 = 0.24 \text{ pu}$$

$$I_q = 1 \cos 14.1 = 0.97 \text{ pu}$$

From Eq. 6.47

$$E_f = 1 \cos 22.8 - 0.24 \times 0.8 = 0.73 \text{ pu}$$

(b) From Eq. 6.52, power due to field excitation

$$P_f = \frac{1 \times 0.73}{0.8} \sin 22.8 = 0.35 \text{ pu}$$

and power due to saliency of the machine

$$P_r = \frac{1^2 \times (0.8 - 0.4)}{2 \times 0.8 \times 0.4} \sin 45.6 = 0.45 \text{ pu}$$

(c) Power output is

$$P_{out} = V_t I_a \cos\phi = 1 \times 1 \times 0.8 = 0.8 \text{ pu}$$

From Eq. 6.52, the maximum power due to saliency of the machine is

$$P_r|_{max} = \frac{1^2 \times (0.8 - 0.4)}{2 \times 0.8 \times 0.4} = 0.63 \text{ pu}$$

The output power is more than the power the machine can develop. Therefore, the machine will lose synchronism.

(d) No-load power is 0.15 pu.

$$0.15 = 0.63 \sin 2\delta$$

or

$$\delta = 6.89°$$

The phasor diagram is shown in Fig. E6.7. With V_t as reference, the q-axis is $\delta = 6.89°$ behind it. The right-angle triangle is formed by V_t, $I_d X_d$, and $I_q X_q$.

$$I_d X_d = V_t \cos\delta = 1 \cos 6.89° = 0.99$$

$$I_d = \frac{0.99}{0.8} = 1.24 \text{ pu}$$

$$I_q X_q = V_t \sin\delta = 1 \sin 6.89° = 0.12$$

$$I_q = \frac{0.12}{0.4} = 0.3 \text{ pu}$$

$$I_a = (1.24^2 + 0.3^2)^{1/2} = 1.276 \text{ pu}$$

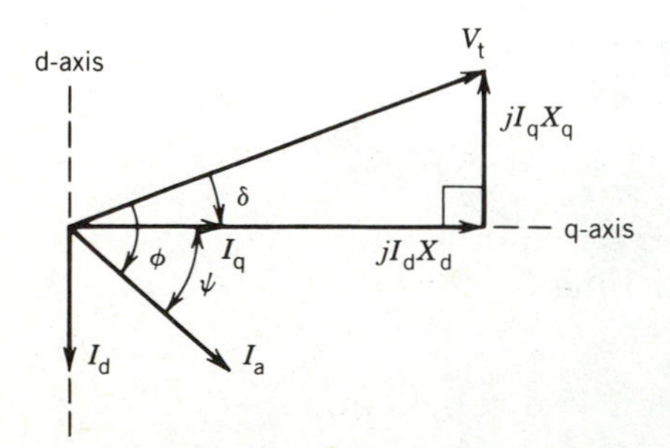

FIGURE E6.7

$$\psi = \tan^{-1}\frac{1.24}{0.3} = 76.4°$$

$$\phi = 76.4° + 6.89° = 83.3°$$

$$\text{Power factor} = \cos 83.3° = 0.117 \quad \text{lagging}$$

EXAMPLE 6.8

A 3ϕ, 12 kV, 15 MVA, 60 Hz, salient pole, synchronous motor is run from a 12 kV, 60 Hz, balanced three-phase supply. The machine reactances are $X_d = 1.2$ pu, $X_q = 0.6$ pu (with the machine rating as base). Neglect rotational losses and armature resistance losses. The machine excitation and load are varied to obtain the following conditions.

(a) Maximum power input is obtained with no field excitation. Determine the value of this power, the armature current, and the power factor for this condition.

(b) Rated power output is obtained with minimum excitation. Determine this minimum value of excitation emf.

Solution

(a) For no field excitation ($E_f = 0$) the machine behaves as a reluctance motor.

$$P_{in} = P_r = \frac{V_t^2(X_d - X_q)}{2X_dX_q} \sin 2\delta$$

For maximum power $2\delta = 90°$ or $\delta = 45°$.

$$P_{in} = \frac{1^2(1.2 - 0.6)}{2 \times 1.2 \times 0.6} = 0.416 \text{ pu}$$

The phasor diagram is drawn in Fig. E6.8.

$$|I_dX_d| = |I_qX_q| = V_t \cos 45° = \frac{1}{\sqrt{2}}$$

$$I_d = \frac{1}{1.2 \times \sqrt{2}} = 0.59 \text{ pu}$$

$$I_q = \frac{1}{0.6 \times \sqrt{2}} = 1.18 \text{ pu}$$

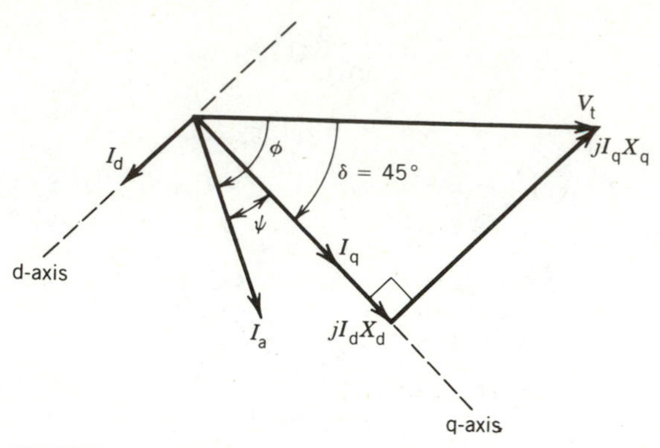

FIGURE E6.8

$$|I_a| = \sqrt{0.59^2 + 1.18^2} = 1.32 \text{ pu}$$

$$\psi = \tan^{-1}\frac{0.59}{1.18} = 26.6°$$

$$\phi = 45 + 26.6 = 71.6°$$

$$\text{Power factor} = \cos 71.6° = 0.32 \quad \text{lagging}$$

(b) From Eq. 6.52

$$P = \frac{E_f \times 1}{1.2}\sin\delta + \frac{1^2(1.2 - 0.6)}{2 \times 1.2 \times 0.6}\sin 2\delta$$

$$= 0.834E_f \sin\delta + 0.416 \sin 2\delta \tag{6.56}$$

From Fig. 6.29, for maximum power

$$\frac{dp}{d\delta} = 0$$

Differentiating Eq. 6.56 with respect to δ and setting it to zero yields

$$0 = 0.834E_f \cos\delta + 0.832\cos 2\delta$$

$$0.834E_f = -0.832\frac{\cos 2\delta}{\cos\delta} \tag{6.57}$$

From Eqs. 6.56 and 6.57

$$P_{\max} = -\frac{0.832\cos 2\delta}{\cos\delta}\sin\delta + 0.416\sin 2\delta$$

$$= -0.832(2\cos^2\delta - 1)\tan\delta + 0.416\sin 2\delta$$

$$= -0.832\sin 2\delta + 0.832\tan\delta + 0.416\sin 2\delta$$

$P_{max} = P_{rated} = 1$ pu

$$1 = -0.832\sin 2\delta + 0.832\tan\delta + 0.416\sin 2\delta$$

or

$$\tan\delta = 1.2 + 0.5\sin 2\delta \tag{6.58}$$

From this equation, by trial and error,

$$\delta = 58.5°$$

From Eq. 6.57

$$E_f = -\frac{0.832}{0.834}\frac{\cos 2 \times 58.5}{\cos 58.5} = 0.87 \text{ pu}$$

6.9.2 DETERMINATION OF X_d AND X_q

As discussed earlier, the d-axis and q-axis reactances are the maximum and minimum values of the armature reactance, respectively, for various rotor positions. These reactances can be measured by the *slip test*. In this test, the rotor of the machine is driven at a speed slightly different from the synchronous speed with the rotor field left open-circuited. The stator is excited by a three-phase source. The rotor must rotate in the same direction as the armature rotating mmf to make the induced voltage in the field winding of low value and frequency. Because the salient pole rotor rotates at a slightly different speed than the stator rotating flux, the latter encounters varying reluctance path. The stator current therefore oscillates, and if the difference between the rotor speed and synchronous speed is small, the pointer in the ammeter connected to the stator terminal will swing back and forth slowly. An oscillogram of the stator current is shown in Fig. 6.30. From the maximum and minimum currents, the reactances can be calculated as follows:

$$X_d = \frac{V_t}{i_{min}/\sqrt{2}} \tag{6.59}$$

$$X_q = \frac{V_t}{i_{max}/\sqrt{2}} \tag{6.60}$$

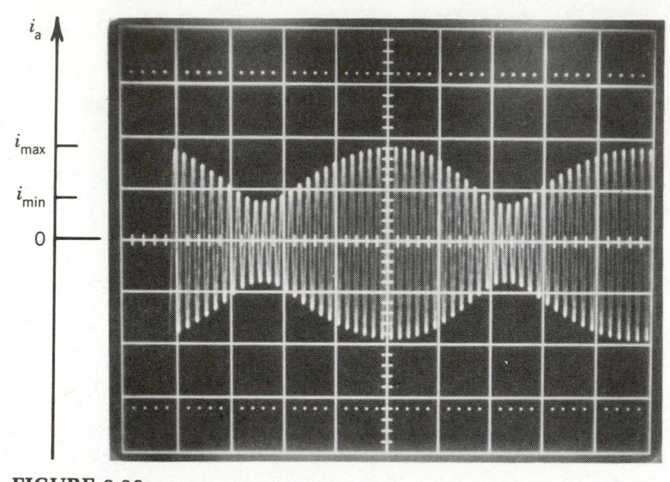

FIGURE 6.30
Armature current during the slip test.

6.10 SPEED CONTROL OF SYNCHRONOUS MOTORS

The speed of a synchronous motor can be controlled by changing the frequency of the power supply. At any fixed frequency the speed remains constant, even for changing load conditions, unless the motor loses synchronism. The synchronous motor is therefore very suitable for accurate speed control and also where several motors have to run in synchronism. A synchronous motor can run at high power factor and efficiency (no power losses due to slip as in the induction motor). At present, it is being increasingly considered for use in variable-speed drives.

Two types of speed control methods are normally in use and are discussed here. In one method the speed is directly controlled by changing the output voltage and frequency of an inverter or cycloconverter. In the other method the frequency is automatically adjusted by the motor speed and the motor is called a "self-controlled" synchronous motor.

6.10.1 FREQUENCY CONTROL

Schematic diagrams for open-loop speed control of a synchronous motor by changing the output frequency and voltage of an inverter or a cycloconverter are shown in Fig. 6.31. The inverter circuit (Fig. 6.31*a*) allows variation of frequency (and hence motor speed) over a wide range, whereas the cycloconverter circuit (Fig. 6.31*b*) permits variation of frequency below one-third of the supply frequency.

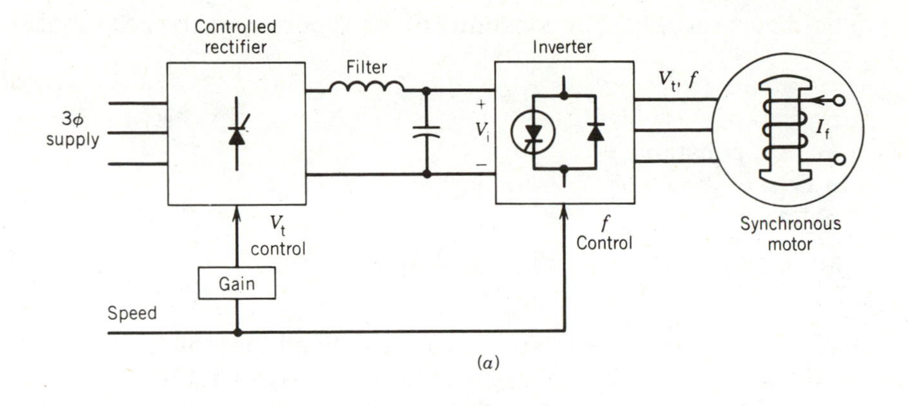

(a)

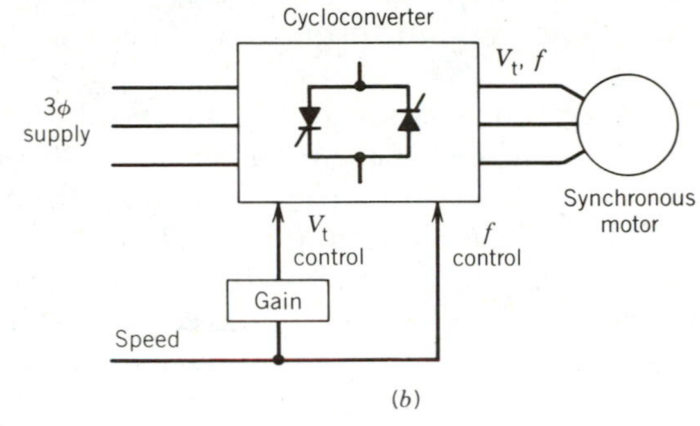

(b)

FIGURE 6.31
Open-loop frequency control.

To obtain the same maximum torque over the whole range of speed variation and also to avoid magnetic saturation in the machine, it is necessary to change the voltage with the frequency. From Eq. 6.24, for a three-phase synchronous machine,

$$P = T\omega_m = \frac{3V_tE_f}{X_s}\sin\delta \tag{6.61}$$

Now,

$$\omega_m = \frac{4\pi f}{p} \tag{6.62}$$

Let

$$X_s = 2\pi f L_s \tag{6.63}$$

If the field current I_f is kept constant, E_f is proportional to speed and so

$$E_f = K_1 f \tag{6.64}$$

where K_1 is a constant.

From Eqs. 6.61 to 6.64,

$$T = K \frac{V_t}{f} \sin \delta \tag{6.65}$$

where K is a constant. A base speed can be defined for which V_t and f are the rated values for the motor. If the ratio V_t/f corresponding to this base speed is maintained at lower speeds by changing voltage with frequency, the maximum torque (i.e., pull-out torque, KV_t/f) is maintained equal to that at the base speed.

The torque–speed characteristic for variable-voltage, variable-frequency (VVVF) operation of the synchronous motor is shown in Fig. 6.32. For regenerative braking of the synchronous motor, the power flow reverses. Note that in the inverter system (Fig. 6.31*a*), because of diodes in the inverter, voltage at the input to the inverter cannot reverse. Therefore, for reverse power flow, current reverses. A dual converter, as shown in Fig. 6.33, is therefore necessary to accept the reversed current flow in the regenerating mode of operation. In a cycloconverter, however, the power flow is reversible. To vary the speed above the base speed, the frequency is increased while maintaining the terminal voltage at the rated value. This, of course, will make the pull-out torque decrease at higher speed range, as shown in Fig. 6.32. In Fig. 6.31*a*, if the controlled rectifier is replaced by a diode-bridge rectifier, the inverter input voltage V_i is constant. The in-

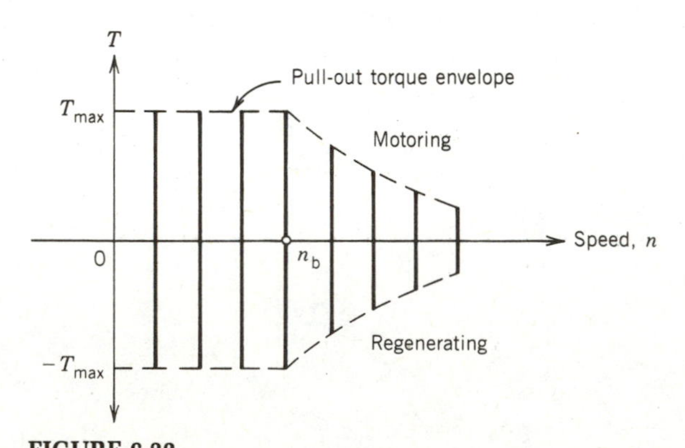

FIGURE 6.32
Torque–speed characteristic of synchronous machines.

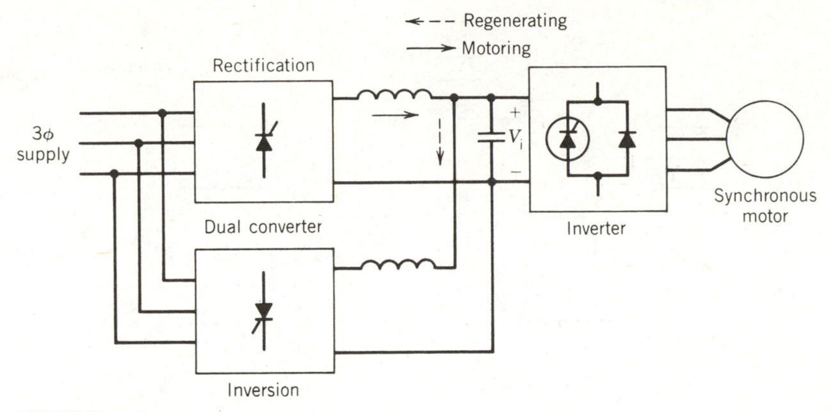

FIGURE 6.33
Controller with reversible power flow.

verter can be a PWM (pulse width-modulated) inverter so that both output voltage and frequency can be changed in the inverter.

Note that if the frequency is suddenly changed or changed at a high rate, the rotor poles may not be able to follow the stator rotating field and the motor will lose synchronism. Therefore, the rate at which the frequency is changed must be restricted. A sudden change in load torque may also cause the motor to lose synchronism. The open-loop drive is therefore not suitable for applications in which load may change suddenly.

6.10.2 SELF-CONTROLLED SYNCHRONOUS MOTOR

A synchronous motor tends to lose synchronism on shock loads. In the open-loop speed control system discussed in the preceding section, if a load is suddenly applied the rotor momentarily slows down, making the torque angle δ increase beyond 90° and leading to loss of synchronism. However, if the rotor position is sensed as the rotor slows down and the information is used to decrease the stator frequency, the motor will stay in synchronism. In such a scheme, the rotor speed will adjust the stator frequency and the drive system is known as a *self-controlled synchronous motor* drive.

The schematic of a self-controlled synchronous motor drive system is shown in Fig. 6.34. Two controlled rectifiers are used, one at the supply end and the other at the machine end. In the motoring mode of operation the supply end rectifier operates in the rectification mode and the machine end rectifier in the inversion mode. The roles of the rectifiers reverse for regenerative braking, in which the power flow reverses. The thyristors in the supply end rectifier are commutated by the supply line voltage and

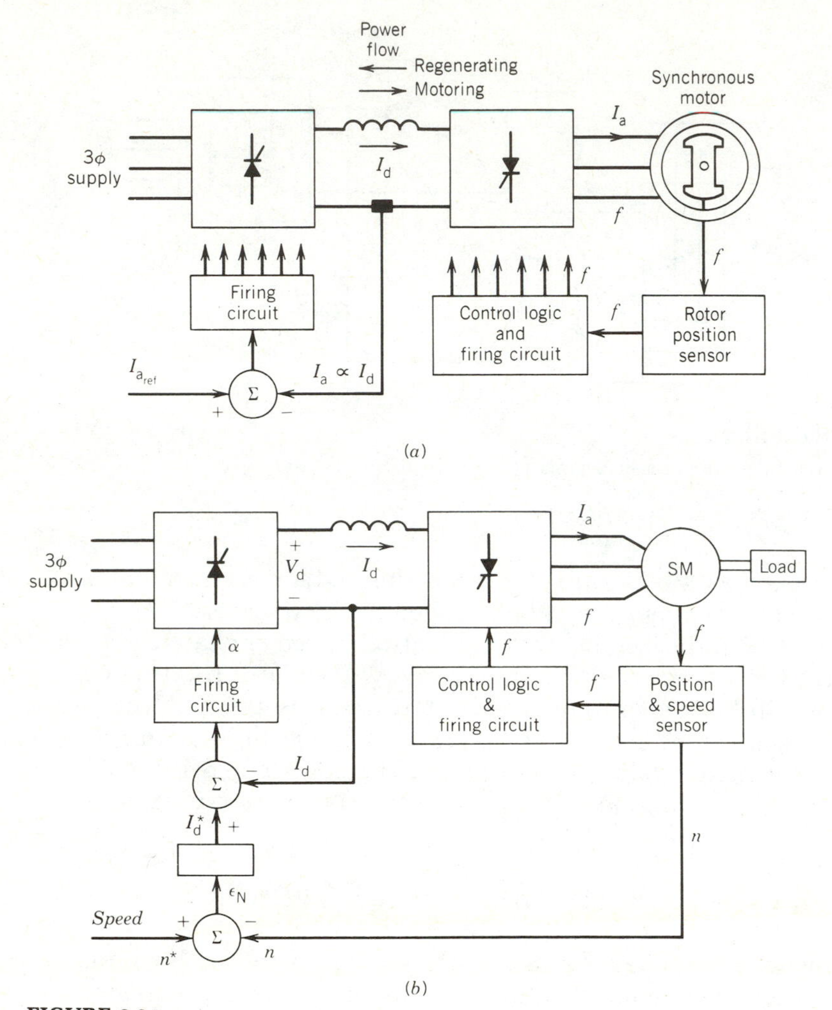

FIGURE 6.34
Self-controlled synchronous motor drive. (*a*) Open-loop control. (*b*) Closed-loop control.

those in the machine end rectifier by the excitation voltage of the synchronous machine.

The rotor position sensor, mounted on the rotor shaft, generates signals having rotor position information. These signals are processed in the control logic circuit and used to fire the thyristors of the machine end rectifier. Therefore, any change in the rotor speed due to change in load will immediately change the frequency of firing of the thyristors and hence adjust the stator frequency at the correct rate to maintain synchronism.

A current loop is implemented around the supply end rectifier to maintain the machine current at the desired value. The dc link current I_d, being proportional to the machine current I_a, is compared with the reference current, and the error signal adjusts the firing of the supply end rectifier to keep the armature current constant at the reference value.

From Eq. 6.29j, the torque depends on the value of the angle β. This angle β can be controlled in the control logic circuit for the machine end rectifier, because the signal from the rotor position sensor defines the position of the field axis and the firing instant of the thyristors defines the position of the armature field axis. If angle β is regulated at a specified value and the field current is kept constant, the torque (and hence the speed) can be directly controlled by the armature current. This current is controlled by the current control loop of the supply end rectifier.

Both rectifiers are simple and inexpensive circuits and, unlike forced commutated inverters, do not require commutation circuits for commutation of the thyristors. As a result, such drive systems can be designed for very high power applications—in the order of a megawatt.

6.10.3 CLOSED-LOOP CONTROL

Notice that in the drive system shown in Fig. 6.34*a*, if the load torque is changed, the speed will change. If the speed is to be maintained constant, the dc link current I_d must be adjusted to satisfy the change in the load torque. This can be achieved by inserting an outer speed loop as shown in Fig. 6.34*b*. The position sensor supplies information about the position of the rotor field as well as the speed of the rotor. If the speed drops because of an increase in load torque on the synchronous motor, the speed error ε_N increases, which in turn increases the current demand I_d^*. Consequently, the firing angle of the controlled rectifier will change to alter the dc link current I_d to produce more torque as required by the increased load torque. The speed will eventually be restored to its initial value.

6.11 APPLICATIONS

There are many applications of synchronous machines. The primary use of the machine is as an ac generator or alternator. It provides more than 99 percent of the electrical energy used by people all over the world. Although significant research and development are under way in pursuit of new types of generators, such as fuel cell, thermoelectric, magnetohydrodynamic (MHD), and solar energy generators, it is conceded that synchronous generators will continue to be the major electrical energy generators for many years to come.

The synchronous motor has a wide range of applications. Its constant-speed operation (even under load variation and voltage fluctuation) and high efficiency (92–96 percent, the highest of all motors) make it most suitable for constant-speed, continuous-running drives such as motor–generator sets, air compressors, centrifugal pumps, blowers, crushers, and many types of continuous-processing mills.

A unique feature of synchronous motors is their power factor control capability. If the motor is overexcited, it draws leading reactive current. The overexcited synchronous motor can be used to compensate for a large number of induction motors that draw lagging reactive current. The leading reactive current drawn by the synchronous motor can improve the plant power factor, while at the same time such motors can act as prime movers for some drives in the plant. An unloaded synchronous motor may be used as a synchronous condenser (overexcited) or reactor (underexcited) to regulate the voltage at the receiving end of a long power transmission line.

A disadvantage of the synchronous motor is that direct current is required for its excitation. However, almost all modern motors are provided with brushless excitation. In this system, a small exciter is mounted on the rotor shaft, the output of which is rectified by shaft-mounted rectifiers rotating with the rotor, thus avoiding the need for slip rings and brushes. A pictorial view of a brushless excitation system is shown in Fig. 6.35.

Because of their higher initial cost compared to induction motors, synchronous motors are not suitable at high speed or below 50 hp in the medium-speed range. For low-speed and high-hp applications, the induction motor is not cheaper than the synchronous motor, because a large amount of iron is required to establish a high air gap flux density, in the neighborhood of 1 tesla. In a synchronous motor, because of separate exci-

FIGURE 6.35
Rotor of a brushless synchronous generator, with rotating diodes. (Courtesy of General Electric Canada Inc.)

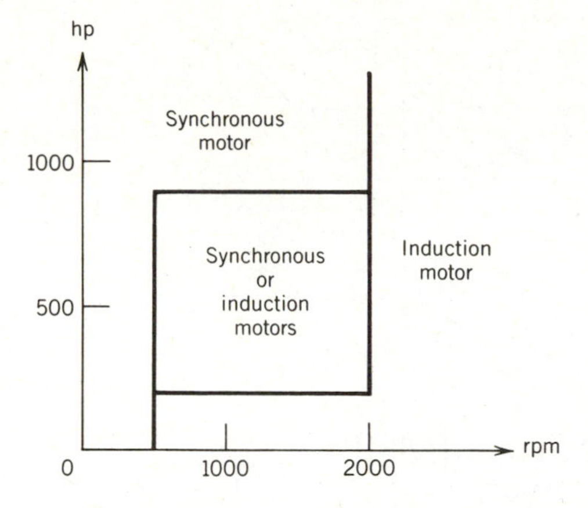

FIGURE 6.36
Application of synchronous and induction motors.

tation by the rotor field, a value almost twice this figure is permissible. Furthermore, the advantages of constant-speed operation, power factor control, and high efficiency may override the disadvantage of higher initial cost for some horsepower and speed ranges, as shown in Fig. 6.36.

In low-speed and high-horsepower drives, some typical applications where synchronous motors are used are large low-head pumps, flour-mill line shafts, ship propulsion, rubber mills and mixers, and pulp grinders.

In variable-speed drive systems, synchronous motors with solid-state control are being considered seriously. The converters in the self-controlled synchronous motor drive system (Fig. 6.34) are simple and inexpensive, because commutation of thyristors is provided by the synchronous machine itself. This system is a serious contender for application as a variable-speed drive. In urban transit system or locomotive drives, dc motors have been used for years. However, dc motors require frequent maintenance and are expensive. Induction motors with inverter control have been tried with limited success. Although induction motors are rugged, the complexity of the inverter control circuit, the relatively poor efficiency, and the poor power factor associated with these motors have not made them attractive. The synchronous motor has fewer maintenance problems than the dc motor and has higher power factor and efficiency than the induction motor. Solid-state power conditioners for synchronous motors can be simpler and less expensive than those required for induction motors. Consequently, the self-controlled synchronous motor drive system (Fig. 6.34) is being considered as a viable alternative for transit system and locomotive drives. A permanent-magnet synchronous motor is as rugged as an induction motor and has great potential for application in variable-speed drive systems.

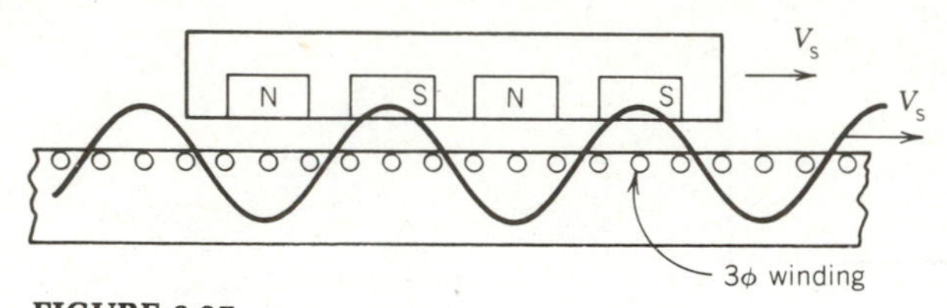

FIGURE 6.37
Linear synchronous motor (LSM).

6.12 LINEAR SYNCHRONOUS MOTOR (LSM)

A linear version of a synchronous motor can produce linear or translational motion. Figure 6.37 shows a schematic diagram of a linear synchronous motor. One member (say the primary) has a three-phase winding, and the other member (say the secondary) has electromagnets, permanent magnets, or superconducting magnets. If the three-phase winding is connected to a three-phase supply, a traveling flux wave will move along the length of the primary. The magnets on the secondary will be synchronously locked with the traveling wave and move at a velocity V_s,

$$V_s = 2T_p f \tag{6.66}$$

where T_p is the pole pitch and f is the frequency of the supply. If powerful magnets such as superconducting magnets are used, the LSM can operate at a higher air gap. It can also operate at a leading power factor and better efficiency (because of no-slip power loss).

The equivalent circuit used for a rotary synchronous motor (Fig. 6.14) can also be used for an LSM. If powerful magnets are used, the field due to currents in the three-phase winding is insignificant compared to the field due to the powerful magnets. Consequently, the armature reaction effect can be neglected in an LSM and the synchronous reactance is due primarily to the leakage reactance.

At present, the linear synchronous motor has not been used as widely as the linear induction motor. However, the LSM has great potential in the field of high-speed transportation,[2] where a large air gap clearance is needed. In many countries research is under way in developing high-speed vehicles, with speeds up to 500 km/hr, using an LSM. Figure 6.38 shows a Japanese test vehicle that may be put into service in the 1990s.

---

[2] P. C. Sen, On Linear Synchronous Motor (LSM) for High Speed Propulsion, *IEEE Transactions on Magnetics*, vol. Mag. 11, no. 5, pp. 1484–1486, 1975.

FIGURE 6.38
Japanese test vehicle using LSM. (Courtesy of Prof. G. E. Dawson.)

PROBLEMS

6.1 The synchronous machine of Example 6.2 is connected to a 3ϕ, 14 kV, 60 Hz infinite bus and draws 5 MW at 0.85 leading power factor.

(a) Determine the values of the stator current (I_a), the excitation voltage (E_f), and the field current (I_f). Draw the phasor diagram.

(b) If the synchronous motor is disconnected from the infinite bus without changing the field current, determine the terminal voltage before the speed decreases.

6.2 A three-phase, 250 hp, 2300 V, 60 Hz, Y-connected nonsalient rotor synchronous motor has a synchronous reactance of 11 ohms per phase. When it draws 165.8 kW the power angle is 15 electrical degrees. Neglect ohmic losses.

(a) Determine the excitation voltage per phase, E_f.

(b) Determine the supply line current, I_a.

(c) Determine the supply power factor.

(d) If the mechanical load is thrown off and all losses become negligible,

i. Determine the new line current and supply power factor.

ii. Draw the phasor diagram for the condition in (i).

iii. By what percent should the field current I_f be changed to minimize the line current?

6.3 A 3ϕ, 2000 kVA, 11 kV, 1800 rpm synchronous generator has a resistance of 1.5 ohms and synchronous reactance of 15 ohms per phase.

(a) The field current is adjusted to obtain the rated terminal voltage at open circuit.

i. Determine the excitation voltage E_f.

ii. If a short circuit is applied across the machine terminals, find the stator current.

(b) The synchronous machine is next connected to an infinite bus. The generator is made to deliver the rated current at 0.8 power factor lagging.

i. Determine the excitation voltage E_f.

ii. Determine the percentage increase in the field current relative to the field current of part (a).

iii. Determine the maximum power the synchronous machine can deliver for the excitation current of part (b). Neglect R_a.

6.4 A 3ϕ, 20 kVA, 208 V, four-pole star-connected synchronous machine has a synchronous reactance of $X_s = 1.5\ \Omega$ per phase. The resistance of the stator winding is negligible. The machine is connected to a 3ϕ, 208 V infinite bus. Neglect rotational losses.

(a) The field current and the mechanical input power are adjusted so that the synchronous machine delivers 10 kW at 0.8 lagging power factor. Determine the excitation voltage (E_f) and the power angle (δ).

(b) The mechanical input power is kept constant but the field current is adjusted to make the power factor unity. Determine the percent change in the field current with respect to its value in part (a).

6.5 A 3ϕ, 2300 V, 60 Hz, 12-pole, Y-connected synchronous motor has 4.5 ohms per phase synchronous reactance and negligible stator winding resistance. The motor is connected to an infinite bus and draws 250 amperes at 0.8 power factor lagging. Neglect rotational losses.

(a) Determine the output power.

(b) Determine the power to which the motor can be loaded slowly without losing synchronism. Determine the torque, stator current, and supply power factor for this condition.

6.6 A 1 MVA, 3ϕ, 2300 V, 60 Hz, 10 pole, star-connected cylindrical-rotor synchronous motor is connected to an infinite bus. The synchronous reactance is 0.8 pu. All losses may be neglected. The synchronous motor delivers 1000 hp and the motor operates at 0.85 power factor leading.

(a) Determine the excitation voltage E_f.

(b) Determine the maximum power and torque the motor can deliver for the excitation current of part (a).

(c) The power output is kept constant at 1000 hp and the field current is decreased. By what factor can the field current of part (a) be reduced before synchronism is lost?

6.7 A 3ϕ cylindrical rotor synchronous machine and a shunt dc machine are mechanically coupled to transfer power from a dc source to an ac source and vice versa. The ratings of the machines are

Synchronous machine:	12 kVA, 208 V
	$X_s = 3.0\ \Omega$
DC machine:	12 kW, 220 V

Neglect all losses.

The dc machine is connected to a 220 V dc bus and the synchronous machine is connected to a 3ϕ, 208 V, 60 Hz bus. The excitation of the synchronous machine is made 1.25 pu.

(a) For zero power transfer, determine the armature current in the dc machine and the current and power factor of the synchronous machine.

(b) Eight kilowatts is transferred from the dc bus to the ac bus through the two machines. What adjustment is necessary? Determine the armature current in the dc machine and the stator current and power factor of the synchronous machine.

(c) Repeat part (b) if 8 kW is transferred from the ac bus to the dc bus.

6.8 A 3ϕ, 4.6 kV, 60 Hz, four-pole, Y-connected synchronous machine has the following current ratings:

Armature current rating = 62.75 A

Field current rating = 15.0 A

Rated voltage synchronous reactance X_s = 1.25 pu

The excitation voltage (E_f) at the rated speed is 4.6 kV (line-to-line) when the field current is 7.5 amps.

(a) Determine the kVA rating of the machine.

(b) Construct the capability curve for the machine for generator operation. Use per-unit values.

(c) Determine the power factor and the power angle for optimum operating conditions; that is, field heating and armature heating are both equal to their allowable maxima.

6.9 A 3ϕ, 100 MVA, 12 kV, 60 Hz salient pole, synchronous machine has X_d = 1.0 pu, X_q = 0.7 pu, and negligible stator resistance. The machine is connected to an infinite bus and delivers 72 MW at 0.9 power factor lagging.

(a) Determine the excitation voltage and the power angle. Draw the phasor diagram with V_t as reference.

(b) Determine the maximum power the synchronous generator can supply if the field current is made zero. Determine the machine current and power factor for this condition. Draw the phasor diagram.

6.10 A 3ϕ, 40 MVA, 11 kV, 60 Hz, salient pole, synchronous machine has X_d = 1.5 pu, X_q = 1.0 pu, and negligible stator resistance. The machine is connected to an infinite bus and the field current is adjusted to make the excitation voltage equal to the bus voltage. Determine the maximum value of the steady-state power that the machine can supply. Find the stator current (I_a) and the power factor at this maximum power condition. Draw the phasor diagram corresponding to this case.

6.11 A 3ϕ salient pole synchronous machine has reactances X_d = 1.2 pu and X_q = 0.6 pu. Neglect armature resistance losses.

(a) The machine operates as a synchronous motor and draws 0.8 pu of power at a power factor of 0.8 leading.

i. Determine the power angle δ and the excitation voltage E_f in pu and draw the phasor diagram.

ii. Determine the power due to excitation and that due to saliency of the machine.

(b) The machine operates as a synchronous generator and delivers 0.8 pu of power at the PF of 0.8 leading. Determine the excitation voltage E_f in pu and the power angle δ.

6.12 (a) A 3ϕ, cylindrical-rotor synchronous machine has $X_s = X_d = 0.9$ pu and negligible stator resistance. The machine delivers rated power to an infinite bus. Determine the minimum value of the excitation voltage in pu that will keep the machine in synchronism.

(b) Now consider a 3ϕ, salient pole synchronous machine that has $X_d = 0.9$ pu, $X_q = 0.6$ pu, and negligible stator resistance. Determine the minimum value of the excitation voltage in pu that will keep the machine in synchronism while delivering rated power to the infinite bus.

6.13 A 3ϕ, 10 MVA, 14 kV, 60 Hz synchronous machine has negligible stator winding resistance and a synchronous reactance of 16.5 ohms per phase. The machine is connected to an infinite bus (3ϕ, 14 kV, 60 Hz) and delivers power to a mechanical load. The rotational losses can be neglected. The magnetization characteristics of the machine are shown in Fig. E6.2.

Write a computer program to study the variation of motor terminal current (I_a) and power factor (PF) with field current (I_f). The program should yield

(a) A computer printout in tabular form showing the variation of I_a, PF with I_f for input power of 5 MW, 10 MW, and 15 MW.

(b) A plot of I_a, PF versus I_f.

SINGLE-PHASE MOTORS

CHAPTER seven

Single-phase motors are small motors, mostly built in the fractional horsepower range. These motors are used for many types of equipment in homes, offices, shops, and factories. They provide motive power for washing machines, fans, refrigerators, lawn mowers, hand tools, record players, blenders, juice makers, and so on. The average home in North America uses a dozen or more single-phase motors. In fact, the number of single-phase (fractional horsepower) motors today far exceeds the number of integral horsepower motors of all types.

Single-phase motors are relatively simple in construction. However, they are not always easy to analyze. Because of the large demand, the market is competitive and the designer uses tricks to save production costs.

Single-phase motors are of three types:

1. Single-phase induction motors.
   The majority of fractional horsepower motors are of the induction type. They are classified according to the methods used to start them and are referred to by names descriptive of these methods. Some common types are resistance-start (split-phase), capacitor-start, capacitor-run, and shaded-pole.

2. Single-phase synchronous motors.
   Motors of the synchronous type run at constant speed and are used in applications such as clocks and turntables where a constant speed is required. Two types of single-phase synchronous motors are in common use: the reluctance type and the hysteresis type.

3. Single-phase series (or universal) motors.
   Motors of the series type can be used with either a dc supply or a single-phase ac supply. These motors provide high starting torque and can operate at high speed. They are widely used in kitchen equipment, portable tools, and vacuum cleaners, where high-speed operation permits high horsepower per unit motor size.

In this chapter the operation of these single-phase motors is described. Methods of analysis for determining their starting and running characteristics are outlined. In some cases, design equations are derived that help in obtaining certain desirable characteristics.

7.1 SINGLE-PHASE INDUCTION MOTORS

Motors of the induction type have cage rotors and a single-phase distributed stator winding. Figure 7.1*a* shows a schematic diagram of a single-phase induction motor. Such a motor inherently does not develop any starting torque and therefore will not start to rotate if the stator winding is connected to an ac supply. However, if the rotor is given a spin or started by auxiliary means, it will continue to run. These starting methods will be discussed later. First, the basic properties of the elementary motor are discussed.

7.1.1 DOUBLE REVOLVING FIELD THEORY

The operation of the single-phase induction motor can be explained and analyzed by the double revolving field theory, as explained in this section.

Rotor at Standstill

First consider that the rotor is stationary and the stator winding is connected to a single-phase ac supply. A pulsating mmf, hence a pulsating flux Φ_s as shown in Fig. 7.1*b*, is established in the machine along the axis of the stator winding. This pulsating stator flux induces current by transformer

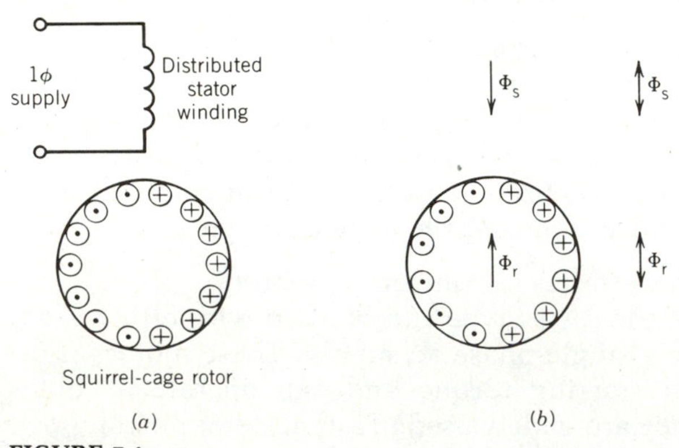

FIGURE 7.1
Single-phase motor. (*a*) Schematic. (*b*) Stator and rotor fluxes.

action in the rotor circuit, which in turn produces a pulsating flux Φ_r acting along the same axis as the stator flux Φ_s. By Lenz's law, these two fluxes tend to oppose each other. As the angle between these fluxes is zero, no starting torque is developed.

Rotor Running

Now assume that the rotor is running. This can be done either by spinning the rotor or by using auxiliary circuits. The single-phase induction motor can develop torque when it is in the running condition, which can be explained as follows.

A pulsating field (mmf or flux) is equivalent to two rotating fields of half the magnitude but rotating at the same synchronous speed in opposite directions. This can be proved either analytically or graphically.

Consider two vectors of equal magnitude $0P$, f moving forward in the anticlockwise direction and b moving backward in the clockwise direction as shown in Fig. 7.2. They rotate at the same speed in opposite directions. Their vector sum $0R$ alternates in magnitude between $+20P$ and $-20P$ and always lies along the same straight line. Moreover, $0R$ is a sine function of time if the vectors rotate at the same constant speed. Therefore, the pulsating field (represented by $0R$) produced by the current in the stator winding may be regarded as the resultant of two rotating fields (represented by f and b) of the same magnitude but rotating in opposite directions. The pulsating stator flux Φ_s, pulsating along the axis of the stator winding, is equivalent to two rotating fluxes Φ_f and Φ_b as shown in Fig. 7.2*b*.

Mathematically, for a sinusoidally distributed stator winding, the mmf along a position θ as shown in Fig. 7.3 is

$$F(\theta) = Ni \cos \theta$$

where N is the effective number of turns of the stator winding.

Let $i = I_{max} \cos \omega t$. Therefore,

$$\begin{aligned} F(\theta, t) &= NI_{max} \cos \theta \cos \omega t \\ &= \frac{NI_{max}}{2} \cos(\omega t - \theta) + \frac{NI_{max}}{2} \cos(\omega t + \theta) \\ &= F_f + F_b \end{aligned}$$

where F_f represents a rotating mmf in the direction of θ, and F_b represents a rotating mmf in the opposite direction. Both of these rotating mmf's produce induction motor torque, although in opposite directions. These component torques and the resultant torque are shown in Fig. 7.4. At standstill, these two torques, forward and backward, are equal in magnitude and therefore the resultant starting torque is zero. At any other speed,

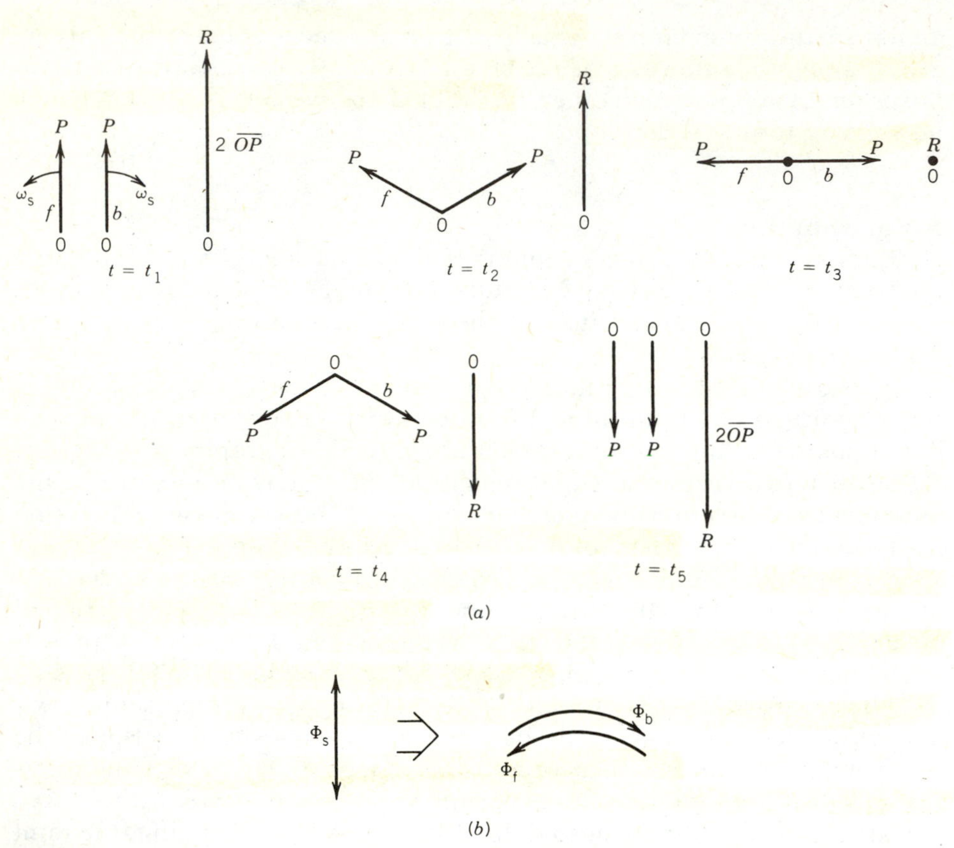

FIGURE 7.2
Pulsating field and rotating field.

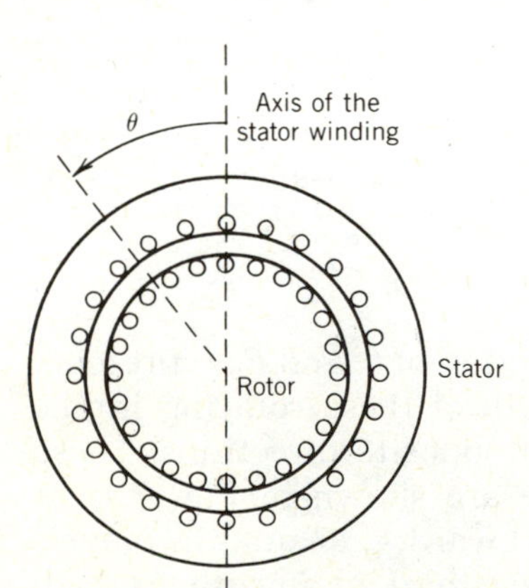

FIGURE 7.3
Cross section of a single-phase induction motor.

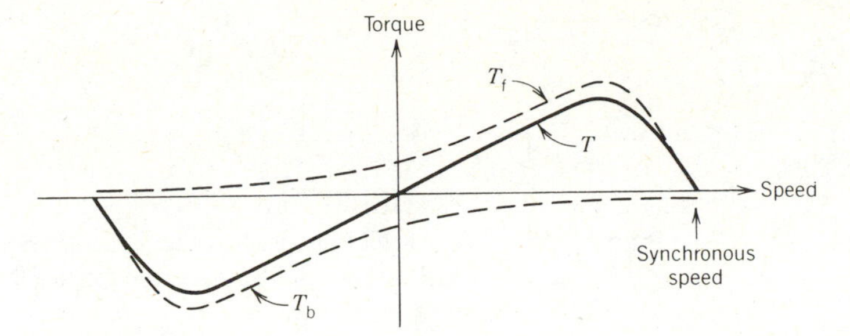

FIGURE 7.4
Torque–speed characteristic of a single-phase induction motor based on constant forward and backward flux waves.

the two torques are unequal and the resultant torque keeps the motor rotating in the direction of rotation.

Slip

Assume that the rotor is rotating in the direction of the forward rotating field at a speed n rpm and the synchronous speed is n_s rpm.

The slip with respect to the forward field is

$$s_f = \frac{n_s - n}{n_s} = s \tag{7.1}$$

The rotor rotates opposite to the rotation of the backward field. Therefore, the slip with respect to the backward field is

$$= \frac{n_s - (-n)}{n_s}$$

$$s_b = \frac{n_s + n}{n_s} = \frac{2n_s - n_s + n}{n_s}$$

$$= 2 - s \tag{7.2}$$

The rotor circuits for the forward- and backward-rotating fluxes are shown in Fig. 7.5. At standstill, the impedances are equal and so are the currents ($I_{2f} = I_{2b}$). Their mmf's affect equally (oppose) the stator mmf's and therefore the rotating forward and backward fluxes in the air gap are equal in magnitude. However, when the rotor rotates, the impedances of the rotor circuits (Fig. 7.5) are unequal and the rotor current I_{2b} is higher (and also at a lower power factor) than the rotor current I_{2f}. Their mmf's, which oppose the stator mmf's, will result in a reduction of the backward-rotating flux. Consequently, as the speed increases, the forward flux in-

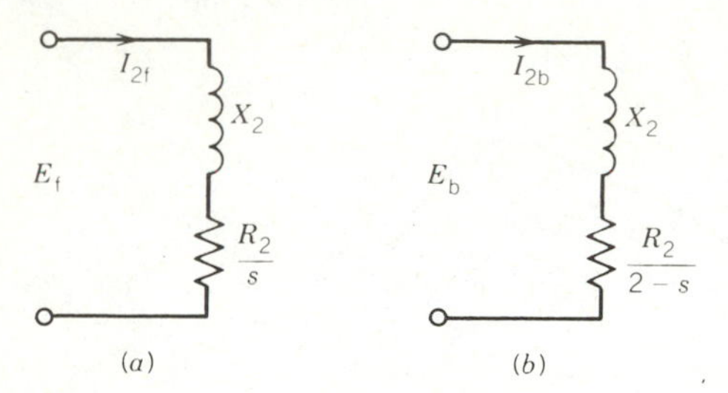

FIGURE 7.5

Rotor equivalent circuits. (*a*) For forward-rotating flux wave. (*b*) For backward-rotating flux wave.

creases while the backward flux decreases; but the resultant flux remains essentially constant to induce voltage in the stator winding, which is almost the same as the applied voltage, if the voltage drops across the winding resistance and the leakage reactance are neglected. Hence, with the rotor in motion, the forward torque increases and the backward torque decreases compared with those shown in Fig. 7.4. The actual torque–speed characteristics are approximately those shown in Fig. 7.6.

Torque Pulsation

In a single-phase motor, instantaneous power pulsates at twice the supply frequency as shown in Fig. 7.7. Consequently, there are torque pulsations at double the stator frequency. The pulsating torque is present in addition to the torque shown in Fig. 7.6. The torque shown on the torque–speed curves of Fig. 7.6 is the time average of the instantaneous torque. The pulsating torque results from the interactions of the oppositely rotating

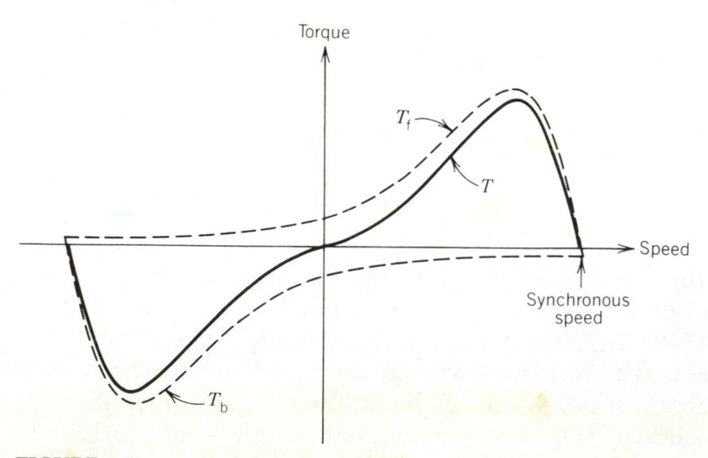

FIGURE 7.6

Actual torque–speed characteristic of a single-phase induction motor taking into account changes in the forward and backward flux waves.

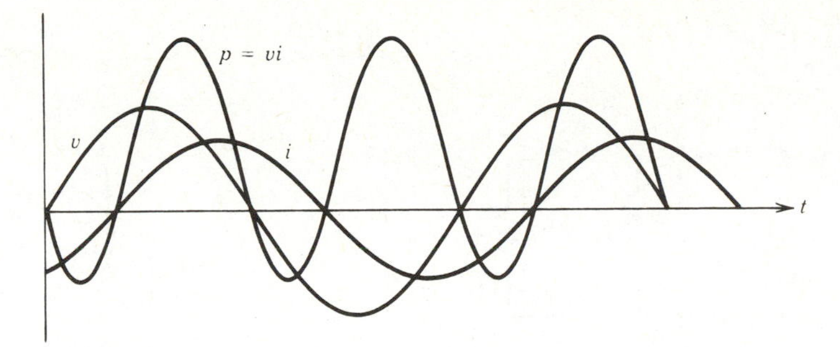

FIGURE 7.7
Waveforms of voltage, current, and power in single-phase induction machine.

fluxes and mmf's, which cross each other at twice the synchronous speed—such as interaction of the forward flux with the backward rotor mmf and of the backward flux with the forward rotor mmf. The interaction of the forward flux with the rotor forward mmf and that of the backward flux with rotor backward mmf produce constant torque.

The pulsating torque produces no average torque but rather produces a humming effect and makes single-phase motors noisier than polyphase motors. The effects of the pulsating torque can be minimized by using elastic mounting, rubber pads, and so forth.

7.1.2 EQUIVALENT CIRCUIT OF A SINGLE-PHASE INDUCTION MOTOR

It was stated earlier that when the stator of a single-phase induction motor is connected to the power supply, the stator current produces a pulsating mmf that is equivalent to two constant-amplitude mmf waves revolving in opposite directions at the synchronous speed. Each of these revolving waves induces currents in the rotor circuits and produces induction motor action similar to that in a polyphase induction machine. This double revolving field theory can be used for the analysis to assess the qualitative and quantitative performance of the single-phase induction motor.

Let us first consider that the rotor is stationary and the stator winding is excited from a single-phase supply. This is equivalent to a transformer with its secondary short-circuited. The equivalent circuit is shown in Fig. 7.8*a*, where

R_1 = resistance of the stator winding

X_1 = leakage reactance of the stator winding

X_{mag} = magnetizing reactance

X_2' = leakage reactance of the rotor referred to the stator

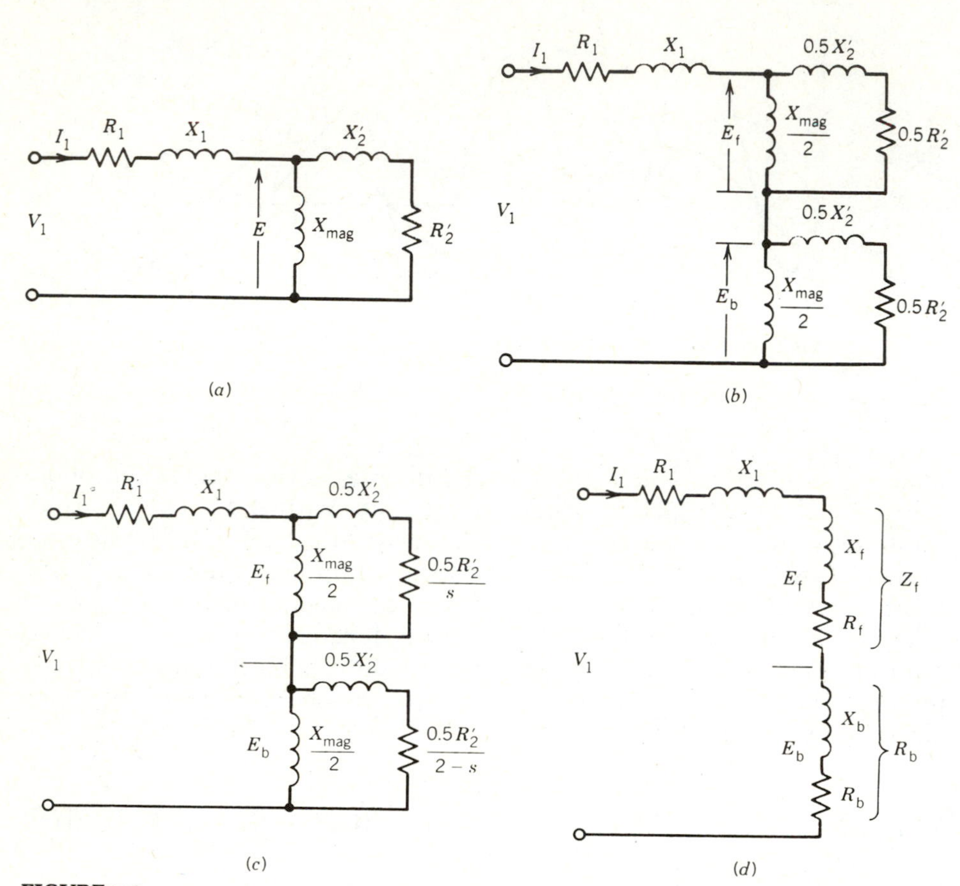

FIGURE 7.8

Equivalent circuits of 1ϕ induction motors. (*a*) and (*b*) → Rotor at standstill. (*c*) and (*d*) → Rotor rotating at slip *s*.

R_2' = resistance of the rotor referred to the stator

V_1 = supply voltage

E = voltage induced in the stator winding (or air gap voltage) by the stationary pulsating air gap flux wave produced by the combined effects of the stator and rotor currents

and

$$E = 4.44fN\Phi \tag{7.3}$$

where Φ is the air gap flux.

According to the double revolving field theory, the equivalent circuit can be split into two halves, as shown in Fig. 7.8*b*, representing the effects of forward and backward fields.

$$E_f = 4.44 f N \Phi_f \tag{7.4}$$

$$E_b = 4.44 f N \Phi_b \tag{7.5}$$

At standstill, as $\Phi_f = \Phi_b$ (these being the revolving air gap fluxes), $E_f = E_b$.

Now consider that the motor is running at some speed in the direction of the forward revolving field, the slip being s. The rotor current induced by the forward field has frequency sf, where f is the stator frequency. As in the polyphase motor, the rotor mmf rotates at the slip rpm with respect to the rotor but at synchronous rpm with respect to the stator. The resultant of the forward stator mmf and the rotor mmf produces a forward air gap flux that induces the voltage E_f. The rotor circuit as reflected in the stator has impedance $j0.5X_2' + 0.5R_2'/s$ as shown in Fig. 7.8*c*.

Now consider the backward-rotating field, which induces current in the rotor circuit at a slip frequency of $(2 - s)f$. The corresponding rotor mmf rotates in the air gap at synchronous speed in the backward direction. The resultant of the backward stator mmf and the rotor mmf produces a backward air gap flux, which induces a voltage E_b. The reflected rotor circuit has impedance $j0.5X_2' + 0.5R_2'/(2 - s)$ as shown in Fig. 7.8*c*. At small slip, the waveform of the rotor current will show a high-frequency component [at $(2 - s)f \simeq 2f$] due to the backward field, superimposed on a low-frequency component (at sf) due to the forward field.

It is obvious from the equivalent circuit that in the running condition, $Z_f > Z_b$, $E_f > E_b$, and therefore the forward air gap flux Φ_f will be greater than the backward air gap flux Φ_b.

The parameters of the circuit in Fig. 7.8*c* can be obtained by performing two tests on the single-phase induction motor, as illustrated by Example 7.1. This equivalent circuit can be used to assess the performance of the motor by computation of the stator current, input power, developed torque, efficiency, and so forth for a particular speed, as illustrated by Example 7.2. To simplify the calculations, the simplified equivalent circuit of Fig. 7.8*d* can be used.

$$Z_f = R_f + jX_f = \frac{j0.5X_{mag}(j0.5X_2' + 0.5R_2'/s)}{0.5R_2'/s + j0.5(X_{mag} + X_2')}$$

$$Z_b = R_b + jX_b = \frac{j0.5X_{mag}[j0.5X_2' + 0.5R_2'/(2 - s)]}{0.5R_2'/(2 - s) + j0.5(X_{mag} + X_2')}$$

The air gap powers due to the forward field and backward field are

$$P_{gf} = I_1^2 R_f \tag{7.6}$$

$$P_{gb} = I_1^2 R_b \tag{7.7}$$

The corresponding torques are

$$T_f = \frac{P_{gf}}{\omega_{syn}} \tag{7.8}$$

$$T_b = \frac{P_{gb}}{\omega_{syn}} \tag{7.9}$$

The resultant torque is

$$T = T_f - T_b = \frac{I_1^2}{\omega_{syn}}(R_f - R_b) \tag{7.10}$$

The mechanical power developed is

$$P_{mech} = T\omega_m \tag{7.11}$$

$$= T\omega_{syn}(1 - s) \tag{7.12}$$

$$= I_1^2(R_f - R_b)(1 - s) \tag{7.13}$$

$$= (P_{gf} - P_{gb})(1 - s) \tag{7.14}$$

The power output is

$$P_{out} = P_{mech} - P_{rot} \tag{7.15}$$

where P_{rot} includes friction and windage losses, and it is assumed that core losses are also included in the rotational losses. The two air gap fields produce currents in the rotor circuit at different frequencies. Therefore the rotor copper loss (i.e., I^2R loss) is the numerical sum of the losses produced by each field.

The rotor copper loss produced by the forward field is

$$P_{2f} = sP_{gf} \tag{7.16}$$

and that produced by the backward field is

$$P_{2b} = (2 - s)P_{gb} \tag{7.17}$$

The total rotor copper loss is

$$P_2 = sP_{gf} + (2 - s)P_{gb} \tag{7.18}$$

The total air gap power is the numerical sum of the air gap powers absorbed from the stator by the two component air gap fields. Thus

$$P_g = P_{gf} + P_{gb} \tag{7.19}$$

EXAMPLE 7.1

The following test data are obtained from a 1/4 hp, 1ϕ, 120 V, 60 Hz, 1730 rpm induction motor.

Stator winding (main) resistance = 2.9 ohms

Blocked rotor (standstill) test: The rotor is prevented from rotating,

$$V = 43 \text{ V}, I = 5 \text{ A}, P = 140 \text{ W}$$

No-load test: Motor is running freely,

$$V = 120 \text{ V}, I = 3.5 \text{ A}, P = 125 \text{ W}$$

(a) Obtain the double revolving field equivalent circuit for the motor.

(b) Determine the rotational loss.

Note: Single-phase induction motors have a main winding and an auxiliary winding on the stator. The auxiliary winding is for starting (discussed in Section 7.1.3). For the standstill rotor test the auxiliary winding is disconnected and only the main winding is connected to the ac supply. For the no-load test, both the main and auxiliary windings are used for starting the motor, but in the running condition the auxiliary winding is automatically disconnected from the supply. Therefore, in both the standstill test and the no-load test, only the main winding is in operation when the test data are taken.

Solution

(a) At standstill, $s = 1$ and the equivalent circuit is shown in Fig. 7.8*b*. In a single-phase induction machine

$$j0.5X_2' + 0.5R_2' \ll 0.5X_{mag}$$

The approximate equivalent circuit at standstill is shown in Fig. E7.1*a*.

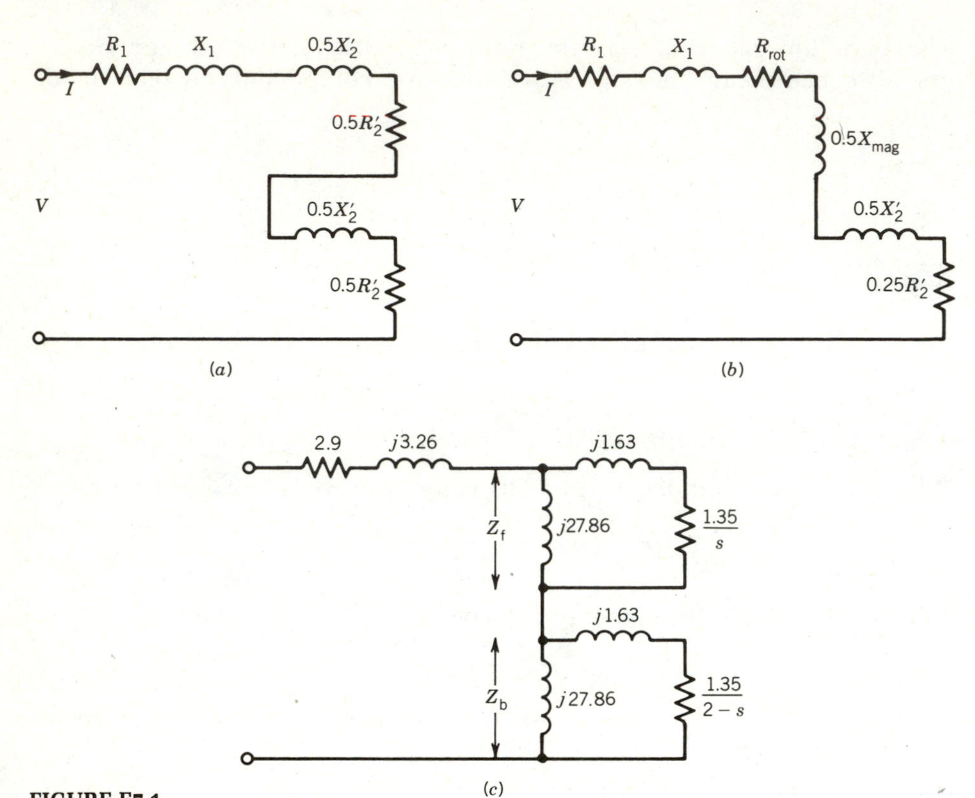

FIGURE E7.1

$$R_1 = 2.9 \text{ ohms}$$

$$P_{BL} = 5^2(2.9 + R_2') = 140 \text{ W}$$

so

$$R_2' = 2.7 \text{ ohms}$$

$$Z_{BL} = 43/5 = 8.6 \text{ ohms}$$

$$8.6^2 = (2.9 + 2.7)^2 + (X_1 + X_2')^2$$

$$X_1 + X_2' = 6.53 \text{ ohms}$$

Assume $X_1 = X_2' = 6.53/2 = 3.26$ ohms.

At no load, the slip is very small. For the circuit shown in Fig. 7.8*c*,

$$j0.5X_2' + 0.5R_2'/s > 0.5X_{mag}$$

$$j0.5X_2' + 0.5R_2'/(2 - s) \simeq j0.5X_2' + 0.25R_2' \ll 0.5X_{mag}$$

The approximate equivalent circuit at no load is shown in Fig. E7.1*b*. Note that a resistance R_{rot} is shown in the circuit to account for the rotational losses.

$$P_{NL} = I^2(R_1 + R_{rot} + 0.25R_2')$$
$$= I^2R_{NL}$$
$$125 = 3.5^2R_{NL}$$
$$R_{NL} = \frac{125}{3.5^2} = 10.2\ \Omega$$

The no-load impedance Z_{NL} from Fig. E7.1*b* is

$$Z_{NL} = [R_{NL}^2 + (0.5X_{mag} + X_1 + 0.5X_2')^2]^{1/2}$$
$$\frac{120}{3.5} = 34.3 = [10.2^2 + (0.5X_{mag} + 3.26 + 1.63)^2]^{1/2}$$
$$X_{mag} = 55.72\ \Omega$$

The equivalent circuit with the parameter values is shown in Fig. E7.1*c*.

(b) From the equivalent circuit at no load (Fig. E7.1*b*),

$$P = I^2(R_1 + 0.25R_2') + P_{rot} = 125 \text{ watts}$$
$$P_{rot} = 125 - 3.5^2(2.9 + 0.675)$$
$$= 81.2 \text{ watts}$$

EXAMPLE 7.2

For the single-phase induction motor of Example 7.1, determine the input current, power, power factor, developed torque, output power, efficiency of the motor, air gap power, and rotor copper loss if the motor is running at the rated speed when connected to a 120 V supply.

Solution

At rated speed,

$$s = \frac{1800 - 1730}{1800} = 0.039$$

From Fig. E7.1*c*,

$$\begin{aligned} Z_f &= R_f + jX_f \\ &= \frac{j27.86[(1.35/0.039) + j1.63]}{1.35/0.039 + j(27.86 + 1.63)} \\ &= 13 + j16.79 \end{aligned}$$

$$\begin{aligned} Z_b &= R_b + jX_b \\ &= \frac{j27.86[1.35/(2 - 0.039) + j1.63]}{1.35/(2 - 0.039) + j(27.86 + 1.63)} \\ &= 0.61 + j1.55 \end{aligned}$$

$$\begin{aligned} Z_{input} &= (R_1 + R_f + R_b) + j(X_1 + X_f + X_b) \\ &= (2.9 + 13 + 0.61) + j(3.26 + 16.79 + 1.55) \\ &= 16.51 + j21.60 = 27.19\angle 52.61^\circ\ \Omega \end{aligned}$$

$$I_{input} = \frac{120}{27.19}\angle -52.61^\circ = 4.41\angle -52.61^\circ \text{ amperes}$$

$$\text{Power factor} = \cos 52.61^\circ = 0.61 \text{ lagging}$$

$$\begin{aligned} P_{input} &= VI\cos\theta \\ &= 120 \times 4.41 \times 0.61 \\ &= 322.81 \text{ W} \end{aligned}$$

Synchronous speed is $\omega_{syn} = 1800 \times 2 \times \pi/60 = 188.5$ rad/sec. From Eq. 7.10, the torque developed is

$$\begin{aligned} T &= \frac{I^2}{\omega_{syn}}(R_f - R_b) \\ &= \frac{4.41^2(13 - 0.61)}{188.5} \\ &= 1.28 \text{ N} \cdot \text{m} \end{aligned}$$

From Eq. 7.12, the mechanical power developed is

$$\begin{aligned} P_{mech} &= T\omega_{syn}(1 - s) \\ &= 1.28 \times 188.5(1 - 0.039) \\ &= 231.87 \text{ W} \end{aligned}$$

$$\begin{aligned} \text{Output power} &= P_{mech} - P_{rot} \\ &= 231.87 - 81.2 \end{aligned}$$

$$P_{out} = 150.67 \text{ watts}$$

$$\text{Efficiency} = \frac{P_{out}}{P_{in}} = \frac{150.67}{322.81}$$

$$= 46.67\%$$

From Eqs. 7.6 and 7.7 the air gap powers due to the forward- and backward-rotating fields are

$$P_{gf} = I^2 R_f = 4.41^2 \times 13 = 252.83 \text{ W}$$

$$P_{gb} = 4.41^2 \times 0.61 = 11.86 \text{ W}$$

The air gap power

$$P_g = 252.83 + 11.86 = 264.69 \text{ W}$$

From Eq. 7.18 the rotor copper loss is

$$P_2 = 0.039 \times 252.83 + (2 - 0.039)11.86$$

$$= 9.86 + 23.26$$

$$= 33.12 \text{ W}$$

7.1.3 STARTING OF SINGLE-PHASE INDUCTION MOTORS

As explained earlier, a single-phase induction motor with one stator winding inherently does not produce any starting torque. In order to make the motor start rotating, some arrangement is required so that the motor produces a starting torque. In the running condition, of course, the motor will produce torque with only one stator winding.

The simplest method of starting a single-phase induction motor is to provide an auxiliary winding on the stator in addition to the main winding and start the motor as a two-phase machine. The two windings are placed in the stator with their axes displaced 90 electrical degrees in space. The impedances of the two circuits are such that the currents in the main and the auxiliary windings are phase-shifted from each other. The motor is equivalent to an unbalanced two-phase motor. However, the result is a rotating stator field that can produce the starting torque. The two windings can be properly designed to make the motor behave as a balanced two-phase motor. This is illustrated in Example 7.3.

In the running condition a single-phase induction motor can develop a torque with only the main winding. Therefore, as the motor speeds up, the auxiliary winding can be taken out of the circuit. In most motors, this is done by connecting a centrifugal switch in the auxiliary circuit. At about

75 percent of the synchronous speed, the centrifugal switch operates and disconnects the auxiliary winding from the supply.

EXAMPLE 7.3

The currents in the main and the auxiliary windings are as follows:

$$i_m = \sqrt{2} I_m \cos \omega t$$

$$i_a = \sqrt{2} I_a \cos(\omega t + \theta_a)$$

The effective numbers of turns for the main and auxiliary windings are N_m and N_a.

The windings are placed in quadrature.

(a) Obtain expressions for the stator rotating mmf wave.

(b) Determine the magnitude and the phase angle of the auxiliary winding current to produce a balanced two-phase system.

Solution

(a) The stator mmf along a position defined by an angle θ (where $\theta = 0°$ defines the axis of the main winding) is contributed by both windings.

$$
\begin{aligned}
F(\theta, t) &= F_m(\theta, t) + F_a(\theta, t) \\
&= N_m i_m \cos \theta + N_a i_a \cos(\theta + 90°) \\
&= N_m \sqrt{2} I_m \cos \omega t \cos \theta + N_a \sqrt{2} I_a \cos(\omega t + \theta_a) \cos(\theta + 90°) \\
&= \sqrt{2} N_m I_m \cos \omega t \cos \theta - \sqrt{2} N_a I_a \sin \theta \cos(\omega t + \theta_a) \\
&= \sqrt{2} N_m I_m \cos \omega t \cos \theta - \sqrt{2} N_a I_a \sin \theta [\cos \omega t \cos \theta_a \\
&\quad - \sin \omega t \sin \theta_a] \\
&= \sqrt{2} N_m I_m \cos \omega t \cos \theta - \sqrt{2} N_a I_a \cos \theta_a \cos \omega t \sin \theta \\
&\quad + \sqrt{2} N_a I_a \sin \theta_a \sin \omega t \sin \theta \\
&= \frac{1}{\sqrt{2}} N_m I_m [\cos(\omega t + \theta) + \cos(\omega t - \theta)] \\
&\quad - \frac{1}{\sqrt{2}} N_a I_a \cos \theta_a [\sin(\omega t + \theta) - \sin(\omega t - \theta)] \\
&\quad + \frac{1}{\sqrt{2}} N_a I_a \sin \theta_a [-\cos(\omega t + \theta) + \cos(\omega t - \theta)]
\end{aligned}
$$

$$F(\theta, t) = \frac{1}{\sqrt{2}}[(N_m I_m - N_a I_a \sin\theta_a)\cos(\omega t + \theta)$$
$$- (N_a I_a \cos\theta_a)\sin(\omega t + \theta)]$$
$$+ \frac{1}{\sqrt{2}}[(N_m I_m + N_a I_a \sin\theta_a)\cos(\omega t - \theta)$$
$$+ (N_a I_a \cos\theta_a)\sin(\omega t - \theta)] \tag{7.20}$$

Terms with $\cos(\omega t - \theta)$ and $\sin(\omega t - \theta)$ form the forward-rotating field. Terms with $\cos(\omega t + \theta)$ and $\sin(\omega t + \theta)$ form the backward-rotating field.

(b) If $N_m I_m = N_a I_a$ and $\theta_a = 90°$ (i.e., the phase difference between I_m and I_a is 90°), it follows from Eq. 7.20 that the backward-rotating mmf vanishes, and the forward-rotating mmf is

$$F_f(\theta, t) = \sqrt{2} N_m I_m \cos(\omega t - \theta)$$

7.1.4 CLASSIFICATION OF MOTORS

Single-phase induction motors are known by various names. The names are descriptive of the methods used to produce the phase difference between the currents in the main and auxiliary windings. Some of the commonly used types of single-phase induction motors are described here.

Split-Phase Motors

A schematic diagram of the split-phase induction motor is shown in Fig. 7.9*a*. The auxiliary winding has a higher resistance-to-reactance ratio than the main winding, so the two currents are out of phase as shown in Fig. 7.9*b*. The high resistance-to-reactance ratio is usually obtained by using finer wire for the auxiliary winding. This is permissible because the auxiliary winding is in the circuit only during the starting period. The centrifugal switch cuts it out at about 75 percent of the synchronous speed.

The typical torque–speed characteristic of this motor is shown in Fig. 7.9*c*. This motor has low to moderate starting torque, which depends on the two currents and the phase angle between them (Eq. 7.27). The starting torque can be increased by inserting a series resistance in the auxiliary winding.

Capacitor-Start Motors

Higher starting torque can be obtained if a capacitor is connected in series with the auxiliary winding as shown in Fig. 7.10*a*. This increases the phase

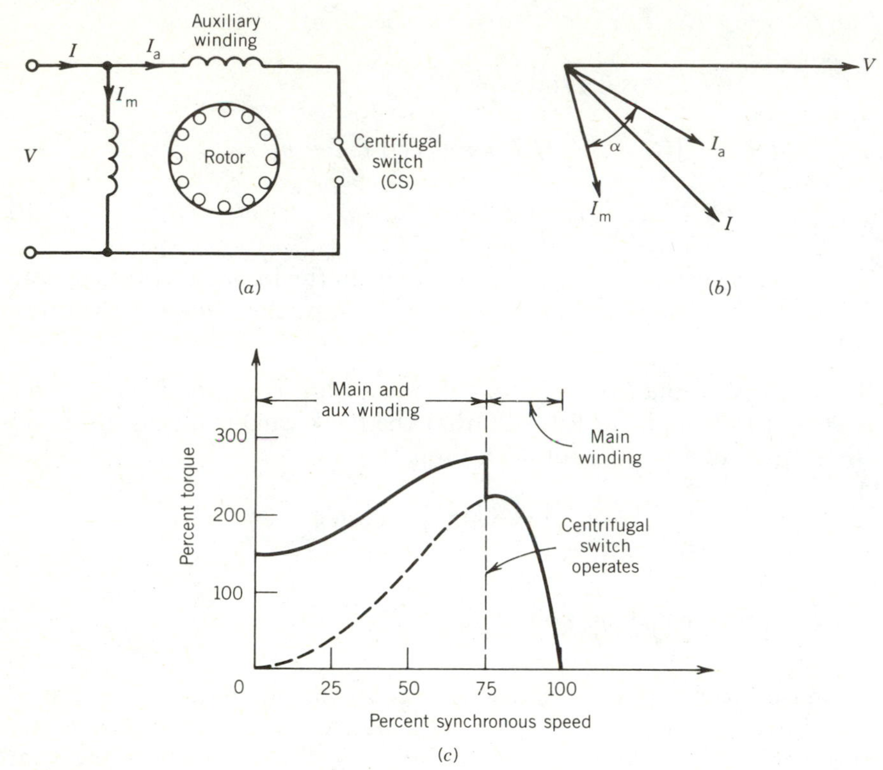

FIGURE 7.9
Split-phase (resistance-start) induction motor.

angle between the winding currents as shown in Fig. 7.10*b*. The torque–speed characteristic is shown in Fig. 7.10*c*. The capacitor is an added cost. A typical capacitor value for a 0.5 hp motor is 300 μF. Because the capacitor is in the circuit only during the starting period, it can be an inexpensive ac electrolytic type. High starting torque is the outstanding feature of this arrangement.

Capacitor-Run Motors

In this motor, as shown in Fig. 7.11*a*, the capacitor that is connected in series with the auxiliary winding is not cut out after starting and is left in the circuit all the time. This simplifies the construction and decreases the cost because the centrifugal switch is not needed. The power factor, torque pulsation, and efficiency are also improved because the motor runs as a two-phase motor. The motor will run more quietly.

The capacitor value is of the order of 20–50 μF and because it operates continuously, it is an ac paper oil type. The capacitor is a compromise

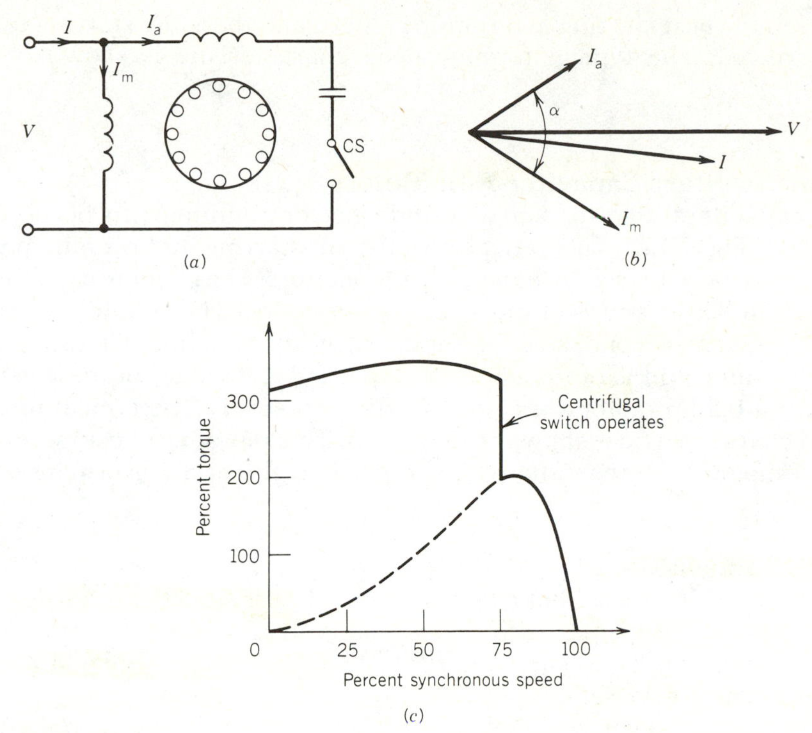

FIGURE 7.10
Capacitor-start induction motor.

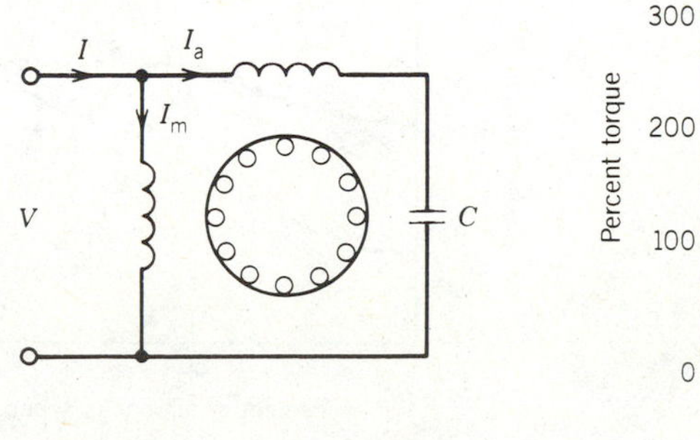

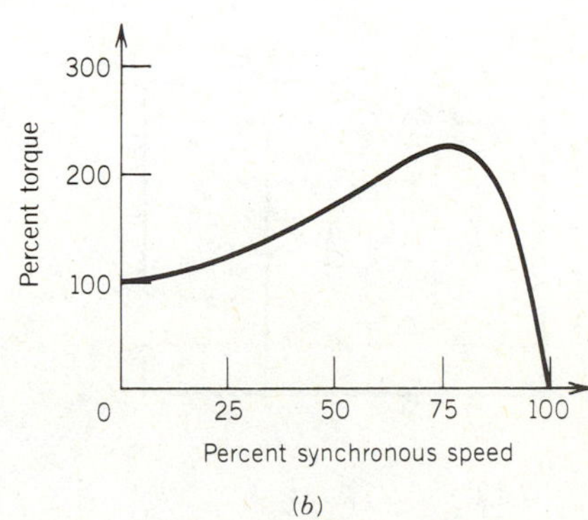

FIGURE 7.11
Capacitor-run induction motor.

between the best starting and running values and therefore starting torque is sacrificed. The typical torque–speed characteristic is shown in Fig. 7.11*b*.

Capacitor-Start Capacitor-Run Motors

Two capacitors, one for starting and one for running, can be used, as shown in Fig. 7.12*a*. Theoretically, optimum starting and running performance can be achieved by having two capacitors. The starting capacitor C_s is larger in value and is of the ac electrolytic type. The running capacitor C_r, permanently connected in series with the starting winding, is of smaller value and is of the paper oil type. Typical values of these capacitors for a 0.5 hp motor are $C_s = 300\ \mu F$, $C_r = 40\ \mu F$. The typical torque–speed characteristic is shown in Fig. 7.12*b*. This motor is, of course, expensive compared to others; however, it provides the best performance.

Shaded-Pole Motors

These motors have a salient pole construction. A shaded band consisting of a short-circuited copper turn, known as a shading coil, is used on one portion of each pole, as shown in Fig. 7.13*a*. The main single-phase winding is wound on the salient poles. The result is that the current induced in the shading band causes the flux in the shaded portion of the pole to lag the flux in the unshaded portion of the pole. Therefore the flux in the shaded portion reaches its maximum after the flux in the unshaded portion reaches its maximum. This is equivalent to a progressive shift of the flux from the unshaded to the shaded portion of the pole. It is similar to a rotating field moving from the unshaded to the shaded portion of the pole. As a result, the motor produces a starting torque.

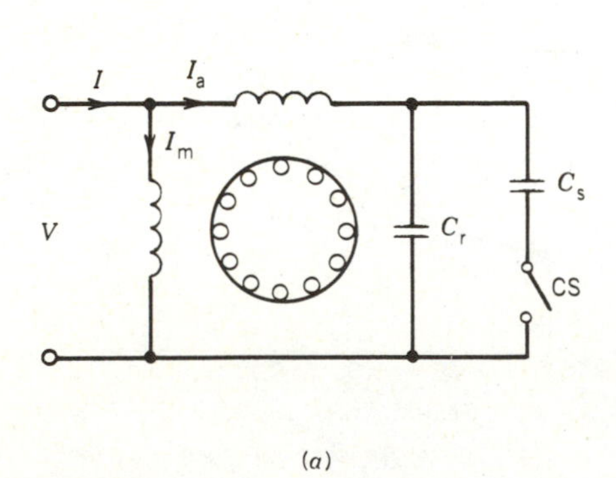

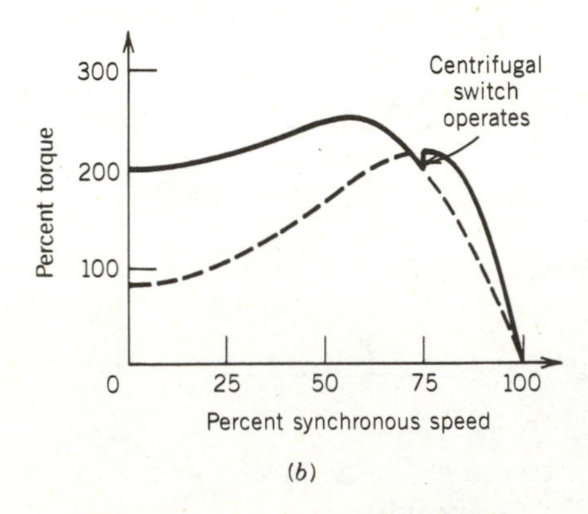

FIGURE 7.12
Capacitor-start capacitor-run induction motor.

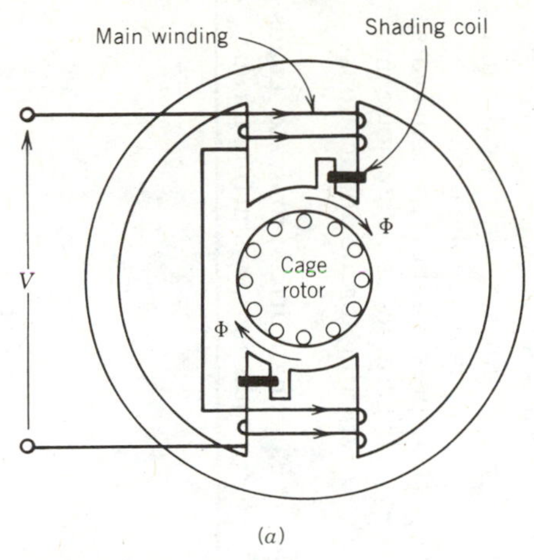

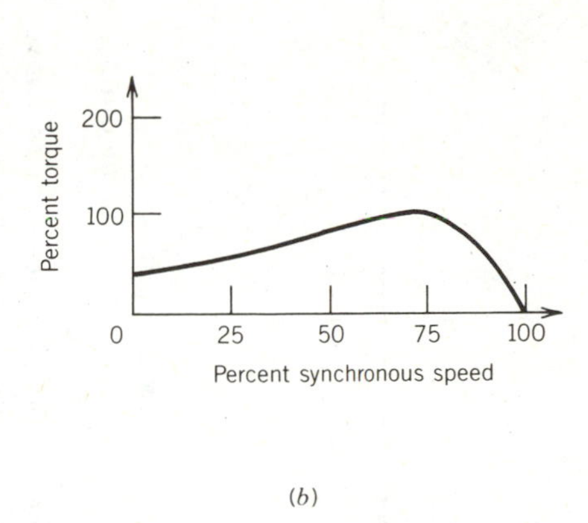

FIGURE 7.13
Shaded-pole induction motor.

The typical torque–speed characteristic is shown in Fig. 7.13*b*. Shaded-pole motors are the least expensive of the fractional horsepower motors and are generally built for low horsepower rating, up to about 1/20 hp.

7.1.5 CHARACTERISTICS AND TYPICAL APPLICATIONS

The main features and the applications of single-phase induction motors are summarized in Table 7.1. Note that for applications below 1/20 hp, shaded-pole motors are invariably used. However, for applications above 1/20 hp, the choice of the motor depends primarily on the starting torque and to some extent on the quietness of operation. If low noise is desired and low starting torque is adequate, such as for driving fans or blowers, capacitor-run motors can be chosen. If low noise is to be combined with high starting torque, as may be required for a compressor or refrigerator drive, then the expensive capacitor-start, capacitor-run motor is better. If the compressor is located in a noisy environment, the choice should be a capacitor-start motor, which will be less expensive.

7.2 STARTING WINDING DESIGN

The main purpose of the starting (auxiliary) winding is to develop a starting torque. However, the starting winding can be designed to provide maximum starting torque or to optimize starting torque per ampere of the starting current. In this section a design procedure is outlined to achieve these objectives. First an expression for the starting torque is derived.

TABLE 7.1
Single-Phase Induction Motors: Characteristics and Applications

Type of Motor	Torque as % of Rated Torque		Rated Load		Horsepower Range	Approx. Comparative Price (%)	Applications
	Starting	Breakdown	Power Factor	Efficiency			
Split-phase (resistance-start)	100–250	Up to 300	50–65	55–65	1/20–1	100	Fans, blowers, centrifugal pumps, washing machines, etc. Loads requiring low or medium starting torque
Capacitor-start	250–400	Up to 350	50–65	55–65	1/8–1	125	Compressors, pumps, conveyors, refrigerators, air-conditioning equipment, washing machines, and other hard-to-start loads
Capacitor-run	100–200	Up to 250	75–90	60–70	1/8–1	140	Fans, blowers, centrifugal pumps, etc. Low noise applications
Capacitor-start, capacitor-run	200–300	Up to 250	75–90	60–70	1/8–1	180	Compressors, pumps, conveyors, refrigerators etc. Low noise and high starting torque applications
Shaded-pole	40–60	140	25–40	25–40	1/200–1/20	60	Fans, hair driers, toys, etc. Loads requiring low starting torque

Starting Torque

Consider the starting of the single-phase induction motor. The two stator windings and the currents flowing in them are shown in Fig. 7.14*a*. The cage rotor can be represented by an equivalent two-phase winding, represented by the coils a–b and c–d. Assume that each of these coils has an effective number of turns N_2, resistance R_2, and reactance X_2 (at the stator frequency f). The current flowing through the main winding produces flux that induces voltage e_{2m} (by transformer action) and current i_{2m} in the a–b coil of the rotor. The current i_{2m} flows in such a direction as to oppose flux Φ_m. Similarly, flux Φ_a in the auxiliary winding induces voltage e_{2a} and current i_{2a} in the c–d coil of the rotor.

$$E_{2m} = 4.44 f N_2 \Phi_m \tag{7.21}$$

$$E_{2a} = 4.44 f N_2 \Phi_a \tag{7.22}$$

The current i_{2m} lags e_{2m} and i_{2a} lags e_{2a} by an angle θ_2, where

$$\cos \theta_2 = \frac{R_2}{(R_2^2 + X_2^2)^{1/2}} = \frac{R_2}{Z_2} \tag{7.23}$$

The various fluxes, currents, and voltages are shown in the phasor diagram of Fig. 7.14*b*. Note that a torque is developed through the interaction of Φ_m and I_{2a} and it acts in the clockwise direction. Torque is also developed through the interaction of Φ_a and I_{2m} and it acts in the anticlockwise direction. No torques are developed through the interaction of Φ_m and i_{2m} or Φ_a and i_{2a}.

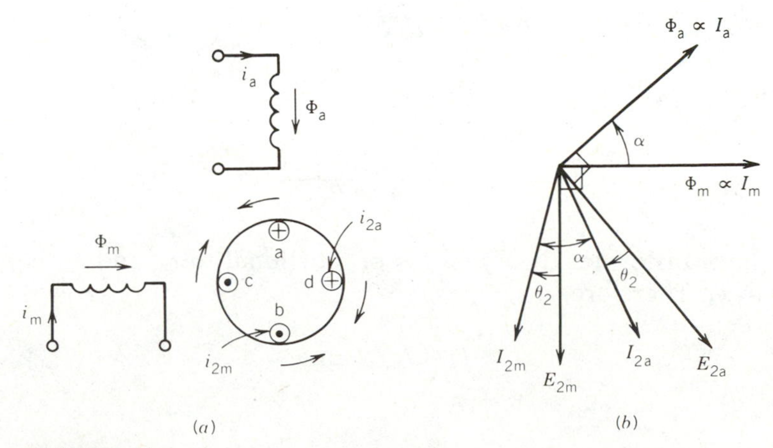

FIGURE 7.14
Starting condition.

$$T_m \propto \Phi_m I_{2a} \cos\underline{/\Phi_m, I_{2a}}$$
$$\propto \Phi_m I_{2a} \cos(90 + \theta_2 - \alpha)$$

Torque developed by Φ_a and I_{2m} is

$$T_a \propto \Phi_a I_{2m} \cos\underline{/\Phi_a, I_{2m}}$$
$$\propto \Phi_a I_{2m} \cos(90 + \theta_2 + \alpha)$$

Net starting torque is

$$T_s \propto \{\Phi_m I_{2a} \cos(90 + \theta_2 - \alpha) - \Phi_a I_{2m} \cos(90 + \theta_2 + \alpha)\}$$

but

$$\frac{\Phi_m}{\Phi_a} = \frac{E_{2m}}{E_{2a}} = \frac{I_{2m}}{I_{2a}}$$

or

$$\Phi_m I_{2a} = \Phi_a I_{2m}$$

Therefore,

$$T_s \propto I_{2m} \Phi_a \sin\alpha \cos\theta_2 \tag{7.24}$$

From Eq. 7.22,

$$\Phi_a = \frac{E_{2a}}{4.44 f N_2} = \frac{I_{2a} Z_2}{444 f N_2} \tag{7.25}$$

From Eqs. 7.24, 7.25, and 7.23,

$$T_s \propto \frac{I_{2m} I_{2a} R_2 \sin\alpha}{4.44 f N_2}$$

From the transformer theory, I_{2m} is proportional to I_m and I_{2a} is proportional to I_a. Therefore,

$$T_s \propto I_m I_a \sin\alpha \tag{7.26}$$

or

$$T_s = K I_m I_a \sin\alpha \tag{7.27}$$

This is a very useful expression. It indicates that the starting torque depends on the magnitudes of currents in the main and the auxiliary winding and the phase difference between these currents. An expression for the starting torque is derived later (Eq. 7.51) based on double revolving field theory and the equivalent circuit.

7.2.1 DESIGN OF SPLIT-PHASE (RESISTANCE-START) MOTORS

In split-phase motors, the main winding is designed to satisfy the running operation of the motor, whereas the auxiliary winding is designed so that, operating in conjunction with the main winding, it produces the desired starting torque without excessive starting current. A convenient approach is to assume a number of turns for the starting winding and calculate the value of the starting winding resistance for the desired starting torque. If this does not yield the optimum design for starting torque and current, a range of values for the starting winding turns can be tried until an optimum design is obtained.

Maximum Starting Torque

If the number of turns (N_a) for the starting winding is specified, the resistance in the auxiliary winding can be determined so as to maximize the starting torque. For the standstill condition, the motor can be represented by the circuit shown in Fig. 7.15*a*. Let

$$Z_m = R_m + jX_m, \quad \text{impedance of the main winding}$$

$$Z_a = R_a + jX_a, \quad \text{impedance of the auxiliary winding}$$

The phasor diagram for the standstill condition of the motor is shown in Fig. 7.15*b*. The phasor I_m (= $0A$) lags V by θ_m. For a particular value of R_a, $I_a = AC$ and $I = I_m + I_a = 0C$. Note that I_m remains fixed and I_a will change if R_a changes. If R_a is infinitely large, I_a is zero, and the input current I is the same as the current I_m. If R_a is zero, $I_a = |V|/X_a$ and I_a will lag V by 90° as represented by the phasor AB in Fig. 7.15*b*. The locus of I_a and the input current I is a semicircle having diameter $AB = |V|/X_a$.

From Eq. 7.26, since I_m is fixed,

$$T_s \propto I_a \sin\alpha$$

$$\propto \text{length } CK \text{ in Fig. 7.15}b$$

For maximum starting torque, the operating point is D, midway between A and E, for which $I_a \sin\alpha$ (= DK') is maximum. The phasor diagram for the maximum starting condition is drawn in Fig. 7.15*c*.

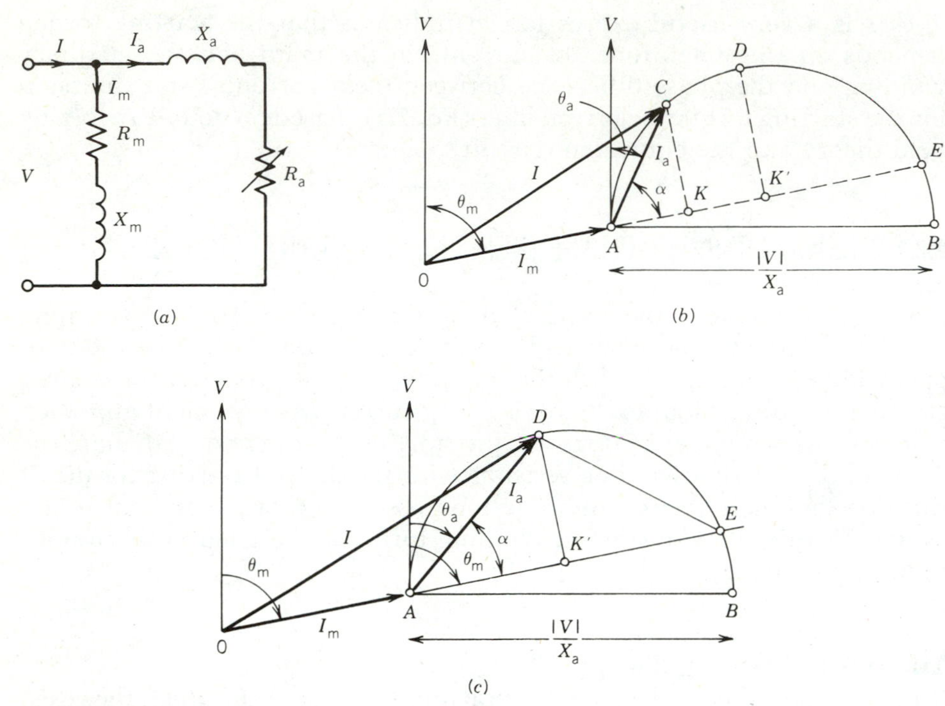

FIGURE 7.15
Split-phase induction motor at starting condition.

Obviously, $AD = DE$, $AK' = K'E$, and

$$\theta_a = \frac{\theta_m}{2} = \alpha$$

Therefore,

$$\cot \theta_a = \cot \frac{\theta_m}{2}$$

$$= \frac{\cos(\theta_m/2)}{\sin(\theta_m/2)}$$

$$= \frac{1 + \cos \theta_m}{\sin \theta_m}$$

or

$$\frac{R_a}{X_a} = \frac{1 + R_m/|Z_m|}{X_m/|Z_m|} = \frac{R_m + |Z_m|}{X_m}$$

$$R_a = \frac{X_a}{X_m}(R_m + |Z_m|) \tag{7.28}$$

$$= \left|\frac{N_a}{N_m}\right|^2 (R_m + |Z_m|) \tag{7.29}$$

The starting winding current is

$$|I_a| = \frac{|V|}{|Z_a|} = \frac{|V|}{(R_a^2 + X_a^2)^{1/2}}$$

$$= \frac{|V|}{[(N_a/N_m)^4(R_m + |Z_m|)^2 + (N_a/N_m)^4 X_m^2]^{1/2}}$$

$$= \frac{|V|}{(N_a/N_m)^2[(R_m + |Z_m|)^2 + X_m^2]^{1/2}} \tag{7.30}$$

For a particular value of turns N_a for the starting winding, Eq. 7.29 (or 7.28) gives the value of R_a for maximum starting torque and Eq. 7.30 gives the starting winding current. A range of values for N_a can be tried and an optimum design can be achieved. If I_a increases, I (= $I_m + I_a$) will increase too, as seen from Fig. 7.15*b* or 7.15*c*.

7.2.2 DESIGN OF CAPACITOR-START MOTORS

For the starting condition, the capacitor-start motor (Fig. 7.10) can be represented by the circuit shown in Fig. 7.16*a*. The phasor diagram for the

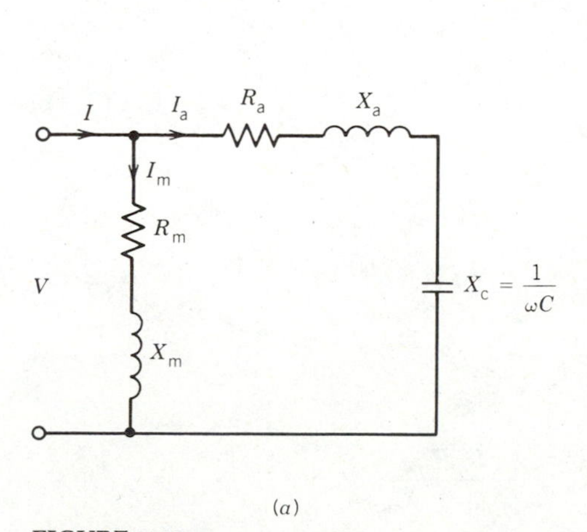

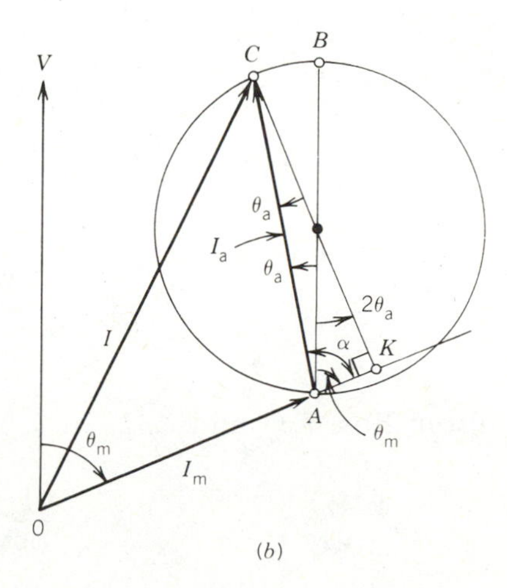

FIGURE 7.16
Capacitor-start induction motor at starting condition.

standstill condition is shown in Fig. 7.16*b*. The main winding current I_m = (OA) lags V by θ_m. The auxiliary winding current I_a (= AC) leads V by θ_a. The starting current is $I = I_m + I_a = OC$. If X_c is infinitely large, I_a is zero and $I = I_m = OA$. If $X_c = X_a$, I_a is a maximum, is equal to $|V|/R_a$, and is in phase with the supply voltage V as represented by the vertical line AB in Fig. 7.16*b*. The locus of I_a is the semicircle ACB having diameter $AB = |V|/R_a$.

Since I_m is fixed,

$$\begin{aligned} T_s &\propto I_a \sin \alpha \\ &\propto \text{length } CK \text{ in Fig. 7.16}b \end{aligned}$$

The length CK is maximum when it passes through the center of the circle as shown in Fig. 7.16*b*. Note that the phasor diagram in Fig. 7.16*b* is drawn for the maximum starting torque condition. From the geometry of the diagram,

$$\theta_a = \frac{90° - \theta_m}{2}$$

Now

$$\begin{aligned} \tan \theta_a &= \left(\frac{1 - \cos 2\theta_a}{1 + \cos 2\theta_a}\right)^{1/2} \\ &= \left[\frac{1 - \cos(90 - \theta_m)}{1 + \cos(90 - \theta_m)}\right]^{1/2} = \left(\frac{1 - \sin \theta_m}{1 + \sin \theta_m}\right)^{1/2} \\ &= \left[\frac{1 - (X_m/|Z_m|)}{1 + (X_m/|Z_m|)}\right]^{1/2} = \left(\frac{|Z_m| - X_m}{|Z_m| + X_m}\right)^{1/2} \\ &= \frac{R_m}{|Z_m| + X_m} \end{aligned} \tag{7.31}$$

Also,

$$\tan \theta_a = \frac{X_c - X_a}{R_a} \tag{7.32}$$

From Eqs. 7.31 and 7.32,

$$X_c = \frac{1}{\omega C} = X_a + \frac{R_a R_m}{|Z_m| + X_m}$$

or

$$C = \frac{1}{\omega \cdot \left(X_a + \frac{R_a R_m}{|Z_m| + X_m}\right)} \tag{7.33}$$

For a given starting winding, the value of C, given by Eq. 7.33, when connected in series with the starting winding will yield maximum starting torque.

Maximum Starting Torque per Ampere of Starting Current

If maximizing the starting torque is the sole criterion, the value of C can be selected by using Eq. 7.33. However, this may not be the best design for the motor. Maximizing the starting torque per ampere of starting current is perhaps the most desirable criterion.

The phasor diagram for the starting condition is shown in Fig. 7.17. The starting current is represented by $0C$ and the starting torque is represented by CK. The ratio $CK/0C$ (i.e., starting torque per ampere of the starting current) is maximum when $0C$ is tangential to the circle $ACBD$, which is the locus of I_a and I. Note that the phasor diagram of Fig. 7.17 is drawn for the condition that $0C$ is tangent to the circle $ACBD$ whose center is F. Now

$$\begin{aligned} 0CF &= 90^\circ \\ 0AF &= 180^\circ - \theta_m \\ |I|^2 &= 0C^2 \\ &= 0F^2 - AF^2 \\ &= 0A^2 + AF^2 - 20A\,AF\cos\underline{/0AF} - AF^2 \end{aligned}$$

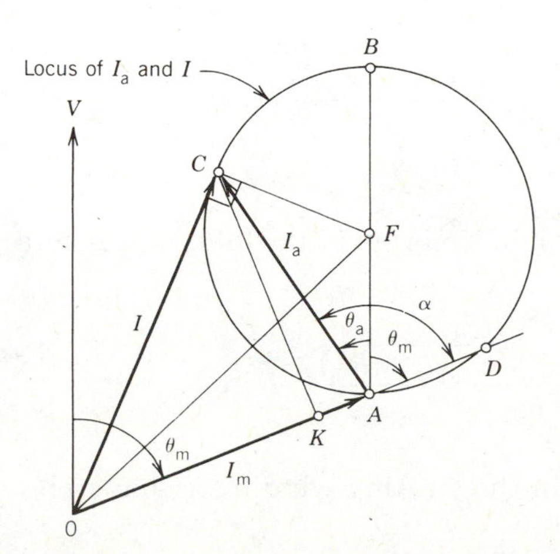

FIGURE 7.17
Phasor diagram for maximum starting torque per ampere of starting current in a capacitor-start induction motor.

$$= OA^2 + 2OA\,AF\cos\theta_m$$

$$= |I_m|^2 + 2|I_m|\frac{|V|}{2R_a}\frac{R_m}{Z_m}$$

$$= |I_m|^2\left(1 + \frac{R_m}{R_a}\right)$$

$$\frac{|I|^2}{|I_m|^2} = \frac{R_a + R_m}{R_a} \tag{7.34}$$

Also,

$$I_m = \frac{V}{Z_m}$$

$$I_a = \frac{V}{Z_a}$$

$$I = I_m + I_a = V\frac{Z_a + Z_m}{Z_m Z_a}$$

so

$$\frac{I}{I_m} = \frac{Z_a + Z_m}{Z_a}$$

$$\frac{|I|^2}{|I_m|^2} = \frac{|Z_a + Z_m|^2}{|Z_a|^2} \tag{7.35}$$

From Eqs. 7.34 and 7.35,

$$\frac{R_a + R_m}{R_a} = \frac{|Z_a + Z_m|^2}{|Z_a|^2}$$

$$= \frac{(R_a + R_m)^2 + (X_m - X_A)^2}{R_a^2 + X_A^2} \tag{7.36}$$

where $X_A = X_c - X_a$.

Equation 7.36 is a quadratic equation in X_A. From it, the following result is obtained:

$$X_A = \frac{-X_m R_a \pm |Z_m|\sqrt{R_a(R_a + R_m)}}{R_m}$$

If I_a has to lead V, the net reactance in the starting winding X_A has to be positive. Hence,

$$X_A = X_c - X_a = \frac{-X_m R_a + |Z_m|\sqrt{R_a(R_a + R_m)}}{R_m}$$

$$X_c = \frac{1}{\omega C} = X_a + \frac{-X_m R_a + |Z_m|\sqrt{R_a(R_a + R_m)}}{R_m} \tag{7.37}$$

The value of C obtained from Eq. 7.37 will maximize the starting torque per ampere of starting current.

EXAMPLE 7.4

A four-pole, single-phase, 120 V, 60 Hz induction motor gave the following *standstill* impedances when tested at rated frequency:

Main winding: $Z_m = 1.4 + j4.0$ ohms

Auxiliary winding: $Z_a = 3 + j6.0$ ohms

(a) Determine the value of external resistance to be inserted in series with the auxiliary winding to obtain maximum starting torque as a resistor split-phase motor.

(b) Determine the value of the capacitor to be inserted in series with the auxiliary winding to obtain maximum starting torque as a capacitor-start motor.

(c) Determine the value of the capacitor to be inserted in series with the auxiliary winding to obtain maximum starting torque per ampere of the starting current as a capacitor-start motor.

(d) Compare the starting torques and starting currents in parts (a), (b), and (c) expressed as per unit of the starting torque without any external element in the auxiliary circuit, when operated at 120 V, 60 Hz.

Solution

(a) From Eq. 7.28,

$$R_a = \frac{X_a}{X_m}(R_m + |Z_m|)$$

$$= \frac{6}{4}(1.5 + \sqrt{1.5^2 + 4^2})$$

$$= 8.66 \text{ ohms}$$

External resistance to be added is $8.66 - 3 = 5.66$ ohms.

(b) From Eq. 7.33,

$$C = \frac{1.0}{2\pi 60\left(6 + \dfrac{3 \times 1.5}{4 + \sqrt{1.5^2 + 4^2}}\right)} \times 10^6\ \mu\text{F}$$

$$= 405\ \mu\text{F}$$

(c) From Eq. 7.37,

$$X_c = 6 + \frac{-4 \times 3 + \sqrt{(1.5^2 + 4^2)}\sqrt{3(3 + 1.5)}}{1.5}$$

$$= 6 + \frac{-12 + 15.69}{1.5}$$

$$= 8.46 \text{ ohms}$$

$$C = \frac{1.0}{2\pi 60 \times 8.46} \times 10^6\ \mu\text{F} = 313.54\ \mu\text{F}$$

(d) If I_m is fixed, the starting torque is

$$T_s \propto I_a \sin\alpha = KI_a \sin\alpha$$

Split-phase motor with no external resistance:

$$I_a = \frac{120}{3 + j6}$$

$$= 17.88\angle -63.43^\circ$$

$$I_m = \frac{120}{1.5 + j4}$$

$$= 28.1\angle -69.44^\circ$$

$$\alpha = 69.44^\circ - 63.43^\circ$$

$$= 6.01^\circ$$

$$T_s = KI_a \sin\alpha = K17.88 \sin 6.01^\circ = 1.87K$$

$$I = 28.1\angle -69.44^\circ + 17.88\angle -63.43^\circ$$

$$= 45.92\angle -67.1^\circ \text{ A}$$

Split-phase motor with external resistance:

$$I_a = \frac{120}{8.66 + j6.0}$$

$$= 11.39\underline{/-34.72^\circ} \text{ amperes}$$

$$\alpha = 69.44^\circ - 34.72^\circ$$

$$= 34.72^\circ$$

$$T_s = KI_a \sin \alpha = K11.39 \sin 34.72^\circ = 6.49K$$

$$I = 28.1\underline{/-69.44^\circ} + 11.39\underline{/-34.72^\circ} = 38.02\underline{/-59.6^\circ} \text{ A}$$

Capacitor-start motor with C = *405 μF:*

$$X_c = 1/(2 \times \pi \times 60 \times 405 \times 10^{-6})$$

$$= 6.55 \text{ ohms}$$

$$I_a = \frac{120}{3 + j6 - j6.55}$$

$$= 39.34\underline{/10.4^\circ}$$

$$\alpha = 69.44 + 10.4^\circ = 79.84^\circ$$

$$T_s = KI_a \sin \alpha = K39.34 \sin 79.84^\circ = 38.72K$$

$$I = 28.1\underline{/-69.44^\circ} + 39.34\underline{/10.4^\circ} = 52.22\underline{/-21.56^\circ} \text{ A}$$

Capacitor-start motor with C = *313.54 μF:*

$$X_c = 8.46 \text{ ohms}$$

$$I_a = \frac{120}{3 + j6 - j8.46}$$

$$= 30\underline{/39.35^\circ}$$

$$\alpha = 69.44^\circ + 39.35^\circ$$

$$= 108.79^\circ$$

$$T_s = KI_a \sin \alpha = K30 \sin 108.79^\circ = 28.4K$$

$$I = 28.1\underline{/-69.44^\circ} + 30\underline{/39.35^\circ} = 33.86\underline{/-12.43^\circ} \text{ A}$$

The comparison is shown in the following table.

Type	I (A)	T_s	T_s (pu)	I (pu)	T_s/I
1. Split-phase without external element	45.9	$1.87K$	1	1	1
2. Split-phase with resistance for maximum starting torque	38	$6.49K$	3.47	0.83	4.2
3. Capacitor-start for maximum starting torque	52.2	$38.72K$	20.71	1.14	18.2
4. Capacitor-start motor for maximum starting torque per ampere of starting current	33.9	$28.4K$	15.19	0.74	20.5

7.3 EQUIVALENT CIRCUIT OF A CAPACITOR-RUN MOTOR

It was pointed out in Section 7.1.4 that there are three types of capacitor motors: capacitor-start, capacitor-run, and capacitor-start capacitor-run. In the latter two types, the auxiliary winding stays in operation all the time and the motor operates as a two-phase induction motor. An equivalent circuit can be derived for the capacitor-run motor based on the double revolving field theory.

The main winding and the auxiliary winding are excited by currents i_m and i_a, as shown in Fig. 7.18. The main winding flux Φ_m can be resolved into two revolving fluxes Φ_{fm} (forward revolving) and Φ_{bm} (backward revolving). Similarly, the auxiliary winding flux Φ_a is resolved into two

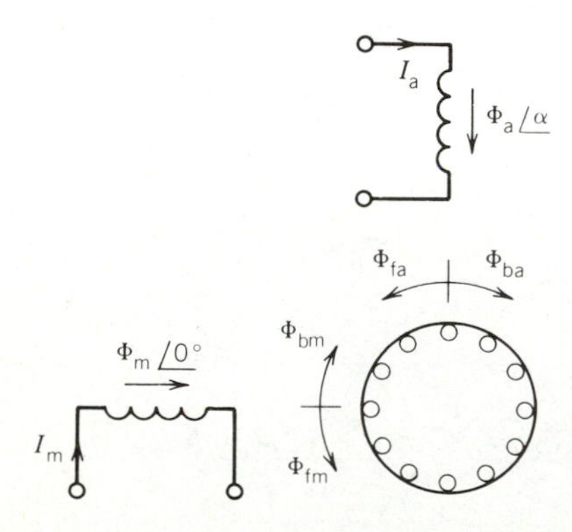

FIGURE 7.18
Revolving fields in the capacitor-run induction motor.

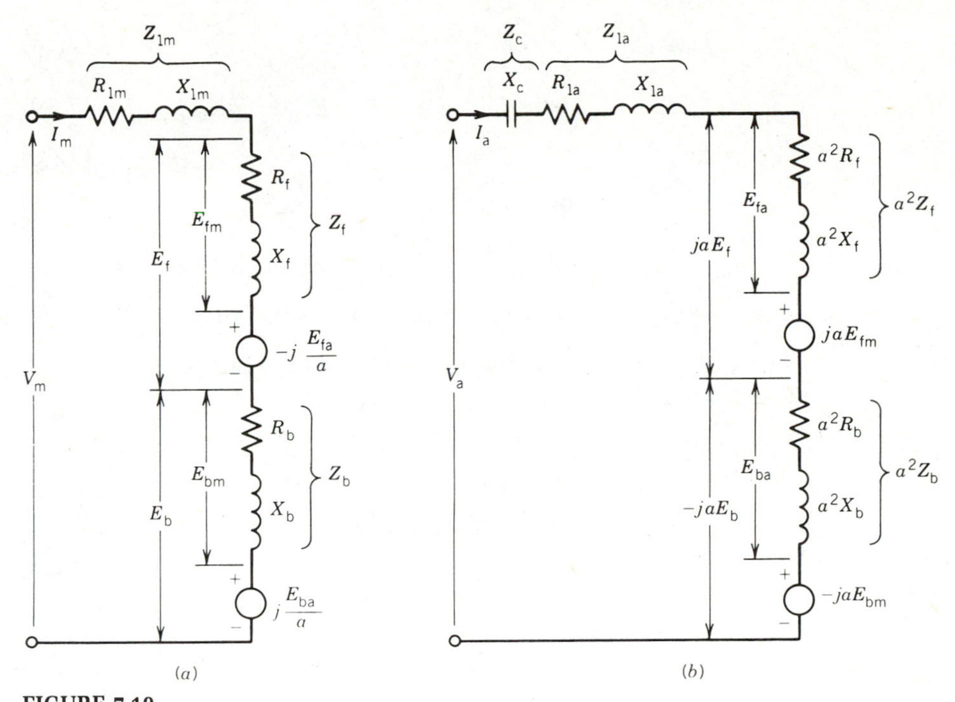

FIGURE 7.19
Equivalent circuit of a capacitor-run induction motor, $a = N_a/N_m$.

revolving fluxes Φ_{fa} and Φ_{ba}. These four revolving fluxes all induce voltages in the two windings.

The main winding can be represented by the equivalent circuit shown in Fig. 7.19*a*, where E_{fm} and E_{bm} are the voltages induced by its own fluxes Φ_{fm} and Φ_{bm}, respectively. The voltages induced, $-j\,E_{fa}/a$ and $j\,E_{ba}/a$ ($a = N_a/N_m$), in the main winding by the respective fluxes Φ_{fa} and Φ_{fb} of the auxiliary winding are shown as internal voltages.

The forward-rotating flux Φ_{fa} of the auxiliary winding induces a voltage E_{fa} in the auxiliary winding. From Fig. 7.18 this flux Φ_{fa} will also induce a voltage in the main winding that will peak $\pi/2$ radian later. If a is the turns ratio of the auxiliary and main winding, the induced voltage is $-j\,E_{fa}/a$, where $-j$ represents a phase lag of $\pi/2$ radian. Similarly, the flux Φ_{ba} will induce a voltage E_{ba} in the auxiliary winding and a voltage $j\,E_{ba}/a$ in the main winding that will peak $\pi/2$ radian earlier.

The auxiliary winding is represented by an equivalent circuit as shown in Fig. 7.19*b*, where the internal voltages jaE_{fm} and $-jaE_{bm}$ are the voltages induced in the auxiliary winding by the revolving fluxes Φ_{fm} and Φ_{bm} of the main winding.

The following are the voltage and current equations for the two windings:

$$V_m = I_m(Z_{1m} + Z_f + Z_b) - j\frac{E_{fa}}{a} + j\frac{E_{ba}}{a} \tag{7.38}$$

$$V_a = I_a(Z_c + Z_{1a} + a^2Z_f + a^2Z_b) + jaE_{fm} - jaE_{bm} \tag{7.39}$$

$$V_m = V_a \tag{7.40}$$

$$I_s = I_m + I_a = \text{input current} \tag{7.41}$$

where $Z_{1m} = R_{1m} + jX_{1m}$ is the leakage impedance of the main winding

$Z_{1a} = R_{1a} + jX_{1a}$ is the leakage impedance of the auxiliary winding

$Z_c = -jX_c$ is the capacitor impedance connected in series with the auxiliary winding

Now

$$E_{fa} = I_a a^2 Z_f \tag{7.42}$$

$$E_{ba} = I_a a^2 Z_b \tag{7.43}$$

$$E_{fm} = I_m Z_f \tag{7.44}$$

$$E_{bm} = I_m Z_b \tag{7.45}$$

From Eqs. 7.42 to 7.45 and 7.38 and 7.39,

$$V_m = (Z_{1m} + Z_f + Z_b)I_m - ja(Z_f - Z_b)I_a \tag{7.46}$$

$$V_a = ja(Z_f - Z_b)I_m + (Z_c + Z_{1a} + a^2Z_f + a^2Z_b)I_a \tag{7.47}$$

Equations 7.46 and 7.47 can be solved to obtain the winding currents I_m and I_a.

Torque

The torque developed by the machine can be expressed as the difference between the forward torque and the backward torque.

$$T = T_f - T_b = \frac{P_{gf} - P_{gb}}{\omega_{syn}} \tag{7.48}$$

From Fig. 7.19

$$P_{gf} = \text{Re}(E_f I_m^* + jaE_f I_a^*)$$

$$P_{gb} = \text{Re}(E_b I_m^* - jaE_b I_a^*)$$

$$P_{gf} - P_{gb} = \text{Re}[(E_f - E_b)I_m^* + ja(E_f + E_b)I_a^*] \tag{7.49}$$

Equation 7.49 can be simplified to the following form:

$$P_{gf} - P_{gb} = (|I_m|^2 + |aI_a|^2)(R_f - R_b) + 2a|I_a||I_m|(R_f + R_b)\sin(\theta_a - \theta_m) \tag{7.50}$$

where $I_m = |I_m|\angle\theta_m$

$I_a = |I_a|\angle\theta_a$

Note that the foregoing analysis is valid for starting and other operating conditions of single-phase induction motors as long as both main and auxiliary windings stay in operation.

For starting, slip $s = 1$ and $R_f = R_b$. From Eqs. 7.48 and 7.50 the starting torque is

$$T_{st} = \frac{2a|I_a||I_m|(R_f + R_b)}{\omega_{syn}}\sin(\theta_a - \theta_m) \tag{7.51}$$

$$= KI_aI_m \sin\alpha \tag{7.52}$$

Note that Eq. 7.52 is the same as Eq. 7.27 derived earlier.

EXAMPLE 7.5

A single-phase 120 V, 60 Hz, four-pole capacitor-run motor has the following equivalent circuit parameters:

$$X_{1m} = 2.0\ \Omega, \quad R_{1m} = 1.5\ \Omega, \quad R_2' = 1.5\ \Omega$$

$$X_{1a} = 2.0\ \Omega, \quad R_{1a} = 1.5\ \Omega, \quad X_2' = 2.0\ \Omega$$

$$X_{mag} = 48\ \Omega$$

Running capacitor $C = 30\ \mu F$

Turns ratio $a = N_a/N_m = 1$

(a) Draw an equivalent circuit based on the double revolving field theory when the motor is running at a slip of 0.05.

(b) Determine the total starting current and the starting torque of the motor at rated voltage.

(c) Determine the value of the starting capacitor to be inserted in parallel with the running capacitor to maximize the starting torque per ampere of starting current.

(d) Determine the maximum starting torque per ampere.

Solution

(a) The equivalent circuit for slip 0.05 is shown in Fig. E7.5*a*.

$$X_c = \frac{1}{2\pi fc} = \frac{10^6}{2\pi 60 \times 30} = 88.4\ \Omega$$

(b) At start $s = 1$.

$$\begin{aligned} Z_f = Z_b &= j24//(0.75 + j1.0) \\ &= 0.69 + j0.98 \\ &= 1.2\underline{/54.85^\circ}\ \Omega \end{aligned}$$

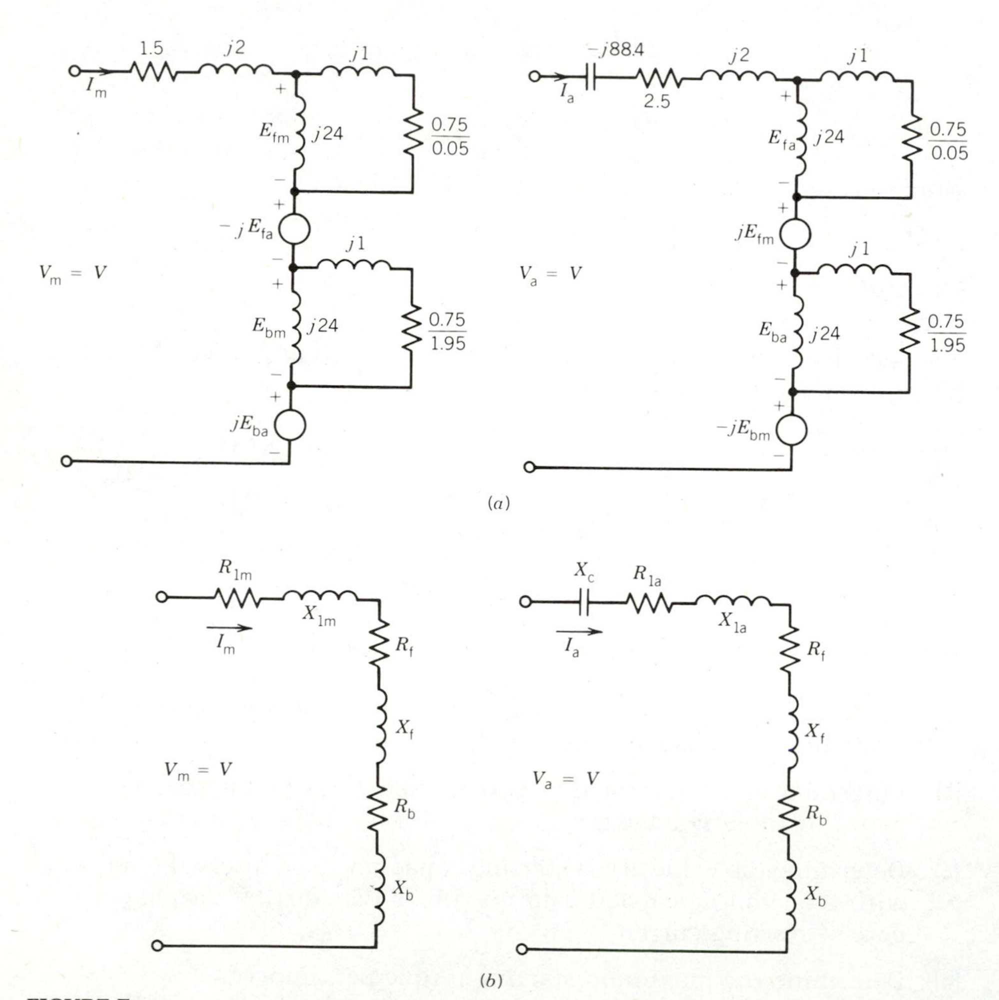

FIGURE E7.5

Also $E_{fm} = E_{bm}$ and $E_{fa} = E_{ba}$. The equivalent circuit for the starting condition is shown in Fig. E7.5*b*. Therefore,

$$I_m = \frac{V}{Z_{1m} + Z_f + Z_b}$$

$$= \frac{120\underline{/0°}}{1.5 + j2 + 2(0.69 + j0.98)}$$

$$= 24.5\underline{/-54°} \text{ A}$$

$$I_a = \frac{V}{Z_{1a} - jX_c + a^2(Z_f + Z_b)}$$

$$= \frac{120\underline{/0°}}{2.5 + j2 - j88.4 + 2(0.69 + j0.98)}$$

$$= 1.42\underline{/87.4°} \text{ A}$$

The starting current is

$$I_s = 24.5\underline{/-54°} + 1.42\underline{/87.4°}$$

$$= 23.4\underline{/-51.82°} \text{ A}$$

From Eq. 7.51 the starting torque is

$$T_s = \frac{2|I_m \mid I_a|(R_f + R_b)\sin(\theta_a - \theta_m)}{\omega_{syn}}$$

$$= \frac{2(24.5)(1.42)2 \times 0.69 \sin(87.4 + 54)}{1800 \times 2\pi/60}$$

$$= 0.318 \text{ N} \cdot \text{m}$$

(c) Equation 7.37 will be used to determine the total capacitance.

$Z_m = R_m + jX_m$ = input impedance of the main winding at start

$$= Z_{1m} + Z_f + Z_b$$

$$= 1.5 + j2 + 2(0.69 + j0.98)$$

$$= 2.88 + j3.96$$

$$= 4.9\underline{/-54°}\ \Omega$$

$Z_a = R_a + jX_a$ = input impedance of the auxiliary winding (excluding capacitors) at start

$$= 2.5 + j2.0 + 2(0.69 + j0.98)$$

$$= 3.88 + j3.96$$

From Eq. 7.37

$$X_c = 3.96 - \left[\frac{3.96 \times 3.88 - 4.9\sqrt{(3.88 + 2.88)3.88}}{2.88}\right]$$

$$= 7.34\ \Omega$$

$$C = \frac{1}{\omega X_c} = \frac{10^6}{377 \times 7.34}\ \mu\text{F}$$

$$= 361.5\ \mu\text{F}$$

The external capacitor C_s to be added in parallel with the running capacitor is

$$C_s = 361.54 - 30 = 331.54\ \mu\text{F}$$

(d) With the starting capacitor and running capacitor in series with the auxiliary winding,

$$I_a = \frac{V}{Z_a - jX_c(\text{total})}$$

$$= \frac{120\angle 0^\circ}{3.88 + j3.96 - j7.34}$$

$$= 23.33\angle 41^\circ\ \text{A}$$

$$I_m = 24.5\angle -54^\circ\ \text{A}$$

The starting current is

$$I_s = 24.5\angle -54^\circ + 23.33\angle 41^\circ$$

$$= 32.33\angle -8^\circ\ \text{A}$$

From Eq. 7.51, the starting torque is

$$T_s = \frac{2(24.5)(23.33)2(0.69)\sin(41^\circ + 54^\circ)}{1800 \times 2\pi/60}$$

$$= 8.35\ \text{N} \cdot \text{m}$$

Maximum starting torque per ampere of input current:

$$= \frac{8.35}{32.33} = 0.258\ \text{N} \cdot \text{m/A}$$

7.4 SINGLE-PHASE SERIES (UNIVERSAL) MOTORS

Single-phase series motors can be used with either a dc source or a single-phase ac source and therefore are called universal motors. They are widely used in fractional horsepower ratings in many domestic appliances such as portable tools, drills, mixers, and vacuum cleaners and usually are light in weight and operate at high speeds (1500 to 10,000 rpm). Large ac series motors in the range of 500 hp are used for traction applications.

Universal motors are mostly operated from a single-phase ac source. Therefore, both the stator and rotor structures are made of laminated steel to reduce core losses and eddy current.

Figure 7.20 shows a schematic diagram of the series motor. The armature current i_a flowing through the series field produces the d-axis flux Φ_d and flowing through the armature winding produces the q-axis flux Φ_q. If eddy current is neglected, both Φ_d and Φ_q are in phase with i_a.

DC Excitation

The behavior with dc excitation was discussed in Section 4.4.2. The torque developed and voltage induced are given by

$$T = K_a \mathbf{\Phi_d} I_a \tag{7.53}$$

$$E_a = K_a \mathbf{\Phi_d} \omega_m \tag{7.53a}$$

If magnetic linearity is assumed

$$T = K_{sr} I_a^2 \tag{7.54}$$

$$E_a = K_{sr} I_a \omega_m \tag{7.54a}$$

The torque–speed characteristic is shown in Fig. 4.55, indicating high torque at low speed and low torque at high speed.

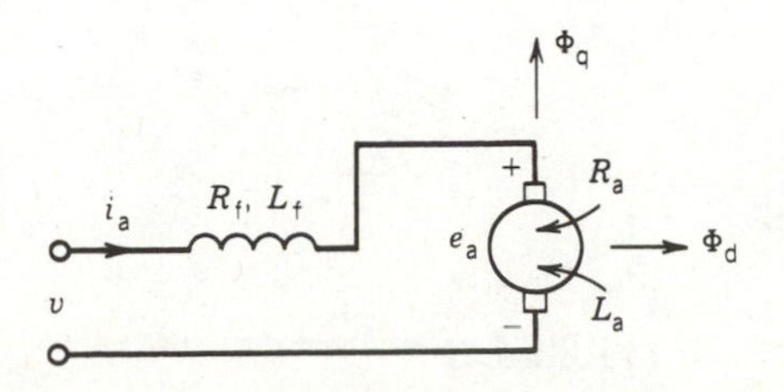

FIGURE 7.20
AC series motor.

AC Excitation

If eddy current effects are neglected, the current i_a and the flux Φ_d are in phase. Let

$$i_a = I_{am} \cos \omega t \tag{7.55}$$

$$\Phi_d = \Phi_{dm} \cos \omega t \tag{7.55a}$$

The back emf is

$$e_a = K_a \Phi_d \omega_m = K_a \Phi_{dm} \omega_m \cos \omega t \tag{7.56}$$

The rms value of the back emf is

$$E_a = K_a \frac{\Phi_{dm}}{\sqrt{2}} \omega_m = K_a \mathbf{\Phi_d} \omega_m \tag{7.56a}$$

where $\mathbf{\Phi_d}$ is the rms value of the d-axis flux.

Note from Eqs. 7.55 and 7.56 that e_a and i_a are in phase. The instantaneous torque is

$$T = K_a \Phi_d i_a = K_a \Phi_{dm} I_{am} \cos^2 \omega t$$

$$T = K_a \frac{\Phi_{dm}}{2} I_{am}(1 + \cos 2\omega t) \tag{7.57}$$

Figure 7.21 shows the variation of i_a, Φ_d, e_a, and T with time. Note that although the current reverses in alternate half-cycles, the instantaneous torque is unidirectional and therefore makes the motor run in the direction of rotation. The instantaneous torque, however, varies at twice the supply frequency, and this fluctuating torque makes the motor operation noisy.

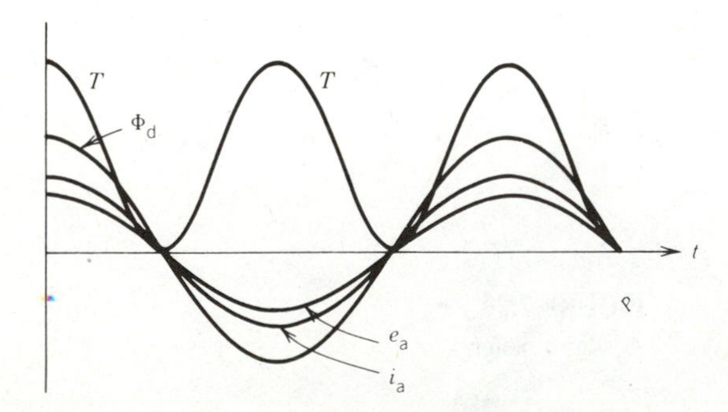

FIGURE 7.21 Voltage, current, flux, and torque waveforms.

From Eq. 7.57 the average torque developed is

$$T = K_a \frac{\Phi_{dm} I_{am}}{2} = K_a \mathbf{\Phi_d} I_a \tag{7.58}$$

where $\mathbf{\Phi_d}$ is the rms value of the d-axis flux and I_a is the rms value of the motor current.

If magnetic linearity is assumed

$$T = K_{sr} I_a^2 \tag{7.58a}$$

$$E_a = K_{sr} I_a \omega_m \tag{7.58b}$$

The electromagnetic (or mechanical) power developed is

$$P_{mech} = E_a I_a \tag{7.59}$$

The developed torque can also be obtained as

$$T = \frac{E_a I_a}{\omega_m} \tag{7.60}$$

The voltage equation for ac excitation is

$$V = I_a(R_f + R_a) + I_a j(X_f + X_a) + E_a$$

The phasor diagram is shown in Fig. 7.22.

DC versus AC Excitation

If it is assumed that the armature current for dc excitation and the rms value of the armature current for ac excitation are the same, the ratio of the back emf's is

$$\frac{E_{a(dc)}}{E_{a(ac)}} = \frac{K_a \mathbf{\Phi}_{d(dc)} \omega_{m(dc)}}{K_a \mathbf{\Phi}_{d(ac)} \omega_{m(ac)}} \simeq \frac{\omega_{m(dc)}}{\omega_{m(ac)}} \tag{7.61}$$

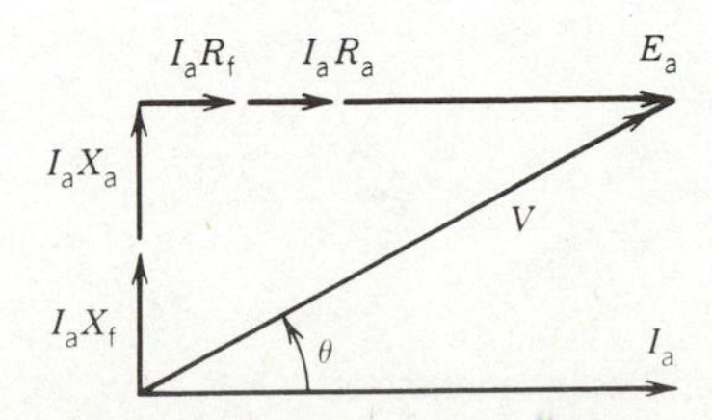

FIGURE 7.22
Phasor diagram for ac series motor (uncompensated).

Note that if magnetic saturation occurs with ac excitation, $\Phi_{d(ac)}$ will be slightly less than $\Phi_{d(dc)}$. Now

$$\frac{E_{a(dc)}}{E_{a(ac)}} = \frac{V - I_a(R_f + R_a)}{V \cos\theta - I_a(R_f + R_a)}$$

$$= \frac{1 - (I_a/V)(R_f + R_a)}{\cos\theta - (I_a/V)(R_f + R_a)}$$

Since $I_a(R_f + R_a)/V << 1$,

$$\frac{E_{a(dc)}}{E_{a(ac)}} \simeq \frac{1}{\cos\theta} \tag{7.62}$$

$$> 1 \tag{7.63}$$

It can be concluded from Eq. 7.63 that for the same terminal voltage and armature current (i.e., same torque) the speed will be lower for ac excitation. The torque–speed characteristics for both dc and ac excitations are shown in Fig. 7.23. Note that ac excitation produces pulsating torque, poor power factor, and lower speed. The latter two undesirable effects are caused by the reactance voltage drop produced by X_f and X_a.

Compensated Motor

A compensating coil (Fig. 7.24) can be connected in series with the armature and will produce flux in opposition to the q-axis flux Φ_q produced by i_a flowing through the armature. The net inductance of the armature winding and the compensating winding is

$$L_{eff} = L_a + L_c - 2M$$

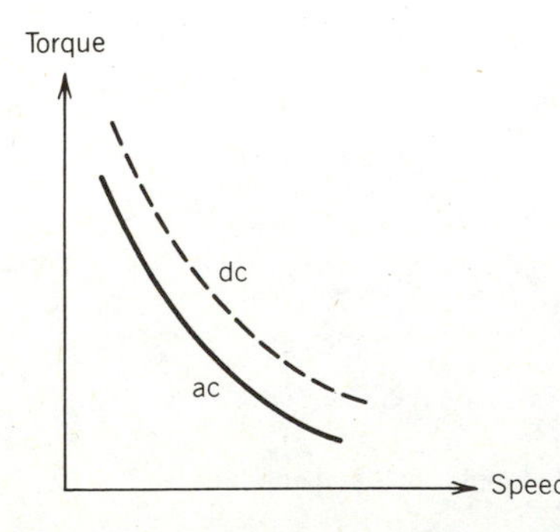

FIGURE 7.23
Torque–speed characteristics in series motors.

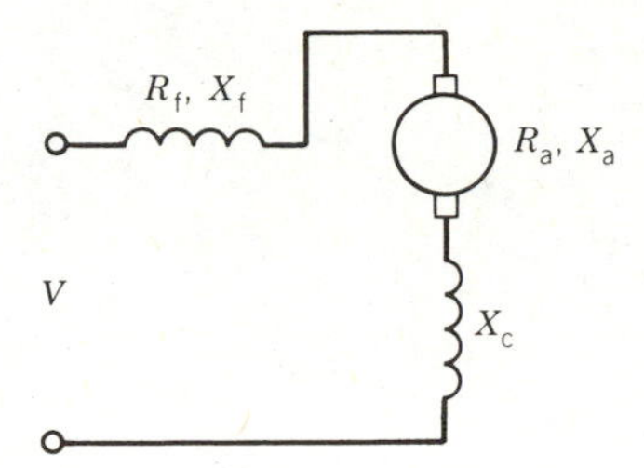

FIGURE 7.24
Compensated series motor.

where L_a is the inductance of the armature winding

L_c is the inductance of the compensating winding

M is the mutual inductance between L_a and L_c

It is possible to make $L_{eff} << L_a$. The phasor diagram for the compensated series motor is shown in Fig. 7.25.

Note that the compensating winding adds additional resistance R_c in the circuit. However, it greatly reduces the effect of the armature reactance X_a. The net result is increased E_a (hence speed), decreased power factor angle θ (hence increased power factor), and increased efficiency. As discussed in Section 4.3.5, a decrease in the q-axis flux due to compensation will improve the commutation of current.

Alternative Design for Compensation Coil

As shown in Fig. 7.26, a shorted compensation coil can be installed in the q-axis so that induced current in this coil can oppose the q-axis flux Φ_q produced by i_a. This coil is predominantly inductive (i.e., high L/R ratio). Compensation by this arrangement is possible with ac excitation only.

EXAMPLE 7.6

A 120 V, 60 Hz, $\frac{1}{4}$ hp universal motor runs at 2000 rpm and takes 0.6 ampere when connected to a 120 V dc source. Determine the speed, torque, and power factor of the motor when it is connected to a 120 V, 60 Hz supply and is loaded to take 0.6 (rms) ampere of current.

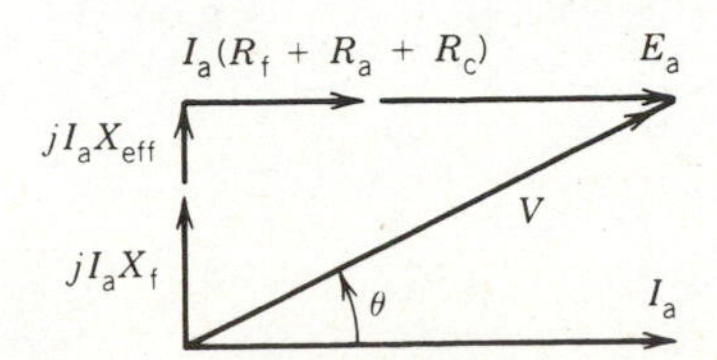

FIGURE 7.25
Phasor diagram for compensated series motor.

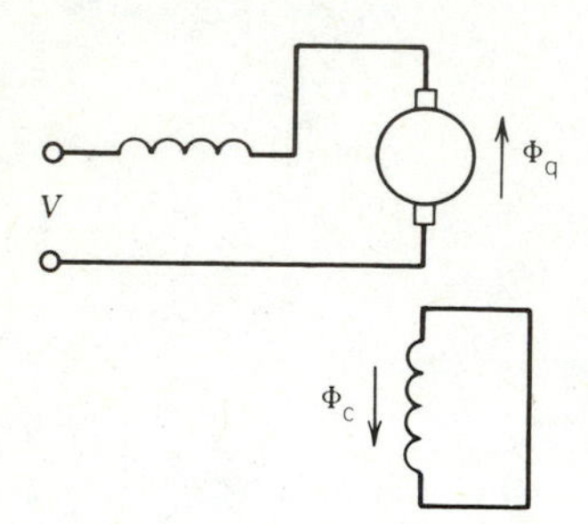

FIGURE 7.26
Inductive compensation.

The resistance and inductance measured at the terminals of the machine are 20 Ω and 0.25 H, respectively.

Solution

DC operation:

$$E_a|_{dc} = 120 - 0.6 \times 20 = 108 \text{ V}$$

AC excitation:

$$X = 2\pi fL = 2\pi 60 \times 0.25 = 94.25 \ \Omega$$

From the phasor diagram shown in Fig. E7.6

$$E_a|_{ac} + I_a R = \sqrt{V^2 - (I_a X)^2}$$

$$\begin{aligned} E_a|_{ac} &= -0.6 \times 20 + \sqrt{120^2 - (0.6 \times 94.25)^2} \\ &= -12 + 105.84 \\ &= 93.84 \text{ V} \end{aligned}$$

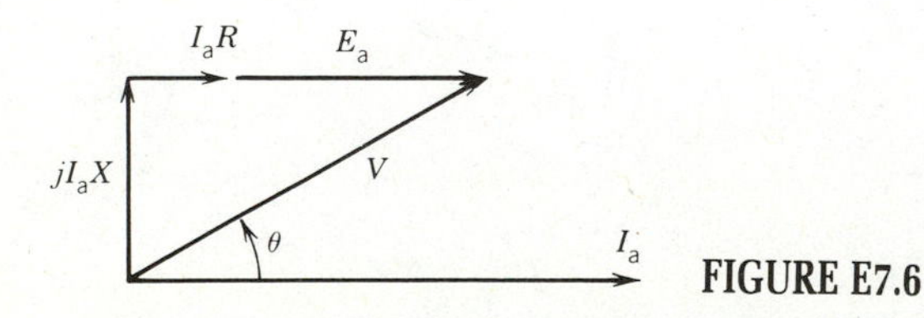

FIGURE E7.6

Assuming the same flux for the same current (i.e., 0.6 A dc and 0.6 A rms),

$$\frac{E_a|_{dc}}{E_a|_{ac}} = \frac{\text{rpm}_{dc}}{\text{rpm}_{ac}} = \frac{n_{dc}}{n_{ac}}$$

$$n_{ac} = 2000 \times \frac{93.84}{108}$$

$$= 1737.78 \text{ rpm}$$

$$\text{Power factor, } \cos\theta = \frac{E_a + I_a R_a}{V}$$

$$= \frac{93.84 + 12}{120}$$

$$= 0.88 \text{ lag}$$

Mechanical power developed is

$$P_{mech} = E_a I_a = 93.84 \times 0.6 = 56.3 \text{ W}$$

Torque developed is

$$T = \frac{P_{mech}}{\omega_m} = \frac{56.3}{1737.78 \times 2\pi/60} = 0.309 \text{ N} \cdot \text{m}$$

7.5 SINGLE-PHASE SYNCHRONOUS MOTORS

The three-phase synchronous motors discussed in Chapter 6 are usually large machines of the order of several hundred kilowatts or megawatts. However, many low-power applications require constant speed. Single-phase synchronous motors of small ratings are ideally suited to such applications as clocks, timers, and turntables. Two types of small synchronous motors are widely used, *reluctance motors* and *hysteresis motors*. These motors do not require dc field excitation, nor do they use permanent magnets. Therefore they are simple in construction.

7.5.1 RELUCTANCE MOTORS

A single-phase synchronous reluctance motor is essentially the same as the single-phase induction motor discussed in Section 7.1 except that some saliency is introduced in the rotor structure by removing some rotor teeth at the appropriate places to provide the required number of poles. Figure 7.27*a* shows the four-pole structure of the rotor for a four-pole reluctance-type synchronous motor. The squirrel-cage bars and end rings are left intact so that the reluctance motor can start as an induction motor. In Section 6.9.1 it was shown that if the motor rotates at synchronous speed, the saliency of the motor will cause a reluctance torque to be developed. This torque arises from the tendency of the rotor to align itself with the rotating field.

The stator of the single-phase reluctance motor has a main winding and an auxiliary (starting) winding. When the stator is connected to a single-phase supply the motor starts as a single-phase induction motor. At a speed of about 75 percent of the synchronous speed a centrifugal switch disconnects the auxiliary winding and the motor continues to speed up as a single-phase motor with the main winding in operation. When the speed is close to the synchronous speed the rotor tends to align itself with the synchronously rotating forward air gap flux wave and eventually snaps into synchronism and continues to rotate at synchronous speed. The performance of the motor will be affected, however, by the torque of the backward-rotating field, and this effect will be similar to an additional shaft load. Figure 7.27*b* shows the typical torque–speed characteristic of the single-phase reluctance motor.

Large-power reluctance motors of integral horsepower rating are invariably of the three-phase type. The reluctance motor has a low power factor because it requires a large amount of reactive current for its excitation. The absence of dc excitation in the rotor greatly reduces the maximum

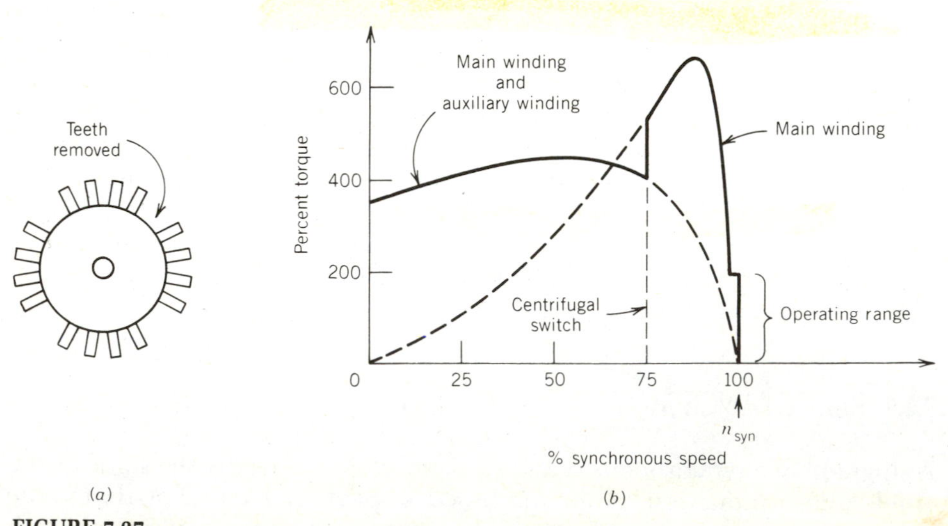

FIGURE 7.27
Torque–speed characteristics of reluctance motor.

torque, as discussed in Section 6.9.1. A reluctance motor is therefore several times larger than a synchronous motor with dc excitation having the same horsepower and speed ratings. However, in some applications these disadvantages may be offset by simplicity of construction (no slip rings, no brushes, and no dc field winding), low cost, and practically maintenance-free operation.

7.5.2 HYSTERESIS MOTORS

Hysteresis motors use the hysteresis property of magnetic materials to produce torque. The rotor has a ring of special magnetic material such as magnetically hard steel, cobalt, or chromium mounted on a cylinder of aluminum or other nonmagnetic material. The stator windings are distributed windings to produce a sinusoidal space distribution of flux. The stator windings are normally the capacitor-run type. The capacitor is chosen to make the two stator windings behave as much like a balanced two-phase system as possible. When the stator windings are connected to a single-phase supply a rotating field is produced, revolving at synchronous speed. This revolving field induces eddy currents in the rotor and, because of hysteresis, the magnetization of the rotor lags behind the inducing revolving field. In Fig. 7.28*a* the axes SS′ and RR′ of the stator and rotor flux waves are displaced by the hysteresis lag angle δ. As long as the rotor speed is less than the synchronous speed, the rotor material is subjected to a repetitive hysteresis cycle at slip frequency. The angle δ depends on the hysteresis loop and is independent of the rate at which the rotor materials are subjected to these hysteresis loops. A constant torque is therefore developed up to the synchronous speed, as shown in Fig. 7.28*b*. As the rotor approaches synchronous speed, the frequency of the eddy currents de-

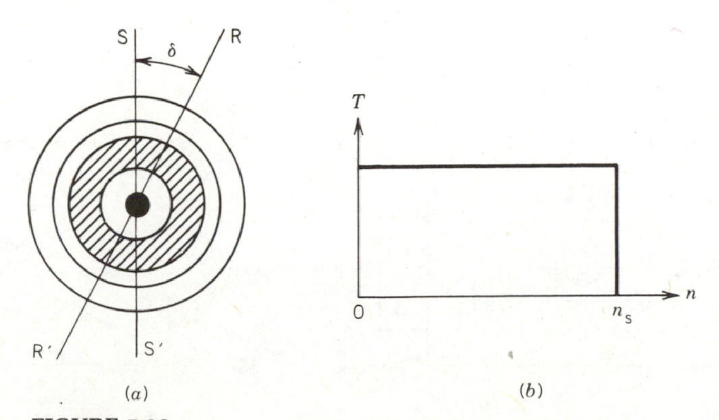

FIGURE 7.28
Hysteresis motor. (*a*) Stator and rotor field. (*b*) *T–n* characteristic.

creases, and at synchronous speed the rotor materials become permanently magnetized in one direction as a result of the high retentivity of the rotor material.

The constant torque–speed characteristic is one of the advantages of a hysteresis motor. Because of this feature it can synchronize any load that it can accelerate, no matter how great the inertia is. On the other hand, a reluctance motor must "snap" its load into synchronism from an induction motor torque–speed characteristic. The hysteresis motor is quiet and smooth-running because of the smooth rotor periphery. However, a high-torque hysteresis motor of good quality is more expensive than a reluctance motor of the same horsepower rating.

7.6 SPEED CONTROL

In many applications of single-phase motors speed must be varied over a certain range. For example, the speed of juice makers, blenders, and hand tools, is often changed. A convenient and economical way of achieving speed control is to control the voltage applied to the motor terminals. In the classical method, shown in Fig. 7.29*a*, speed is changed by changing the value of an external resistance connected in series with the motor. This method is easy to implement, but the power loss in the resistance, its physical size, and the problems of durability and maintenance of the resistance are some of the disadvantages of this method. Recently, a solid-state controller, as shown in Fig. 7.29*b*, has been widely used to vary the speed. Because most single-phase motors (induction or series) are of fractional horsepower rating, a triac (Fig. 7.30*a*) can be used to control the voltage in both positive and negative half-cycles. If the firing angle α is changed, it changes the value of the rms voltage applied to the motor terminals (see

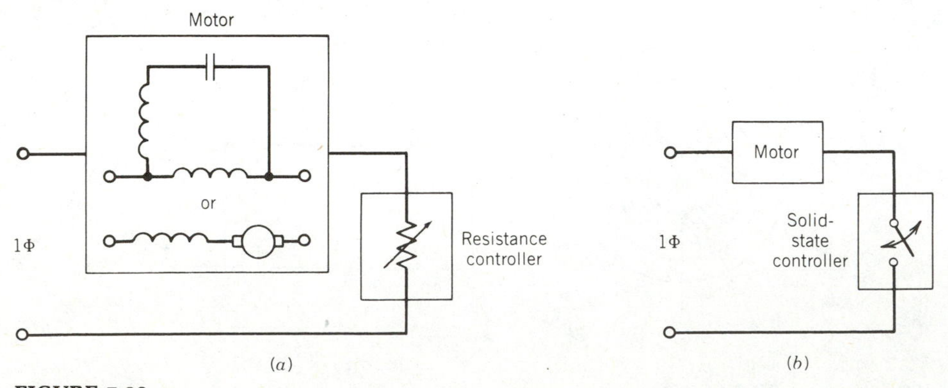

FIGURE 7.29
Speed control of single-phase motors. (*a*) Resistance control. (*b*) Solid-state control.

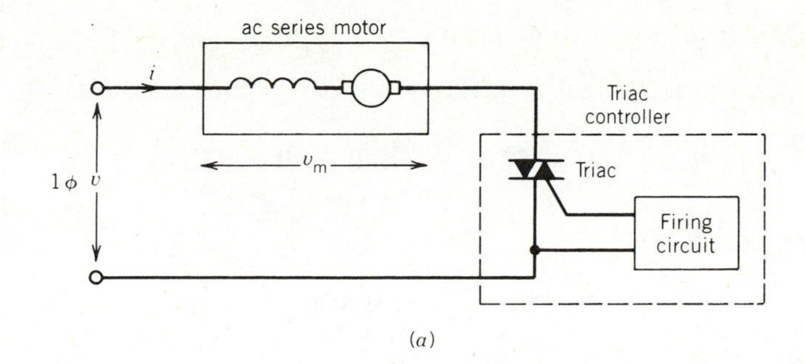

(a)

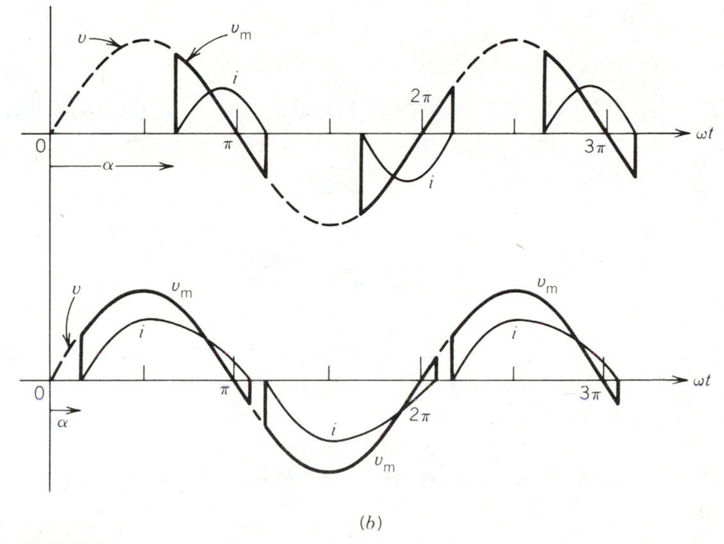

(b)

FIGURE 7.30
Speed control using triacs. (*a*) Circuit. (*b*) Waveforms.

Chapter 10, Section 10.3, on ac voltage controllers). This process is illustrated by the waveforms in Fig. 7.30*b*. At low motor terminal voltage, the firing angle α is large and the input current is very distorted (i.e., highly nonsinusoidal) as shown in Fig. 7.30*b*. The harmonic current increases the heating of the motor.

PROBLEMS

7.1 The nameplate specifications for a single-phase capacitor-start induction motor are 1ϕ, 110 V, $\frac{1}{2}$ hp, 1720 rpm, 8.0 A, 60 Hz. The following test data are obtained for this motor.

Stator main winding resistance = 2.0 Ω

Blocked motor test (auxiliary winding disconnected),

$$V = 52\ \text{V}, I = 8.0\ \text{A}, P = 255\ \text{W}$$

No-load test,

$$V = 110\ \text{V}, I = 4.5\ \text{A}, P = 100\ \text{W}$$

(a) Obtain the double revolving field equivalent circuit for the motor.

(b) Determine the no-load rotational loss.

7.2 A 1ϕ, $\frac{1}{4}$ hp, 115 V, 1725 rpm, 60 Hz, four-pole, capacitor-start induction motor has the following equivalent circuit parameters for the main winding.

$$R_1 = 2.2\ \Omega, \quad R_2' = 3.5\ \Omega$$

$$X_1 = 2.5\ \Omega, \quad X_2' = 2.5\ \Omega, \quad X_{mag} = 60\ \Omega$$

The core loss at 115 V is 20 W and the friction and windage loss is 15 W. The motor is connected to a 115 V, 60 Hz supply and runs at a slip of 0.04. While running, the starting winding remains disconnected. Determine the speed, input current, power factor, input power, developed torque, output power, efficiency, and rotor copper loss.

7.3 The motor in Examples 7.1 and 7.2 runs at rated speed. Determine the ratio of the forward flux to the backward flux.

7.4 A 1ϕ, 120 V, 60 Hz split-phase induction motor has the following standstill impedances.

Main winding: $Z_m = 2.8 + j4.8$

Auxiliary winding: $Z_a = 8 + j6$

Determine the value of the capacitor to be connected in series with the auxiliary winding and the turns ratio a ($= N_a/N_m$) to produce a pure forward mmf wave.

7.5 A single-phase, 120 V, 60 Hz, four-pole, split-phase induction motor gave the following blocked rotor test data:

	V	I	P
Main winding	32	4	80
Auxiliary winding	40	4	128

(a) Determine the standstill impedances of the main and auxiliary windings.

(b) Determine the value of the resistances to be added in series with the auxiliary winding to obtain maximum starting torque.

(c) Compare the starting torques and starting currents with and without the added resistance in the auxiliary winding circuit if the motor is connected to a 120 V, 60 Hz supply.

7.6 A single-phase, 120 V, 60 Hz, four-pole, split-phase induction motor has the following standstill impedances.

Main winding:	$Z_m = 5 + j6.25$
Auxiliary winding:	$Z_a = 8 + j6$

(a) Determine the value of capacitance to be added in series with the auxiliary winding to obtain maximum starting torque.

(b) Compare the starting torques and starting current with and without the added capacitance in the auxiliary winding circuit when operated from a 120 V, 60 Hz supply.

7.7 A four-pole, 115 V, 60 Hz, 1710 rpm, capacitor-start single-phase induction motor has been designed to produce maximum starting torque per unit starting current. The motor has the following parameters.

$$R_{1m} = 1.5 \text{ ohms}, \qquad X_{1m} = 2.6 \text{ ohms}$$

$$R_{1a} = 2.5 \text{ ohms}, \qquad X_{1a} = 2.5 \text{ ohms}$$

$$X_{mag} = 40 \text{ ohms}$$

$$R_2' = 1.0 \text{ ohms}, \qquad X_2' = 1.6 \text{ ohms}$$

$$N_a/N_m = 1$$

Capacitor in series with auxiliary winding = 375 μF

(a) Draw the equivalent circuit for the motor under starting conditions. Determine the value of the starting torque.

(b) Draw the equivalent circuit for the motor when it runs at the rated (i.e., full-load) speed. Determine the torque developed at this speed.

(c) Determine the ratio of the starting torque to the torque developed at the rated speed.

7.8 A single-phase, 120 V, 60 Hz, four-pole, split-phase induction motor has the following equivalent circuit parameters:

$$R_{1m} = 1.5\ \Omega, \qquad R_{1a} = 2.5\ \Omega, \qquad R_2' = 1.0\ \Omega$$

$$X_{1m} = 2.5\ \Omega, \qquad X_{1a} = 2.5\ \Omega, \qquad X_2' = 1.5\ \Omega$$

$$X_{mag} = 40\ \Omega$$

(a) Determine the standstill impedances (Z_m, Z_a) of the windings.

(b) Determine the starting torque and the starting current of the motor if it is started from rated voltage mains as a resistor split-phase motor.

(c) Determine the value of the capacitor to be connected in series with the auxiliary winding to produce maximum starting torque per ampere of starting current. Determine the value of the starting torque and starting current.

(d) Compare the starting torque per ampere of starting current for cases (b) and (c).

7.9 Determine the operating power factor, output power, and efficiency for the following single-phase motors when operated from a 120 V, 60 Hz supply at 1728 rpm. Assume the rotational loss to be 40 W.

(a) A four-pole, capacitor-start, single-phase induction motor with the following main winding equivalent circuit parameters:

$$R_{1m} = 1.2\ \Omega, \qquad X_{1m} = 1.9\ \Omega, \qquad X_{mag} = 36\ \Omega$$

$$R_2' = 1.6\ \Omega, \qquad X_2' = 2.0\ \Omega$$

(b) A compensated series motor with the same standstill input impedance as the main winding of the above induction motor. At 1728 rpm, both motors draw the same line current from the 120 V supply. Assume the same rotational loss as in the induction motor.

7.10 A single-phase, 120 V, 60 Hz series motor gave the following standstill impedances:

Without the compensating winding, $Z_1 = 5 + j25$

With the compensating winding, $Z_1 = 5.5 + j3.0$

(a) Uncompensated motor: The uncompensated motor is connected to a 120 V, 60 Hz supply and rotates at 1800 rpm when loaded to draw a current of 1.6 A. The rotational loss is 30 W. Determine the

i. Supply power factor.

ii. Mechanical power developed.

iii. Efficiency.

(b) Compensated motor: The compensated motor is connected to a 120 V, 60 Hz supply and loaded to draw a current of 1.6 A. Determine the

i. Speed of the motor.

ii. Supply power factor.

iii. Mechanical power developed.

iv. Efficiency [assume the same rotational loss as in part (a)].

(c) The uncompensated motor is connected to a 120 V, 60 Hz supply. Determine the starting torque. Assume magnetic linearity, that is, no saturation.

7.11 Write a computer program to study the performance characteristics of the single-phase induction motor of Problem 7.2. For various speeds (1600 rpm to 1795 rpm in steps of 5 rpm) calculate the following:

Input impedance (Z_{in}), input current (I_{in}), input power factor (PF), input power (P_{in}), torque (T), mechanical power developed (P_{mech}), output (shaft) power (P_{out}), air gap power (P_g), rotor copper loss (P_2), and efficiency (Eff)

Assume that rotational losses remain constant over the speed range.

(a) Write a computer flowchart.

(b) Obtain a computer printout for the performance characteristics mentioned above in tabular form.

SPECIAL MACHINES

CHAPTER eight

Large electric machines, dc or ac, are used primarily for continuous energy conversion. However, there are many special applications where continuous energy conversion is not required. For example, robots require position control for movement of the arm from one position to another. The printer of a computer requires that the paper move by steps in response to signals received from the computer. Such applications require special motors of low power rating. The basic principle of operation of these motors is the same as that of other electromagnetic motors. However, their construction, design, and mode of operation may be different. In this chapter the operation of servomotors, synchro motors, and stepper motors is discussed.

8.1 SERVOMOTORS

Servomotors, sometimes called *control motors,* are electric motors that are specially designed and built, primarily for use in feedback control systems, as output actuators. Their power rating can vary from a fraction of a watt up to a few hundred watts. They have a high speed of response, which requires low rotor inertia. These motors are therefore smaller in diameter and longer in length. They normally operate at low or zero speed and thus have a larger size for their torque or power rating than conventional motors of similar rating. They may be used for various applications, such as robots, radars, computers, machine tools, tracking and guidance systems, and process controllers. Both dc and ac servomotors are used at present.

8.1.1 DC SERVOMOTORS

DC servomotors are separately excited dc motors or permanent magnet dc motors. A schematic diagram of a separately excited dc servomotor is shown in Fig. 8.1*a*. The basic principle of operation is the same as that of the con-

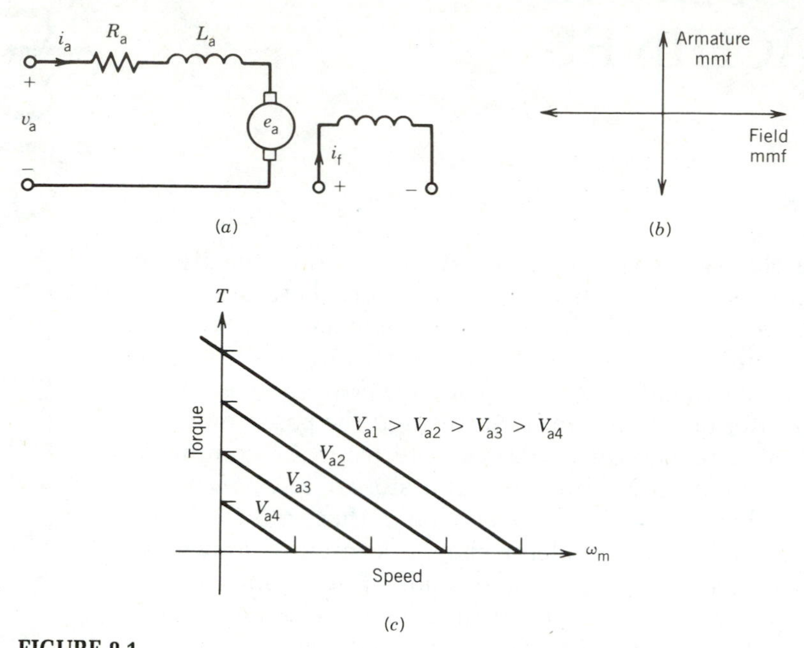

FIGURE 8.1
DC servomotor. (*a*) Schematic diagram. (*b*) Armature mmf and field mmf. (*c*) Torque–speed characteristics.

ventional dc motors discussed in Chapter 4. These dc servomotors are normally controlled by the armature voltage. The armature is designed to have large resistance so that the torque–speed characteristics are linear and have a large negative slope as shown in Fig. 8.1*c*. The negative slope provides viscous damping for the servo drive system. Recall that the armature mmf and excitation field mmf are in quadrature in a dc machine (Fig. 8.1*b*). This provides a fast torque response because torque and flux are decoupled. Therefore, a step change in the armature voltage (or current) results in a quick change in the position or speed of the rotor.

8.1.2 AC SERVOMOTORS

The power rating of dc servomotors ranges from a few watts to several hundred watts. In fact, most high-power servomotors are dc servomotors. At present, ac servomotors are used for low-power applications. AC motors are robust in construction and have lower inertia. However, in general, they are nonlinear and highly coupled machines, and their torque–speed characteristics are not as ideal as those of dc servomotors. Besides, they are low-torque devices compared to dc servomotors of the same size.

Most ac servomotors used in control systems are of the two-phase squirrel-cage induction type. The frequency is normally rated at 60 or 400 Hz; the higher frequency is preferred in airborne systems.

A schematic diagram of a two-phase ac servomotor is shown in Fig. 8.2. The stator has two distributed windings displaced 90 electrical degrees apart. One winding, called the *reference* or *fixed phase* is connected to a constant-voltage source, $V_m \angle 0°$. The other winding, called the *control phase,* is supplied with a variable voltage of the same frequency as the reference phase but is phase-displaced by 90 electrical degrees. The control phase voltage is usually supplied from a servo amplifier. The direction of rotation of the motor depends on the phase relation, leading or lagging, of the control phase voltage with respect to the reference phase voltage.

For balanced two-phase voltages, $|V_a| = |V_m|$, the torque–speed characteristic of the motor is similar to that of a three-phase induction motor. For low rotor resistance this characteristic is nonlinear, as shown in Fig. 8.2*b*. Such a torque–speed characteristic is unacceptable in control systems. However, if the rotor resistance is high the torque–speed character-

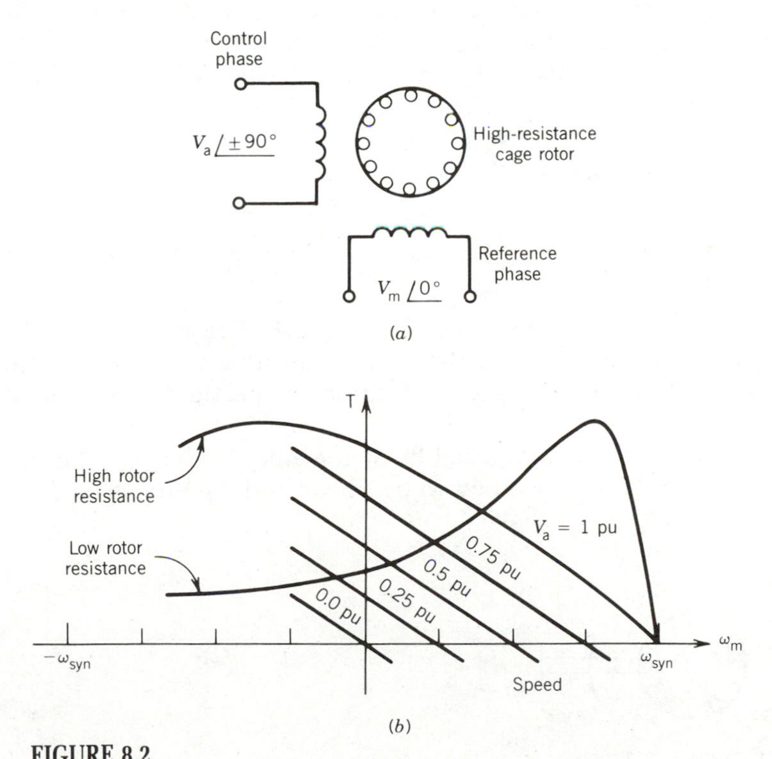

FIGURE 8.2

Two-phase ac servomotor. (*a*) Schematic diagram. (*b*) Torque–speed characteristics.

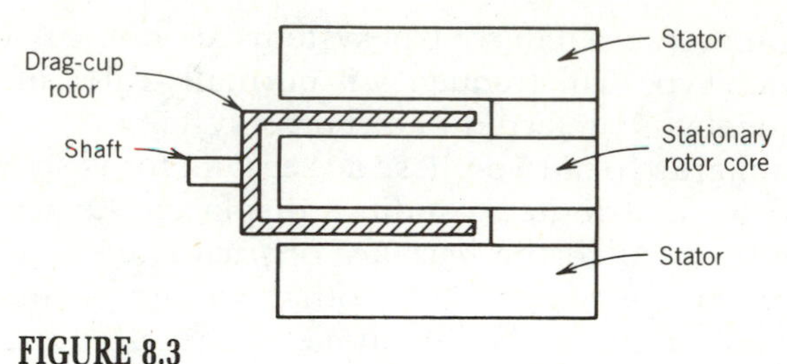

FIGURE 8.3
Drag-cup rotor construction.

istic, as shown in Fig. 8.2*b*, is essentially linear over a wide speed range, particularly near zero speed. To control the machine it is operated with fixed voltage for the reference phase and variable voltage for the control phase. The torque–speed characteristics are essentially linear (high rotor resistance assumed) for various control phase voltages, as shown in Fig. 8.2*b*.

In low-power control applications (below a few watts) a special rotor construction is used to reduce the inertia of the rotor. A thin cup of non-magnetic conducting material is used as the rotor, as shown in Fig. 8.3. Because of the thin conductor, the rotor resistance is high, resulting in high starting torque. A stationary iron core at the middle of the conducting cup completes the magnetic circuit. With this type of construction the rotor is called a *drag-cup rotor*.

8.1.3 ANALYSIS: TRANSFER FUNCTION AND BLOCK DIAGRAM

Consider the servo system shown in Fig. 8.4. The input variable is the control phase voltage V_a and the output variable is either position θ or speed ω_m. Most loads are a combination of inertia J_L and viscous friction F_L.

The torque–speed characteristics of the unbalanced two-phase motor shown in Fig. 8.2*b* are assumed to be linear and equally spaced for equal

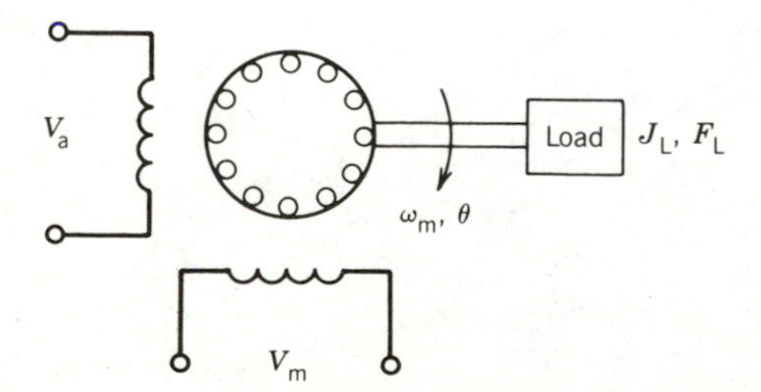

FIGURE 8.4
Servo system using a two-phase motor.

increments of the control phase voltage. The motor torque can be written as

$$T = K_m V_a - F_m \omega_m \tag{8.1}$$

where K_m is the motor torque constant in N · m/volt

F_m is the motor viscous friction in N · m/radian/sec

Note that F_m is just the slope of the torque–speed curves at constant control phase voltage V_a. Also, K_m is the change in torque per unit change in control phase voltage at constant speed.

The equation of motion of the servomotor driving the load is

$$T = K_m V_a - F_m \omega_m = (J_m + J_L) \frac{d\omega_m}{dt} + F_L \omega_m \tag{8.2}$$

where J_L is the load inertia

J_m is the motor inertia

If θ is the angular position of the load

$$\frac{d\theta}{dt} = \omega_m \quad \text{is the speed of the system}$$

Equation 8.2 can also be written as

$$K_m V_a - F_m \frac{d\theta}{dt} = (J_m + J_L) \frac{d^2\theta}{dt^2} + F_L \frac{d\theta}{dt} \tag{8.3}$$

Equations 8.2 and 8.3 can also be written as

$$K_m V_a = (J_m + J_L) \frac{d\omega_m}{dt} + (F_m + F_L)\omega_m \tag{8.4}$$

$$K_m V_a = (J_m + J_L) \frac{d^2\theta}{dt} + (F_m + F_L) \frac{d\theta}{dt} \tag{8.5}$$

Note that the negative slope (F_m) of the torque–speed characteristic of the motor corresponds to viscous friction and therefore provides damping for the system.

Taking the Laplace transforms of Eqs. 8.4 and 8.5,

$$\frac{\omega_m(s)}{V_a(s)} = \frac{K_m/F}{1 + s\tau_m} \tag{8.6}$$

$$\frac{\theta(s)}{V_a(s)} = \frac{K_m/F}{s(1 + s\tau_m)} \tag{8.7}$$

where $F = F_L + F_m$

$J = J_L + J_m$

$\tau_m = J/F$ is the mechanical time constant of the drive system

Equations 8.6 and 8.7 are shown in block diagram forms in Fig. 8.5.

Time Response for a Step Change in Control Phase Voltage: Open-Loop Operation

Consider a step change in the control phase voltage V_a, as shown in Fig. 8.6*a*.

$$V_a(s) = \frac{V}{s}$$

From Eq. 8.6

$$\omega_m(s) = \frac{K_m/F}{1 + s\tau_m}\frac{V}{s}$$

$$= \frac{K_m V}{F}\left(\frac{1}{s} - \frac{1}{s + 1/\tau_m}\right)$$

The corresponding time function is

$$\omega_m(t) = \frac{K_m V}{F}(1 - e^{-t/\tau_m}) \tag{8.8}$$

The steady-state speed is

$$\omega_m(\infty) = \frac{K_m V}{F} \tag{8.9}$$

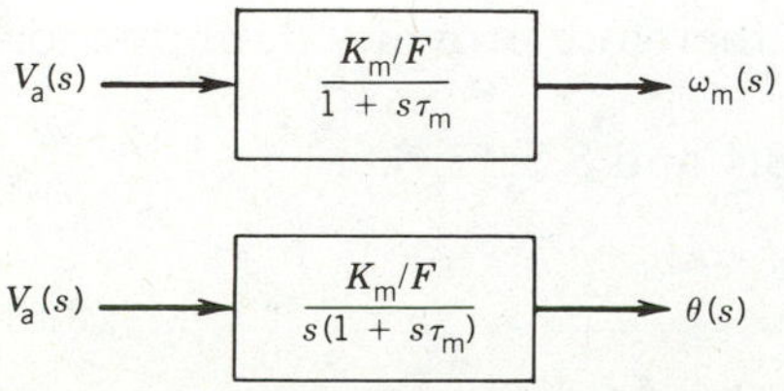

$$V_a(s) \longrightarrow \boxed{\frac{K_m/F}{1 + s\tau_m}} \longrightarrow \omega_m(s)$$

$$V_a(s) \longrightarrow \boxed{\frac{K_m/F}{s(1 + s\tau_m)}} \longrightarrow \theta(s)$$

FIGURE 8.5
Transfer functions.

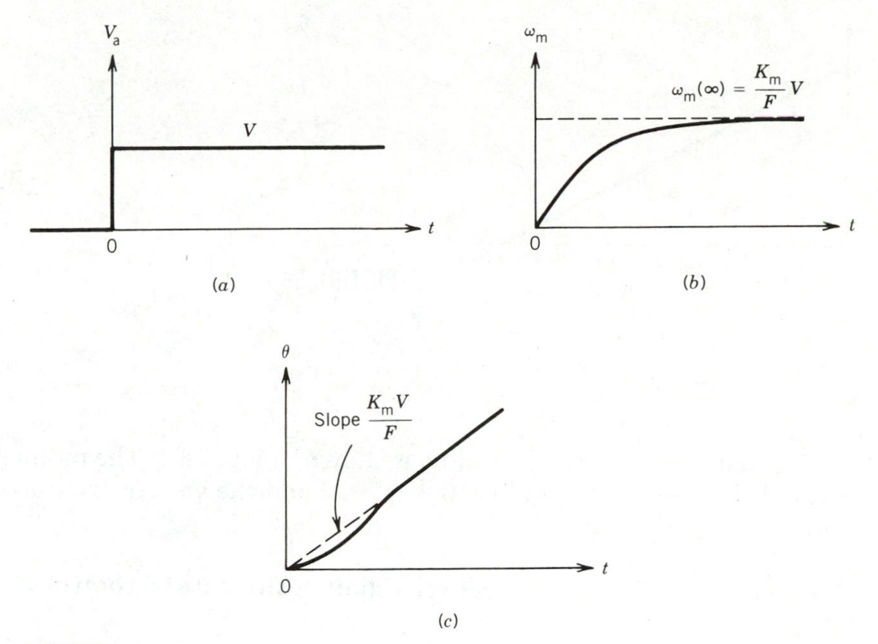

FIGURE 8.6
Step response in a two-phase servo system. (a) Step change in V_a. (b) Response in speed. (c) Response in position.

From Eq. 8.7

$$\theta(s) = \frac{K_m/F}{s(1 + s\tau_m)}\frac{V}{s}$$

$$= \frac{K_m V}{Fs^2} - \frac{K_m V\tau_m}{Fs} + \frac{K_m V\tau_m}{F(s + 1/\tau_m)}$$

The corresponding time function is

$$\theta(t) = \frac{K_m V}{F}t - \frac{K_m V\tau_m}{F} + \frac{K_m V\tau_m}{F}e^{-t/\tau_m} \tag{8.10}$$

The speed response and the position response are shown in Figs. 8.6*b* and 8.6*c*, respectively.

EXAMPLE 8.1

A two-phase servomotor has rated voltage applied to its reference phase winding. The torque–speed characteristic of the motor with $V_a = 115$ V, 60

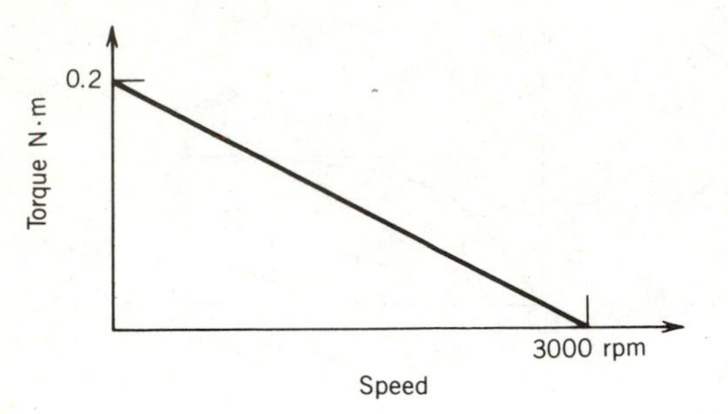

FIGURE E8.1

Hz applied to its control phase winding is shown in Fig. E8.1. The moment of inertia of the motor and load is 10^{-5} kg · m² and the viscous friction of the load is negligible (Fig. 8.4).

(a) Obtain the transfer function between shaft position θ and control voltage V_a.

(b) Obtain an expression for the shaft position due to the application of a step voltage $V_a = 115$ V to the control phase winding.

Solution

(a)

$$K_m = \left.\frac{T}{V_a}\right|_{\omega_m = \text{constant}} = \left.\frac{0.2}{115}\right|_{\omega_m = 0} = 0.00174 \text{ N} \cdot \text{m/V}$$

$$F_m = \left.\frac{T}{\omega_m}\right|_{V_a = \text{constant}} = \frac{0.2}{3000 \times 2\pi/60} = 0.0006366 \text{ N} \cdot \text{m/rad/sec}$$

$$F = F_m + F_L = F_m + 0 = F_m$$

$$J = 10^{-5} \text{ kg} \cdot \text{m}^2$$

$$\tau_m = \frac{J}{F} = \frac{10^{-5}}{0.0006366} = 15.71 \times 10^{-3} \text{ sec}$$

$$\frac{K_m}{F} = \frac{0.00174}{0.0006366} = 2.733$$

From Eq. 8.7

$$\frac{\theta(s)}{V_a(s)} = \frac{2.733}{s(1 + 0.01571s)}$$

(b)

$$V_a(s) = \frac{115}{s}$$

$$\frac{K_m V}{F} = 2.733 \times 115 = 314.3$$

$$\frac{K_m V}{F} \tau_m = 314.3 \times 0.01571 = 4.94$$

From Eq. 8.10

$$\theta(t) = 314.3t - 4.94 + 4.94e^{-t/0.01571}$$

$$\simeq 314.3t$$

Application: Radar Position Control

A typical closed-loop position control system using a two-phase ac servomotor is shown in Fig. 8.7. With this system the position of a radar antenna can be controlled.

Two potentiometers are used as position transducers. The reference potentiometer generates a voltage E_{ref} depending on the desired position command θ_{ref}. The second potentiometer coupled to the shaft of the servomotor produces a voltage E proportional to the output shaft position θ. The difference in the two voltages, E_{error} ($= E_{ref} - E$), is therefore proportional to the position error $\theta_{ref} - \theta$. This error is fed to a servo amplifier, which generates the necessary voltage V_a for the control phase winding of the servomotor to reduce the position error to zero.

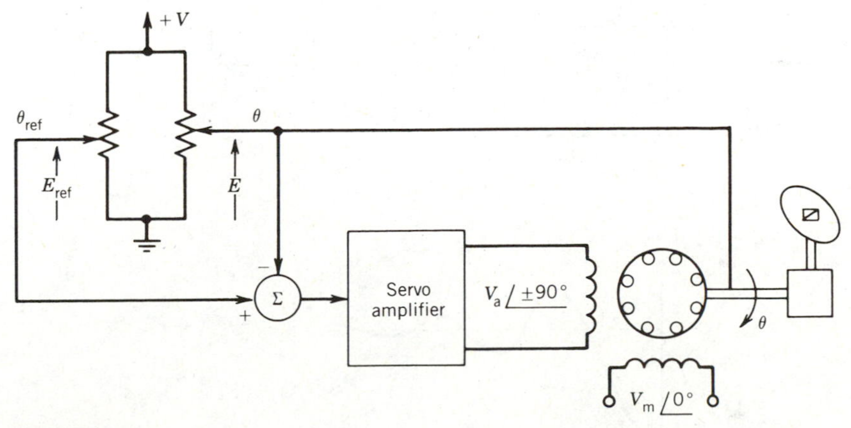

FIGURE 8.7
Radar position control system.

EXAMPLE 8.2

For the position control system shown in Fig. 8.7, let the potentiometer transducers give a voltage of 1 volt per radian of position. The transfer function of the servo amplifier is $G(s) = 10(1 + 0.01571s)/(7 + s)$. Assume that the initial angular position of the radar is zero. The transfer function between the motor control phase voltage V_a and radar position θ is $M(s) = 2.733/s(1 + 0.01571s)$

(a) Derive the transfer function of the system.

(b) For a step change in the command angle of 180° (= π radians) find the time response of the angular position of the antenna.

Solution

(a) The block diagram is shown in Fig. E8.2*a*. This can be simplified to the block diagram shown in Fig. E8.2*b*. From Fig. E8.2*b*

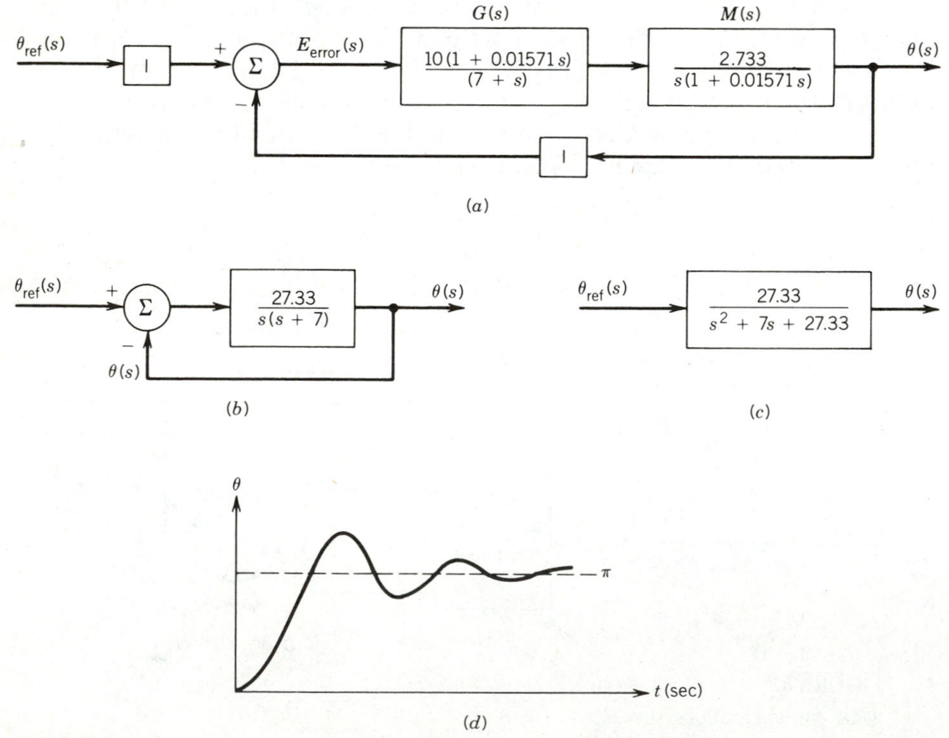

FIGURE E8.2

$$\frac{\theta(s)}{\theta_{\text{ref}}(s)} = \frac{27.33/s(s+7)}{1 + 27.33/s(s+7)} = \frac{27.33}{s^2 + 7s + 27.33}$$

This equation represents a second-order system. The corresponding block diagram is shown in Fig. E8.2*c*.

(b)

$$\theta_{\text{ref}}(s) = \frac{\pi}{s}$$

$$\begin{aligned}
\theta(s) &= \frac{27.33}{s^2 + 7s + 27.33}\,\frac{\pi}{s} \\
&= \pi\,\frac{27.33}{s(s^2 + 7s + 27.33)} \\
&= \pi\,\frac{\omega_n^2}{s(s^2 + 2\xi\omega_n s + \omega_n^2)}
\end{aligned}$$

where $\omega_n = \sqrt{27.33} = 5.228$ rad/sec

$$\xi = \frac{7}{2\omega_n} = \frac{7}{2 \times 5.288} = 0.67$$

The time response is

$$\begin{aligned}
\theta(t) &= \pi\left[1 - \frac{e^{-\xi\omega_n t}}{\sqrt{1 - \xi^2}} \sin(\omega_n \sqrt{1 - \xi^2}\, t + \cos^{-1} \xi)\right] \\
&= \pi[1 - 1.347e^{-3.5t} \sin(3.88t + 48°)] \text{ radian}
\end{aligned}$$

The position response is shown in Fig. E8.2*d*.

8.1.4 THREE-PHASE AC SERVOMOTORS

DC servomotors have dominated the area of high-power servo systems. Recently, however, a great deal of research has been conducted on the use of three-phase squirrel-cage induction motors as servomotors for application in high-power servo systems. A three-phase induction motor is normally a highly nonlinear coupled-circuit device. Many researchers have operated this machine successfully as a linear decoupled machine, similar to a dc machine, using a control method known as *vector control* or *field-oriented control.*[1] In this method the currents in the machine are controlled so that torque and flux are decoupled as in a dc machine. This provides a high speed response and high torque response. In Japan, three-phase in-

[1] W. Leonhard, *Control of Electrical Drives*, Springer-Verlag, New York, 1985.

duction motors with vector control are being increasingly used as servomotors.

8.2 SYNCHROS

Synchros are ac electromagnetic devices that convert a mechanical displacement into an electrical signal. Synchros are widely used in control systems for transmitting shaft position information or for maintaining synchronism between two or more shafts. They are used primarily to synchronize the angular positions of two shafts at different locations where it is not practical to make a mechanical interconnection of the shafts.

There are many types of synchros and a wide variety of applications. In this section only the synchro *control transmitter* (CX), synchro *control receiver* (CR), and synchro *control transformer* (CT) are discussed.

The control transmitter (CX) has a balanced three-phase stator winding similar to the stator winding of a three-phase synchronous machine. The rotor is of the salient pole type using dumbbell construction with a single winding, as shown in Fig. 8.8*a*. If a single-phase ac voltage is applied to the rotor through a pair of slip rings, an alternating flux field is produced along the axis of the rotor. This alternating flux induces voltages in the stator windings by transformer action. If the rotor is aligned with the axis of stator winding 2, flux linkage of this stator winding is maximum, and this position is defined as the *electrical zero* position of the rotor. Figure 8.8*b* shows the rotor position displaced from the electrical zero by the angle α.

The control transformer (CT) has a uniform air gap because of the cylindrical shape of the rotor, as shown in Fig. 8.9*a*. This feature is important

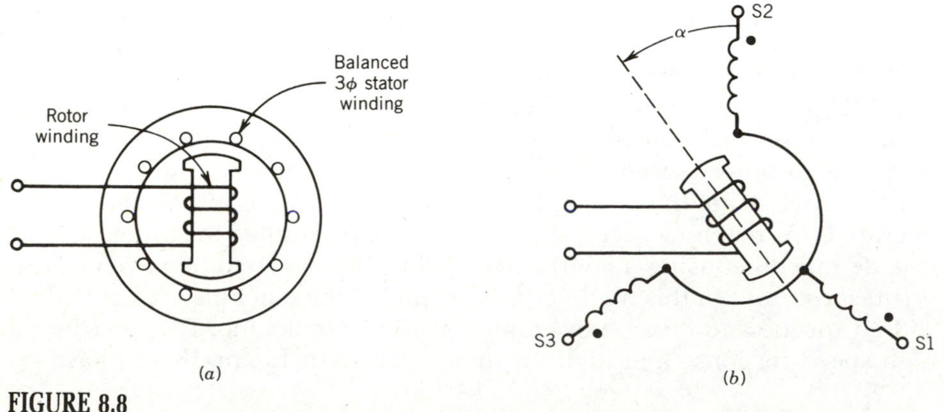

FIGURE 8.8
Synchro control transmitter (CX). (*a*) Construction. (*b*) Schematic diagram.

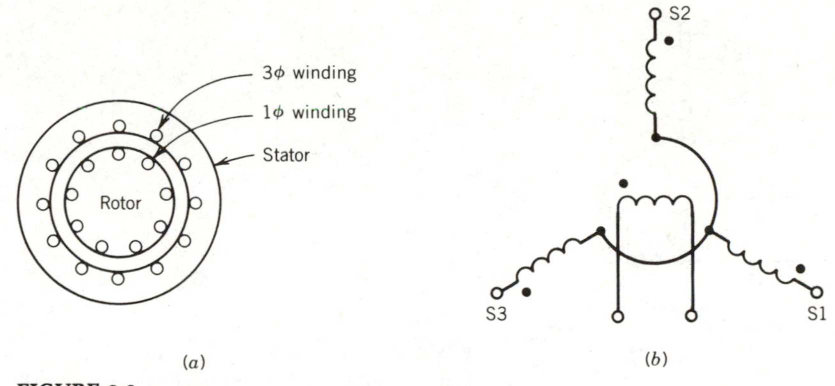

FIGURE 8.9
Synchro control transformer (CT). (*a*) Construction. (*b*) Schematic diagram, rotor winding at electrical zero position.

for the control transformer, because the rotor terminals are normally connected to an amplifier and the latter should see a constant impedance irrespective of rotor position. The *electrical zero* is defined as the position of the rotor that makes the coupling with stator winding 2 zero. This position is shown in Fig. 8.9*b*. The stator has a balanced three-phase winding. However, the impedance per phase is greater in the transformer than in the transmitter. This feature allows several control transformers to be fed from a single control transmitter.

The control receiver (CR) has essentially the same basic structure as the control trnsmitter, that is, three-phase stator winding and single-phase salient pole rotor. However, in the control receiver a mechanical viscous damper is provided on the shaft to permit the receiver rotor to respond without causing the rotor to overshoot its mark.

8.2.1 VOLTAGE RELATIONS

To understand the various applications of synchros, it is necessary to know how the stator phase voltages vary with the rotor displacement. Let us consider the schematic diagram of the control transmitter shown in Fig. 8.10. A single-phase ac voltage is applied to the rotor winding, and the rotor is displaced by an angle α from its electrical zero position and held fast. The rotor voltage is

$$e_r = \sqrt{2}\, E_r \sin \omega t \tag{8.11}$$

Single-phase voltages having the same frequency as the rotor voltage are induced in each stator phase by transformer action. The value of the

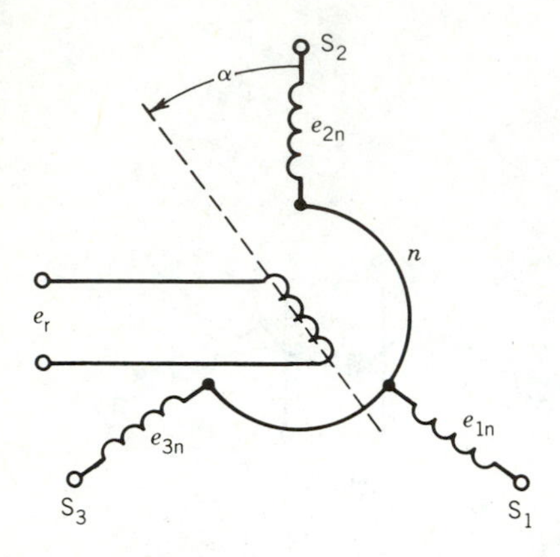

FIGURE 8.10
Voltages in rotor and stator of a synchro transmitter.

induced stator phase voltage depends on the coupling between the stator phase and the rotor winding.

Let

$$a = \frac{\text{effective stator turns}}{\text{effective rotor turns}} \tag{8.12}$$

The voltages induced in the stator phases are

$$e_{1n} = \sqrt{2}\, aE_r \sin \omega t \cos(\alpha + 120°) \tag{8.13}$$

$$e_{2n} = \sqrt{2}\, aE_r \sin \omega t \cos \alpha \tag{8.14}$$

$$e_{3n} = \sqrt{2}\, aE_r \sin \omega t \cos(\alpha - 120°) \tag{8.15}$$

The rms voltages are

$$E_{1n} = aE_r \cos(\alpha + 120°) \tag{8.16}$$

$$E_{2n} = aE_r \cos \alpha \tag{8.17}$$

$$E_{3n} = aE_r \cos(\alpha - 120°) \tag{8.18}$$

The rms line-to-line voltages are

$$E_{12} = E_{1n} - E_{2n} = \sqrt{3}\, aE_r \sin(\alpha - 120°) \tag{8.19}$$

$$E_{23} = E_{2n} - E_{3n} = \sqrt{3}\, aE_r \sin(\alpha + 120°) \tag{8.20}$$

$$E_{31} = E_{3n} - E_{1n} = \sqrt{3}\, aE_r \sin \alpha \tag{8.21}$$

These terminal voltages are shown in Fig. 8.11 as a function of the rotor shaft position. Note that each rotor position corresponds to one unique set

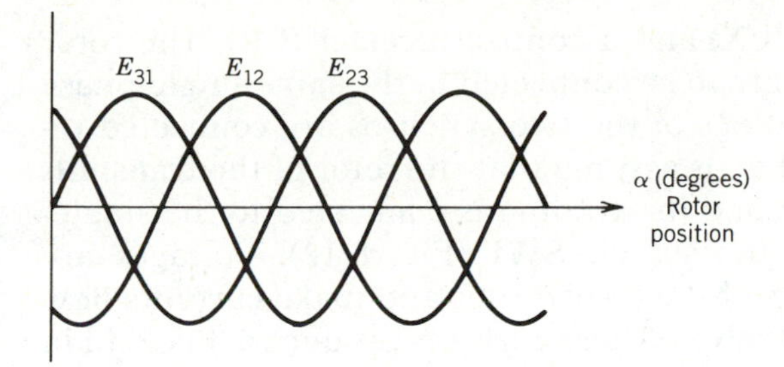

FIGURE 8.11
Variation of terminal voltages in a synchro transmitter as a function of the rotor position.

of stator voltages. This characteristic can be exploited to identify the angular position of the rotor shaft.

8.2.2 APPLICATIONS

Synchros are extensively used in servomechanism and other applications. Two major applications are discussed here.

Torque Transmission

Synchros can be used to transmit torque over a long distance without the use of a rigid mechanical connection. Figure 8.12 illustrates such an arrangement for maintaining alignment of two shafts. The arrangement re-

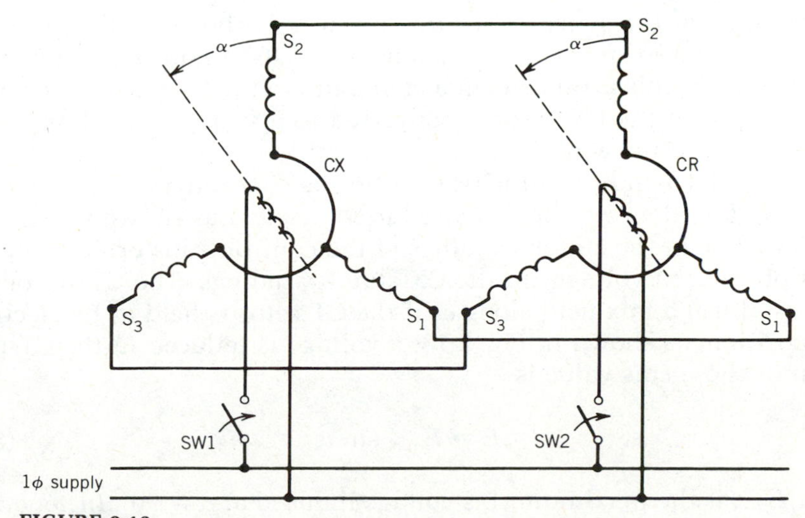

FIGURE 8.12
Synchro transmitter (CX) and synchro receiver (CR) arrangement for shaft alignment.

quires a control transmitter (CX) and a control receiver (CR). The rotor windings of the two synchros are to be connected to the same single-phase ac supply and the stator windings of the two synchros are connected together as shown in Fig. 8.12. Let us assume that the rotor of the transmitter is displaced by an angle α and its winding is connected to the single-phase ac supply by closing the switch SW1 (Fig. 8.12). Voltages are induced in the stator windings of the transmitter and make currents flow in the stator windings of the receiver. These currents produce a flux field in the transmitter whose axis is fixed by the angle α. If the rotor winding of the receiver is now connected to the single-phase supply by closing the switch SW2 (Fig. 8.12), a flux field is created along the axis of the receiver rotor which interacts with the flux field of the stator windings to produce a torque. This torque rotates the receiver rotor, which is free to run, to a position of correspondence with the transmitter rotor, that is, to the same displacement angle α, as shown in Fig. 8.12. Note that at this position the induced stator voltages of the receiver have the same magnitudes and phases as those prevailing in the stator windings of the transmitter. Therefore no current flows in the stator windings and no torque is produced. However, if the transmitter rotor, called the *master,* is displaced to a new position, the receiver rotor, called the *slave,* will take a similar position of correspondence. Note that master–slave roles are not uniquely assigned because a displacement of the receiver rotor will also cause the transmitter rotor to be displaced in similar fashion.

Error Detection

Synchros can be used for error detection in a servo control system. The arrangement of the synchros for this purpose is shown in Fig. 8.13. The synchros required are a control transmitter (CX) and a control transformer (CT). In this arrangement a command in the form of a mechanical displacement of the CX rotor is converted to a voltage signal appearing across the CR rotor winding.

Let the rotor winding of the transmitter be connected to a single-phase supply and let the rotor be displaced to an angle α as shown in Fig. 8.13. Currents flow in the stator windings of the control transformer (CT) as a result of induced voltages in the CX stator windings. The CT stator currents establish a flux field along α. If the CT rotor is held at its electrical zero position, as shown in Fig. 8.13, a voltage is induced in the CT rotor winding whose rms value is

$$E = E_{\max} \sin \alpha \tag{8.22}$$

where $E_{\max}$ is the maximum rms voltage induced at $\alpha = 90°$. In general, if α_x is the position of the CX rotor and α_T is the position of the CT rotor,

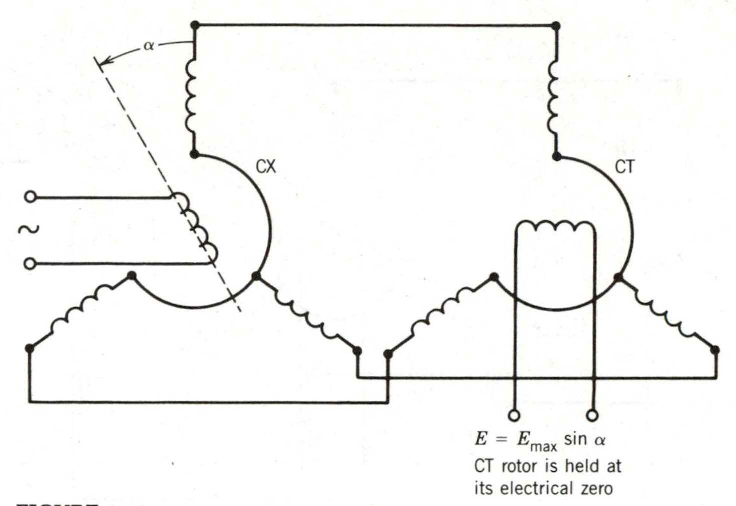

FIGURE 8.13
Synchro arrangement for error detection.

measured from their respective electrical zero positions, the rms voltage available at the CT rotor winding is

$$E = E_{\max} \sin(\alpha_x - \alpha_T) \tag{8.23}$$

The corresponding instantaneous voltage is

$$e = \sqrt{2}\, E_{\max} \sin(\alpha_R - \alpha_T) \sin \omega t \tag{8.24}$$

The application of the synchro error detector in a position servo control system is illustrated in Fig. 8.14. The objective of this servo system is to make an output shaft follow the angular displacement of a reference input shaft as closely as possible. The CX rotor is mechanically connected to an input shaft. As shown in Fig. 8.14, the rotor of the control transformer is mechanically connected to the output shaft and the rotor winding is electrically connected to the input of an amplifier.

The electrical zero positions for CX and CT rotors are 90° apart. Therefore, when the output shaft is 90° from the input shaft position, error voltage e is zero, making the input voltage V_a to the servomotor zero, and the motor does not turn. If the input shaft is moved from this 90° relative position, an error voltage e is produced which, after amplification by the amplifier, will turn the servomotor in a direction such that the output shaft follows the input shaft until the error voltage is zero and the 90° relative position of the input and output shafts is restored.

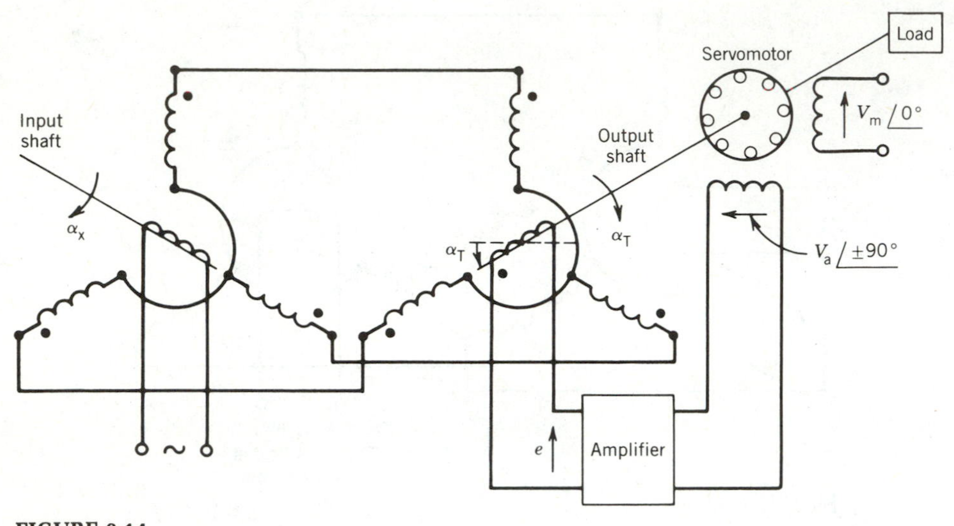

FIGURE 8.14
Servo system using synchros for error detection.

8.3 STEPPER MOTORS

A stepper motor rotates by a specific number of degrees in response to an input electrical pulse. Typical step sizes are 2°, 2.5°, 5°, 7.5°, and 15° for each electrical pulse. The stepper motor is an electromagnetic incremental actuator that can convert digital pulse inputs to analog output shaft motion. It is therefore used in digital control systems. A train of pulses is made to turn the shaft of the motor by steps. Neither a position sensor nor a feedback system is normally required for the stepper motors to make the output response follow the input command. Typical applications of stepper motors requiring incremental motion are printers, tape drives, disk drives, machine tools, process control systems, X–Y recorders, and robotics. Figure 8.15 illustrates a simple application of a stepper motor in

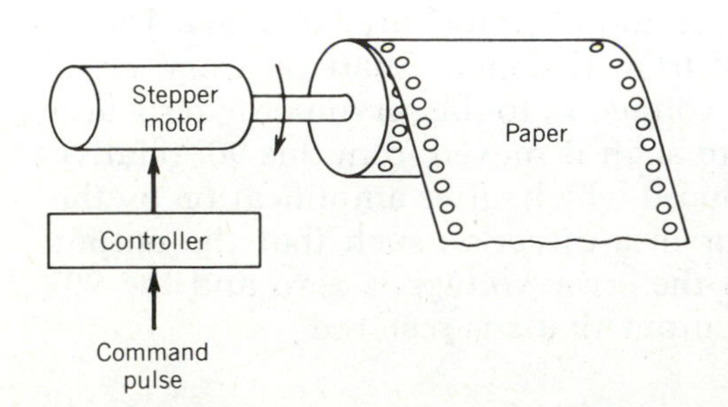

FIGURE 8.15
Paper drive using stepper motor.

the paper drive mechanism of a printer. The stepper motor is directly coupled to the platen so that the paper is driven a certain incremental distance whenever the controller receives a digital command pulse.

Typical resolution of commercially available stepper motors ranges from several steps per revolution to as many as 400 steps per revolution and even higher. Stepper motors have been built to follow signals as rapid as 1200 pulses per second with power ratings up to several horsepower.

Two types of stepper motors are widely used: (1) the variable-reluctance type and (2) the permanent magnet type.

8.3.1 VARIABLE RELUCTANCE STEPPER MOTOR

A variable reluctance stepper motor can be of the single-stack type or the multiple-stack type.

Single-Stack Stepper Motor

A basic circuit configuration of a four-phase, two-pole, single-stack, variable reluctance stepper motor is shown in Fig. 8.16. When the stator phases are excited with dc current in proper sequence, the resultant air gap field steps around and the rotor follows the axis of the air gap field by virtue of reluctance torque. This reluctance torque is generated because of the tendency of the ferromagnetic rotor to align itself along the direction of the resultant magnetic field.

Figure 8.17 shows the mode of operation for a 45° step in the clockwise direction. The windings are energized in the sequence A, A + B, B, B + C, and so forth, and this sequence is repeated. When winding A is excited, the rotor aligns with the axis of phase A. Next, both windings A and B are

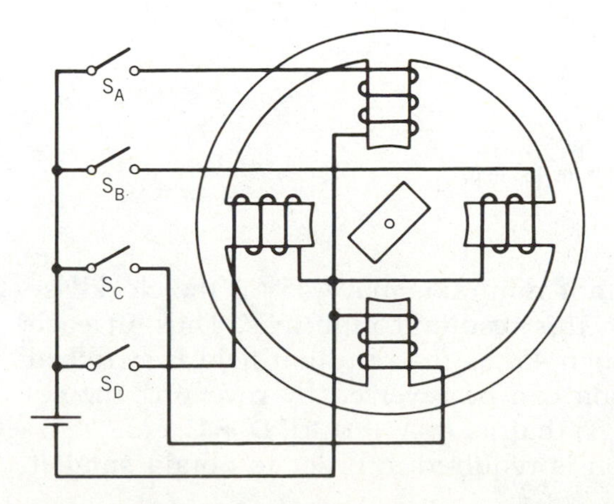

FIGURE 8.16
Basic circuit for a four-phase, two-pole stepper motor.

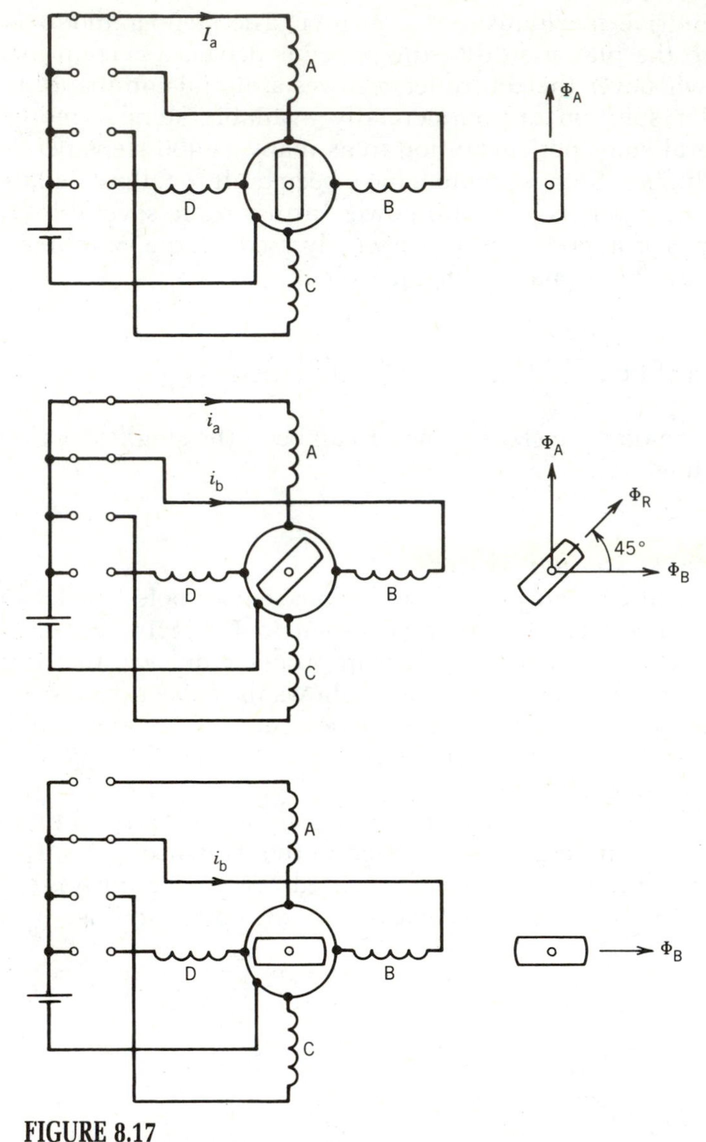

FIGURE 8.17

Operating modes of stepper motor for 45° step.

excited, which makes the resultant mmf axis move 45° in the clockwise direction. The rotor aligns with this resultant mmf axis. Thus, at each transition the rotor moves through 45° as the resultant field is switched around. The direction of rotation can be reversed by reversing the sequence of switching the windings, that is, A, A + D, D, D + C, etc.

A multipole rotor construction is required in order to obtain smaller step sizes. The construction of a four-phase, six-pole stepper motor is shown in Fig. 8.18. When phase A winding is excited, pole P_1 is aligned with the axis of phase A as shown in Fig. 8.18. Next, phase A and phase B windings are excited. The resultant mmf axis moves in the clockwise direction by 45° and pole P_2, nearest to this new resultant field axis, is pulled to align with it. The motor therefore steps in the anticlockwise direction by 15°. Next, phase A winding is de-excited and the excitation of phase B winding pulls pole P3 to align with the axis of phase B. Therefore, if the windings are excited in the sequence A, A + B, B, B + C, C, . . . , the rotor rotates in steps of 15° in the anticlockwise direction.

Multistack Stepper Motor

Multistack variable reluctance-type stepper motors are widely used to give smaller step sizes. The motor is divided along its axial length into magnetically isolated sections ("stacks") and each of these sections can be excited by a separate winding ("phase"). Three-phase arrangements are most common, but motors with up to seven stacks and phases are available.

Figure 8.19 shows the longitudinal cross section (i.e., parallel to the shaft) of a three-stack variable reluctance stepper motor. The stator of each stack has a number of poles. Figure 8.20 shows an example with four poles. Adjacent poles are wound in the opposite sense, and this produces four main flux paths as shown in Fig. 8.20. Both stator and rotor have the same number of teeth (12 in Fig. 8.20*a*). Therefore, when a particular phase is excited, the position of the rotor relative to the stator in that stack is accurately defined, as shown in Fig. 8.20*a*. The rotor teeth in each stack are aligned, whereas the stator teeth have a different orientation between stacks as shown in the developed diagram of rotor and stator teeth in Fig. 8.20*b*. Therefore, when stack A is energized, the rotor and stator teeth in stack A are aligned but those in stacks B and C are not aligned, as shown in

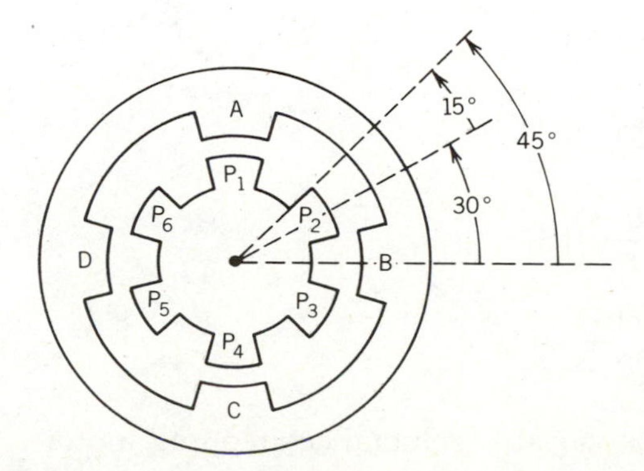

FIGURE 8.18
Multiple stepper motor for 15° step.

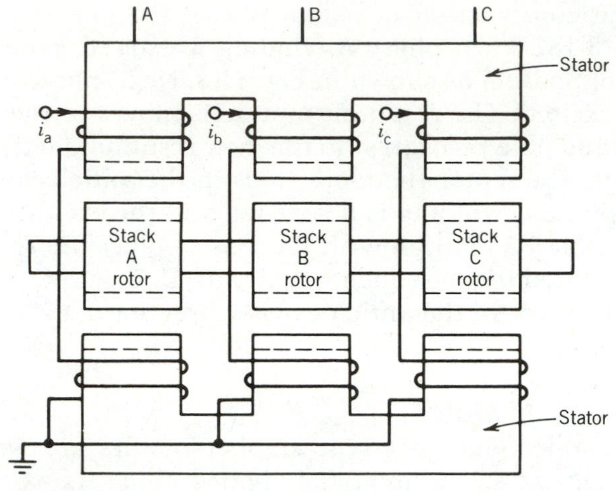

FIGURE 8.19
Cross section of a three-stack, variable-reluctance stepper motor parallel to the shaft.

Fig. 8.20*b*. Next, if excitation is changed from stack A to stack B, the stator and rotor teeth in stack B are aligned. This new alignment is made possible by a rotor movement in the clockwise direction; that is, the motor moves one step as a result of changing excitation from stack A to stack B. Another step motion in the clockwise direction can be obtained if excitation is changed from stack B to stack C. Another change of excitation from stack C to stack A will once more align the stator and rotor teeth in stack A. However, during this process (A → B → C → A) the rotor has moved one rotor tooth pitch, that is, the angle between adjacent rotor teeth. Let x be the number of rotor teeth and N the number of stacks or phases. Then

$$\text{Tooth pitch} \qquad \tau_p = \frac{360^\circ}{x} \tag{8.25}$$

$$\text{Step size} \qquad \Delta\theta = \frac{360^\circ}{xN} \tag{8.26}$$

For the motor illustrated in Fig. 8.20,

$$\tau_p = \frac{360^\circ}{12} = 30^\circ$$

$$\Delta\theta = \frac{360^\circ}{12 \times 3} = 10^\circ$$

Typical step sizes for the multistack variable reluctance stepping motor are in the range 2 to 15 degrees.

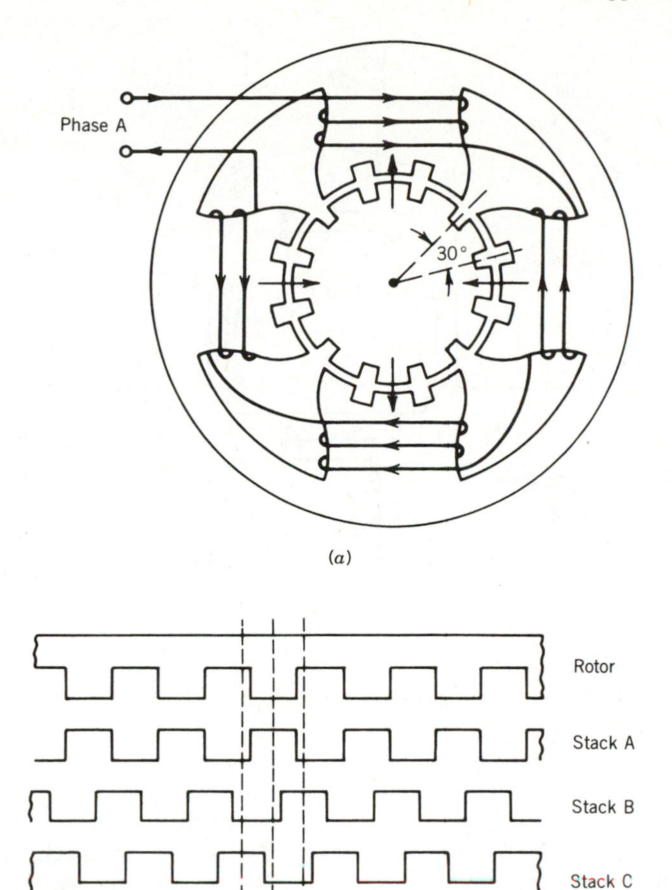

FIGURE 8.20
Teeth position in a four-pole, three-stack, variable-reluctance stepper motor. (*a*) Phase A excited. Rotor and stator teeth are aligned. (*b*) Developed diagram for rotor and stator teeth for phase A excitation.

8.3.2 PERMANENT MAGNET STEPPER MOTOR

The permanent magnet stepper motor has a stator construction similar to that of the single-stack variable reluctance type, but the rotor is made of a permanent magnet material. Figure 8.21 shows a two-pole, permanent magnet stepper motor. The rotor poles align with two stator teeth (or poles) according to the winding excitation. Figure 8.21 shows the alignment if phase A winding is excited. If the excitation is switched to phase B,

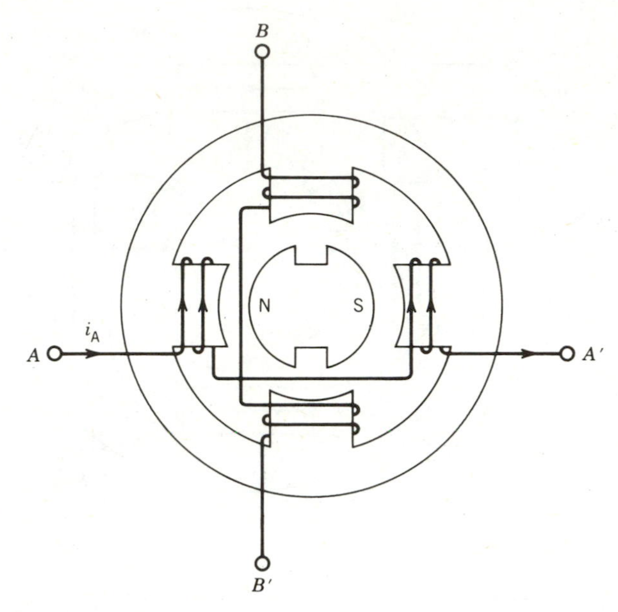

FIGURE 8.21
Permanent magnet stepper motor.

the rotor moves by a step of 90°. Note that current polarity is important in the permanent magnet stepper motor, because it decides the direction in which the motor will move. Figure 8.21 illustrates the rotor position for positive current in phase A. A switch over to positive current in phase B winding will produce a clockwise step, whereas a negative current in phase B winding will produce an anticlockwise step. It is difficult to make a small permanent magnet rotor with a large number of poles, and therefore stepper motors of this type are restricted to larger step sizes in the range 30 to 90 degrees.

Permanent magnet stepper motors have higher inertia and therefore slower acceleration than variable reluctance stepper motors. The maximum step rate for permanent magnet stepper motors is 300 pulses per second, whereas it can be as high as 1200 pulses per second for variable reluctance stepper motors. The permanent magnet stepper motor produces more torque per ampere stator current than the variable reluctance stepper motor.

Hybrid stepper motors are also commercially available in which the rotor has an axial permanent magnet at the middle and ferromagnetic teeth at the outer sections as shown in Fig. 8.22. Smaller step sizes can be obtained from these motors, but they are more expensive than the variable reluctance-type stepper motors.

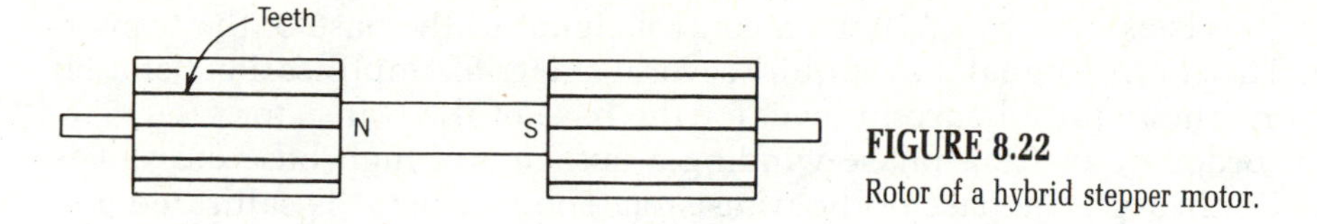

FIGURE 8.22
Rotor of a hybrid stepper motor.

8.3.3 DRIVE CIRCUITS

The command signals for a stepper motor are normally obtained from low-power logic circuits that are built with TTL or CMOS digital integrated circuits (ICs). The driving current available is either 20 mA at 5 V (TTL) or 1 mA at 5–15 V (CMOS). However, a typical variable reluctance stepper motor producing a torque of 1.2 N · m has a rated winding excitation of 5 V and 3A. Therefore power amplification stages are required between the low-power command signals and the high-power stepper motors.

Variable reluctance stepper motors require more than two phases (three phases are typical). The phase currents need only be switched on or off and current polarity is irrelevant for torque production. Permanent magnet stepper motors require two phases and the current polarity is important.

Unipolar Drive Circuit

Figure 8.23 shows a simple unipolar drive circuit suitable for a three-phase variable reluctance stepper motor. Each phase winding is excited by a separate drive circuit. The main switching device is a transistor. A phase

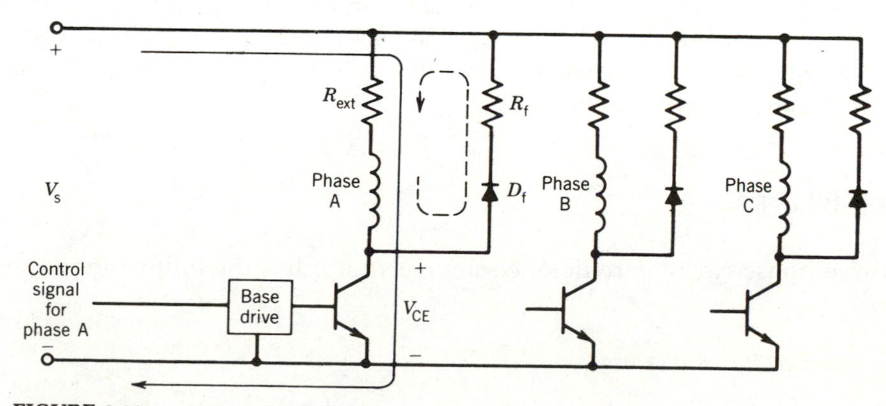

FIGURE 8.23
Unipolar drive circuit for a three-phase variable reluctance stepper motor.

winding is excited by applying a control signal to the base of the transistor. The control signal may require several stages of amplification before it attains the required current level for the base of the transistor.

In order to excite a phase winding a sufficiently high base current is passed through the base of the transistor. The transistor is saturated and its collector–emitter path behaves like a short-circuit. The supply voltage V_s appears across the phase winding and the resistance R_{ext} connected in series with the winding. The dc supply voltage V_s is chosen so that it produces the rated current I in the winding.

$$V_s = I(R_w + R_{ext}) \tag{8.27}$$

where R_w is the phase winding resistance. The phase winding has a large inductance and therefore the electrical time constant (ratio of inductance to resistance) is large. As a result, buildup of current in the phase winding to its rated value is slow, causing unsatisfactory operation of the motor at high stepping rates. The addition of the external resistance R_{ext} decreases the electrical time constant, thereby speeding up the current buildup.

When the base drive current is removed to switch off the transistor, a large induced voltage will appear across the transistor if the winding current is suddenly interrupted. The large voltage may permanently damage the transistor. This possibility is avoided by providing an alternative path for the phase winding current—known as a freewheeling path. Therefore, when the transistor is switched off the phase winding current will continue to flow in the freewheeling diode D_f and a freewheeling resistance R_f. The maximum voltage across the transistor occurs at the instant of switchoff and is

$$V_{CE(max)} = V_s + IR_f \tag{8.28}$$

Subsequently, the phase current will decay in the closed circuit formed by the phase winding, D_f, R_f, and R_{ext}. The magnetic energy stored in the phase inductance at turnoff of the transistor is dissipated in the resistances of this closed circuit.

EXAMPLE 8.3

A three-phase variable reluctance stepper motor has the following parameters:

$R_w = 1\Omega$

$L_w = 30$ mH, average phase winding inductance

$I = 3$A, rated winding current

Design a simple unipolar drive circuit such that the electrical time constant is 2 msec at phase turn-on and 1 msec at turnoff. The stepping rate is 300 steps per second.

Solution

The turn-on time constant

$$\tau_{on} = \frac{L_w}{R_w + R_{ext}}$$

$$R_w + R_{ext} = \frac{30}{2} = 15\ \Omega$$

$$R_{ext} = 15 - 1 = 14\ \Omega$$

This resistance must be able to dissipate the power lost when rated current flows through the phase winding continuously, namely

$$P_{Rext} = 3^2 \times 14 = 126\ \mathrm{W}$$

The required dc supply voltage, from Eq. 8.27, is

$$V_s = 3 \times 15 = 45\ \mathrm{V}$$

The turnoff time constant

$$\tau_{off} = \frac{L_w}{R_w + R_{ext} + R_f}$$

$$R_w + R_{ext} + R_f = \frac{30}{1} = 30\ \Omega$$

$$R_f = 30 - 15 = 15\ \Omega$$

Energy stored in the phase winding at turnoff $= \frac{1}{2}L_w I^2$

$$= \tfrac{1}{2} \times 30 \times 10^{-3} \times 3^2\ \mathrm{J}$$

$$= 0.135\ \mathrm{J}$$

This energy is dissipated in R_f, R_{ext}, and R_w. Since $R_f = R_{ext} + R_w$ (= 15 Ω) the energy dissipated in R_f is 0.0675 J.

Stepping rate = 300 steps/sec

Number of turnoffs in each phase = 100

Average power dissipated in R_f = 100 × 0.0675 W = 6.75 W

When the transistor conducts, the reverse voltage across the diode D_f is $V_s = 45$ V. The peak current of the freewheeling diode is 3A, which is the phase winding current at the instant the transistor turns off.

From Eq. 8.28,

$$V_{CE(max)} = 45 + 3 \times 15 = 90 \text{ V}$$

Current rating of the transistor is 3 A.

Bipolar Drive Circuit

Figure 8.24 shows one phase of a bipolar drive circuit suitable for a permanent magnet or hybrid-type stepper motor. The transistors are switched in pairs according to the current polarity required for the phase winding. For example, transistors T_1 and T_2 are turned on simultaneously so that current can flow from left to right in the phase winding as shown in Fig. 8.24. If transistors T_3 and T_4 are turned on simultaneously current will flow in the opposite direction.

The four diodes D_1 to D_4 connected in antiparallel with the switching transistors provide the paths for the freewheeling currents. For example, when T_1 and T_2 are switched on, current flows from dc supply to T_1, phase winding (left to right), T_2, and back to dc supply. When T_1 and T_2 are switched off (by removing their base currents) current in the phase winding cannot decay instantaneously because of winding inductance. The current therefore flows through diodes D_3 and D_4 to the dc supply, as shown in Fig. 8.24 by dashed lines. Note that when current flows through D_3 and D_4 to the dc supply some of the energy stored in the phase winding induc-

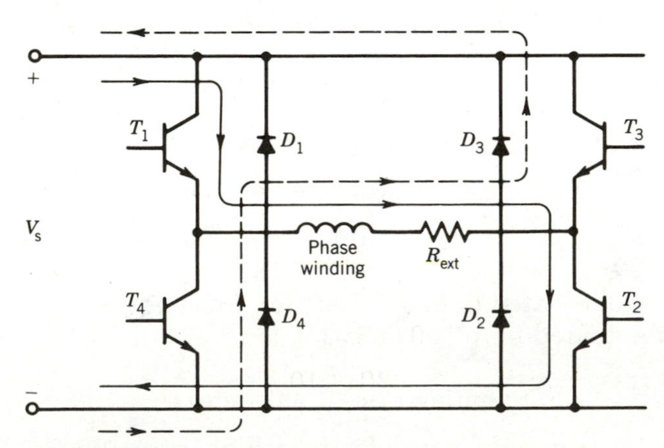

FIGURE 8.24
One phase of a bipolar drive circuit.

tance at turnoff (of the transistors) is returned to the supply. This improves the overall system efficiency and is a significant advantage of the bipolar drive circuit over the unipolar drive circuit. Most large stepper motors (greater than 1 kW), including variable reluctance types, are operated from bipolar drive circuits. Of course, bipolar drive circuits require more switching devices and are therefore more expensive than unipolar drive circuits.

Note that the freewheeling currents in the bipolar drive circuit decay more rapidly than in the unipolar drive circuit, because the dc supply opposes them. Consequently no additional freewheeling resistance is necessary in the bipolar drive circuit.

EXAMPLE 8.4

A stepper motor driven by a bipolar drive circuit has the following parameters.

Winding inductance (average) $L_w = 30$ mH

Rated current = 3 A

Total resistance in each phase $R = 15\ \Omega$

DC supply = 45 V

When transistors are turned off, determine the

(a) Time taken by the phase current to decay to zero.

(b) Proportion of the stored inductive energy returned to the dc supply.

Solution

(a) The equivalent circuit at turnoff is shown in Fig. E8.4. The current can be considered to have two components.

(i) One component of current is the initial current 3 A, which decays in L_w and R, with zero supply voltage, that is, $i_1 = 3e^{-t/\tau}$ where $\tau = L_w/R = 30/15$ msec = 2 msec.

(ii) The other component of current is i_2, which is produced by the supply voltage V_s, assuming no initial current: $i_2 = -3(1 - e^{-t/\tau})$.

Hence the net current is

$$i = 3e^{-t/\tau} - 3(1 - e^{-t/\tau})$$

$$= -3 + 6e^{-t/\tau}$$

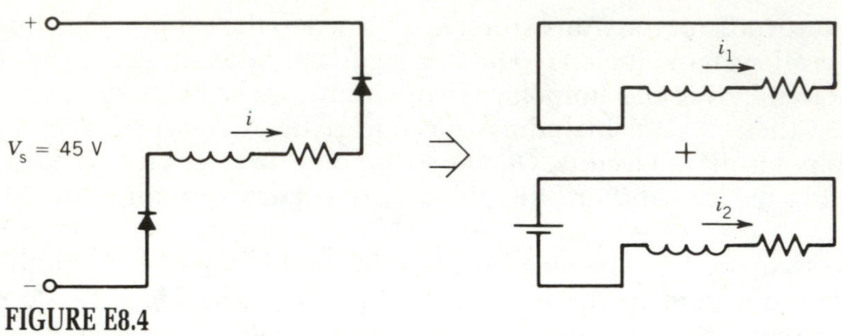

FIGURE E8.4
Equivalent circuits at turnoff. $i(0) = 3$ A.

Let i fall to zero in time t_1.

$$0 = -3 + 6e^{-t_1/\tau}$$

$$\text{or} \quad e^{-t_1/\tau} = \tfrac{3}{6}$$

$$\text{or} \quad \frac{t_1}{\tau} = 0.7$$

$$\text{or} \quad t_1 = 0.7 \times 2 \text{ msec} = 1.4 \text{ msec}$$

(b) Energy returned to the supply, W_s:

$$W_s = \int_0^{t_1} V_s i \, dt = \int_0^{t_1} 45(-3 + 6e^{-t/\tau}) \, dt$$

$$= \int_0^{t_1} -135 \, dt + \int_0^{t_1} 270e^{-t/\tau} \, dt$$

$$= -135t\Big|_0^{t_1} + 270(-\tau)e^{-t/\tau}\Big|_0^{t_1}$$

$$= -135t_1 - 270\tau \cdot \left|e^{-t/\tau}\right|_0^{t_1}$$

$$= -135t_1 - 270\tau(e^{-t_1/\tau} - 1)$$

Now $t_1 = 1.4 \times 10^{-3}$ sec and $\tau = 2 \times 10^{-3}$ sec. Therefore,

$$W_s = -135 \times 1.4 \times 10^{-3} - 270 \times 2 \times 10^{-3}(0.5 - 1)$$

$$= 0.081 \text{ joules}$$

$$= 81 \text{ mJ}$$

$$\text{Stored energy} = \tfrac{1}{2}L_w I^2$$
$$= \tfrac{1}{2} \times 30 \times 3^2 \text{ mJ}$$
$$= 135 \text{ mJ}$$

Proportion of energy returned to supply

$$= \frac{81}{135} \times 100\%$$
$$= 60\%$$

PROBLEMS

8.1 A closed-loop speed control system using a two-phase ac servomotor is shown in Fig. P8.1. The transfer function between speed and control phase voltage is

$$\frac{\omega(s)}{V_a(s)} = \frac{2.733}{1 + 0.0157s}$$

The servo amplifier is a PI (proportional–integral) type of controller whose transfer function is

$$G(s) = K_p + \frac{K_i}{s} = \frac{K_p s + K_i}{s}$$

(a) Derive the transfer function for the speed control system $\omega_m(s)/\omega_m^*(s)$.

(b) If $K_p/K_i = 0.0157$, derive the transfer function for the system. Determine the values of K_p and K_i if the time constant of the speed response is 1.0 second.

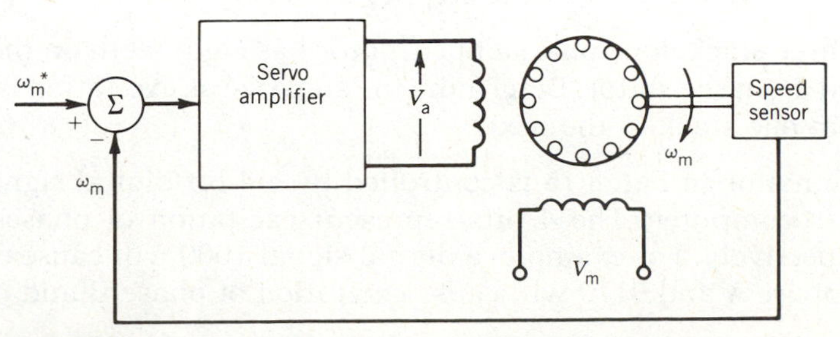

FIGURE P8.1

8.2 For the position control system shown in Fig. 8.7, the transfer function of the servo amplifier is

$$G(s) = \frac{5(1 + 0.02s)}{s + 5}$$

The transfer function between the motor control phase voltage V_a and shaft position θ is

$$M(s) = \frac{2}{s(1 + 0.02s)}$$

(a) Derive the transfer function of the system.

(b) For a step change in the command angle of 180°, find the time response of the speed and position of the shaft. Sketch the responses.

8.3 For the synchro transmitter of Fig. 8.10, the rms voltage induced in winding S_2 is 50 V at the electrical zero position. When the rotor is at 30° from electrical zero position in a counterclockwise direction,

(a) Determine the rms voltages in the windings.

(b) Determine the rms voltages between lines.

8.4 For the synchro system of Fig. 8.13, the maximum rms voltage induced in the rotor of the CT is 10 V. The rotor of the CT is held at the electrical zero position. Determine the voltage induced in the rotor of the CT in response to a displacement of the rotor of the CX from its electrical zero position by 20°.

8.5 Consider the multipole stepper motor shown in Fig. 8.18. Determine the sequence of excitation for a 30° step.

8.6 A single-stack, four-phase (stator) multipole stepper motor is required to produce an 18° step motion. Determine the number of rotor poles and the sequence of excitation of the stator phases. Draw a cross-sectional view of the stepper motor.

8.7 A three-stack, four-pole stepper motor has eight teeth on the rotor as well as on the stator. Determine the step size as excitation is changed from one stack to the next.

8.8 The motor in Fig. 8.16 is controlled by a 4-bit digital signal from a microcomputer. The 4 bits represent excitation of phases A to D, respectively. For example, a digital signal 1000 will cause excitation of phase A and 0110 will cause excitation of phase B and phase C.

(a) Write a table for the 4-bit digital signals for 45° step rotation. Show the angle of rotation and phases excited.

(b) Continuous sequencing of the digital signals of part (a) causes the motor rotate at constant speed. Determine the number of signals per second (i.e., nibbles/sec) if the motor rotates at 720 rpm.

8.9 A 3ϕ variable reluctance stepper motor has the following parameters:

$$R_w = 2.0\ \Omega$$

$$L_w = 50\ \text{mH} \quad \text{(average)}$$

$$I = 5\ \text{A} \quad \text{(rated winding current)}$$

Each phase is controlled by a unipolar drive circuit. The resistance connected in series with the winding is $R_{ext} = 10\ \Omega$ and that with the freewheeling diode is $R_f = 5\ \Omega$.

(a) Determine the electrical time constants (τ_{on} and τ_{off}) at turn-on and turnoff of a phase.

(b) Determine the value of the supply voltage V_s.

(c) Determine the voltage and current ratings of the transistor and diode.

(d) If phase current conducts for $3(\tau_{on} + \tau_{off})$, determine the maximum value of the stepping rate (steps per second) of the motor.

8.10 A stepper motor is driven by bipolar drive circuits. The stepper motor has the following parameters:

$$L_w = 50\ \text{mH}$$

$$R_w = 2\ \Omega$$

$$I_{rated} = 5.0\ \text{A}$$

Determine the supply voltage (V_s) and external resistance (R_{ext}) to be connected in series with the phase winding such that rated current flows when the transistor is on and phase current decays to zero in 1.0 msec when the transistor is off.

TRANSIENTS AND DYNAMICS

CHAPTER nine

In earlier chapters the steady-state operation and performance of dc and ac machines have been discussed. However, when a disturbance is applied, the machine behavior can be quite different. A transient period of readjustment occurs between the initial and final steady-state operating conditions. In many applications it is necessary to know the behavior of the machine (i.e., its response to the disturbance) during this transient period. In this chapter both the electrical transient behavior and the mechanical transient behavior (the dynamics) of dc and ac machines are studied. The study of transients and of dynamic behavior is quite complex and simplifying assumptions are frequently made.

9.1 DC MACHINES

DC machines can be controlled with ease and are used in applications requiring control of speed over a wide range or applications requiring precise control of other variables such as position in servo drives. The following assumptions are made to reduce the complexity of the analysis when the dc machine is used as a system component.

1. Magnetic saturation is neglected. This assumption implies that inductances are independent of currents.
2. The field mmf is assumed to act along the d-axis and the armature mmf to act along the q-axis. Consequently, there is no mutual inductance between the field circuit and the armature circuit. A further consequence is that there is no demagnetizing effect due to armature reaction.

We first study electrical transients in a dc generator. Following this analysis we study the dynamic behavior of a dc motor. For this purpose the machine behavior is described by equations, and transfer functions relating the output variables to the input variables are derived to obtain the response of the system resulting from a sudden input change.

9.1.1 SEPARATELY EXCITED DC GENERATOR

A schematic representation of a separately excited dc generator is shown in Fig. 9.1. The armature inductance is represented by the q-axis inductance L_{aq}, because the armature mmf acts along the q-axis. The basic equations for the dc machine are

$$e_a = K_a \Phi \omega_m \tag{9.1}$$

$$T = K_a \Phi i_a \tag{9.2}$$

If magnetic linearity is assumed,

$$e_a = K_f i_f \omega_m \tag{9.3}$$

$$T = K_f i_f i_a \tag{9.4}$$

1. *Field circuit transient.* Let us first consider the electrical transients in a separately excited dc generator resulting from changes in the excitation (i.e., in the field circuit voltage). The armature circuit is open-circuited and the generator is assumed to be running at a constant speed ω_m. From Fig. 9.1, the voltage equation for the field circuit, after the switch SW is closed, is

$$V_f = R_f i_f + L_f \frac{di_f}{dt} \tag{9.5}$$

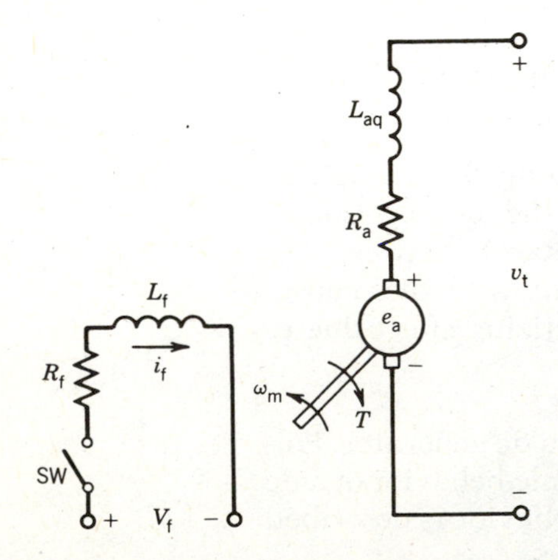

FIGURE 9.1
Schematic representation of a separately excited dc generator.

The Laplace transform of Eq. 9.5 with zero initial conditions is

$$V_f(s) = R_f I_f(s) + L_f s I_f(s) = I_f(s)(R_f + sL_f) \tag{9.6}$$

The transfer function relating the field current to the field voltage is

$$\frac{I_f(s)}{V_f(s)} = \frac{1}{R_f + sL_f} = \frac{1}{R_f(1 + s\tau_f)} \tag{9.7}$$

where $\tau_f = L_f/R_f$ is the time constant of the field circuit.

The generated voltage in the armature circuit, from Eq. 9.3, is

$$e_a = K_f i_f \omega_m = K_g i_f \tag{9.8}$$

where $K_g = (K_f \omega_m)$ is the slope of the linear portion of the magnetization curve, at speed ω_m, representing e_a plotted against i_f. The Laplace transform of Eq. 9.8 is

$$E_a(s) = K_g I_f(s) \tag{9.9}$$

From Eqs. 9.7 and 9.9 the transfer function relating the armature (generated) voltage to the field circuit voltage is

$$\frac{E_a(s)}{V_f(s)} = \frac{E_a(s)}{I_f(s)} \cdot \frac{I_f(s)}{V_f(s)} = \frac{K_g}{R_f(1 + s\tau_f)} \tag{9.10}$$

Equations 9.7, 9.9, and 9.10 are represented in block diagram form in Fig. 9.2. The time domain response corresponding to the transfer function of Eq. 9.10 (for a step change of V_f) is

$$e_a(t) = \frac{K_g V_f}{R_f}(1 - e^{-t/\tau_f}) \tag{9.11}$$

$$= E_a(1 - e^{-t/\tau_f}) \tag{9.12}$$

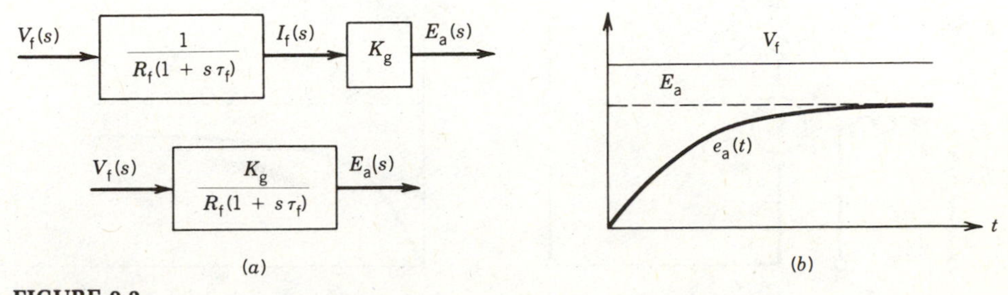

FIGURE 9.2
Field circuit transients. (a) Block diagram. (b) Response.

where $E_a = e_a(\infty) = K_g V_f / R_f = K_g I_f$ is the steady-state generated voltage

$I_f = V_f / I_f$ is the steady-state field current

The response is shown in Fig. 9.2*b*. It is a first-order response with time constant τ_f. The field circuit time constant τ_f is quite large and varies in the range 0.1 to 2 seconds.

2. *Armature circuit transient.* Let us now consider an electrical transient in the armature circuit. In Fig. 9.3*a* the load, consisting of resistance R_L and inductance L_L, is connected to the armature terminal by closing the switch SW at $t = 0$. It is assumed that the armature rotates at constant speed and that the field current also stays constant. After the switch is closed,

$$E_a = R_a i_a + L_{aq}\frac{di_a}{dt} + R_L i_a + L_L\frac{di_a}{dt} \tag{9.13}$$

or

$$E_a = (R_a + R_L)i_a + (L_{aq} + L_L)\frac{di_a}{dt} \tag{9.14}$$

or

$$E_a = R_{at} i_a + L_{at}\frac{di_a}{dt} \tag{9.15}$$

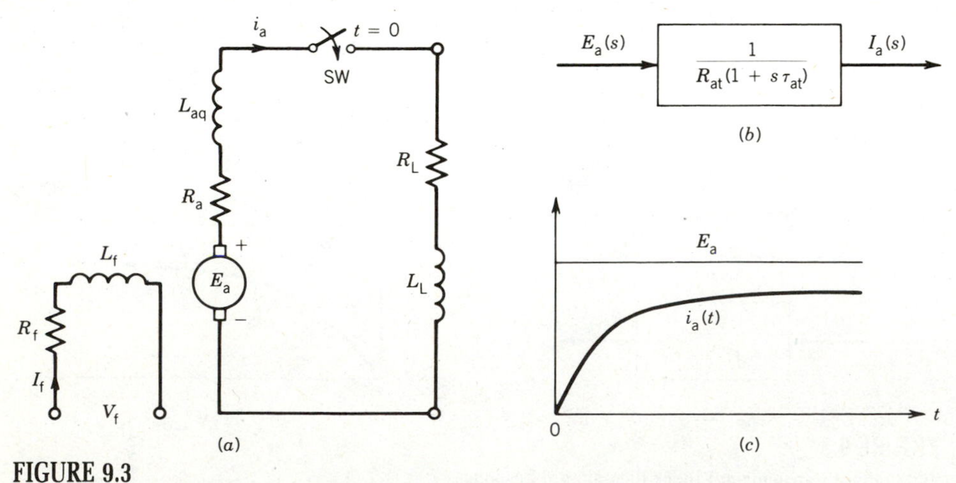

FIGURE 9.3
Armature circuit transient. (*a*) Schematic circuit. (*b*) Block diagram. (*c*) Response.

where $R_{at} = R_a + R_L$ is the total resistance in the armature circuit
$L_{at} = L_{aq} + L_L$ is the total inductance in the armature circuit

The Laplace transform of Eq. 9.15 is

$$E_a(s) = R_{at}I_a(s) + L_{at}sI_a(s) \tag{9.16}$$

The transfer function is

$$\frac{I_a(s)}{E_a(s)} = \frac{1}{R_{at}(1 + s\tau_{at})} \tag{9.17}$$

where $\tau_{at} = L_{at}/R_{at}$ is the armature circuit time constant.

A block diagram representation of the transfer function of Eq. 9.17 is shown in Fig. 9.3*b*. The time domain response is

$$i_a(t) = \frac{E_a}{R_{at}}(1 - e^{-t/\tau_{at}}) \tag{9.18}$$

The response is shown in Fig. 9.3*c*. This is also a first-order response with time constant τ_{at}. Normally, τ_{at} is low and therefore armature current i_a builds up quickly.

From Eqs. 9.10 and 9.17 the total transfer function relating the armature current to the field circuit voltage is

$$\frac{I_a(s)}{V_f(s)} = \frac{I_a(s)}{E_a(s)} \cdot \frac{E_a(s)}{V_f(s)} = \frac{K_g}{R_f R_{at}(1 + s\tau_f)(1 + s\tau_{at})} \tag{9.19}$$

The corresponding block diagram representation is shown in Fig. 9.4*a*.

For a step change of voltage in the field circuit

$$V_f(s) = \frac{V_f}{s} \tag{9.20}$$

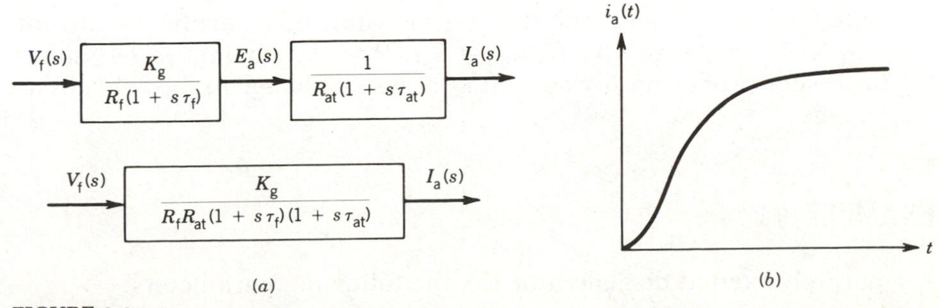

FIGURE 9.4
Field and armature circuit transient. (*a*) Block diagram. (*b*) Response.

From Eqs. 9.19 and 9.20,

$$I_a(s) = \frac{K_g V_f}{R_f R_{at} s(1 + s\tau_f)(1 + s\tau_{at})} \tag{9.21}$$

or

$$I_a(s) = \frac{K_g V_f}{R_f R_{at} \tau_f \tau_{at} s(s + 1/\tau_f)(s + 1/\tau_{at})} \tag{9.22}$$

or

$$I_a(s) = \frac{A}{s(s + 1/\tau_f)(s + 1/\tau_{at})} \tag{9.23}$$

where $A = K_g V_f / R_f R_{at} \tau_f \tau_{at}$

or

$$I_a(s) = \frac{A_1}{s} + \frac{A_2}{s + 1/\tau_f} + \frac{A_3}{s + 1/\tau_{at}} \tag{9.24}$$

where $A_1 = \left.\frac{A}{(s + 1/\tau_f)(s + 1/\tau_{at})}\right|_{s=0} = A\tau_f\tau_{at}$

$$A_2 = \left.\frac{A}{s(s + 1/\tau_{at})}\right|_{s=-1/\tau_f}$$

$$A_3 = \left.\frac{A}{s(s + 1/\tau_f)}\right|_{s=-1/\tau_{at}}$$

The time domain response of i_a is

$$i_a(t) = A_1 + A_2 e^{-t/\tau_f} + A_3 e^{-t/\tau_{at}} \tag{9.25}$$

Note that A_1 represents the steady-state value of the armature current, that is, $A_1 = i_a(\infty) = (K_g V_f)/(R_f R_{at}) = K_g I_f / R_{at} = E_a / R_{at}$. Figure 9.4*b* shows the response of armature current i_a for a step change in the field circuit voltage.

EXAMPLE 9.1

A separately excited dc generator has the following parameters:

$$R_f = 100\ \Omega, \qquad L_f = 25\ \text{H}$$

$$R_a = 0.25\ \Omega, \qquad L_{aq} = 0.02\ \text{H}$$

$$K_g = 100\ \text{V} \quad \text{per field ampere at rated speed}$$

(a) The generator is driven at rated speed and a field circuit voltage V_f = 200 V is suddenly applied to the field winding.

(i) Determine the armature generated voltage as a function of time.

(ii) Determine the steady-state armature voltage.

(iii) Determine the time required for the armature voltage to rise to 90 percent of its steady-state value.

(b) The generator is driven at rated speed and a load consisting of R_L = 1 Ω and L_L = 0.15 H in series is connected to the armature terminals. A field circuit voltage V_f = 200 V is suddenly applied to the field winding. Determine the armature current as a function of time.

Solution

(a) Field circuit time constant $\tau_f = 25/100 = 0.25$ sec.

(i) From Eq. 9.11,

$$e_a(t) = \frac{100 \times 200}{100}(1 - e^{-t/0.25})$$

$$= 200(1 - e^{-4t})$$

(ii)

$$e_a(\infty) = 200\ \text{V}$$

(iii)

$$0.9 \times 200 = 200(1 - e^{-4t})$$

$$t = 0.575\ \text{sec}$$

(b)

$$\tau_f = 0.25\ \text{sec}$$

$$\tau_{at} = \frac{0.15 + 0.02}{1 + 0.25} = 0.136\ \text{sec}$$

From Eq. 9.22,

$$I_a(s) = \frac{100 \times 200}{100 \times 1.25 \times 0.25 \times 0.136 s(s + 4)(s + 7.35)}$$

$$= \frac{4705.88}{s(s + 4)(s + 7.35)}$$

$$= \frac{A_1}{s} + \frac{A_2}{s + 4} + \frac{A_3}{s + 7.35}$$

$$\text{where} \quad A_1 = \left.\frac{4705.88}{(s + 4)(s + 7.35)}\right|_{s=0} = 160$$

$$A_2 = \left.\frac{4705.88}{s(s + 7.35)}\right|_{s=-4} = -351$$

$$A_3 = \left.\frac{4705.88}{s(s + 4)}\right|_{s=-7.35} = 191$$

From Eq. 9.25,

$$i_a(t) = 160 - 351e^{-4t} + 191e^{-7.35t}$$

9.1.2 DC MOTOR DYNAMICS

DC motors are extensively used in applications where precise control of speed and torque is required over a wide range. A common method of control is the use of a separately excited dc motor with constant field excitation. The speed is controlled by changing the voltage applied to the motor terminals. We now investigate how the speed of the motor responds to changes in the terminal voltage. The study involves electrical transients in the armature circuit and mechanical transients in the mechanical system driven by the motor.

A separately excited dc motor system is shown in Fig. 9.5*a*. Assuming magnetic linearity, the basic motor equations are

$$T = K_f i_f i_a = K_m i_a \tag{9.26}$$

$$e_a = K_f i_f \omega_m = K_m \omega_m \tag{9.27}$$

where $K_m = K_f i_f$ is a constant, which is also the ratio e_a/ω_m, e_a being the generated voltage corresponding to the field current i_f at the speed ω_m.

The Laplace transforms of Eqs. 9.26 and 9.27 are

$$T(s) = K_m I_a(s) \tag{9.28}$$

$$E_a(s) = K_m \omega_m(s) \tag{9.29}$$

In Fig. 9.5*a* let the switch SW be closed at $t = 0$. After the switch is closed,

$$V_t = e_a + R_a i_a + L_{aq} \frac{di_a}{dt} \tag{9.30}$$

From Eqs. 9.27 and 9.30

$$V_t = K_m\omega_m + R_a i_a + L_{aq}\frac{di_a}{dt} \tag{9.31}$$

The Laplace transform of Eq. 9.31 for zero initial conditions is

$$V_t(s) = K_m\omega_m(s) + R_a I_a(s) + L_{aq}sI_a(s) \tag{9.32}$$

or

$$V_t(s) = K_m\omega_m(s) + I_a(s)R_a(1 + s\tau_a) \tag{9.33}$$

where $\tau_a = L_{aq}/R_a$ is the electrical time constant of the armature.

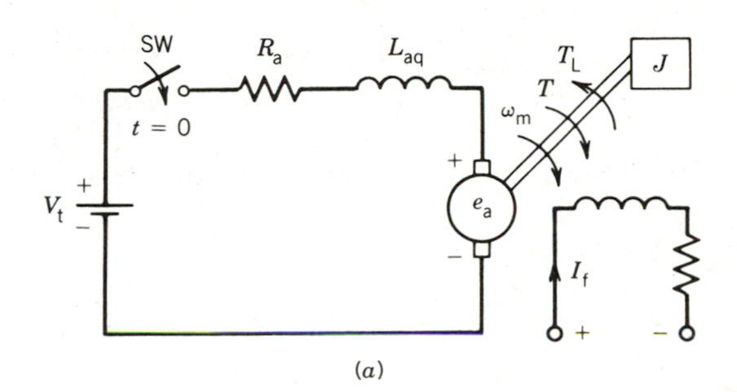

(a)

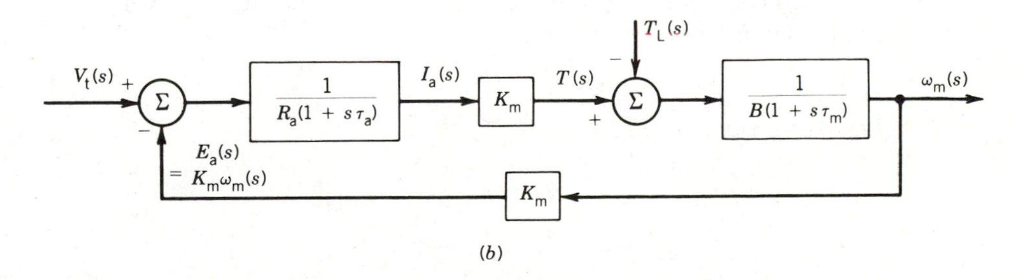

(b)

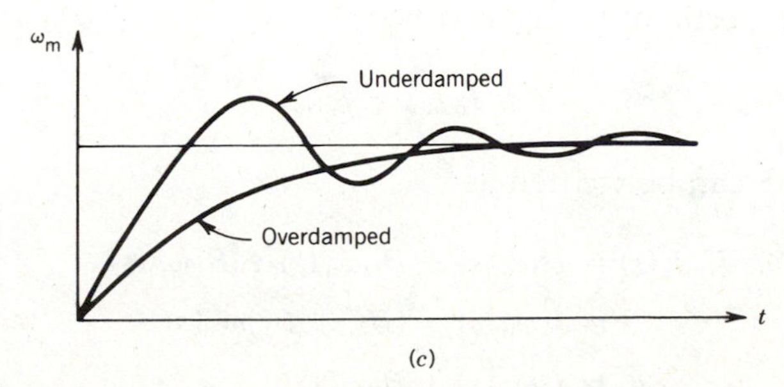

(c)

FIGURE 9.5
Separately excited dc motor. (a) Schematic diagram. (b) Block diagram representation. (c) Response.

The dynamic equation for the mechanical system is

$$T = K_m i_a = J\frac{d\omega_m}{dt} + B\omega_m + T_L \tag{9.34}$$

where J is the combined polar moment of inertia of the load and the rotor of the motor, B is the equivalent viscous friction constant of the load and motor, and T_L is the mechanical load torque. The term $B\omega_m$ represents the rotational loss torque of the system.

The Laplace transform of Eq. 9.34 is

$$T(s) = K_m I_a(s) = Js\omega_m(s) + B\omega_m(s) + T_L(s) \tag{9.35}$$

From Eqs. 9.35 and 9.28,

$$\omega_m(s) = \frac{T(s) - T_L(s)}{B(1 + sJ/B)} = \frac{K_m I_a(s) - T_L(s)}{B(1 + s\tau_m)} \tag{9.36}$$

where $\tau_m = J/B$ is the mechanical time constant of the system. From Eqs. 9.29 and 9.33,

$$I_a(s) = \frac{V_t(s) - E_a(s)}{R_a(1 + s\tau_a)} = \frac{V_t(s) - K_m\omega_m(s)}{R_a(1 + s\tau_a)} \tag{9.37}$$

A block diagram representation of Eqs. 9.36 and 9.37 is shown in Fig. 9.5*b*.

Let us consider a few special cases:

1. *Load torque proportional to speed.*

$$T_L \propto \omega_m$$
$$= B_L\omega_m \tag{9.38}$$

Let the total inertia of the system be

$$J = J_{motor} + J_{load} \tag{9.38a}$$

Equation 9.35 can be written as

$$K_m I_a(s) = Js\omega_m(s) + B_m\omega_m(s) + B_L\omega_m(s) \tag{9.39}$$
$$= Js\omega_m(s) + (B_m + B_L)\omega_m(s) \tag{9.39a}$$
$$= Js\omega_m(s) + B\omega_m(s) \tag{9.39b}$$

The load therefore increases the viscous friction of the mechanical system. From Eqs. 9.33 and 9.39b,

$$V_t(s) = K_m\omega_m(s) + \frac{BR_a}{K_m}(1 + s\tau_m)(1 + s\tau_a)\omega_m(s) \tag{9.40}$$

$$\frac{\omega_m(s)}{V_t(s)} = \frac{1}{K_m + (BR_a/K_m)(1 + s\tau_m)(1 + s\tau_a)} \tag{9.40a}$$

The speed response due to a step change in the terminal voltage V_t is a second-order response because of the two time constants τ_m and τ_a. The response can be underdamped or overdamped depending on the values of these time constants and the other parameters K_m, B, and R_a. Two typical responses are shown in Fig. 9.5*c*.

2. $L_{aq} = 0$. If the armature circuit inductance is neglected, the electrical time constant τ_a is zero. From Eq. 9.40a, the transfer function becomes

$$\frac{\omega_m(s)}{V_t(s)} = \frac{1}{K_m + (R_aB/K_m)(1 + s\tau_m)} \tag{9.41}$$

or

$$\frac{\omega_m(s)}{V_t(s)} = \frac{K_m}{K_m^2 + R_aB} \cdot \frac{1}{1 + s\tau'_m} \tag{9.42}$$

where

$$\tau'_m = \frac{R_aB}{K_m^2 + R_aB}\tau_m < \tau_m \tag{9.43}$$

3. $B = 0$, *inertia load.* If the viscous friction is zero, Eq. 9.39b becomes

$$K_mI_a(s) = Js\omega_m(s) \tag{9.44}$$

From Eqs. 9.33 and 9.44,

$$V_t(s) = \frac{K_m\omega_m(s) + Js\omega_m(s)R_a(1 + s\tau_a)}{K_m}$$

or

$$\frac{\omega_m(s)}{V_t(s)} = \frac{1}{K_m + (JR_a/K_m)s(1 + s\tau_a)} \tag{9.45}$$

4. *Supply disconnected.* Let us now investigate what happens if the supply is suddenly disconnected, that is, the switch SW in Fig. 9.5*a* is opened at $t = 0$. The dynamic equation for the mechanical system is

$$T = K_m i_a = J\frac{d\omega_m}{dt} + B\omega_m = 0 \tag{9.46}$$

or

$$B\omega_m = -J\frac{d\omega_m}{dt} \tag{9.47}$$

The Laplace transform of Eq. 9.47 is

$$B\omega_m(s) = -J[s\omega_m(s) - \omega_{m0}] \tag{9.48}$$

where ω_{m0} is the initial speed. From Eq. 9.48,

$$\omega_m(s) = \frac{J\omega_{m0}}{B + sJ} = \frac{\omega_{m0}}{(s + B/J)}$$

or

$$\omega_m(s) = \frac{\omega_{m0}}{s + 1/\tau_m} \tag{9.49}$$

The time domain response of speed is

$$\omega_m(t) = \omega_{m0}e^{-t/\tau_m} \tag{9.50}$$

The speed decreases exponentially with time constant τ_m. The deceleration of speed is shown in Fig. 9.6. The intersection of the initial slope on the time axis represents the mechanical time constant τ_m $(= J/B)$.

EXAMPLE 9.2

A separately excited dc motor has the following parameters:

$$R_a = 0.5\ \Omega, \qquad L_{aq} \simeq 0, \qquad B \simeq 0$$

The motor generates an open-circuit armature voltage of 220 V at 2000 rpm and with a field current of 1.0 ampere.

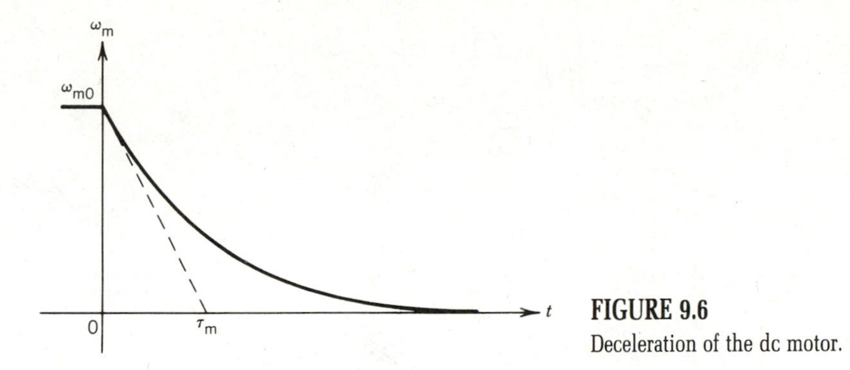

FIGURE 9.6
Deceleration of the dc motor.

The motor drives a constant load torque $T_L = 25$ N · m. The combined inertia of motor and load is $J = 2.5$ kg · m^2. With field current $I_f = 1.0$ A, the armature terminals are connected to a 220 V dc source.

(a) Derive expressions for speed (ω_m) and armature current (i_a) as a function of time.

(b) Determine the steady-state values of the speed and armature current.

Solution

(a)

$$E_a = K_m\omega_m$$

$$K_m = \frac{220}{(2000/60) \times 2\pi} = 1.05 \text{ V/rad/sec}$$

$$V_t = e_a + i_a R_a = K_m\omega_m + i_a R_a$$

$$T = K_m i_a = J\frac{d\omega_m}{dt} + T_L$$

From the last two equations,

$$\begin{aligned} V_t &= K_m\omega_m + R_a\left(\frac{J}{K_m}\frac{d\omega_m}{dt} + \frac{T_L}{K_m}\right) \\ &= K_m\omega_m + \frac{R_a J}{K_m}\frac{d\omega_m}{dt} + \frac{R_a T_L}{K_m} \\ &= 1.05\omega_m + \frac{0.5 \times 2.5}{1.05}\frac{d\omega_m}{dt} + \frac{0.5 \times 25}{1.05} \\ &= 1.05\omega_m + 1.19\frac{d\omega_m}{dt} + 11.9 \end{aligned}$$

$$V_t(s) = \frac{220}{s} = 1.05\omega_m(s) + 1.19s\omega_m(s) + \frac{11.9}{s}$$

$$\omega_m(s) = \frac{220 - 11.9}{s(1.05 + 1.19s)}$$

$$= \frac{174.874}{s(s + 0.8824)}$$

$$= \frac{198.2}{s} - \frac{198.2}{s + 0.8824}$$

$$\omega_m(t) = 198.2(1 - e^{-0.8824t})$$

$$i_a = \frac{V_t - K_m\omega_m}{R_a}$$

$$= \frac{220 - 1.05\omega_m}{0.5}$$

$$= 440 - 2.1 \times 198.2(1 - e^{-0.8824t})$$

$$= 23.8 + 416.2e^{-0.8824t}$$

(b) Steady-state speed is $\omega_m(\infty) = 198.2$ rad/sec.
Steady-state current is $I_a = i_a(\infty) = 23.8$ A.

9.2 SYNCHRONOUS MACHINES

Synchronous machines are used primarily as generators, either supplying power to an individual load or connected to an infinite bus. A disturbance may occur in a synchronous machine in various ways. An accidental short circuit may occur between line and ground, between line and line, or between all three lines. A disturbance may also be caused by the sudden application of a load to the machine. Any kind of disturbance will cause electrical and mechanical transients. The machine may even lose synchronism because of a disturbance. Transient phenomena in a synchronous machine are inherently very complex. The study of synchronous machine transients and dynamics has been a formidable challenge to power system engineers for many years. The general subject is so broad and complicated that many books have been written and many courses offered on this topic.

In this section we provide only a basic understanding of transient phenomena in a synchronous machine. We consider two particular cases: (1) a sudden three-phase short circuit at the stator terminals and (2) mechanical transients caused by a sudden load change. The development of an understanding will be based primarily on physical or semi-intuitive reasoning. A rigorous analysis of this complex subject matter is beyond the scope of this book.

9.2.1 THREE-PHASE SHORT CIRCUIT

Short Circuit on an Open-Circuited Synchronous Generator

Figure 9.7 shows a schematic representation of a three-phase synchronous machine. The rotor is rotating at some speed ω_m and the field current I_f generates an open-circuit voltage E_f in each phase. If the stator terminals are now shorted, a large transient current will flow through them. However, when the transient dies down, the steady-state short-circuit current is

$$I_{sc} = \frac{E_f}{X_s} \tag{9.51}$$

If $E_f = 1$ pu and $X_s = 1$ pu (typically) the steady-state short-circuit current is 1 pu. This is a good feature of the synchronous machine. If the short circuit is sustained, it will not damage the machine. However, at the instant the short is applied, the armature current can be very high—as high as 5 to 10 pu. To determine the circuit breaker rating or relay setting of the protective system, a prior knowledge of the armature current during the transient period is essential.

Before the short is applied, the flux linkages of the field winding and damper (or amortisseur) winding are constant. No current is present in the damper winding because it rotates at the same speed as the field winding. The only current present is the dc current in the field winding. However, when the short is applied, armature current flows whose mmf directly opposes the mmf of the field winding. The flux linkages of both field winding and damper winding are affected. To maintain the component fluxes constant at their initial values, induced components of current will flow in both field winding and damper winding. This phenomenon can be explained as follows. Consider a ring of conductor, as shown in Fig. 9.8, having an inductance L. If a magnetic field is suddenly applied, the loop resists change in the flux linkage. Consequently, a current i is induced in the loop,

$$i = \frac{\Phi}{L} \tag{9.52}$$

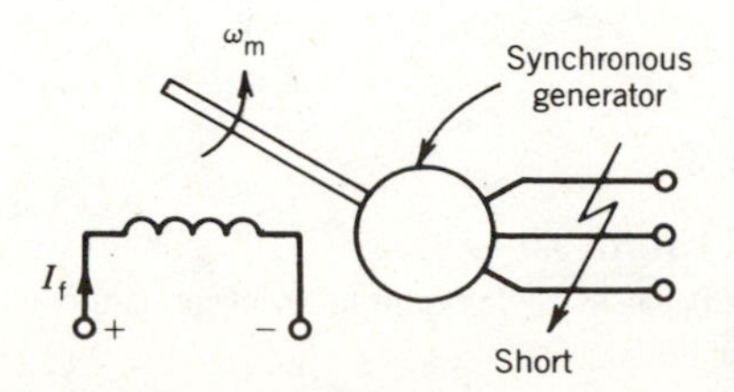

FIGURE 9.7
Schematic diagram of a synchronous generator with short on stator terminals.

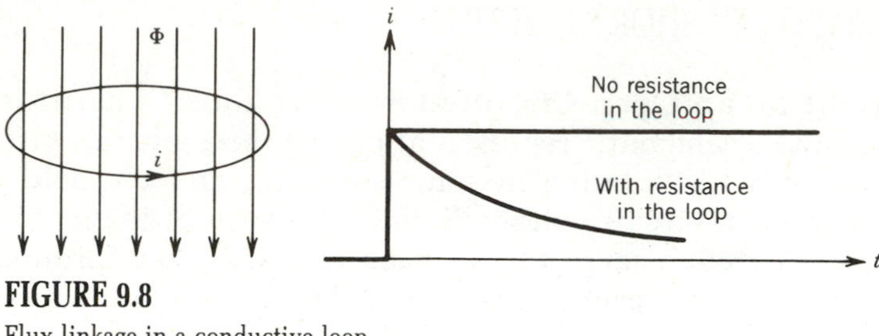

FIGURE 9.8
Flux linkage in a conductive loop.

where Φ is the applied flux on the loop. If the loop has no resistance, the current is sustained, thereby maintaining the initial zero flux linkage. If the loop has resistance, the current decays, thereby allowing the flux linkage to change.

The induced currents in the field winding and damper winding decay because of the resistances in these circuits. The change in the field current resulting from an armature short circuit is shown in Fig. 9.9. These induced currents are equivalent to an increase of the field excitation, and therefore a large current will flow in the armature circuit immediately after the short circuit is applied.

In a salient pole machine the damper winding is placed in the rotor pole faces. Solid cylindrical rotor machines do not generally have damper windings. However, during transient periods, currents that are induced directly in the rotor body produce essentially the same effects as the damper currents in a salient pole machine.

Figure 9.10*a* shows the trace of short-circuit current in a stator phase resulting from a three-phase short suddenly applied at the stator terminals. This symmetrical trace can be obtained oscillographically if the short is applied at the instant when the preshort flux linkage of the phase is zero. The envelope of the wave is shown in Fig. 9.10*b*. The envelope shows three distinct periods: the *subtransient period,* lasting only the first few cycles, during which current decreases very rapidly; the *transient period,* lasting a relatively longer time, during which the current decrease is more moderate; and finally the *steady-state period,* during which the cur-

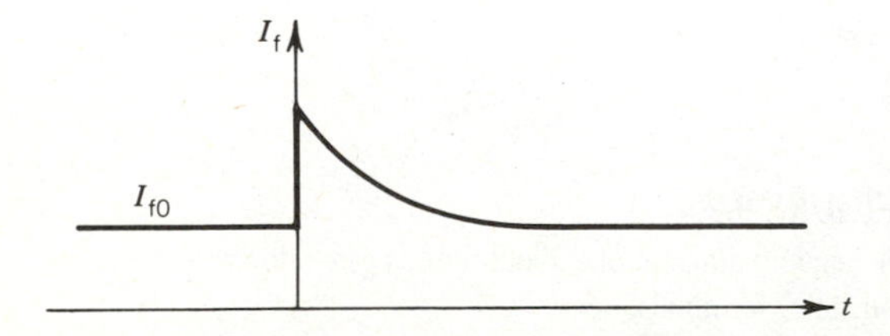

FIGURE 9.9
Change in field current following armature short circuit.

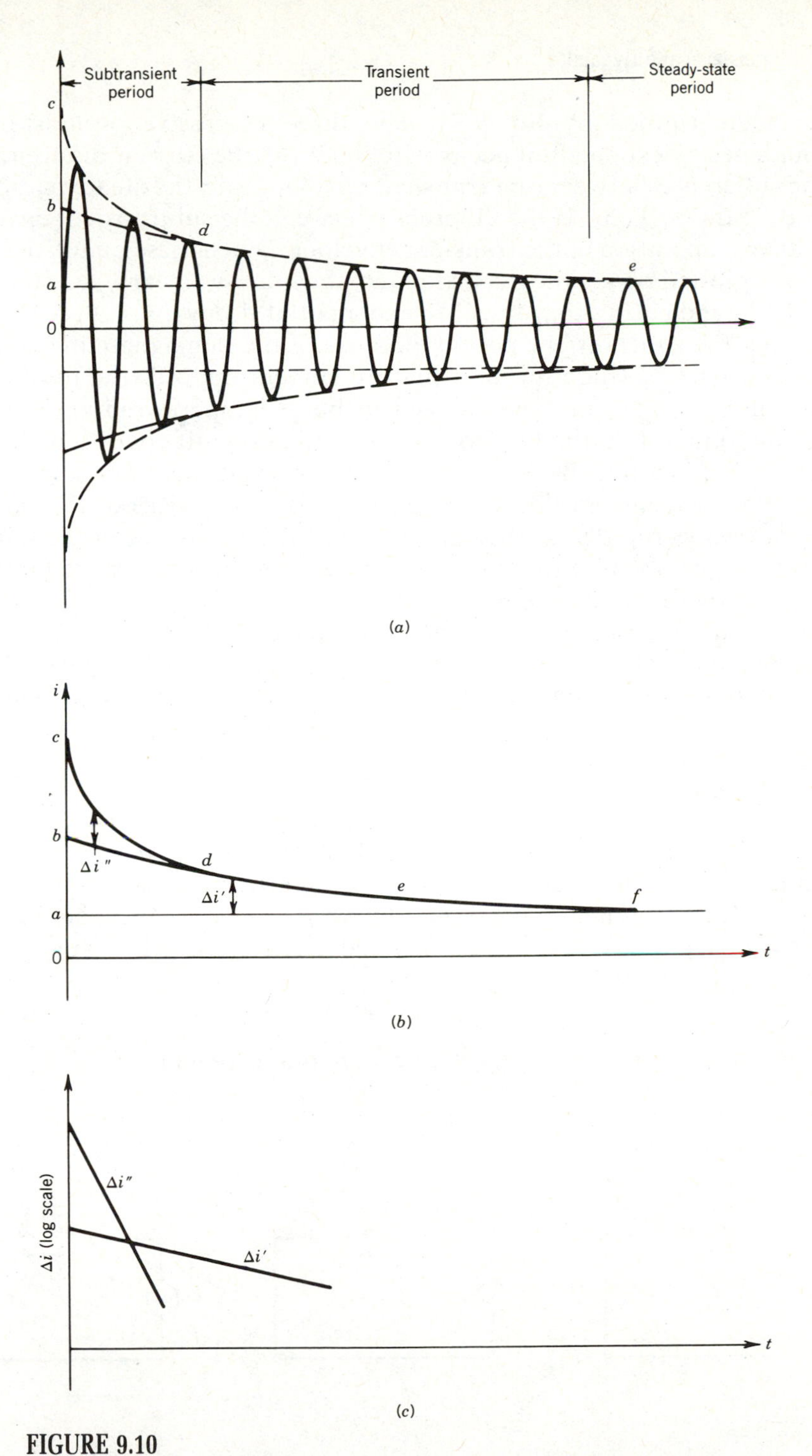

FIGURE 9.10

Armature short circuit. (*a*) Symmetrical armature current. (*b*) Envelope of current. (*c*) Decay of current difference.

rent is determined by Eq. 9.51. The three successive periods merge through nearly exponential decays. In Fig. 9.10*b* the current difference $\Delta i'$ is the difference between the transient envelope and the steady-state amplitude. Similarly, $\Delta i''$ is the difference between the subtransient envelope and an extrapolation of the transient envelope. When these quantities ($\Delta i'$, $\Delta i''$) are plotted on semilog coordinates, they decay linearly, as shown in Fig. 9.10*c*, indicating that they are exponential decays.

During the subtransient period, because of the demagnetizing effect of armature current (the mmf of armature current opposes the mmf of the field winding), currents are induced in both damper winding and field winding to maintain the flux constancy of the prefault condition. This, in effect, is similar to a large increase in rotor excitation, and therefore a large armature current flows during the subtransient period. The damper current decays rapidly because of the small time constant of the damper circuit. The behavior of the stator current during this period is determined primarily by the damper current.

During the transient period, the damper current has decayed to zero. The behavior of the armature current during this period is determined by the field winding current, which decays with a larger time constant.

Short-Circuit Current

The armature current can be determined for the various periods by using appropriate reactances and time constants. During short circuit, mmf's act along the d-axis. The equivalent circuits during the various periods of short circuit are shown in Fig. 9.11. The d-axis synchronous reactances for the various periods can be determined from Figs. 9.10*b* and 9.11 as follows:

$$X_{\mathrm{d}} = \sqrt{2}\,\frac{E_{\mathrm{f}}}{0a}, \qquad \text{steady-state d-axis reactance} \tag{9.53}$$

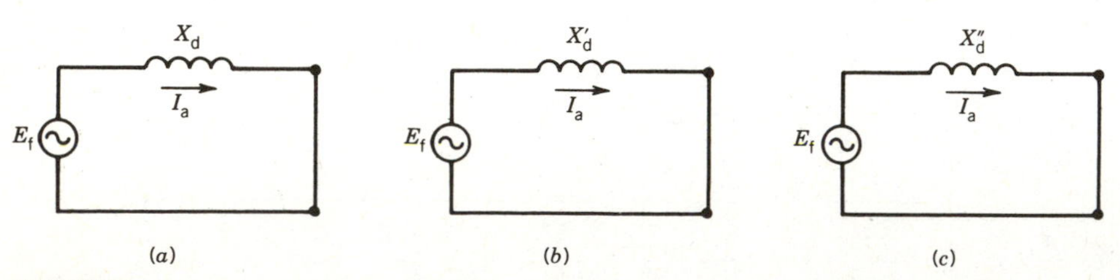

FIGURE 9.11

Equivalent circuits for calculation of fault current during various periods of armature short circuit (open-circuited generator). (*a*) Steady-state short circuit. (*b*) Transient period (I_a decays). (*c*) Subtransient period (I_a decays).

$$X_d' = \sqrt{2}\,\frac{E_f}{0b}, \qquad \text{d-axis transient reactance} \tag{9.54}$$

$$X_d'' = \sqrt{2}\,\frac{E_f}{0c}, \qquad \text{d-axis subtransient reactance} \tag{9.55}$$

The short-circuit current of an armature phase is

$$\begin{aligned} i_{sc} &= [\text{varying amplitude}] \sin \omega t \\ &= [0a + (0b - 0a)e^{-t/T_{do}'} + (0c - 0b)e^{-t/T_{do}''}] \sin \omega t \end{aligned} \tag{9.56}$$

or

$$= \sqrt{2}\left[\frac{E_f}{X_d} + \left(\frac{E_f}{X_d'} - \frac{E_f}{X_d}\right)e^{-t/T_{do}'} + \left(\frac{E_f}{X_d''} - \frac{E_f}{X_d'}\right)e^{-t/T_{do}''}\right] \sin \omega t \tag{9.57}$$

where T_{d0}' is the time constant during the transient period and is so defined that it determines the decay of the transient envelope *bde*

T_{d0}'' is the time constant during the subtransient period and is so defined that it determines the decay of the subtransient envelope *cd*

DC Component

The symmetrical wave of Fig. 9.10*a* is a special case rather than a general case. The more usual short-circuit currents are shown in Fig. 9.12. These currents are not symmetrical about the zero-current axis and definitely show the dc components responsible for the offset waves. A symmetrical wave, as shown in Fig. 9.10*a*, can be obtained by replotting the offset wave with the dc component subtracted from it.

The dc component in the short-circuit armature current is due to the flux linkage of a phase at the instant the short is applied. If the flux linkage of a phase is zero at the instant the short circuit is applied, no dc component is required to maintain the flux linkage at that zero value and the short-circuit current wave for that phase is symmetrical. On the other hand, if the flux linkage of a phase is nonzero at the instant the short circuit is applied, the dc component must appear in the current of that phase to keep the flux linkage constant. The dc component decays with the armature time constant.

$$I_{dc} = I_{dc0}e^{-t/T_a} \tag{9.58}$$

where T_a is the armature time constant and is so defined that it determines the decay of the dc component.

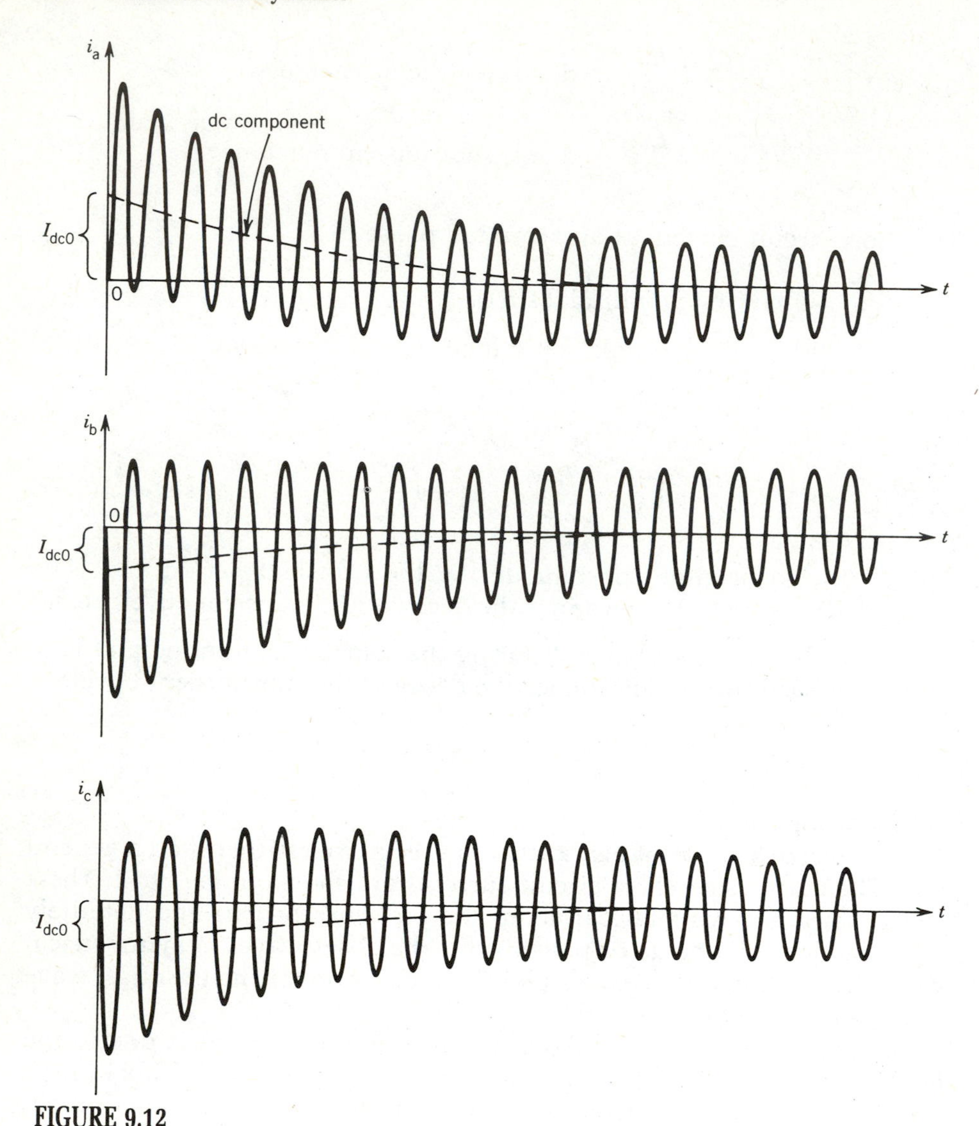

FIGURE 9.12
Three-phase short-circuit currents of a synchronous generator.

Note that in a three-line system the sum of the dc components in the three phases at any instant is zero. The short-circuit armature current with the dc offset is

$$i_{sc} = \sqrt{2}\left[\frac{E_f}{X_d} + \left(\frac{E_f}{X'_d} - \frac{E_f}{X_d}\right)e^{-t/T'_{do}} + \left(\frac{E_f}{X''_d} - \frac{E_f}{X'_d}\right)e^{-t/T''_{do}}\right]\sin\omega t + I_{dc0}e^{-t/T_a} \tag{9.59}$$

The largest dc component occurs in a phase current when the flux linkage of that phase is maximum at the instant the short circuit is applied. The largest possible dc component is equal to the amplitude of the subtransient current $t = 0$.

$$(I_{dc0})_{max} = \sqrt{2}\,\frac{E_f}{X''_d} \tag{9.60}$$

The dc components in the stator phases establish a stationary field in the air gap, and this induces voltage and current at the fundamental frequency in the synchronously rotating rotor circuits. Figure 9.13 shows this ac component superimposed on the field current immediately following the short circuit at the armature terminals.

Typical values of the machine constants of a salient pole synchronous generator are

$$X_d = 1.0 \text{ pu}, \qquad T'_{d0} = 6.0 \text{ sec}$$
$$X'_d = 0.3 \text{ pu}, \qquad T''_{d0} = 0.06 \text{ sec}$$
$$X''_d = 0.2 \text{ pu}, \qquad T_a = 0.15 \text{ sec}$$

Short Circuit on a Loaded Synchronous Generator

Let us now consider a three-phase generator that is delivering power to a load or to an infinite bus. If a short circuit is applied across the machine terminals, the short-circuit armature current will pass through a subtransient period and a transient period and finally will settle down to a steady-state condition. When the short is applied, the machine reactance changes from X_d to X''_d. The excitation voltages must also change to satisfy the initial condition of flux linkage constancy.

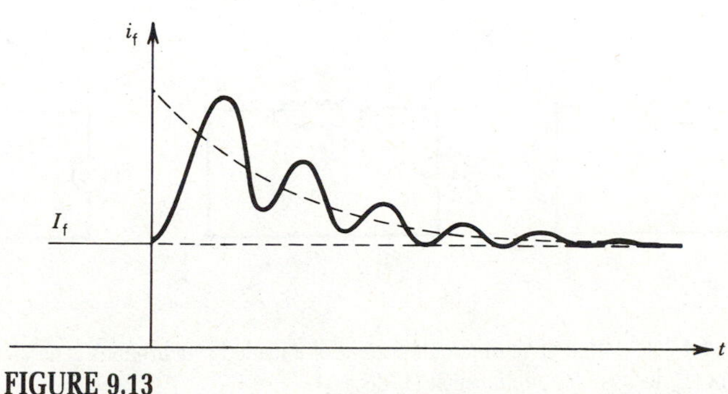

FIGURE 9.13
Field current following a three-phase armature short circuit.

The equivalent circuits during the three periods of the short circuit are shown in Fig. 9.14. These equivalent circuits are models for the synchronous machine after the short circuit is applied. The voltages E_i, E_i', and E_i'' are the internal voltages and can be computed from the prefault condition as follows:

$$E_i(= E_f) = V_t + I_a jX_d, \quad \text{voltage behind synchronous reactance before fault} \tag{9.61}$$

$$E_i' = V_t + I_a jX_d', \quad \text{voltage behind transient reactance before fault} \tag{9.62}$$

$$E_i'' = V_t + I_a jX_d'', \quad \text{voltage behind subtransient reactance before fault} \tag{9.63}$$

where I_a is the prefault steady-state current. The short-circuit current is

$$i_{sc} = \sqrt{2}\left[\frac{E_i}{X_d} + \left(\frac{E_i'}{X_d'} - \frac{E_i}{X_d}\right)e^{-t/T_d'} + \left(\frac{E_i''}{X_d''} - \frac{E_i'}{X_d'}\right)e^{-t/T_d''}\right]\sin \omega t + I_{dc0}e^{-t/T_a} \tag{9.64}$$

where

$$T_d' \simeq \frac{X_d'}{X_d} T_{d0}' \tag{9.65}$$

$$T_d'' \simeq \frac{X_d''}{X_d'} T_{d0}'' \tag{9.66}$$

A rigorous analysis of the phenomena resulting from the application of a short circuit (short-circuiting) to a loaded synchronous generator is quite complex because it involves detailed study of many coupled circuits with initial currents. The discussion above is a simplistic analysis of a very

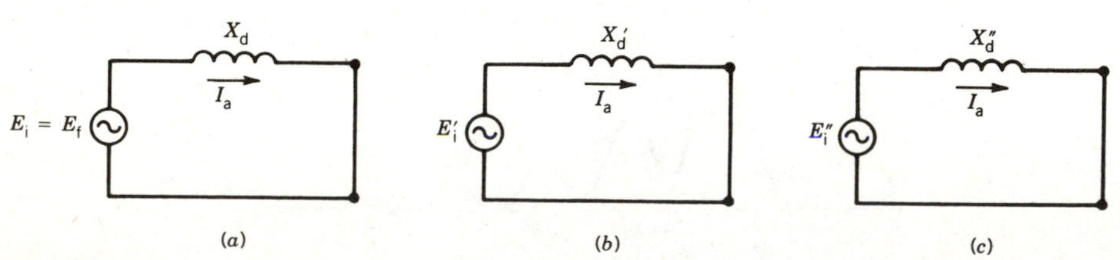

FIGURE 9.14
Equivalent circuit for calculation of short-circuit current of a loaded synchronous generator. (a) Steady state. (b) Transient (I_a decays). (c) Subtransient (I_a decays).

complex problem. However, this simplified analysis provides results that are reasonably accurate.

EXAMPLE 9.3

A 3ϕ, 50 MVA, 15 kV, 60 Hz synchronous generator has the following parameters:

$$X_d = 0.9 \text{ pu}, \quad X'_d = 0.4 \text{ pu}, \quad X''_d = 0.2 \text{ pu}$$

$$T'_{d0} = 4 \text{ sec}, \quad T''_{d0} = 0.6 \text{ sec}, \quad T_a = 0.2 \text{ sec}$$

The generator is delivering full load to the infinite bus at 0.9 lagging power factor. A three-phase short circuit suddenly occurs at the machine terminals.

(a) Determine the prefault values of the voltages behind the reactances.

(b) Determine the initial value of the maximum possible dc offset current in the machine.

(c) Obtain an expression for the machine fault current as a function of time. Consider the maximum possible dc offset current in the machine.

(d) Determine the rms value of the machine fault current at $t = 0.1$ sec.

Solution

(a) Prefault condition: $V_t = 1 \underline{/0^\circ}$ pu, $I_a = 1 \underline{/-25.8^\circ}$ pu.

$$\begin{aligned} E_i &= V_t + I_a j X_d \\ &= 1\underline{/0^\circ} + 1\underline{/-25.8^\circ} \times 0.9\underline{/90^\circ} \\ &= 1.61 \text{ pu} \end{aligned}$$

$$\begin{aligned} E'_i &= 1 + 1\underline{/-25.8^\circ} \times 0.4\underline{/90^\circ} \\ &= 1.23 \text{ pu} \end{aligned}$$

$$\begin{aligned} E''_i &= 1 + 1\underline{/-25.8^\circ} \times 0.2\underline{/90^\circ} \\ &= 1.1 \text{ pu} \end{aligned}$$

(b)

$$I_{dc0(\max)} = \sqrt{2}\frac{E''_i}{X''_d} = \sqrt{2} \times \frac{1.1}{0.2} = 7.78 \text{ pu}$$

(c)
$$T'_d = \frac{X'_d}{X_d} T'_{d0} = \frac{0.4}{0.9} \times 4 = 1.78 \text{ sec}$$

$$T''_d = \frac{X''_d}{X'_d} T''_{d0} = \frac{0.2}{0.4} \times 0.6 = 0.3 \text{ sec}$$

$$i_{sc}(t) = \sqrt{2}\left[\frac{1.61}{0.9} + \left(\frac{1.23}{0.4} - \frac{1.61}{0.9}\right) e^{-t/1.78} + \left(\frac{1.1}{0.2} - \frac{1.23}{0.4}\right) e^{-t/0.3}\right] \sin \omega t + 7.78e^{-t/0.2}$$

$$= \sqrt{2}(1.79 + 1.29e^{-0.562t} + 2.42e^{-3.3t}) \sin \omega t + 7.78e^{-5t}$$

(d)
$$i_{sc}\big|_{t=0.1 \text{ sec}} = \sqrt{2}(1.79 + 1.29e^{-0.562\times 0.1} + 2.42e^{-3.3\times 0.1}) \sin \omega t + 7.78e^{-5\times 0.1}$$

$$= \sqrt{2}(4.75) \sin \omega t + 4.72$$

$$I_{sc}\big|_{t=1.0 \text{ sec}} = (4.75^2 + 4.72^2)^{1/2}$$

$$= 6.7 \text{ pu}$$

9.2.2 DYNAMICS: SUDDEN LOAD CHANGE

A sudden change in the operating condition of a synchronous machine connected to a power system may result in loss of synchronism. For a synchronous generator the most severe disturbance arises if a short circuit accidentally occurs across the machine terminals. For a synchronous motor, a disturbance may arise from a sudden application of load torque to the shaft. It is important to predict the ability of a synchronous machine to remain in synchronism after a disturbance occurs. In this section we study the dynamic behavior of a synchronous machine resulting from a load torque disturbance and discuss some techniques for predicting whether or not the machine will stay in synchronism after the disturbance is applied.

Steady-State Stability Limit

The power and torque developed by a three-phase synchronous machine connected to a power system are given by the following expressions.

$$P = \frac{3V_t E_f}{X_d} \sin \delta \tag{9.67}$$

$$T = \frac{3V_t E_f}{\omega_{syn} X_d} \sin \delta \tag{9.68}$$

or

$$T = T_{max} \sin \delta \tag{9.69}$$

The T–δ characteristic is shown in Fig. 9.15. If a load torque T_L is slowly applied on the shaft, the torque angle δ will increase. The load torque can be increased to the value T_{max}, for which $\delta = 90°$. If the load torque is further increased, the machine will lose synchronism because the torque developed by the machine will be less than the load torque. The load torque, therefore, can be slowly increased to the maximum value of the torque the machine can develop, which is called the *steady-state stability limit or static stability limit.*

Dynamic Stability

If a load torque T_L is applied suddenly, the machine may lose synchronism even if the load torque is less than the maximum torque the machine can develop. The maximum value of the load torque that can be applied suddenly without losing synchronism can be ascertained from the dynamic behavior of the synchronous machine.

Consider an unloaded synchronous motor connected to a power system. If losses are neglected, the angle δ is zero. Now a load torque is suddenly applied to the shaft of the synchronous motor. The motor slows down, increasing the torque angle δ. As δ increases, the machine develops torque to meet the load torque. As the machine slows down, the kinetic energy of the moving mass provides the load torque. The torque angle δ reaches the value δ_L at which load torque is same as the torque developed by the machine, as shown in Fig. 9.16*a*. However, because of inertia, δ increases

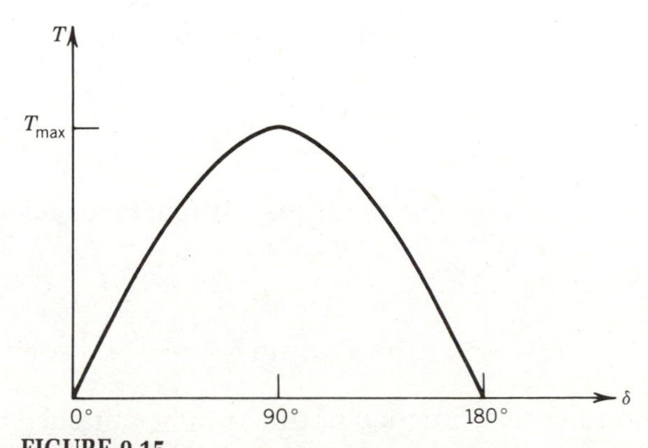

FIGURE 9.15
T–δ characteristic of a synchronous machine.

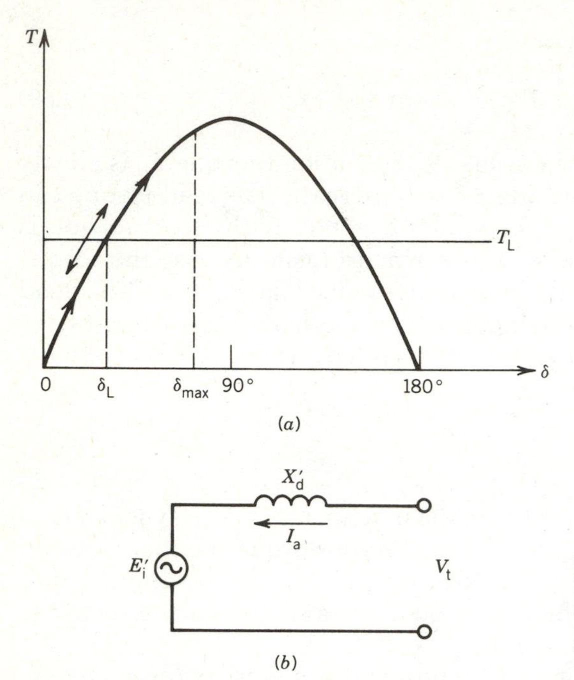

FIGURE 9.16
Sudden load disturbance on a synchronous motor. (*a*) T–δ characteristic. (*b*) Equivalent circuit during disturbance.

beyond δ_L and the machine develops more torque than required for the load. The deceleration decreases, and δ reaches a maximum value δ_{max} and then swings back. The torque angle oscillates around δ_L. Because of damping in the system, the torque angle will settle down to the required value δ_L.

The subtransient period lasts a very short time, say less than three cycles. However, the oscillation of δ lasts for several cycles, say 20 or more. It can be assumed that the transient equivalent circuit of Fig. 9.16*b* provides a satisfactory representation of the electrical system. The T–δ relationship, shown graphically in Fig. 9.16*a*, is given by

$$T = \frac{3V_t E'_i}{X'_d} \sin \delta \tag{9.70}$$

where E'_i is the voltage behind the transient reactance before the disturbance and is determined by the equation $E'_i = V_t - I_a jX'_d$. Equation 9.70 can also be written as

$$T = T_{max} \sin \delta \tag{9.71}$$

The oscillation in δ as a function of time can be obtained by solving the differential equation that describes the dynamics of the system. The torque balance equation for a synchronous motor is

$$T = T_a + T_d + T_L \tag{9.72}$$

where $T = T_{max} \sin \delta$ is the torque developed by the motor

$T_a = K_j \, d^2\delta/dt^2$ is the acceleration torque

$T_d = K_d \, d\delta/dt$ is the damping torque

T_L is the load torque

Equation 9.72 can be written as

$$T_{max} \sin \delta = K_j \frac{d^2\delta}{dt^2} + K_d \frac{d\delta}{dt} + T_L \tag{9.73}$$

This is a nonlinear equation and can be solved by numerical methods to determine the oscillation in δ as a function of time.

Equal-Area Method

In most cases we are only interested in knowing whether synchronism is restored, that is, whether the angle δ settles down to a steady operating value after the disturbance occurs. A graphical approach known as the equal-area method will enable us to determine this.

Let us consider the specific case of a synchronous motor having the torque–angle curve of Fig. 9.17, based on Eq. 9.70. Assume that the motor is initially unloaded and therefore that the operating point is at the origin of the curve. Now a load torque T_L is suddenly applied. The rotor decelerates toward $\delta = \delta_L$, where the torque produced by the machine equals the load torque T_L. During this deceleration, kinetic energy is removed from the rotating mass. The area *0AB* represents this energy. At $\delta = \delta_L$, the speed is somewhat less than the synchronous speed. The angle δ continues to increase, making the machine produce more torque than required for the load. This makes the motor accelerate and restore its lost kinetic energy. At $\delta = \delta_{max}$, the speed reaches its synchronous value. Therefore the total change in kinetic energy is zero. Thus, the areas *0AB* and *BCD* are equal.

At $\delta = \delta_{max}$ the machine is being accelerated, and when δ swings back to δ_L the speed is above synchronous value. The angle δ swings beyond δ_L and, if damping is neglected, reaches its initial value and repeats the oscillation. In practice, the oscillation in δ damps out and δ settles down to the value δ_L.

The equal-area method provides the following information:

- An easy means of finding the maximum angle of swing.
- An estimate of whether synchronism will be maintained.

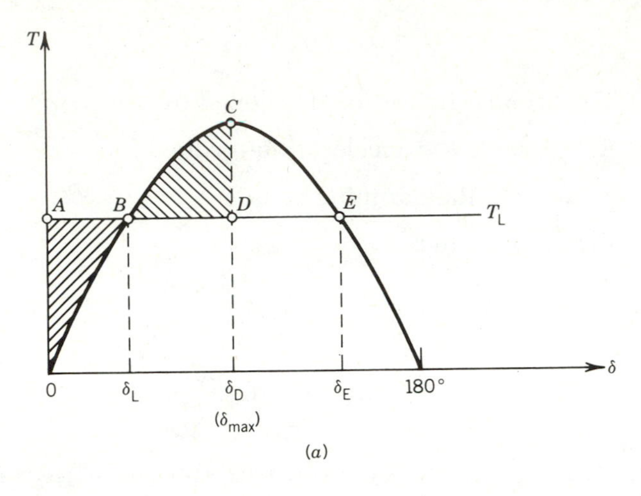

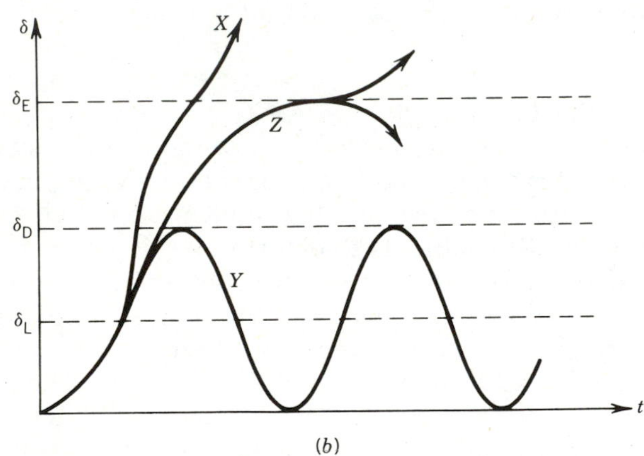

FIGURE 9.17
Swing curves for sudden load torque disturbance (no damping).

- The maximum amount of disturbance that can be allowed without losing synchronism.

If the load disturbance is such that the area *BCEDB* in Fig. 9.17*a* is less than the area *0AB*, synchronism will never be restored and δ will continue to increase with time, as shown by curve *X* in Fig. 9.17*b*. On the other hand, if area *BCEDB* is less than *0AB* synchronism will be restored and the δ–t curve will follow curve *Y* in Fig. 9.17*b*. If the areas *BCEDB* and *0AB* are equal, the system will remain in unstable equilibrium, δ following curve *Z* of Fig. 9.17*b*.

The maximum value of the load torque $T_{L(\max)}$ that can be applied sud-

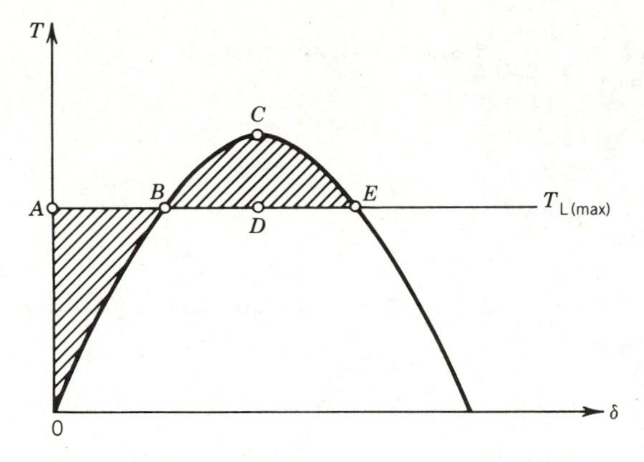

FIGURE 9.18
Dynamic stability limit.

denly without losing synchronism is called the *dynamic stability limit,* as shown in Fig. 9.18.

Note that when the transient period is over, the maximum torque developed by the machine is determined by E_f and X_d and is smaller than the maximum torque developed during the transient period. Therefore, if the load disturbance is greater than the maximum value of torque for the steady-state condition, the system may stay in synchronism during the transient period but not during the steady state.

EXAMPLE 9.4

A 3ϕ synchronous machine has the following parameters:

$$X_d = 0.8 \text{ pu}, \qquad X'_d = 0.3 \text{ pu}$$

The field current of the synchronous machine is adjusted to produce an open-circuit voltage of 1 pu and the machine is synchronized to an infinite bus.

(a) Determine the maximum per-unit torque that can be applied slowly without losing synchronism.

(b) Determine the maximum per-unit torque that can be applied suddenly after initial synchronization without losing synchronism. Can this torque be sustained for long?

Solution

(a) For gradually changing torque, use the steady-state equivalent circuit.

$$T = \frac{|V_t \mid E_f|}{\omega_{syn} X_d} \sin \delta = \frac{1 \times 1}{1 \times 0.8} \sin \delta = 1.25 \sin \delta \text{ pu}$$

$$T_{max} = 1.25 \text{ pu}$$

$$\left(\text{Note: in pu } T = P = \frac{|V_t \mid E_f|}{X_d} \sin \delta.\right)$$

(b) For rapidly changing conditions, use the transient equivalent circuit. Before load torque was applied,

$$V_t = E_f; \qquad \text{therefore } I_a = 0$$

Hence,

$$E_i' = V_t - I_a j X_d' = V_t = 1 \text{ pu}$$

$$T = \frac{|V_t \mid E_i'|}{\omega_{syn} X_{d'}} \sin \delta = \frac{1 \times 1 \sin \delta}{1 \times 0.3} = 3.33 \sin \delta \text{ pu}$$

The T–δ relation is shown in Fig. E9.4. The area A is

$$A = T_1 \delta_1 - \int_0^{\delta_1} 3.33 \sin \delta \, d\delta$$

$$= T_1 \delta_1 - 3.33(1 - \cos \delta_1)$$

The area B is

$$B = \int_{\delta_1}^{\pi - \delta_1} 3.33 \sin \delta \, d\delta - T_1(\pi - 2\delta_1)$$

$$= 3.33(2 \cos \delta_1) - T_1(\pi - 2\delta_1)$$

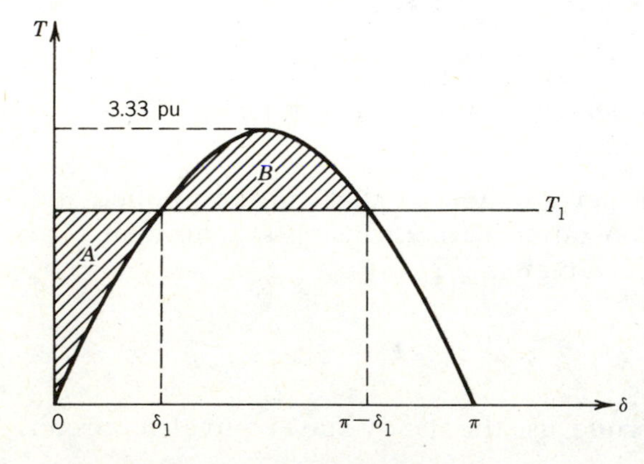

FIGURE E9.4

If area A = area B, then

$$T_1(\pi - \delta_1) - 3.33 - 3.33 \cos \delta_1 = 0 \tag{E9.4a}$$

Also,

$$T_1 = 3.33 \sin \delta_1 \tag{E9.4b}$$

From Eqs. E9.4a and E9.4b,

$$(\pi - \delta_1) \sin \delta_1 - \cos \delta_1 = 1$$

Solving for δ_1,

$$\delta_1 = 46.5°$$

Therefore,

$$\begin{aligned} T_1 &= 3.33 \sin 46.5° \\ &= 2.42 \text{ pu} \end{aligned}$$

This torque cannot be sustained for a long time, because in the steady state the maximum torque the machine can produce is only 1.25 pu.

9.3 INDUCTION MACHINES

The transient behavior of induction machines can be studied by following the same approach used for synchronous machines.

Short-Circuit Transients

During a fault condition, the cage bars in the rotor of the induction machine behave in essentially the same way as the amortisseur winding on the pole faces of a synchronous machine. Therefore, a large subtransient current will flow in the induction machine following a short circuit across the terminals. Further, because of the absence of any field winding circuit in an induction machine, the long transient period of a synchronous machine will be absent. Therefore, short-circuit current in an induction machine will be large but will decay quickly, as shown in Fig. 9.19. Note that the current decay is characterized by one time constant, the subtransient time constant.

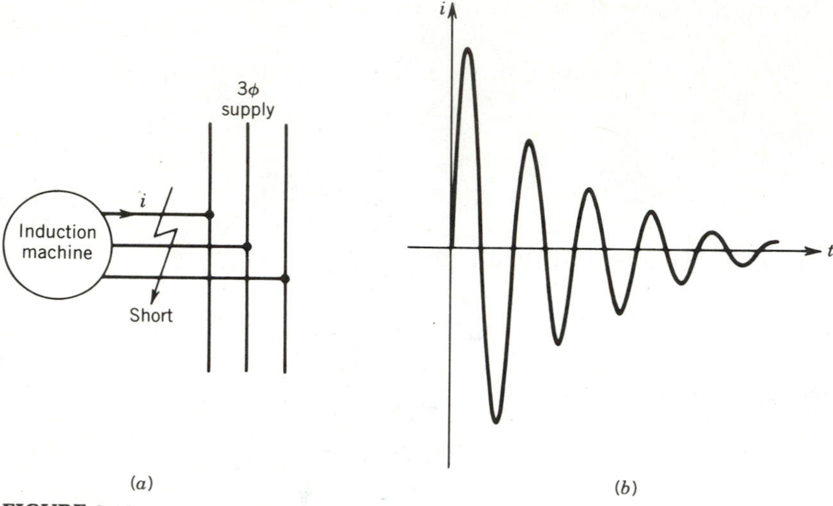

FIGURE 9.19

Symmetrical three-phase short on an induction machine. (a) Schematic diagram. (b) Waveform of short-circuit current in a phase (dc component not considered).

Dynamics

The induction machine has no synchronization problem. It can develop torque at speeds other than the synchronous speed. The common dynamic problems are associated with starting and stopping the machine and also with mechanical load change on the motor shaft.

Consider the case of starting an induction motor by connecting it directly to the power system. The initial starting current is high—it can be as high as three to eight times its rated value.

The motor current decays as the motor speeds up. Figure 9.20*a* shows the steady-state torque–speed characteristic of the motor. Also shown is the steady-state load–torque characteristic of a typical load. The initial electrical transients decay very quickly, and also, because of the high inertia of the motor and its load, the change in speed during one cycle of the supply frequency is very small. The electrical system can be considered to be in a slowly varying steady state during the starting period, and the steady-state torque–speed characteristic can be assumed to represent performance under starting conditions with sufficient accuracy. In other words, at any speed during the motor start-up the electrical system is assumed to be in quasi-steady-state condition.

From Fig. 9.20*a* the torque difference $T - T_L$ is the accelerating torque.

$$J\frac{d\omega_m}{dt} = T - T_L \tag{9.74}$$

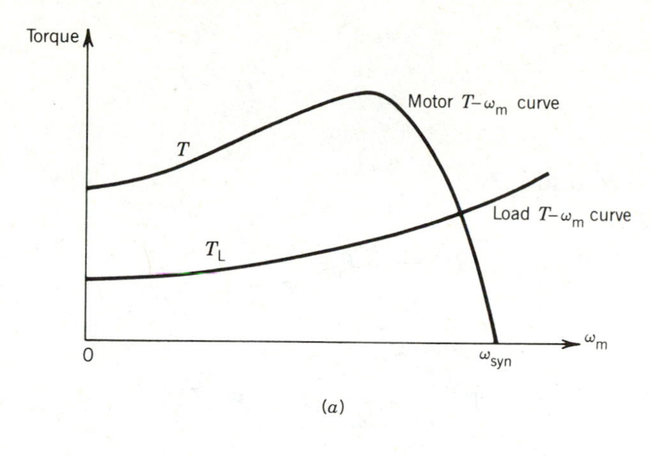

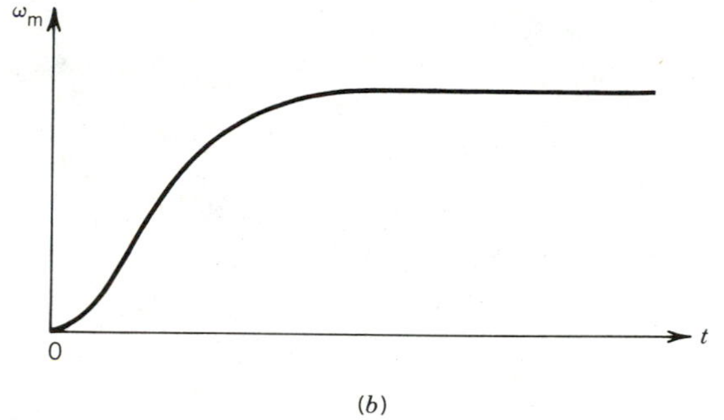

FIGURE 9.20

Starting transient in an induction motor. (a) Torque–speed curves for motor and load. (b) Speed–time curve during starting.

or

$$\omega_m = \frac{1}{J}\int_0^t (T - T_L)\,dt \tag{9.75}$$

The integration in Eq. 9.75 can be performed graphically or by use of numerical methods. A typical speed–time curve for the starting of an induction machine is shown in Fig. 9.20*b*. In practice, it may take about 15 seconds for an induction motor to reach the steady-state speed.

Speed transients also occur if the mechanical load on the motor shaft is changed suddenly. Let us consider that an unloaded induction motor is running near the synchronous speed and suddenly a load torque T_L is applied to the motor shaft. Near the synchronous speed the torque–speed curve is assumed to be linear and is

$$T = Ks \tag{9.76}$$

$$= K \frac{\omega_{syn} - \omega_m}{\omega_{syn}} \tag{9.77}$$

From Eqs. 9.74 and 9.77

$$J \frac{d\omega_m}{dt} = K - \frac{K\omega_m}{\omega_{syn}} - T_L \tag{9.78}$$

or

$$J \frac{d\omega_m}{dt} + \frac{K}{\omega_{syn}} \omega_m = K - T_L \tag{9.79}$$

The speed–time solution is

$$\omega_m = \omega_{syn} - \frac{T_L \omega_{syn}}{K} (1 - e^{-t/\tau_m}) \tag{9.80}$$

where the mechanical time constant is

$$\tau_m = \frac{J\omega_{syn}}{K} \tag{9.81}$$

9.4 TRANSFORMER; TRANSIENT INRUSH CURRENT

In normal steady-state operation the exciting current of a transformer is usually very low—less than 5 percent of rated current. However, at the moment when a transformer is connected to the power system, a large *inrush current* will flow in the transformer during the transient period. This current may be as high as 10 to 20 times the rated current. Knowledge of this large inrush current is important in determining the maximum mechanical stresses that could occur in the transformer windings and also in designing the protective system for the transformer.

The magnitude of the inrush current depends on the instant of the voltage wave at which the transformer is connected to the power supply. Consider a transformer whose core is initially unmagnetized. The transformer primary winding is now connected to a supply voltage

$$v = \sqrt{2}\, V \sin \omega t \tag{9.82}$$

If we neglect the core losses and the primary-winding resistance,

$$v = N \frac{d\Phi}{dt} \tag{9.83}$$

$$\Phi = \frac{1}{N}\int v\, dt \tag{9.84}$$

Consider two cases, as follows:

1. *The transformer is connected when the voltage is maximum.* The voltage and flux variations for this situation are shown in Fig. 9.21. Note that there is no transient in flux and that the time variation of flux is

$$\Phi = \Phi_{max} \sin(\omega t - 90°) \qquad \text{for } \omega t > 90° \tag{9.85}$$

   where

$$\Phi_{max} = \frac{\sqrt{2}\, V}{\omega N} \tag{9.86}$$

   The magnetizing current can be found from the B–H characteristic of the transformer core and is also shown in Fig. 9.21. No inrush current will flow and the system is in steady state from the start.

2. *The transformer is connected when the voltage is zero.* From Eqs. 9.82 and 9.84 the flux is given by

$$\begin{aligned}\Phi &= \frac{\sqrt{2}\, V}{N}\int_0^t \sin \omega t\, dt \\ &= \frac{\sqrt{2}\, V}{\omega N}(1 - \cos \omega t) \\ &= \Phi_{max} - \Phi_{max} \cos \omega t \end{aligned} \tag{9.87}$$

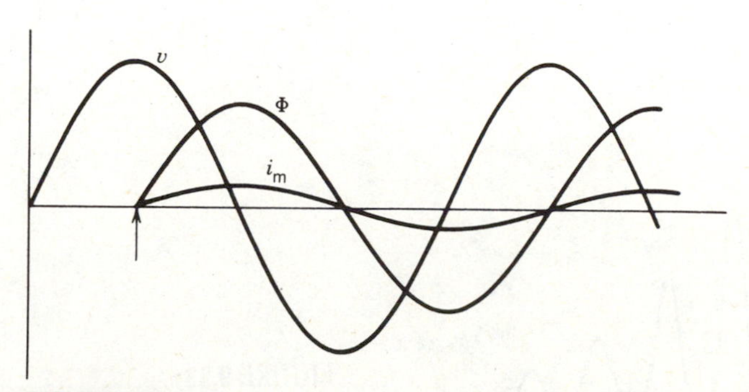

FIGURE 9.21
Transformer inrush current. Transformer connected to supply at the instant of maximum voltage.

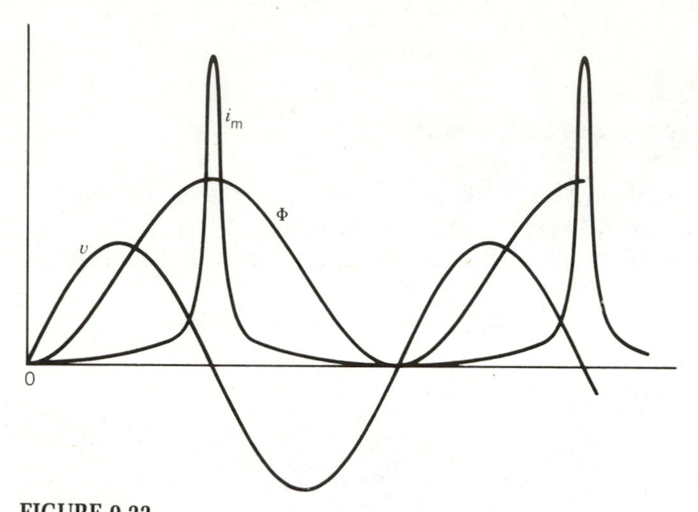

FIGURE 9.22
Transformer inrush current. Transformer connected to supply at the instant of zero voltage.

The time variations of voltage, flux, and magnetizing current are shown in Fig. 9.22. The peak flux has doubled and the corresponding peak magnetizing current is very large because of core saturation.

In practice, because of winding resistance, the large inrush current will decay rapidly, as shown in Fig. 9.23.

In a three-phase transformer, there is always an inrush current, because even if the voltage is a maximum for one phase at the instant the transformer is connected to the power supply, it is not maximum for the other phases.

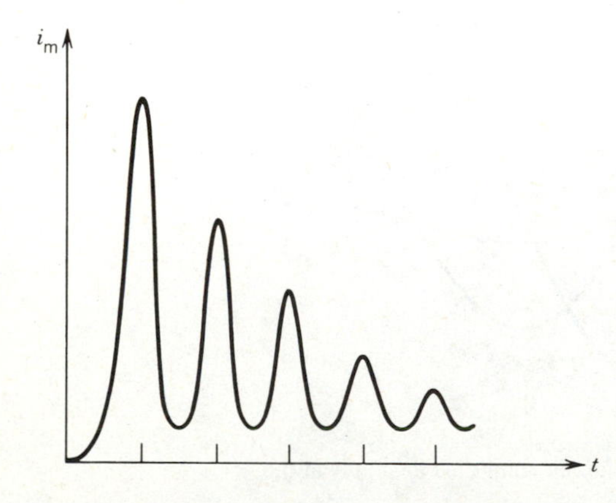

FIGURE 9.23
Effect of winding resistance on transformer inrush current.

PROBLEMS

9.1 A separately excited dc generator has the following parameters:

$$R_f = 110\ \Omega, \qquad L_f = 40\ \text{H}, \qquad R_a = 0.2\ \Omega, \qquad L_{aq} = 10\ \text{mH}$$

$$K_g = 100\ \text{V/field ampere at 1000 rpm}$$

The generator is driven at the rated speed of 1200 rpm, and the field current is adjusted at 2 A. The armature is then suddenly connected to a load consisting of a resistance of 1.8 ohms and an inductance of 10 mH connected in series.

(a) Determine the load terminal voltage as a function of time.

(b) Determine the steady-state value of the load terminal voltage.

(c) Determine the torque as a function of time.

9.2 A separately excited dc motor has the following parameters:

$$R_a = 0.4\ \Omega, \qquad L_{aq} \simeq 0, \qquad K_m = 2\ \text{V/rad/sec}$$

The motor is connected to a load whose torque is proportional to the speed.

$$J = J_{motor} + J_{load} = 2.5\ \text{kg} \cdot \text{m}^2$$

$$B = B_{motor} + B_{load} = 0.25\ \text{kg} \cdot \text{m}^2/\text{sec}$$

The field current is maintained constant at its rated value. A voltage $V_t = 200$ V is suddenly applied across the motor armature terminals.

(a) Obtain an expression for the motor speed as a function of time.

(b) Determine the steady-state speed.

(c) Determine the time required for the motor to reach 95 percent of the steady-state speed.

9.3 A separately excited dc motor has the following parameters.

$$R_a = 0.5\ \Omega, \quad L_{aq} = 0, \quad B = 0, \quad J = 0.1\ \text{kg} \cdot \text{m}^2$$

The rotational loss is negligible.

The motor is used to drive an inertia load of 1.0 kg · m². With the rated field current and an armature terminal voltage of 100 V the motor and the load have a steady-state speed of 1500 rpm. At a certain time the armature terminal voltage is suddenly increased to 120 V.

(a) Obtain an expression for the speed of the motor–load system as a function of time.

(b) Determine the speed 1 second after the step increase in the terminal voltage.

(c) Determine the final steady-state speed of the motor.

9.4 A separately excited dc motor has the following parameters.

$$R_a = 0.4\ \Omega, \qquad L_{aq} = 0, \qquad K_f = 1$$
$$B = 0, \qquad J = 4.5\ \text{kg} \cdot \text{m}^2$$

The motor operates at no load with $V_t = 220$ V and $I_f = 2$ A. Rotational losses are negligible.

The motor is intended to be stopped by plugging, that is, by reversal of its armature terminal voltage ($V_t = -220$ V).

(a) Determine the no-load speed of the motor.

(b) Obtain an expression for the motor speed after plugging.

(c) Determine the time taken for the motor to reach zero speed.

9.5 A motor–generator set consists of a dc generator and a dc motor whose armatures are connected in series. The generator is driven at the rated speed and the motor field current is kept constant at its rated value. The machines have the following parameters.

Generator	**Motor**
$R_a = 0.3\ \Omega$	$R_a = 0.6\ \Omega$
$K_g = K_f\omega_m = 100$ V/A	$K_m = K_f I_f = 1.1$ N · m/A

$$R_f = 100\ \Omega$$
$$L_f = 50\ \text{H}$$

The rotational losses and the armature inductances are negligible. The motor is coupled to an inertia load and the combined inertia of the motor and load is $J = 1.75$ kg · m².

Derive an expression for the motor speed subsequent to the application of a step voltage of 50 V to the generator field circuit.

9.6 A 3ϕ, 30 MVA, 13.8 kV, 60 Hz synchronous machine has the following parameters:

$$X_d = 0.8\ \text{pu}, \qquad X'_d = 0.35\ \text{pu}, \qquad X''_d = 0.2\ \text{pu}$$
$$T'_{do} = 2.5\ \text{sec}, \qquad T''_{do} = 0.07\ \text{sec}, \qquad T_a = 0.25\ \text{sec}$$

The synchronous machine is rotated at the rated speed and its field current is adjusted to produce an open-circuit voltage of 1.0 pu. A 3ϕ short circuit is suddenly applied across the machine terminals.

(a) Determine the maximum possible dc offset current.

(b) Derive an expression for the short-circuit current in a phase as a function of time. Consider the maximum possible dc offset current in the phase.

(c) The fault is cleared at $t = 0.2$ sec. Determine the rms value of the fault current at the instant of clearing the fault (use the result of part b).

9.7 A 3ϕ, 500 MVA, 23 kV, 60 Hz, salient pole synchronous machine has the following reactances:

$$X_d = 1.0 \text{ pu}, \qquad X'_d = 0.4 \text{ pu}, \qquad X''_d = 0.3 \text{ pu}$$

The armature resistance is negligibly small. The machine delivers rated MVA to the infinite bus at unity power factor. A sudden three-phase short circuit occurs at the machine terminals.

(a) Determine the maximum possible rms ac component of fault current from the machine.

(b) Determine the maximum possible dc component of fault current from the machine.

(c) Determine the maximum possible fault current (rms) from the machine.

(d) If $T'_d = 1.5$ sec and $T''_d = 0.03$ sec, $T_a = 0.2$ sec, determine the rms value of the fault current at 0.5 sec after the fault.

9.8 A 3ϕ, 200 MVA, 23 kV, 60 Hz synchronous generator has the following parameters: $X_d = 1.2$ pu, $X'_d = 0.4$ pu, $X''_d = 0.25$ pu, $T'_d = 1.2$ sec, $T''_d = 0.025$ sec, $T_a = 0.15$ sec. The generator delivers rated power to an infinite bus at unity power factor. A three-phase short circuit suddenly occurs at the machine terminals.

(a) Determine the rms value of the short-circuit current as a function of time. Neglect the dc offset current.

(b) Determine the rms value of the short circuit current as $t = 0$ (i.e., at the instant the short circuit occurs), $t = 0.5$ sec, and $t = 5$ sec.

9.9 The synchronous machine in Problem 9.8 is operated as a motor and delivers rated load torque. The field current is adjusted to make the input power factor unity. Determine how much additional load torque can be added suddenly without losing synchronism. Can the new load torque be sustained for long?

CHAPTER ten

POWER SEMICONDUCTOR CONVERTERS

The configurations of basic electric machines (dc, induction, and synchronous) discussed in earlier chapters have remained essentially the same for the past several decades and will most likely remain so for many years in the future. However, the techniques for controlling these machines have recently changed in a significant way. For example, series dc motors are used to propel subway cars. The speed of these cars has been controlled for many years by inserting resistances in series with the dc motors as shown in Fig. 10.1*a*. In recent years solid-state choppers (which can convert a fixed voltage dc into a variable voltage dc) have been used for this purpose, as shown in Fig. 10.1*b*. Solid-state control provides smoother control and higher efficiency. Other electric machines can also be controlled by using the appropriate converters. The following are the various types of converters that are frequently used to control electric machines.

AC Voltage Controller (AC to AC). An ac voltage controller converts a fixed voltage ac to a variable voltage ac. It can be used to control the speed of an induction motor (voltage control method) and for smooth induction motor starting.

Controlled Rectifier (AC to DC). A controlled rectifier converts a fixed voltage ac to a variable voltage dc. It is used primarily to control the speed of dc motors, such as those used in rolling mills.

Chopper (DC to DC). A chopper converts a fixed voltage dc to a variable voltage dc. It is used primarily to control the speed of dc motors.

Inverter (DC to AC). An inverter converts a fixed voltage dc to a fixed (or variable) voltage ac with variable frequency. It can be used to control the speed of ac motors.

Cycloconverter (AC to AC). A cycloconverter converts a fixed voltage and fixed frequency ac to a variable voltage

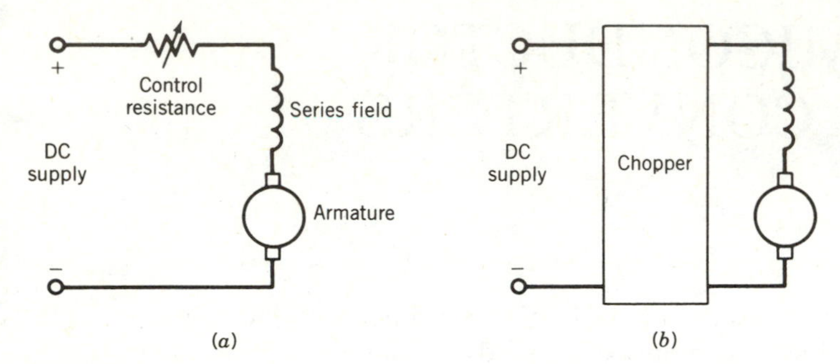

FIGURE 10.1
Speed control of dc series motors. (a) Resistance control. (b) Chopper control.

and variable (lower) frequency ac. It can be used to control the speed of ac motors.

In this chapter the input–output characteristics of these converters are discussed. High-power semiconductor devices are used in these converters to function as on–off switches. The characteristics of these devices are discussed first.

10.1 POWER SEMICONDUCTOR DEVICES

The power semiconductor devices that are generally used in converters can be grouped as follows:

- Thyristors (SCR)
- Power transistors
- Diode rectifiers

These devices are operated in the switching mode so that losses are reduced and conversion efficiency is improved. The disadvantage of switching mode operation is the generation of harmonics and radio-frequency interference (RFI). In this section the external electrical characteristics of these devices are discussed briefly. No attempt is made to describe the physics of operation of these devices and methods for fabricating them, which are well covered in the literature.

10.1.1 THYRISTOR (SCR)

The thyristor, also known as a silicon-controlled rectifier (SCR), has been widely used in industry for more than two decades for power conversion and control. The thyristor has a four-layer p–n–p–n structure with three terminals, anode (A), cathode (K), and gate (G), as shown in Fig. 10.2. The anode and cathode are connected to the main power circuit. The gate terminal carries a low-level gate current in the direction from gate to cathode. The thyristor operates in two stable states: on or off.

Volt–Ampere Characteristics

The terminal volt–ampere characteristics of a thyristor are shown in Fig. 10.3. With zero gate current ($i_g = 0$), if a forward voltage is applied across the device (i.e., anode positive with respect to cathode) junctions J_1 and J_3 are forward biased while junction J_2 remains reverse biased, and therefore the anode current is a small leakage current. If the anode-to-cathode forward voltage reaches a critical limit, called the *forward breakover voltage*, the device switches into high conduction. If gate currents are applied, the forward breakover voltage is reduced. For a sufficiently high gate current, such as i_{g3}, the entire forward blocking region is removed and the device behaves as a diode. When the device is conducting, the gate current can be removed and the device remains in the on state. If the anode current falls below a critical limit, called the *holding current* I_h, the device returns to its forward blocking state.

If a reverse voltage is applied across the device (i.e., anode negative with respect to cathode), the outer junctions J_1 and J_3 are reverse biased and the central junction J_2 is forward biased. Therefore only a small leakage current flows. If the reverse voltage is increased, then at a critical breakdown level (known as the *reverse breakdown voltage*), an avalanche will occur at J_1 and J_3 and the current will increase sharply. If this current is not limited to a safe value, power dissipation will increase to a dangerous level that will destroy the device.

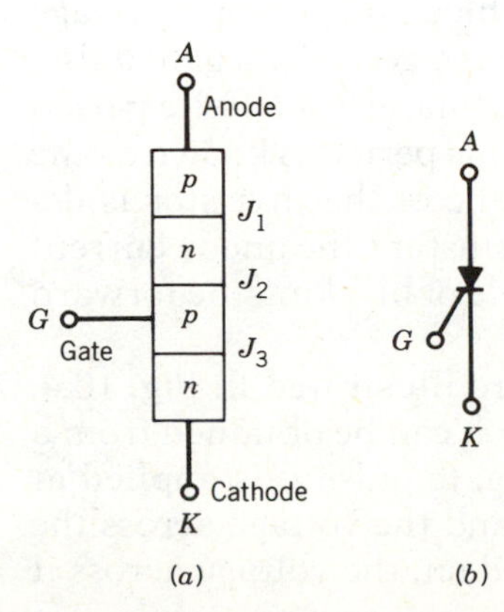

FIGURE 10.2 Thyristor (SCR). (*a*) Structure. (*b*) Symbol.

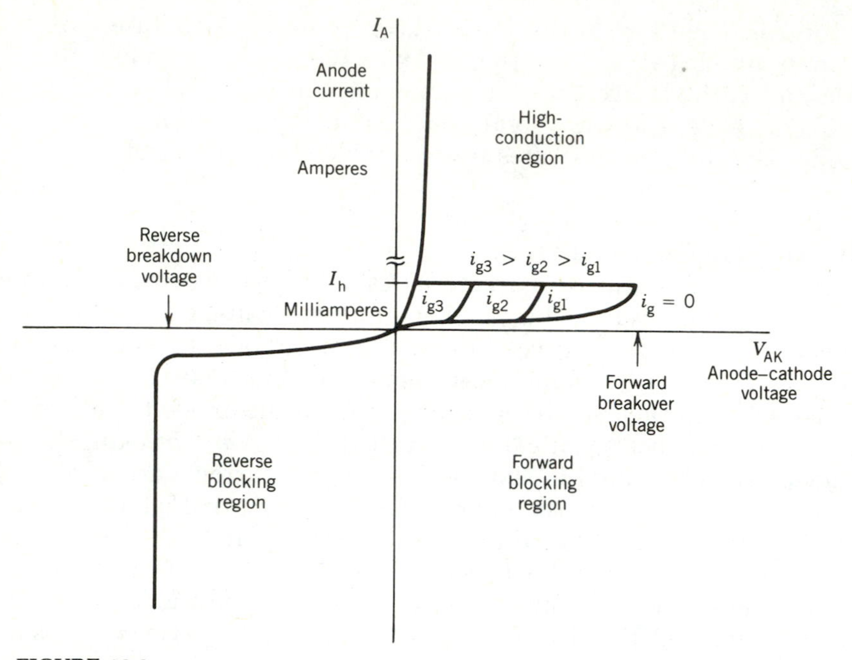

FIGURE 10.3
Terminal volt–ampere characteristics of a thyristor (SCR).

Switching Characteristics

If a thyristor is forward biased and a gate pulse is applied, the thyristor switches on. However, once the thyristor starts conducting an appreciable forward current, the gate has no control on the device. The thyristor will turn off if the anode current becomes zero, called *natural commutation*, or is forced to become zero, called *forced commutation*.

However, if a forward voltage is applied immediately after the anode current is reduced to zero, the thyristor will not block the forward voltage and will start conducting again although it is not triggered by a gate pulse. It is therefore necessary to keep the device reverse biased for a finite period before a forward anode voltage can be applied. This period is known as the turnoff time, t_{off}, of the thyristor. The turnoff time of the thyristor is defined as the minimum time interval between the instant the anode current becomes zero and the instant the device is capable of blocking the forward voltage.

The switching characteristics of a thyristor are illustrated in Fig. 10.4. The thyristor is turned on by a gate pulse i_g, which can be obtained from a firing circuit as shown in Fig. 10.5*a*. When the gate pulse i_g is applied at instant t_1 (Fig. 10.4) anode current I_A builds up and the voltage across the device (V_{AK}) falls. When the device is fully turned on, the voltage across it

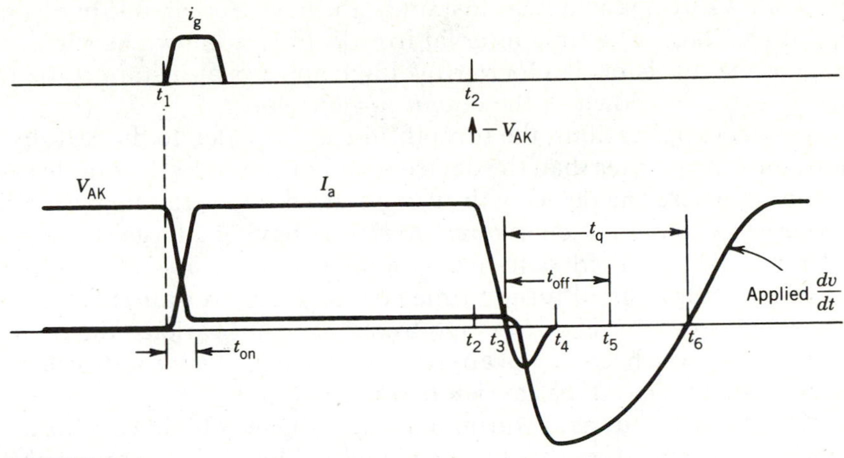

FIGURE 10.4
Switching characteristics of a thyristor (SCR).

is quite small (typically 1 to 2.5 V, the higher voltage drop for higher-current devices) and for all practical purposes the device behaves as a short circuit. The device switches on very quickly, the turn-on time t_{on} typically being 1 to 3 μsec. Typically, the width of the gate pulse is in the range 10 to 50 μsec and its amplitude in the range 20 to 200 mA.

If the current through the thyristor is required to be switched off at a desired instant t_2 (Fig. 10.4), it is momentarily reverse biased by making the cathode positive with respect to the anode (i.e., V_{AK} is negative). For this forced commutation, a commutation circuit as shown in Fig. 10.5*b* is required. In most commutation circuits a precharged capacitor is momentarily connected across the conducting thyristor to reverse-bias it. If the device is reverse biased, its current falls, becomes zero at t_3, then reverses,

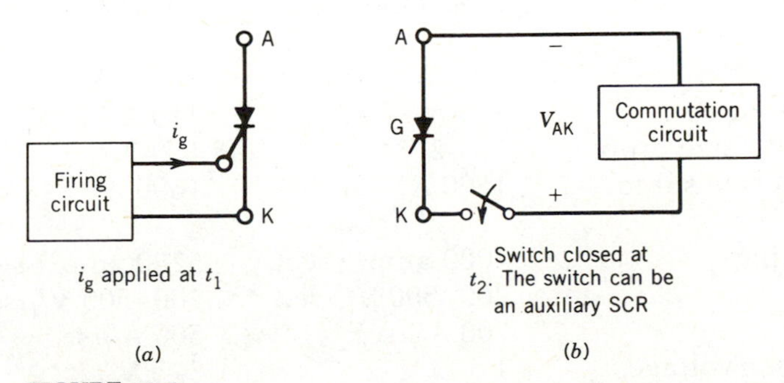

FIGURE 10.5
Thyristor turn-on and turn-off. (*a*) Turn-on by i_g. (*b*) Turn-off by applying $-V_{AK}$.

and becomes zero again at t_4. At instant t_5, the device is capable of blocking a forward voltage. The time interval from t_3 to t_5 is known as the *turnoff time* t_{off} of the thyristor. If a forward voltage appears at instant t_6 the time interval t_3 to t_6 is known as the *circuit turnoff time,* t_q.

In practical applications, the turnoff time t_q, provided to the SCR by the circuit, must be greater than the device turnoff time t_{off} by a suitable safety margin; otherwise the device will turn on at an undesired instant, a process known as *commutation failure.* Thyristors having a large turnoff time (50–100 μsec) are called *slow-switching or phase control-type* thyristors, and those having a small turnoff time (10–50 μsec) are called *fast-switching or inverter-type* thyristors. In high-frequency applications, the required circuit turnoff time becomes an appreciable portion of the total cycle time, and therefore inverter-type thyristors must be used.

Note that during thyristor turn-on, if the voltage is high, current is low and vice versa. Therefore the turn-on switching loss is low. During thyristor turnoff also, if the reverse current is small, the turnoff switching loss is low. The low switching loss in a thyristor is a significant advantage, particularly for high-frequency applications.

Table 10.1 summarizes a typical data sheet for two types of thyristors. Manufacturers can provide more detailed information on request. Thyris-

TABLE 10.1
Thyristor Data Sheet

	Phase Control Type (Slow Switching) G.E. C150 Series	**Inverter Type (Fast Switching) G.E. C158 Series**
Repetitive peak forward and reverse voltages	500 V (C150E) 800 V (C150N) 1000 V (C150P)	500 (C158E) 800 (C158N) 1000 (C158P)
Reverse and forward leakage current	10–20 mA	10–15 mA
Holding current	20 mA	100 mA
RMS current	100 A	110 A
Forward voltage drop	1.2–2 V	1.2–2 V
Peak one-cycle surge (60 Hz)	1500 A	1600 A
I^2t for fusing	7000 amp² · sec	5200 amp² · sec
dv/dt	200–500 V/μsec	200–500 V/μsec
di/dt	500 A/μsec	500 A/μsec
Gate trigger voltage	1.5–3 V	3–5 V
Gate trigger current	50–200 mA	80–300 mA
Turn-on time	3 μsec	2 μsec
Turnoff time	100 μsec	20 μsec

tors with power ratings as high as 4000 V, 2000 A, 20–100 μsec or 3000 V, 2000 A, 10–20 μsec are presently being used in industry.

Protection

If the current in a thyristor rises at too high a rate, that is, high *di/dt*, the device can be destroyed. Some inductance must be present or inserted in series with the thyristor so that *di/dt* is below a safe limit specified by the manufacturer.

A thyristor may turn on (without any gate pulse) if the forward voltage is applied too quickly. This is known as *dv/dt turn-on* and it may lead to improper operation of the circuit. A simple *R–C* snubber, as shown in Fig. 10.6, is normally used to limit the *dv/dt* of the applied forward voltage.

10.1.2 TRIAC

A triac can be considered as an integration of two SCRs in inverse-parallel. The circuit symbol and volt–ampere characteristics of a triac are shown in Fig. 10.7. When terminal T_1 is positive with respect to terminal T_2 and the device is fired by a positive gate current ($+i_g$), it turns on. Also, when terminal T_2 is positive with respect to terminal T_1 and the device is fired by a negative gate current ($-i_g$), the device turns on.

A triac is frequently used in many low-power applications such as juice makers, blenders, and vacuum cleaners, etc. It is economical and easy to control compared to two SCRs connected antiparallel. However, a triac has a lower *dv/dt* capability and a longer turnoff time. It is not available in high voltage and current ratings.

10.1.3 GTO (GATE-TURN-OFF) THYRISTOR

A GTO thyristor can be turned on by a single pulse of positive gate current (like a thyristor), but in addition it can be turned off by a pulse of negative

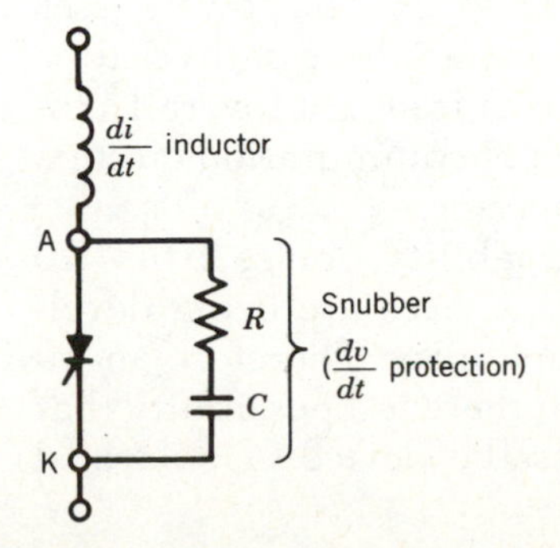

FIGURE 10.6
Thyristor protection for di/dt and dv/dt.

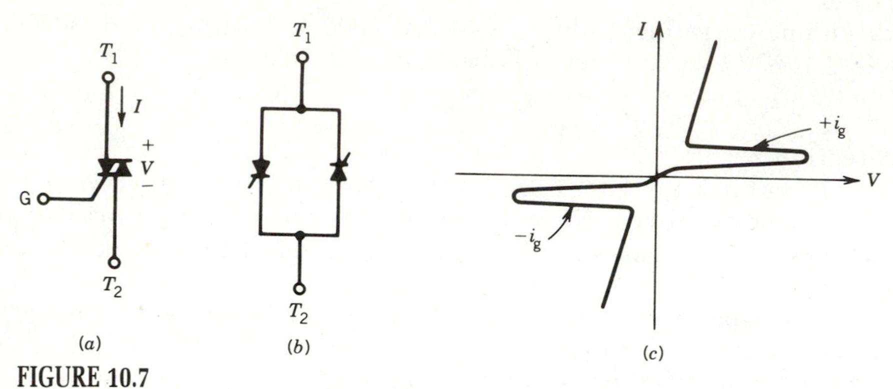

FIGURE 10.7
Triac (a) Symbol. (b) Equivalent circuit. (c) Volt–ampere characteristic.

gate current. Both on-state and off-state operation of the device are therefore controlled by the gate current.

A symbol for the GTO thyristor frequently used in North America is shown in Fig. 10.8*a*. The switching characteristics of the GTO thyristor are shown in Fig. 10.8*b*. The turn-on process is the same as that of a thyristor. The turnoff characteristics are somewhat different. When a negative voltage is applied across the gate and cathode terminals, the gate current i_g rises. When the gate current reaches its maximum value, I_{GR}, the anode current begins to fall, and the voltage across the device, V_{AK}, begins to rise. The fall time of I_A is abrupt, typically less than 1 μsec. Thereafter the anode current changes slowly, and this portion of the anode current is known as the *tail current*.

The ratio (I_A/I_{GR}) of the anode current I_A (prior to turnoff) to the maximum negative gate current I_{GR} required for turnoff is low, typically between 3 and 5. For example, a 2500 V, 1000 A GTO typically requires a peak negative gate current of 250 A for turnoff.

Note that during turnoff both voltage and current are high. Therefore switching losses are somewhat higher in GTO thyristors. Consequently GTOs are restricted to operate at or below a 1 kHz switching frequency. If the spike voltage V_p is large, the device may be destroyed. The power losses in the gate drive circuit are also somewhat higher than those of thyristors. However, since no commutation circuits are required, the overall efficiency of the converter is improved. Elimination of commutation circuits also results in a smaller and less expensive converter.

GTOs may have no reverse-voltage blocking capability, or else little—20 percent of the forward breakover voltage. New devices are being developed having higher reverse-voltage blocking capability. Therefore an inverse diode must be used, as shown in Fig. 10.9, if there is a possibility that appreciable reverse voltage may appear across the device. A polarized

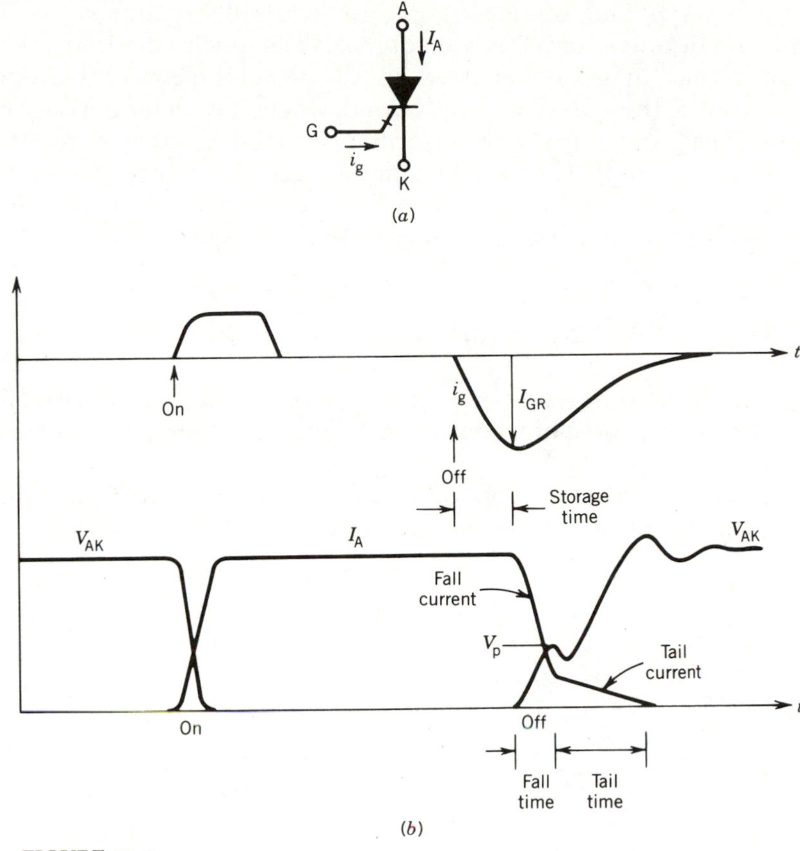

FIGURE 10.8
GTO thyristor. (a) Symbol. (b) Switching characteristics.

snubber consisting of a diode, capacitor, and resistor as shown in Fig. 10.9 is used for the following purposes:

- During the fall time of the turnoff process the device current is diverted (known as current snubbing) to the snubber capacitor (charging it up).
- The snubber limits the dv/dt across the device during turnoff.

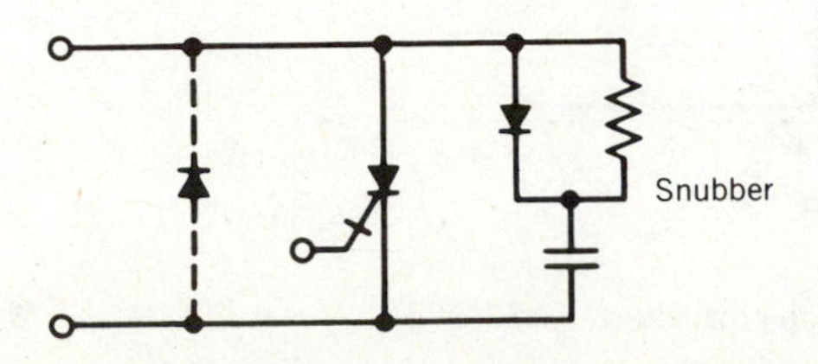

FIGURE 10.9
GTO with antiparallel diode and snubber circuit.

Although GTOs and thyristors became available at almost the same time, the development of GTOs did not receive as much attention as that of thyristors. The Japanese persisted in the development of high-power GTOs. Recently, these devices have been developed with large voltage and current ratings and improved performance (4500 V, 2000 A, 5–10 μsec GTOs are being used). They are becoming increasingly popular in power control equipment, and it is predicted that GTOs will replace thyristors where forced commutation is necessary, as in choppers and inverters.

10.1.4 POWER TRANSISTOR (BJT)

A transistor is a three-layer *p–n–p* or *n–p–n* semiconductor device having two junctions. This type of transistor is known as a bipolar junction transistor (BJT).

The structure and the symbol of an *n–p–n* transistor are shown in Fig. 10.10. The three terminals of the device are called the collector (C), the

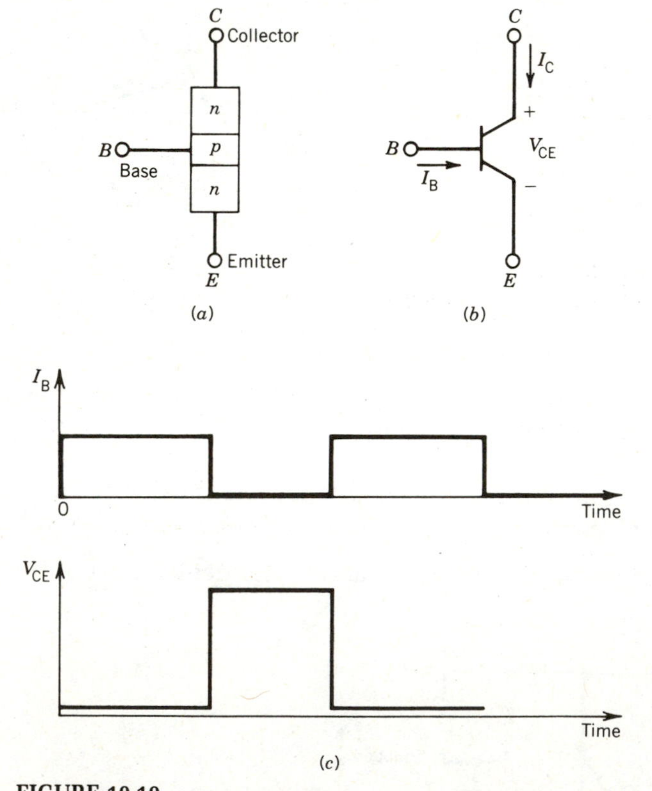

FIGURE 10.10
Transistor. (*a*) Structure. (*b*) Symbol. (*c*) Switching operation of a transistor.

base (B), and the emitter (E). The collector and emitter terminals are connected to the main power circuit, and the base terminal is connected to a control signal.

Like thyristors, transistors can also be operated in the switching mode. If the base current I_B is zero, the transistor is in an off state and behaves as an open switch. On the other hand, if the base is driven hard, that is, if the base current I_B is sufficient to drive the transistor into saturation, then the transistor behaves as a closed switch. This type of operation is illustrated in Fig. 10.10*c*.

The transistor is a current-driven device. The base current determines whether it is in the on state or the off state. To keep the device in the on state there must be sufficient base current.

Transistors with high voltage and current ratings are known as *power transistors*. The current gain (I_C/I_B) of a power transistor can be as low as 10, although it is higher than that of a GTO thyristor. For example, a base current of 10 amperes may be required for 100 amperes of collector current. High current gain can be obtained from a Darlington connected transistor pair, as shown in Fig. 10.11. The pair can be fabricated on one chip, or two discrete transistors can be physically connected as a Darlington transistor. Current gains in the hundreds can be obtained in a high-power Darlington transistor. Power transistors switch on and switch off much faster than thyristors. They may switch on in less than 1 μsec and turn off in less than 2 μsec. Therefore, power transistors can be used in applications where the frequency is as high as 100 kHz. These devices are, however, very delicate. They fail under certain high-voltage and high-current conditions. They should be operated within specified limits, known as the safe operating are (SOA).

The SOA is partitioned into four regions, as shown in Fig. 10.12, defined by the following limits:

- Peak current limit (*ab*)
- Power dissipation limit (*bc*)
- Secondary breakdown limit (*cd*)
- Peak voltage limit (*de*)

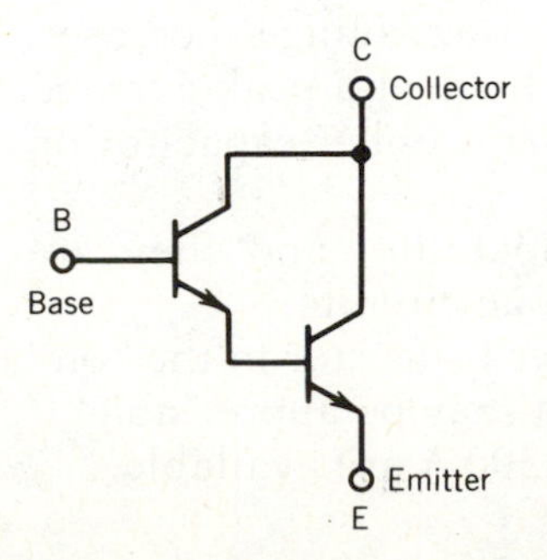

FIGURE 10.11
Darlington transistor.

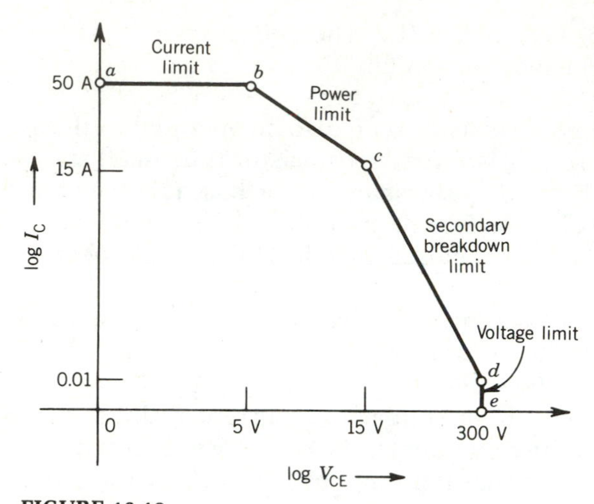

FIGURE 10.12
Safe operating area (SOA) of a power transistor.

If high voltage and high current occur simultaneously during turnoff, a hot spot is formed and the device fails by thermal runaway, a phenomenon known as *secondary breakdown.*

Polarized snubbers are used with power transistors to avoid the simultaneous occurrence of peak voltage and peak current. Figure 10.13 shows the effects of the snubber circuit on the turnoff characteristics of a power transistor. A chopper circuit with an inductive load is considered.

If no snubber circuit is used and the base current is removed to turn off the transistor, the voltage across the device, V_{CE}, first rises, and when it reaches the dc supply voltage (V_d) the collector current (I_C) falls. The power dissipation (P) during the turnoff interval is also shown in Fig. 10.13 by the dashed line. Note that in these idealized waveforms, the peaks of V_{CE} and I_C occur simultaneously, and this may lead to secondary breakdown failure.

If the snubber circuit is used and base current is removed to turn off the transistor, the collector current is diverted to the capacitor. The collector current, therefore, decreases as the collector–emitter voltage increases, avoiding the simultaneous occurrence of peak voltage and peak current. Figure 10.13 also shows the effect of the size of the snubber capacitor on the turnoff characteristics.

Transistors do not have reverse blocking capability, and they are shunted by antiparallel diodes if they are used in ac circuits.

Because base current is required to keep a power transistor in the "on" condition, the power loss in the base drive circuit may be appreciable.

Power transistors of ratings as high as 1000 V, 500 A are available.

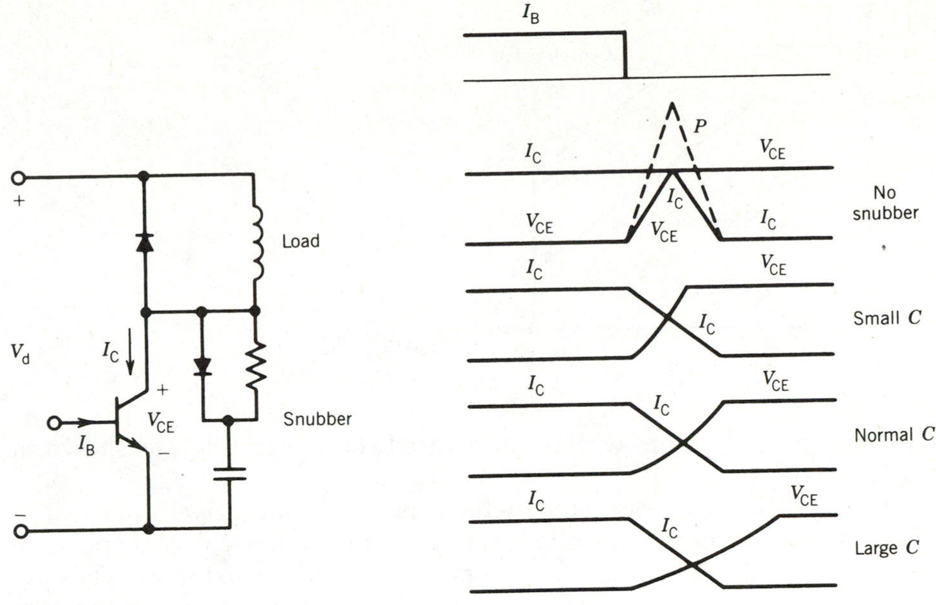

FIGURE 10.13
Effects of snubber capacitor on turn-off characteristics of a power transistor.

10.1.5 POWER MOSFET

The MOSFET (metal oxide semiconductor field effect transistor) is a very fast switching transistor that has shown great promise for applications involving high frequency (up to 1 MHz) and low power (up to a few kilowatts). There are other trade names for this device, such as HEXFET (International Rectifier), SIMMOS (Siemens), and TIMOS (Motorola).

The circuit symbol of the MOSFET is shown in Fig. 10.14. The three terminals are called drain (D), source (S), and gate (G). The current flow is from drain to source. The device has no reverse-voltage blocking capabil-

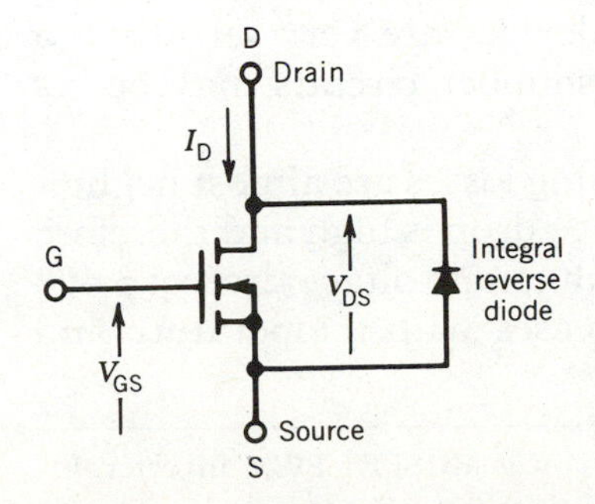

FIGURE 10.14
Power MOSFET.

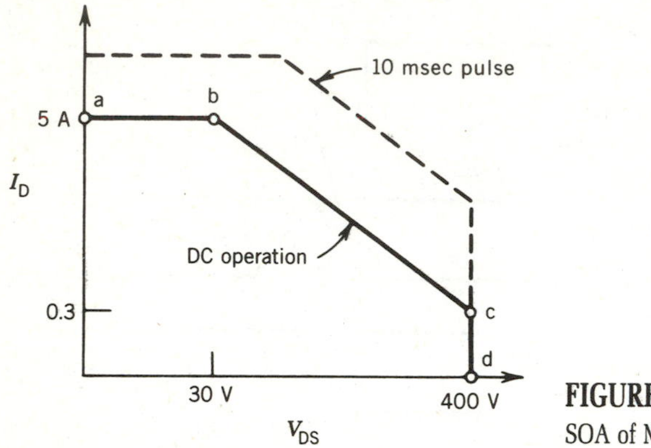

FIGURE 10.15
SOA of MOSFET.

ity and it always comes with an integrated reverse rectifier, as shown in Fig. 10.14.

Unlike a bipolar transistor (which is a current-driven device), a MOSFET is a voltage-controlled majority carrier device. With positive voltage applied to the gate (i.e., V_{GS} positive), the transistor switches on. The gate is isolated by a silicon oxide (SO_2) layer, and therefore the gate circuit input impedance is extremely high. This feature allows a MOSFET to be driven directly from CMOS or TTL logic. The gate drive current is therefore very low—it can be less than 1 milliamp.

The MOSFET has a positive temperature coefficient of resistance and the possibility of secondary breakdown is almost nonexistent. If local heating occurs, the effect of the positive temperature coefficient of resistance forces the local concentrations of current to be distributed over the area, thereby avoiding the creation of local hot spots. The safe operating area of a MOSFET is shown in Fig. 10.15. It is bounded by three limits: the current limit (*ab*), the power dissipation limit (*bc*), and the voltage limit (*cd*). The SOA can be increased for pulse operation of the device, shown dashed in Fig. 10.15.

The switching characteristics of the MOSFET are similar to those of the BJT. However, MOSFETs switch on and off very fast, in less than 50 nanoseconds. Because MOSFETs can switch under high voltage and current conditions (i.e., practically no secondary breakdown), no current snubbing is required during turnoff. However, these devices are very sensitive to voltage spikes appearing across them, and snubber circuits may be required to suppress voltage spikes.[1]

MOSFETs switch very fast and their switching losses are almost negligible. However, conduction (i.e., on-state) voltage drop is high and therefore conduction loss is high. For example, the conduction voltage drop of a 400 V device is 4 V at 10 A, and this drop increases with temperature and current.

---

[1] B. Nair and P. C. Sen, Voltage Clamp Circuits for a Power MOSFET PWM Inverter, IA-IEEE Transactions, Vol. IA-23, No. 5, pp 911–920, 1987.

MOSFETs are still not available in high power ratings. MOSFETs with ratings of 500 V, 10 A, 50 nsec are available. These devices can be used in parallel for higher current ratings.

10.1.6 DIODE

A diode is a two-layer *p–n* semiconductor device. The structure of a diode and its symbol are shown in Fig. 10.16*a* and *b*. High-power diodes are silicon rectifiers that can operate at high junction temperatures.

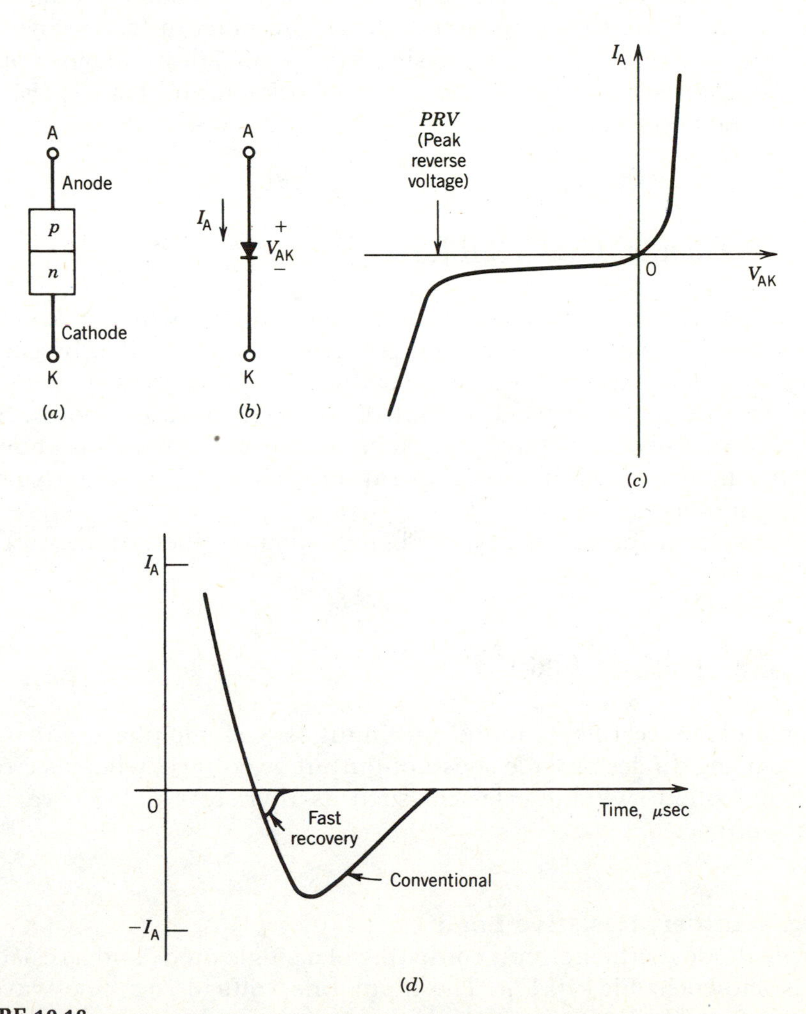

FIGURE 10.16
Diode rectifier. (*a*) Structure. (*b*) Symbol. (*c*) V–I characteristics. (*d*) Reverse recovery characteristics.

The voltage–current characteristic of a diode is shown in Fig. 10.16*c*. If a reverse voltage is applied across the diode, it behaves essentially as an open circuit. If a forward voltage is applied, it starts conducting and behaves essentially as a closed switch. It can provide uncontrolled ac-to-dc power rectification. The forward voltage drop when it conducts current is in the range of 0.8 to 1 V. Diodes with ratings as high as 4000 V and 2000 A are available.

Following the end of forward conduction in a diode, a reverse current flows for a short time. The device does not attain its full blocking capability until the reverse current ceases. The time interval during which reverse current flows is called the rectifier recovery time. During this time, charge carriers stored in the diode at the end of forward conduction are removed. The recovery time is in the range of a few microseconds (1–5 μsec) in a conventional diode to several hundred nanoseconds in fast-recovery diodes. This recovery time is of great significance in high-frequency applications. The recovery characteristics of conventional and fast-recovery diodes are shown in Fig. 10.16*d*.

10.2 CONTROLLED RECTIFIERS

A controlled rectifier converts ac power to dc power, which is known as *rectification*. The output voltage and power can be controlled by controlling the instants at which the semiconductor devices switch. Thus controlled rectifiers can be used to control the speed of a dc motor. Some controlled rectifiers can convert dc power to ac power, which is known as *inversion*. This inversion mode of operation is used for regenerative braking of dc motors.

In this section the various types of controlled rectifier circuits are discussed.

10.2.1 SINGLE-PHASE CIRCUITS

In single-phase rectifier circuits the input is a single-phase ac supply. Circuits using diodes provide constant-output dc voltage, whereas circuits using controlled switching devices such as thyristors provide variable-output voltage.

Diode Rectifier, Resistive Load

A simple diode rectifier circuit consisting of a single diode and a resistance load is shown in Fig. 10.17*a*. The input line voltage is a sine wave, as shown in Fig. 10.17*b*. During the positive half-cycle, that is, $0 < \omega t < \pi$, the

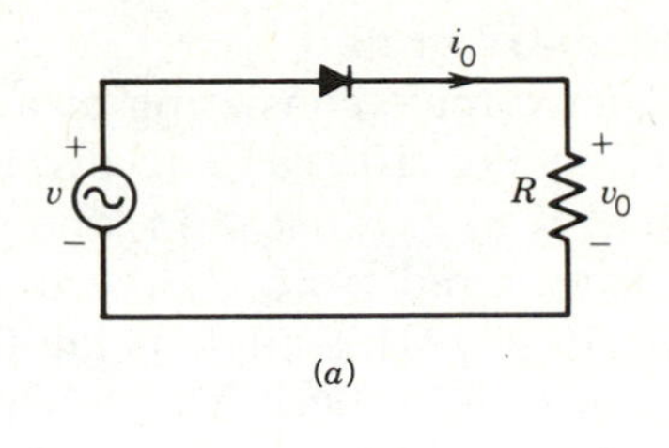

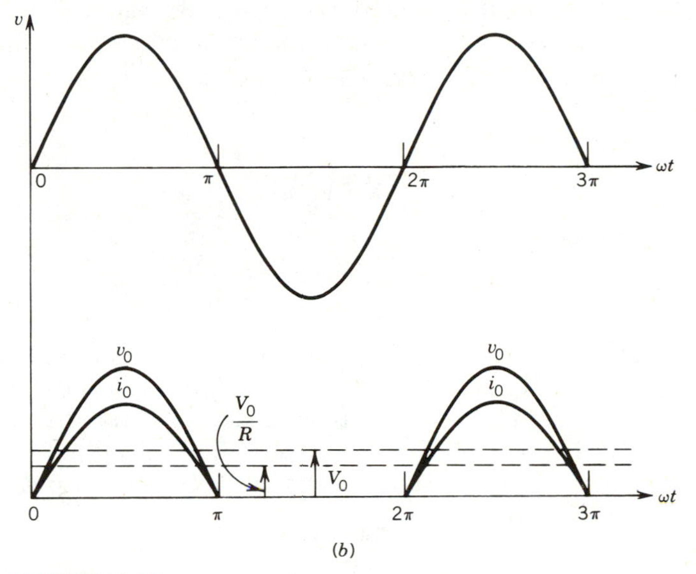

FIGURE 10.17

Single-phase diode rectifier circuit. (a) Circuit. (b) Waveforms.

diode conducts and acts like a closed switch (the diode is assumed to be ideal) connecting the supply to the load (i.e., $v_0 = v$). The load current is $i_0 = v_0/R$ and, because R is a constant, v_0 and i_0 have the same waveforms (Fig. 10.17*b*). During the negative half-cycle the diode acts like an open switch and conducts zero current.

Let $v = \sqrt{2}\, V_p \sin \omega t$. The average value of the load voltage is

$$V_0 = \frac{1}{2\pi} \int_0^{\pi} v_0 \, d(\omega t)$$

$$= \frac{1}{2\pi} \int_0^{\pi} \sqrt{2}\, V_p \sin \omega t \, d(\omega t)$$

$$= \frac{\sqrt{2}\, V_p}{\pi} \tag{10.1}$$

Thyristor Rectifier, Resistive Load

A simple thyristor rectifier circuit consisting of a single thyristor and a resistance load is shown in Fig. 10.18*a*. The thyristor is forward biased during the intervals $0 < \omega t < \pi$, $2\pi < \omega t < 3\pi$, and so forth. A gate pulse is applied at an angle α (measured from the zero crossing of the supply voltage) as shown in Fig. 10.18*b*. This angle is known as the *firing angle* of the thyristor. The thyristor current becomes zero at $\omega t = \pi, 3\pi$, and so on and the thyristor conducts from α to π, $2\pi + \alpha$ to 3π, and so on. During the interval when the thyristor conducts, known as *conduction interval,* the load voltage is the same as the supply voltage, $v_0 = v$. The average value of the load voltage is

$$V_0 = \frac{1}{2\pi}\int_\alpha^\pi \sqrt{2}\, V_p \sin \omega t \, d(\omega t)$$

$$= \frac{V_p}{\sqrt{2}\,\pi}(1 + \cos \alpha) \qquad (10.2)$$

The firing angle α can be changed from zero to π, which will change the output voltage. Note that at $\alpha = 0$, $V_0 = \sqrt{2}\, V_p/\pi$, which is the same as the voltage obtained from the diode rectifier (Eq. 10.1). If the thyristor is fired at $\alpha = 0$, the thyristor circuit behaves like a diode circuit. This reference for the firing angle, by convention, results in the largest output voltage for the thyristor rectifier.

Thyristor Rectifier, Reactive Load

Most practical loads have resistance (R) and inductance (L). For example, the armature of a dc motor load has resistance and inductance. The field circuit of a dc motor is highly inductive. A thyristor rectifier circuit with a load consisting of R and L is shown in Fig. 10.19*a*. The thyristor is fired at a firing angle α, meaning that it starts to conduct at $\omega t = \alpha$. The inductance in the load forces the current to lag the voltage and therefore the current decays to zero at $\omega t = \beta$ instead of $\omega t = \pi$, which would have been the case if the load were purely resistive. The waveform of the load current (i_0) is shown in Fig. 10.19*b*. During the conduction interval (α to β), $v_0 = v$. The waveform of the output voltage (v_0) is also shown in Fig. 10.19*b*.

Thyristor Full Converter

The waveforms of the load current and load voltage in a single-thyristor rectifier circuit, as shown in Fig. 10.19*b*, contain a significant amount of ripple. The single-thyristor rectifier circuit is, therefore, not suitable for speed control of dc motors. A full-converter circuit, shown in Fig. 10.20*a*, consists of four thyristors (S_1 to S_4) and is used for the speed control of dc motors.

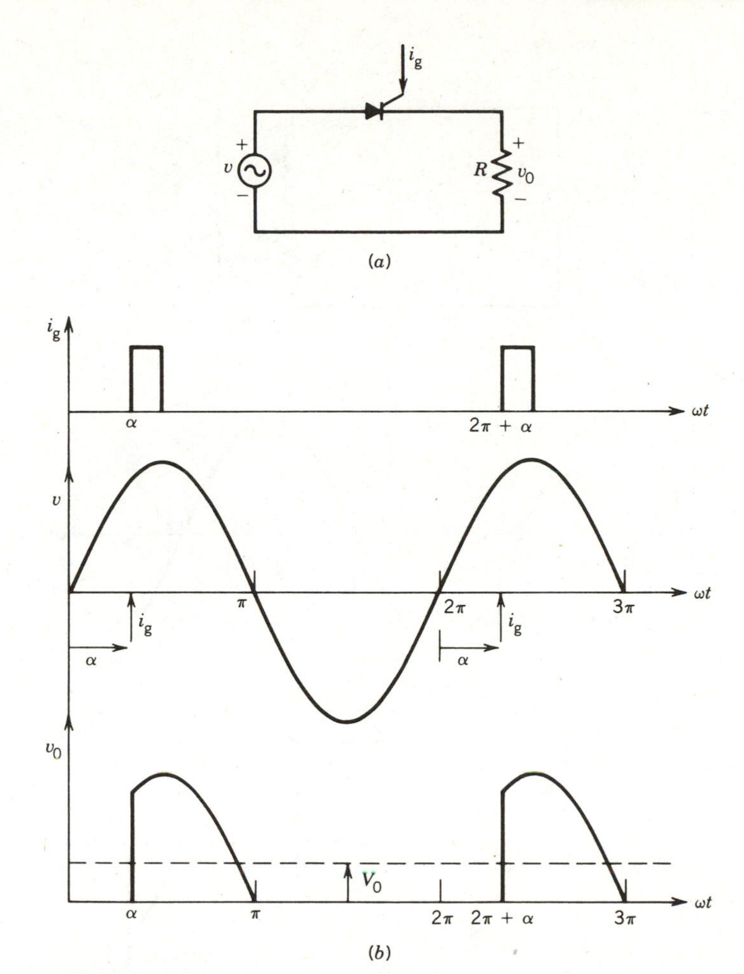

FIGURE 10.18
Thyristor rectifier with resistive load. (a) Circuit. (b) Waveforms.

1. *Resistive load.* Thyristors S_1 and S_2 are forward biased during the positive half-cycle of the input voltage, and thyristors S_3 and S_4 are forward biased during the negative half-cycle. Let S_1 and S_2 be fired at α. During the conduction interval of S_1 and S_2, the output voltage is the same as the input voltage, $v_O = v$. The load current $i_O = v_O/R$ has the same waveform as the load voltage v_O. Therefore the current through S_1 and S_2 becomes zero at $\omega t = \pi$, and they turn off (natural commutation). Thyristors S_3 and S_4 are fired at $\pi + \alpha$. During their conduction interval, the input supply is connected to the load and $v_O = -v$. Current through S_3 and S_4 becomes zero at $\omega t = 2\pi$ and they turn off.

   Thyristors S_1 and S_2 are fired again at $\omega t = 2\pi + \alpha$ and S_3 and S_4 at $\omega t = 3\pi + \alpha$, and the process continues. The resulting load voltage

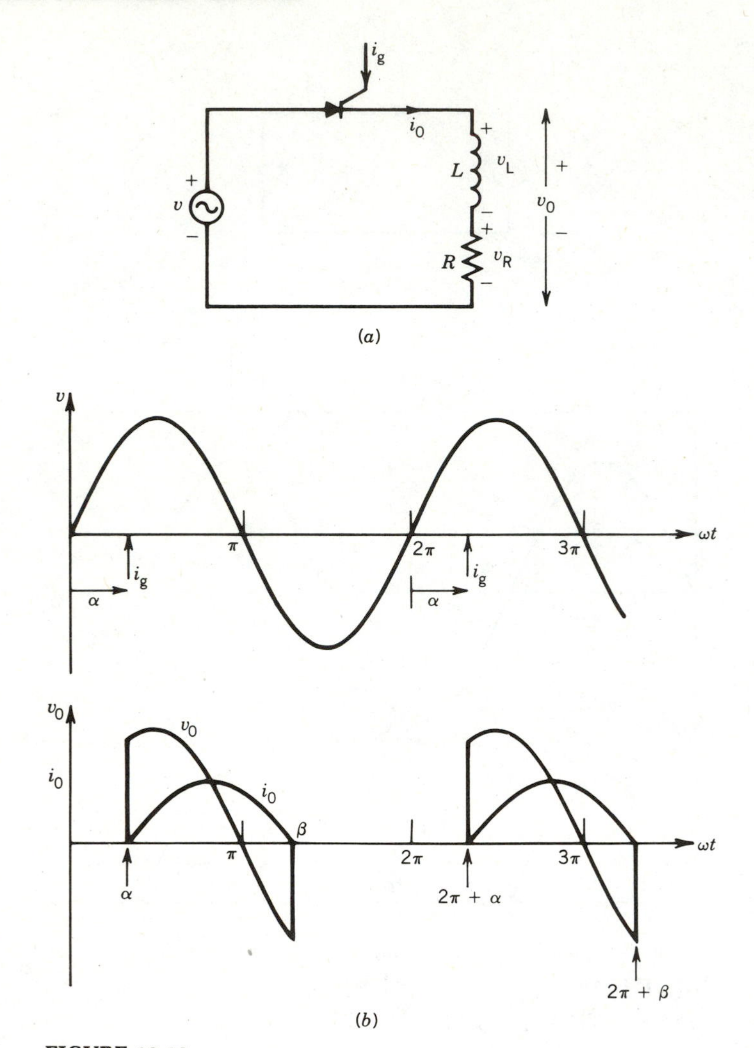

FIGURE 10.19
Thyristor rectifier with inductive load. (a) Circuit. (b) Waveforms.

waveform is shown in Fig. 10.20*b*. Note from the waveforms of Figs. 10.19*b* and 10.20*b* that the full converter has less ripple than the single-thyristor rectifier circuit. The ripple frequency (f_r) and input supply frequency (f_i) are related as follows:

$$f_r = f_i \quad \text{for a single-thyristor rectifier}$$

$$f_r = 2f_i \quad \text{for a full converter}$$

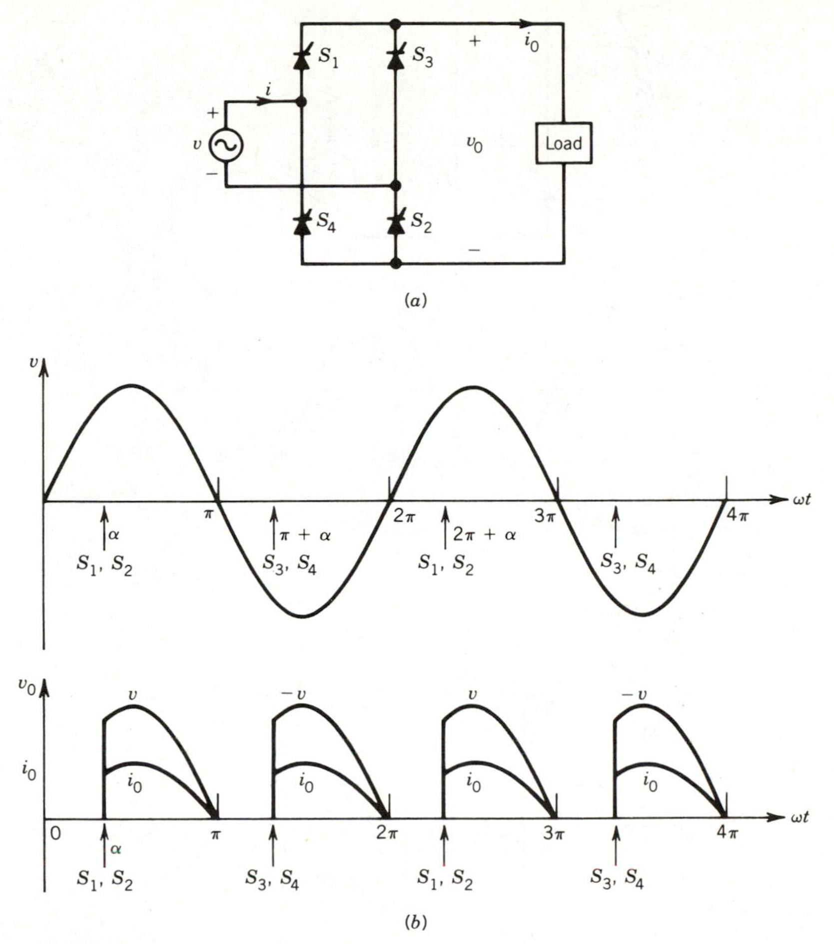

FIGURE 10.20
Thyristor full converter with R load. (a) Circuit. (b) Waveforms.

2. *DC motor load.* A full converter with a dc motor load is shown in Fig. 10.21a. Let us assume that sufficient inductance is present in the dc armature circuit to ensure that motor current is continuous (i.e., present all the time). The motor current i_0 flows from the supply through S_1 and S_2 for one half-cycle and through S_3 and S_4 for the next half-cycle.

   As shown in Fig. 10.21b, thyristors S_1 and S_2 conduct the motor current during the interval $\alpha < \omega t < (\pi + \alpha)$ and connect the motor to the supply ($v_0 = v$). At $\pi + \alpha$, thyristors S_3 and S_4 are fired. The supply voltage appears immediately across thyristors S_1 and S_2 as a reverse-bias voltage and turns them off. This is called *line commutation.* The

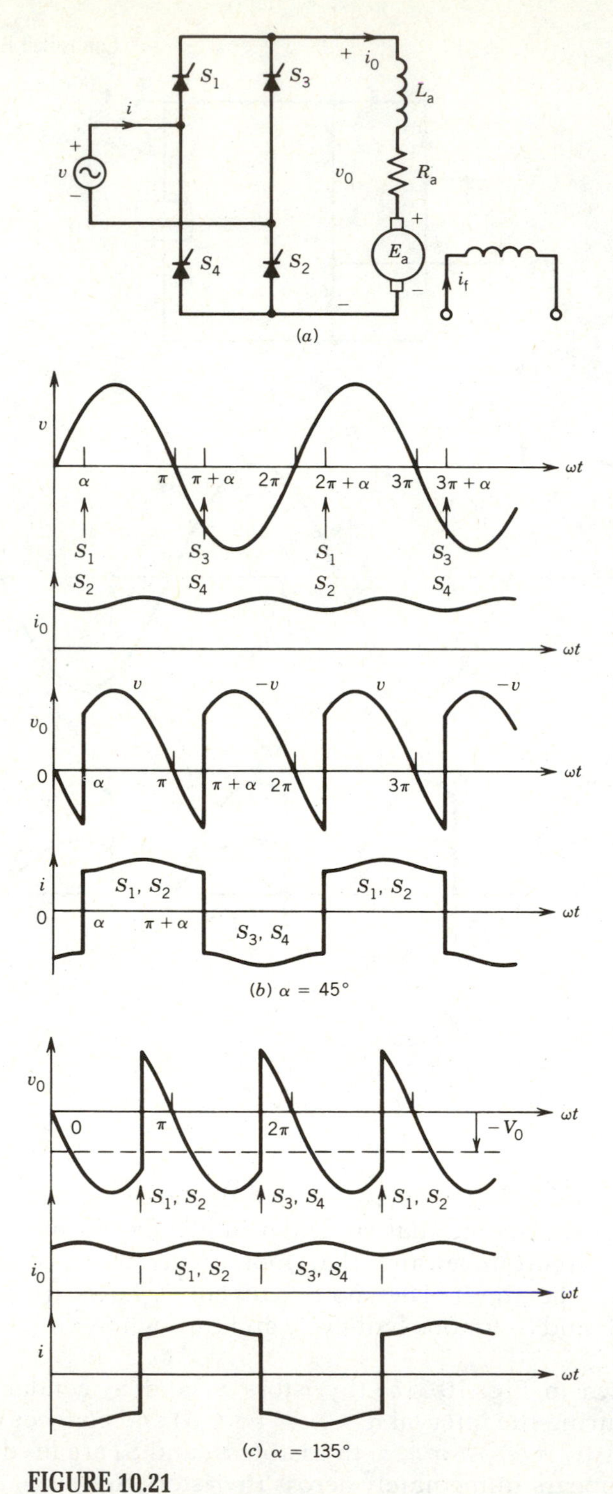

FIGURE 10.21

Thyristor full converter with dc motor load. (*a*) Circuit. (*b*, *c*) Waveforms.

motor current i_a, which was flowing from the supply through S_1 and S_2, is transferred to S_3 and S_4. Thyristors S_3 and S_4 conduct the motor current during the interval $\pi + \alpha < \omega t < 2\pi + \alpha$ and connect the motor to the supply ($v_0 = -v$).

In Fig. 10.21*c* voltage and current waveforms are shown for $\alpha > 90°$. From the waveform of the output voltage, it can be seen that its average value is

$$V_0 = \frac{1}{\pi}\int_{\alpha}^{\pi+\alpha} v_0 \, d(\omega t)$$

$$= \frac{1}{\pi}\int_{\alpha}^{\pi+\alpha} \sqrt{2}\, V_p \sin \omega t \, d(\omega t)$$

$$= \frac{2\sqrt{2}}{\pi} V_p \cos \alpha \tag{10.3}$$

The inductance (L_a) does not sustain any average voltage. Therefore,

$$V_0 = I_0 R_a + E_a \tag{10.4}$$

where I_0 is the average motor current

E_a is the armature back emf, which is constant if the speed and field current are constant

The variation of motor terminal voltage (V_0) as a function of the firing angle (α), based on Eq. 10.3, is shown in Fig. 10.22. For firing angles in the range $0° < \alpha < 90°$, the average output voltage is positive. Since the current can flow only in one direction in the load circuit because of the thyristors, the power ($V_0 I_0$) is positive; that is, power flow is from the input ac supply to the dc machine and the dc machine operates as a motor. For firing angles in the range $90° < \alpha < 180°$, the output voltage is negative and therefore the power ($V_0 I_0$) is negative; that is, power flow is from the dc machine to the ac supply. This is known as *inversion operation* of the converter, and this mode of operation is used for regenerative braking of the motor. Note that for inversion operation, the polarity of the motor back emf (E_a) must be negative. It can be reversed by reversing the field current (i_f) so that the dc machine behaves as a dc generator.

Two full converters can be connected back to back, as shown in Fig. 10.23. This arrangement is known as the *dual-converter* connection. If one converter is used, it causes motor current to flow in one direction. If the other converter is used, the motor current reverses and so does the speed. The dual converter provides virtually instantaneous reversal of current through the dc motor and therefore provides fast reversal of motor speed.

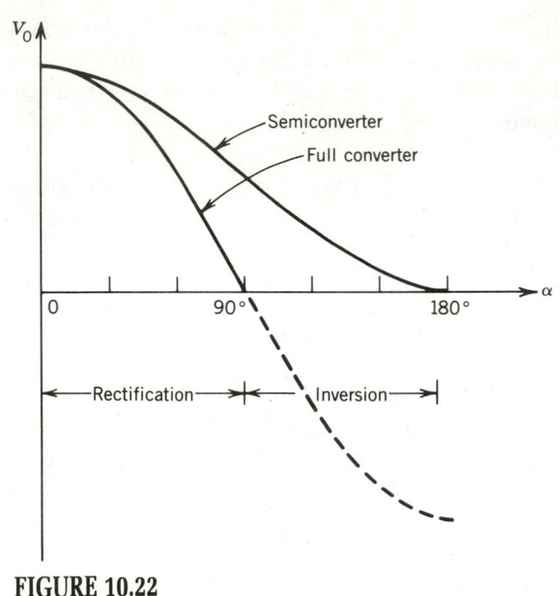

FIGURE 10.22
Converter output characteristics for continuous load current.

Thyristor Semiconverter

A thyristor semiconverter that can be used for speed control of a dc motor consists of thyristors and diodes. A semiconverter consisting of two thyristors and three diodes is shown in Fig. 10.24*a*. Thyristors S_1 and S_2 are fired at α and $\pi + a$, respectively, as shown in Fig. 10.24*b*. The motor is connected to the input supply for the period $\alpha < \omega t < \pi$ through S_1 and D_2, and the load voltage v_0 is the same as the input supply voltage v. Beyond π, v_0 tends to reverse as the input voltage changes polarity. As v_0 tends to reverse, the diode D_{FW} (known as the *freewheeling diode*) becomes forward-biased and starts to conduct. The motor current i_a, which was flowing from the supply through S_1, is transferred to D_{FW} (i.e., S_1 commutates) and freewheels through it. The output terminals are shorted through the freewheeling diode during the interval $\pi < \omega t < (\pi + \alpha)$, making $v_0 = 0$. At $\omega t =$

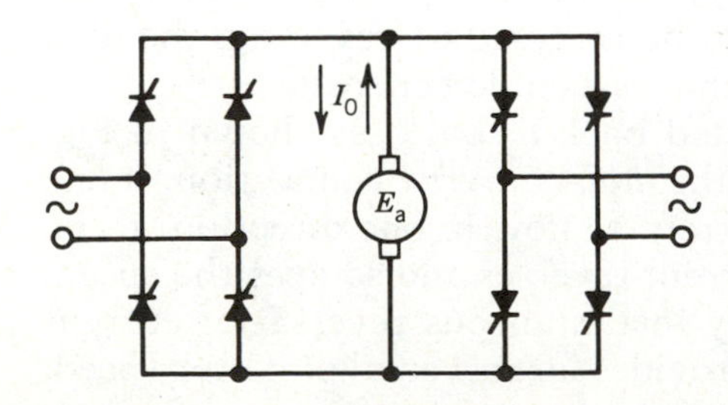

FIGURE 10.23
Dual converter.

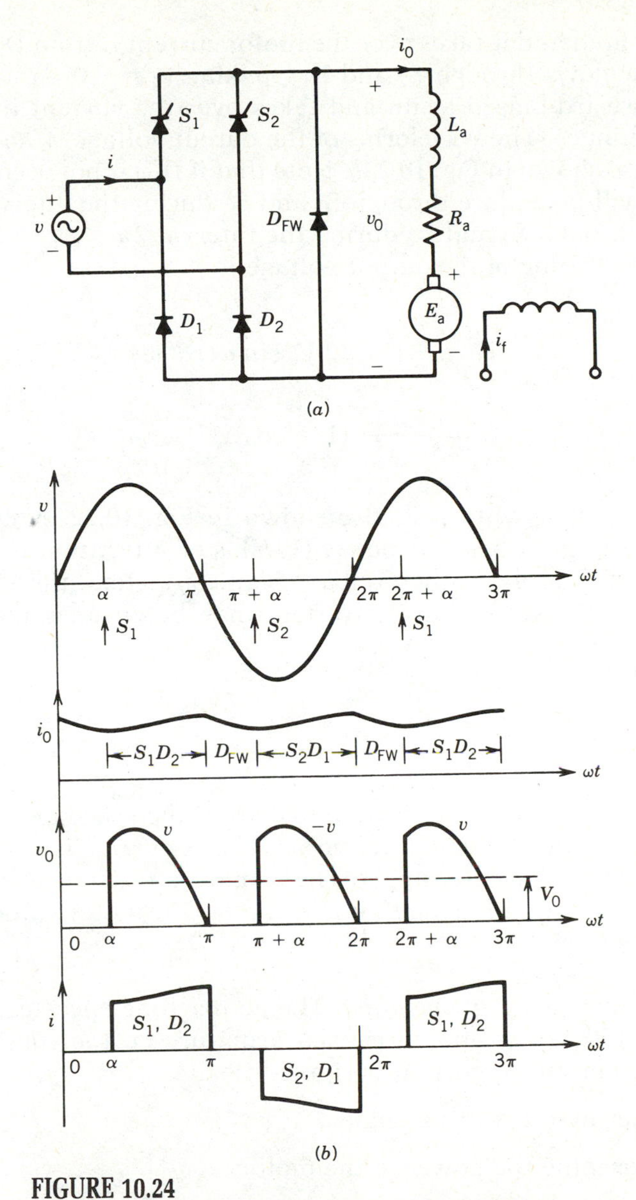

FIGURE 10.24
Thyristor semiconverter with dc motor load. (a) Circuit. (b) Waveforms.

$\pi + \alpha$, S_2 is fired and it takes over the motor current i_a from D_{FW}. The load current now flows through S_2 and D_1, making $v_0 = -v$. At $\omega t = 2\pi$, D_{FW} becomes forward-biased again and takes over the current from S_2. The process continues. The waveforms of the output voltage v_0 and input current i are also shown in Fig. 10.24*b*. Note that if D_{FW} is not used, freewheeling action will take place through S_1 and D_1 during the interval $\pi < \omega t < \pi + \alpha$ and through S_2 and D_2 during the interval $2\pi < \omega t < 2\pi + \alpha$.

The average value of the output voltage is

$$V_0 = \frac{1}{\pi} \int_{\alpha}^{\pi} (\sqrt{2}\, V_p \sin \omega t)\, d(\omega t)$$

$$= \frac{\sqrt{2}\, V_p}{\pi} (1 + \cos \alpha) \tag{10.5}$$

The variation of V_0 with α is also shown in Fig. 10.22. Note that V_0 is always positive and therefore power ($V_0 I_0$) is positive; that is, power flow is from the ac supply to the dc load. Semiconverters, therefore, do not invert power. However, semiconverters may be cheaper than full converters.

EXAMPLE 10.1

A single-phase full converter is used to control the speed of a 5 hp, 110 V, 1200 rpm, separately excited dc motor. The converter is connected to a single-phase 120 V, 60 Hz supply. The armature resistance is $R_a = 0.4\ \Omega$ and armature circuit inductance is $L_a = 5$ mH. The motor voltage constant is $K\Phi = 0.09$ V/rpm.

1. *Rectifier (or motoring) operation.* The dc machine operates as a motor, runs at 1000 rpm, and carries an armature current of 30 amperes. Assume that motor current is ripple-free.
   - (a) Determine the firing angle α.
   - (b) Determine the power to the motor.
   - (c) Determine the supply power factor.
2. *Inverter operation (regenerating action).* The polarity of the motor back emf E_a is reversed, say by reversing the field excitation.
   - (a) Determine the firing angle to keep the motor current at 30 amperes when the speed is 1000 rpm.
   - (b) Determine the power fed back to the supply at 1000 rpm.

Solution

Refer to Fig. 10.21.

1. (a)

$$E_a = 0.09 \times 1000 = 90 \text{ V}$$

$$V_0 = E_a + I_0 R_a = 90 + 30 \times 0.4 = 102 \text{ V}$$

From Eq. 10.3

$$120 = \frac{2\sqrt{2} \times 120}{\pi} \cos \alpha$$

$$\alpha = 19.2°$$

(b)

$$P = I_0^2 R_a + E_a I_0 = V_0 I_0$$

$$= 102 \times 30$$

$$= 3060 \text{ W}$$

(c) The supply current has a square waveform with amplitude 30 A ($= I_0$). The rms supply current is

$$I = 30 \text{ A}$$

The supply volt–amperes are

$$S = VI = 120 \times 30 = 3600 \text{ VA}$$

If losses in the converter are neglected, the power from the supply is the same as the power to the motor.

$$P_s = 3060 \text{ W}$$

Thus, the supply power factor is

$$\text{PF} = \frac{P_s}{S} = \frac{3060}{3600} = 0.85$$

2. (a) At the time of polarity reversal the back emf is

$$E_a = 90 \text{ V}$$

From Eq. 10.4

$$V_0 = E_a + I_0 R_a$$
$$= -90 + 30 \times 0.4$$
$$= -90 + 12$$
$$= -78 \text{ V}$$

Now

$$V_0 = \frac{2\sqrt{2} \times 120}{\pi} \cos \alpha = -78 \text{ V}$$

or $\alpha = 136.2°$.

(b) Power from dc machine:

$$P_{dc} = 90 \times 30 = 2700 \text{ W}$$

Power lost in R_a:

$$P_R = 30^2 \times 0.4 = 360 \text{ W}$$

Power fed back to the ac supply:

$$P_s = 2700 - 360 = 2340 \text{ W}$$

10.2.2 THREE-PHASE CIRCUITS

For high-power applications, several kilowatts or more, it is desirable to use three-phase rectifier circuits. Three-phase rectifier circuits provide better load voltage waveforms. Various circuit configurations are used, and some of the important ones are discussed here.

Half-Wave Diode Rectifier

A simple three-phase, half-wave, diode rectifier circuit consisting of three diodes and a resistance load is shown in Fig. 10.25*a*. The phase voltages of the input three-phase supply are shown in Fig. 10.25*b* and can be expressed as follows:

$$v_{AN} = \sqrt{2}\, V_p \sin \omega t \tag{10.6}$$

$$v_{BN} = \sqrt{2}\, V_p \sin(\omega t - 120°) \tag{10.7}$$

$$v_{CN} = \sqrt{2}\, V_p \sin(\omega t + 120°) \tag{10.8}$$

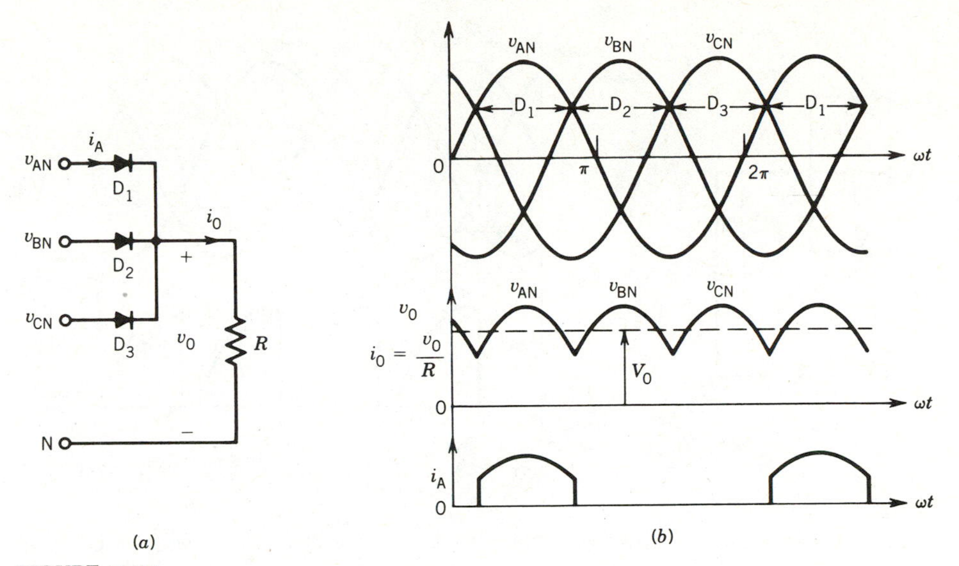

FIGURE 10.25
Three-phase half-wave rectifier circuit with R-load. (a) Circuit. (b) Waveforms.

The diodes conduct in the sequence $D_1, D_2, D_3, D_1, \ldots$. They conduct one at a time for 120° intervals. At any time the diode whose anode is at the highest instantaneous supply voltage will conduct. For example, during the interval $30° < \omega t < 150°$, v_{AN} is higher than both v_{BN} and v_{CN}. Therefore, during this interval, diode D_1 conducts. When D_1 conducts the voltage across D_2 is v_{BA} and that across D_3 is v_{CA}. Diodes D_2 and D_3, therefore, remain reverse-biased. When D_1 conducts, the output load voltage v_0 is the same as the input phase voltage v_{AN}. The conduction interval of the diodes is indicated in Fig. 10.25*b*. The load voltage follows the envelope of the highest instantaneous input supply voltages.

The average value of the output voltage is

$$V_0 = \frac{1}{2\pi/3}\int_{30°}^{150°} \sqrt{2}\, V_p \sin \omega t \, d(\omega t)$$

$$= \frac{3\sqrt{6}}{2\pi} V_p \qquad (10.9)$$

Half-Wave Thyristor-Controlled Rectifier

To control the output voltage, the diodes in Fig. 10.25*a* can be replaced by thyristors, as shown in Fig. 10.26*a*. The circuit of Fig. 10.26*a* will behave like the diode circuit of Fig. 10.25*a* if thyristor S_1 is fired at $\omega t = 30°$, S_2 is

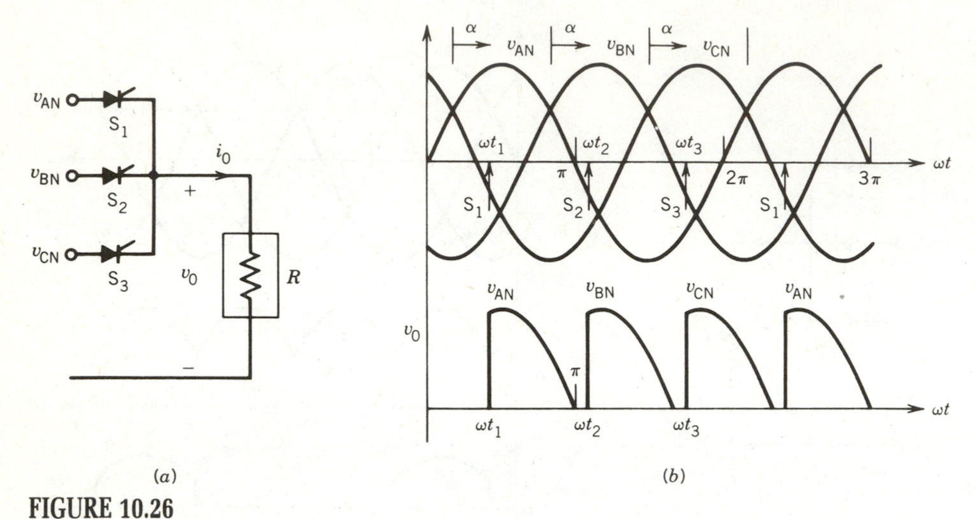

FIGURE 10.26

Three-phase half-wave controlled rectifier circuit with R-load. (*a*) Circuit. (*b*) Waveforms (*R* load).

fired at $\omega t = 150°$, and S_3 is fired at 270°, that is, at the crossing points of the phase voltages. Firings at these instants will also result in maximum output voltage. The reference for the firing angle α is therefore the crossing point of the phase voltages. The firing of the thyristors can be delayed from these crossing points. In other words, the firing angle is measured from the crossing points of the phase voltages. Recall that in single-phase converters, the firing angle was measured from the zero crossing of the input supply voltage.

For a particular firing angle α, the thyristors are fired at the instants (ωt_1, ωt_2, ωt_3) shown in Fig. 10.26*b*. Thyristor S_1 is fired at firing angle α (i.e., at $\omega t = \omega t_1 = 30° + \alpha$) and the output voltage $v_0 = v_{AN}$. The output current i_0 is V_0/R, and becomes zero at $\omega t = \pi$. Thyristor S_1 turns off at this instant. Thyristor S_2 is fired at ωt_2, making $v_0 = v_{BN}$. Thyristor S_2 turns off at $\pi + 2\pi/3$. Thyristor S_3 is fired at ωt_3, making $v_0 = v_{CN}$. The waveform of the output voltage v_0 is shown in Fig. 10.26*b*.

EXAMPLE 10.2

The load in Fig. 10.26*a* consists of a resistance and a very large inductance. The inductance is so large that the output current i_0 can be assumed to be continuous and ripple-free. For $\alpha = 60°$,

(a) Draw the waveforms of v_0 and i_0.

(b) Determine the average value of the output voltage, if phase voltage $V_p = 120$ V.

Solution

(a) The supply voltages and the firing instants of the thyristors are shown in Fig. E10.2*a*. The output current i_0 is constant and is shown in Fig. E10.2*b*.

During the interval $(30° + \alpha) < \omega t < (30° + \alpha + 120°)$ thyristor S_1 conducts the load current and therefore during this interval $v_0 = v_{AN}$, as shown in Fig. E10.2*c*. Similarly, $v_0 = v_{BN}$ when S_2 conducts and $v_0 = v_{CN}$ when S_3 conducts. The output voltage waveform v_0 is shown in Fig. E10.2*c*.

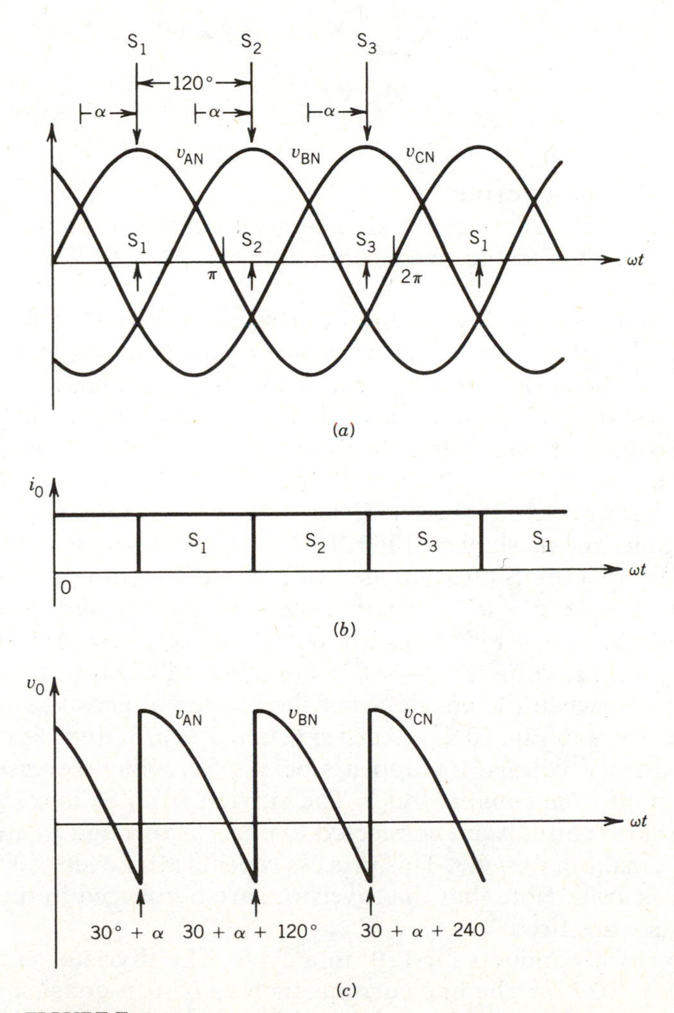

FIGURE E10.2

(b)

$$V_0 = \frac{1}{2\pi/3} \int_{30°+\alpha}^{30°+\alpha+120°} v_{AN}\, d(\omega t)$$

$$= \frac{3}{2\pi} \int_{30°+\alpha}^{30°+\alpha+120°} \sqrt{2}\, V_p \sin \omega t\, d(\omega t)$$

$$= \frac{3\sqrt{6}}{2\pi} V_p \cos \alpha$$

For $V_p = 120$ V and $\alpha = 60°$

$$V_0 = \frac{3\sqrt{6}}{2\pi} \times 120 \times \cos 60°$$

$$= 70.2 \text{ V}$$

Thyristor Full Converter

A three-phase full converter consists of six thyristor switches, as shown in Fig. 10.27*a*. This is the most commonly used controlled-rectifier circuit.

Thyristors S_1, S_3, and S_5 are fired during the positive half-cycle of the voltages of the phases to which they are connected, and thyristors S_2, S_4, and S_6 are fired during the negative cycle of the phase voltages. The references for the firing angles are the crossing points of the phase voltages. The times at which the thyristors fire are marked in Fig. 10.27*b* for $\alpha = 30°$.

Assume that the output current i_0 (i.e., the dc motor current) is continuous and ripple-free as shown in Fig. 10.27*b*. At $\omega t = \pi/6 + \alpha$, S_1 turns on. Prior to this instant S_6 was turned on. Therefore, during the interval $(\pi/6 + \alpha) < \omega t < (\pi/6 + \alpha + \pi/3)$, thyristors S_1 and S_6 conduct the output current and the motor terminals are connected to phase A and phase B, making the output voltage $v_0 = v_{AB} = v_{AN} - v_{BN}$. The output voltage v_0 is the distance between the envelopes of the phase voltages v_{AN} and v_{BN}, as shown by arrows in Fig. 10.27*b*. At $\omega t = \pi/6 + \alpha + \pi/3$, thyristor S_2 is fired and immediately voltage v_{CB} appears across S_6, which reverse-biases it and turns it off *(line commutation)*. The current from S_6 is transferred to S_2. The motor terminals are connected to phase A through S_1 and phase C through S_2, making $v_0 = v_{AC}$. This process repeats after every 60° whenever a thyristor is fired. Note that the thyristors are numbered in the sequence in which they are fired.

Each thyristor conducts for 120° in a cycle. The thyristor current i_{S1} is shown in Fig. 10.27*b*. The line current, such as i_A, is a quasi-square wave (i.e., square wave having 120° pulse width), as shown in Fig. 10.27*b*.

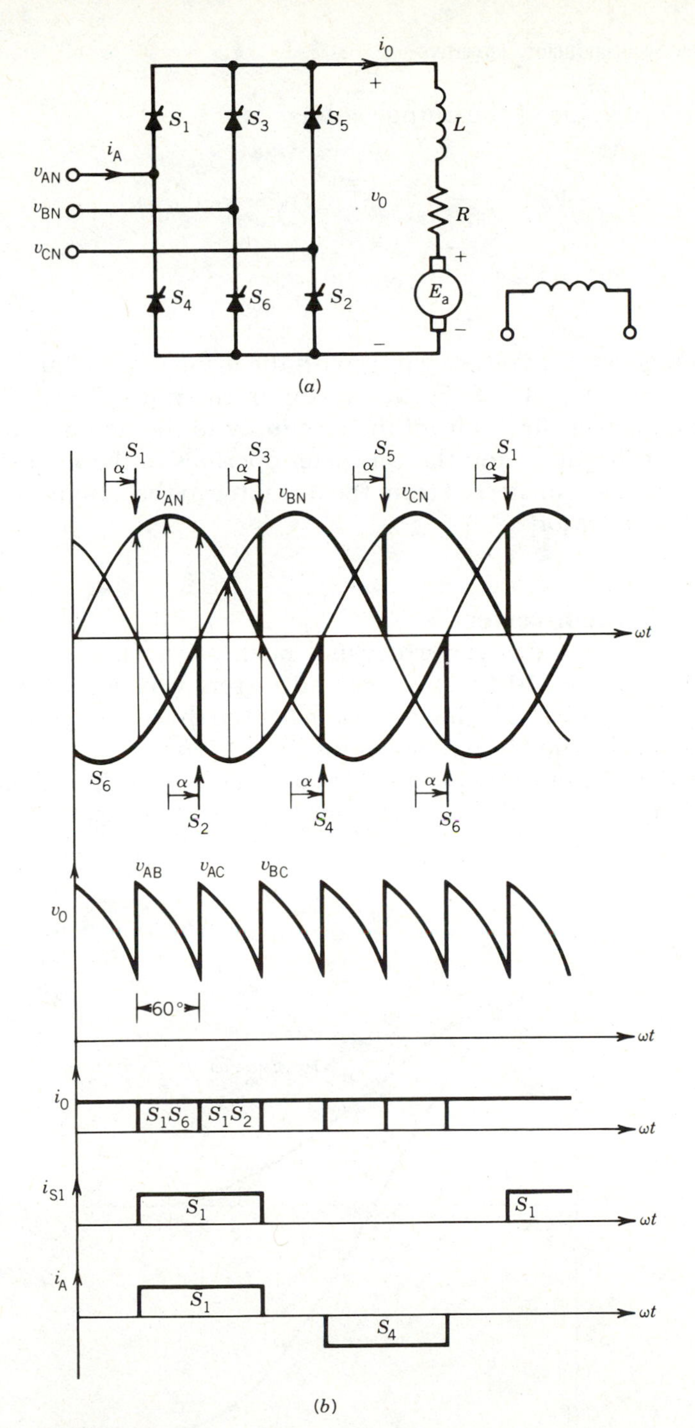

FIGURE 10.27
Thyristor full converter (i_0 assumed continuous).

The average value of the output voltage is

$$V_0 = \frac{1}{2\pi/3} \int_{\pi/6+\alpha}^{\pi/6+\alpha+\pi/3} (v_{AN} - v_{BN})\, d(\omega t)$$

$$= \frac{3\sqrt{6}}{\pi} V_p \cos\alpha \tag{10.10}$$

The average output voltage varies with the firing angle α, and this variation is shown in Fig. 10.28. For α varying in the range $0 < \alpha < 90°$, v_0 is positive and power flow is from the ac supply to the dc motor. For $90° < \alpha < 180°$, v_0 is negative and the converter operates in the inversion mode. The power can be transferred from the dc motor to the ac supply, a process known as *regeneration.*

Thyristor Semiconverter

The three-phase semiconverter consists of three thyristors and three diodes as shown in Fig. 10.29*a*. Voltage and current waveforms are drawn in Fig. 10.29*b* for $\alpha = 90°$. The instants of firing the thyristors and duration of conduction of the diodes are shown in Fig. 10.29*b*.

Assume that the output current i_0 is continuous and ripple-free. At $\omega t = \pi/6 + \alpha$, S_1 turns on and S_1 and D_3 conduct the output current i_0, making

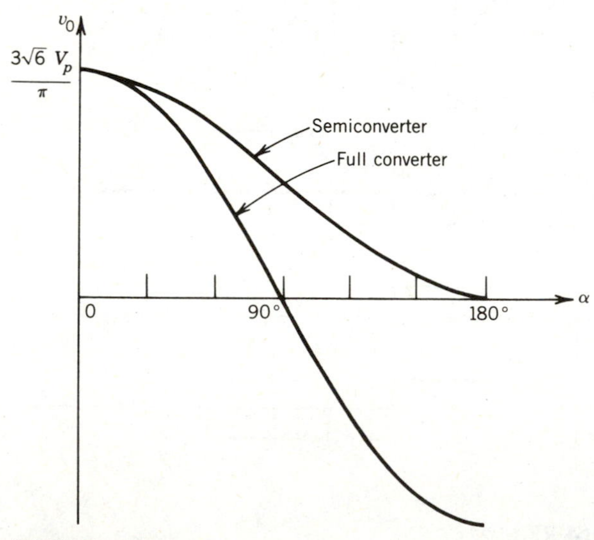

FIGURE 10.28
V_0 versus α in three-phase converters for continuous output current.

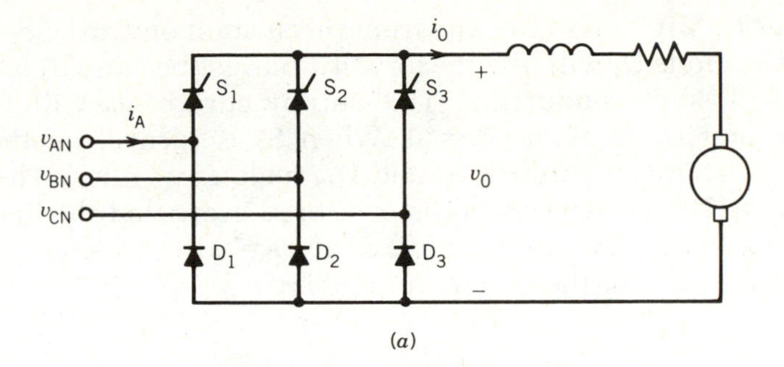

(a)

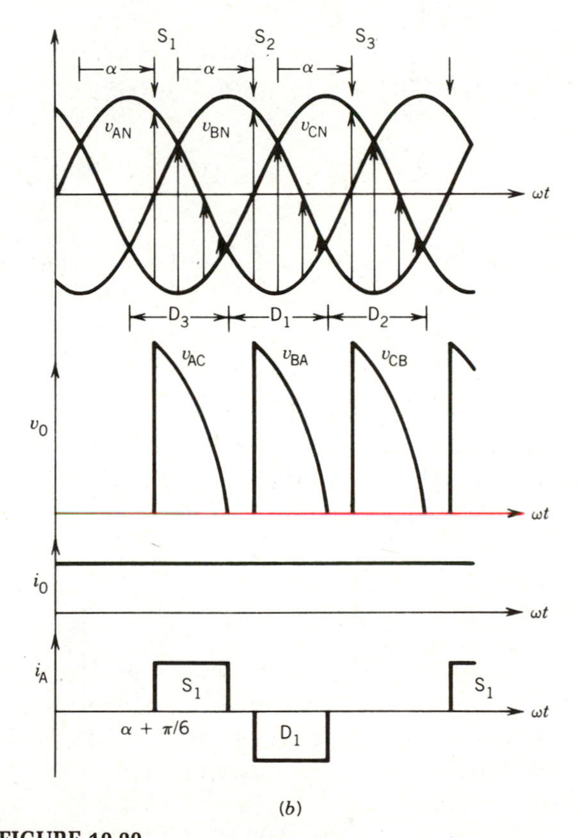

(b)

FIGURE 10.29
Thyristor semiconverter (i_0 ripple-free).

$v_0 = v_{AC}$. At $\omega t = 210°$, v_0 is zero, and from this instant onward v_0 tends to be negative. The diode D_1 will become forward-biased (because S_1 is conducting) and will start conducting. The output current i_0 will freewheel through S_1 and D_1, making $v_0 = 0$. When S_2 is turned on, the output current i_0 will conduct through S_2 and D_1, making $v_0 = v_{BA}$. The process repeats every 120° whenever a thyristor is fired. Note that the line current i_A starts at $\omega t = \pi/6 + \alpha$ and terminates at $\omega t = 210°$.

The average value of the output voltage is

$$V_0 = \frac{1}{2\pi/3} \int_{\pi/6+\alpha}^{\pi+\pi/6} v_{AC}\, d\theta$$

$$= \frac{3}{2\pi} \int \sqrt{6}\, V_p \sin(\theta - 30°)\, d\theta$$

$$= \frac{3\sqrt{6}}{2\pi} V_p(1 + \cos\alpha) \tag{10.10a}$$

The variation of V_0 with α is shown in Fig. 10.28. Note that the output voltage cannot reverse. Hence this converter does not operate in the inversion mode.

Dual Converter[2]

Two full converters can be connected back-to-back to form a dual converter, as shown in Fig. 10.30. Both the voltage V_0 and the current I_0 can reverse in a dual converter.

EXAMPLE 10.3

A 3ϕ full converter is used to control the speed of a 100 hp, 600 V, 1800 rpm, separately excited dc motor. The converter is operated from a 3ϕ, 480 V, 60 Hz supply. The motor parameters are $R_a = 0.1\ \Omega$, $L_a = 5$ mH, $K\Phi = 0.3$ V/rpm ($E_a = K\Phi n$). The rated armature current is 130 A.

1. *Rectifier* (*or motoring*) *operation.* The machine operates as a motor, draws rated current, and runs at 1500 rpm. Assume that motor current is ripple-free.
   - (a) Determine the firing angle.
   - (b) Determine the supply power factor.
2. *Inverter operation.* The dc machine is operated in the regenerative braking mode. At 1000 rpm and rated motor current,

[2] P. C. Sen, *Thyristor DC Drives,* Wiley–Interscience, New York, 1981.

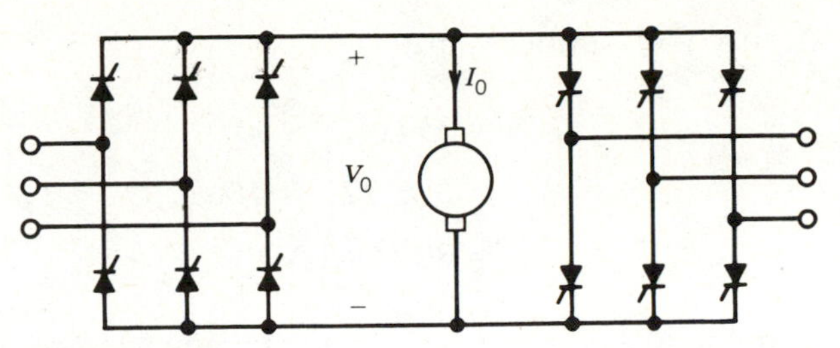

FIGURE 10.30
Three-phase dual converter.

(a) Determine the firing angle.

(b) Determine the power fed back to the supply and the supply power factor.

Solution

1. (a)

$$V_p = \frac{480}{\sqrt{3}} = 277 \text{ V}$$

$$E_a = 0.3 \times 1500 = 450 \text{ V}$$

$$\begin{aligned} V_0 &= E_a + I_0 R_a \\ &= 450 + 130 \times 0.1 \\ &= 463 \text{ V} \end{aligned}$$

From Eq. 10.10

$$463 = \frac{3\sqrt{6} \times 277}{\pi} \cos \alpha$$

$$\alpha = 44.4°$$

(b) Since ripple in the motor current is neglected, from Fig. 10.27 the supply current i_A is a square wave of magnitude 130 A and width 120°. The rms value of the supply current is

$$\begin{aligned} I_A &= \left(\frac{1}{\pi} \times 130^2 \times \frac{2\pi}{3}\right)^{1/2} \\ &= \sqrt{\frac{2}{3}} \times 130 \\ &= 106.1 \text{ A} \end{aligned}$$

The supply volt–amperes are

$$S = 3VI_A$$
$$= 3 \times 277 \times 106.1$$
$$= 88{,}169.1 \text{ VA}$$

Assuming no losses in the converter, the power from the supply P_s is the same as the power input to the motor. Hence,

$$P_s = V_0 I_0$$
$$= 463 \times 130$$
$$= 60{,}190 \text{ W}$$

Therefore, the supply power factor is

$$\text{PF} = \frac{P_s}{S} = \frac{60{,}190}{88{,}169.1} = 0.68$$

2. (a)

$$E_a = 0.3 \times 1000 = 300 \text{ V}$$

For inversion, the polarity of E_a is reversed

$$V_0 = E_a + I_0 R_a$$
$$= -300 + 130 \times 0.1$$
$$= -287 \text{ V}$$

Now,

$$V = \frac{3\sqrt{6} \times 277}{\pi} \cos \alpha = -287 \text{ V}$$
$$\alpha = 116.3°$$

(b) Power from the dc machine (operating as a generator):

$$P_{dc} = 300 \times 130 = 39{,}000 \text{ W}$$

Power lost in R_a: $P_R = 130^2 \times 0.1 = 1690$ W

Power to source: $P_s = 39{,}000 - 1690 = 37{,}310$ W

Supply volt–amperes:

$$S = 88{,}169.1 \text{ VA}$$

Supply power factor:

$$\text{PF} = \frac{37{,}310}{88{,}169.1} = 0.423$$

10.3 AC VOLTAGE CONTROLLERS

AC voltage controllers convert a fixed-voltage ac supply into a variable-voltage ac supply. They are equivalent to autotransformers. AC controllers can be used to control the speed of induction motors.

10.3.1 SINGLE-PHASE AC VOLTAGE CONTROLLERS

The power circuit of a single-phase ac voltage controller supplying an inductive load is shown in Fig. 10.31*a*. Thyristor S_1 is fired at α and thyristor S_2 is fired at $\pi + \alpha$. When S_1 turns on at α, the supply voltage is connected to the load, making $v_0 = v$. The load current i_0 builds up at α and

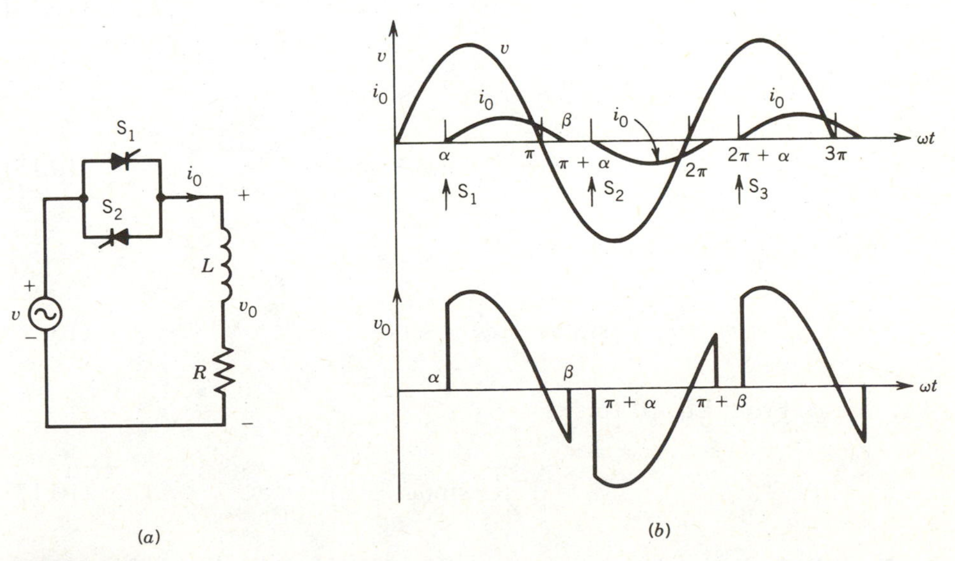

FIGURE 10.31
Single-phase ac voltage controller. (*a*) Circuit. (*b*) Waveforms.

decays to zero at β as shown in Fig. 10.31*b*. When S_2 turns on at $\pi + \alpha$, a negative current pulse flows in the load.

The waveforms of the load voltage v_0 and load current i_0 are shown in Fig. 10.31*b* for a firing angle α.

During the conduction interval, that is, $\alpha < \omega t < \beta$,

$$v_0 = v = Ri_0 + L\frac{di_0}{dt} \tag{10.11}$$

$$\sqrt{2}V_p \sin \omega t = Ri_0 + L\frac{di_0}{dt} \tag{10.12}$$

The load current is

$$i_0 = i_{\text{steady state}} + i_{\text{transient}}$$

$$i_0 = \frac{\sqrt{2}V_p}{Z}\sin(\omega t - \phi) + Ae^{-t/\tau} \tag{10.13}$$

where $Z = \sqrt{R^2 + (\omega L)^2}$

$\phi = \tan^{-1} \omega L/R$

$\tau = L/R$

At $\omega t = \alpha$, $i_0 = 0$. From Eq. 10.13,

$$0 = \frac{\sqrt{2}V_p}{Z}\sin(\alpha - \phi) + Ae^{-(R/\omega L)\alpha} \tag{10.14}$$

From Eq. 10.14

$$A = -\frac{\sqrt{2}V_p}{Z}\sin(\alpha - \phi)e^{(R/\omega L)\alpha} \tag{10.15}$$

From Eqs. 10.13 and 10.15

$$i_0 = \frac{\sqrt{2}V_p}{Z}[\sin(\omega t - \phi) - \sin(\alpha - \phi)e^{(R/\omega L)(\alpha - \omega t)}] \tag{10.16}$$

Let $\alpha = \phi$. From Eq. 10.16,

$$i_0 = \frac{\sqrt{2}V_p}{Z}\sin(\omega t - \phi) \tag{10.17}$$

In Eq. 10.17, the load current i_0 is sinusoidal, which indicates that if the firing angle is the same as the impedance angle (i.e., $\alpha = \phi$), the load current becomes purely sinusoidal. Each thyristor conducts for 180° and full supply voltage appears across the load. Figure 10.32 shows waveforms of load current for two different firing angles. For $\alpha > \phi$, i_0 is nonsinusoidal, and for $\alpha = \phi$, i_0 is sinusoidal. Even for $\alpha < \phi$, i_0 will be sinusoidal in the steady state. The thyristor will be fired at $\omega t = \alpha$, but it will turn on at $\omega t = \phi$, as shown in Fig. 10.32*b*.

10.3.2 THREE-PHASE AC VOLTAGE CONTROLLERS

For high-power loads, such as large-horsepower induction motors driving fans or pumps, three-phase controllers are used. Figure 10.33 shows the power circuits of two types of three-phase ac voltage controllers. In one circuit (Fig. 10.33*a*) the thyristor switches are in the lines and the load is connected in star (or delta). In the other circuit (Fig. 10.33*b*) the thyristor switches are connected in series with the phase loads to form a delta connection.

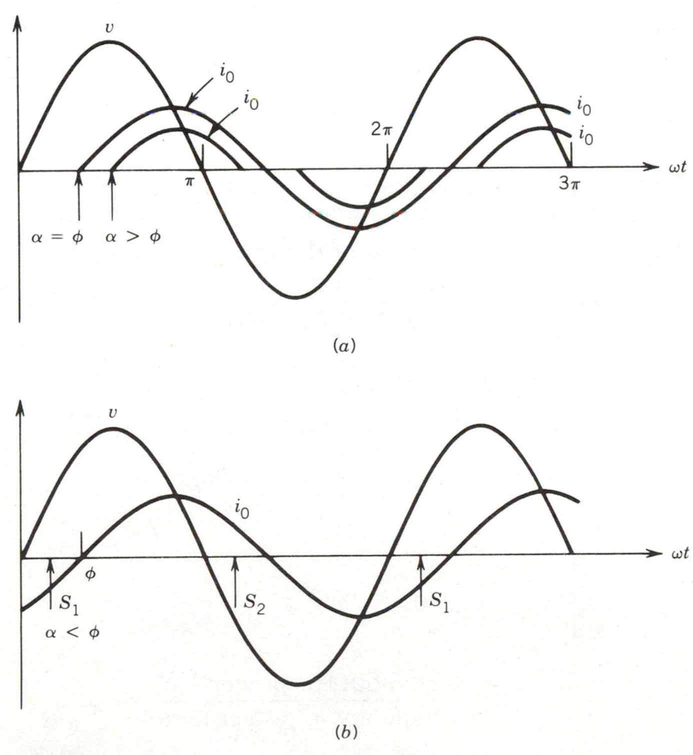

FIGURE 10.32
Load current waveforms at different firing angles.

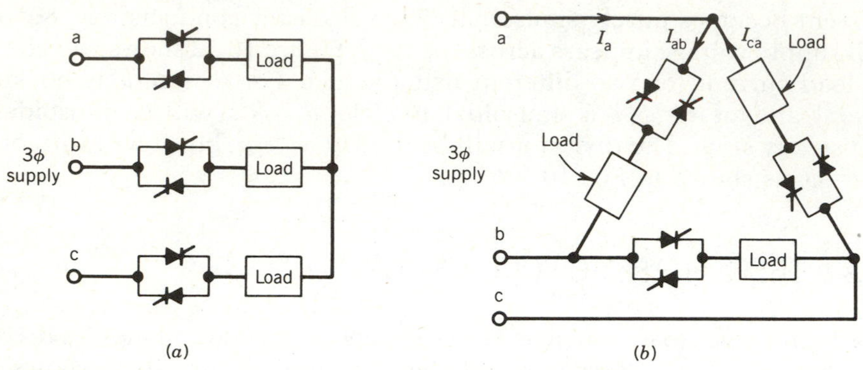

FIGURE 10.33
Three-phase ac voltage controllers. (*a*) Star-connected controller. (*b*) Delta-connected controller.

Analysis of the star-connected three-phase controller is very complex because operation of one phase is dependent on the operation of the other phases. However, the operation of the delta-connected controller can be studied on a per-phase basis because each phase is connected across a known supply voltage.

EXAMPLE 10.4

A three-phase ac voltage controller is used to start and control the speed of a 3ϕ, 100 hp, 460 V, four-pole induction motor driving a centrifugal pump. At full-load output the power factor of the motor is 0.85 and the efficiency is 80 percent. The motor current is sinusoidal. The controller and motor are connected in delta, as shown in Fig. 10.33*b*.

(a) Determine the rms current rating of the thyristors.

(b) Determine the peak voltage rating of the thyristor.

(c) Determine the control range of the firing angle α.

Solution

(a) Input kVA at full load is

$$\begin{aligned} S &= \frac{\text{output power}}{\text{efficiency} \times \text{power factor}} \\ &= \frac{100 \times 0.746}{0.8 \times 0.85} \\ &= 109.71 \text{ kVA} \end{aligned}$$

Input line current I_L:

$$I_L = \frac{109.71 \times 10^3}{\sqrt{3} \times 460} = 137.7 \text{ A}$$

Motor phase current I_p:

$$I_p = \frac{137.7}{\sqrt{3}} = 79.5 \text{ A}$$

Thyristor rms current I_s:

$$I_s = \frac{79.5}{\sqrt{2}} = 56.22 \text{ A}$$

(b) Peak voltage across a thyristor:

$$V_s = \sqrt{2} \times 460 = 650.4 \text{ V}$$

(c)

$$\phi = \cos^{-1} 0.85 = 31.8°$$

The control range is $31.8° < \alpha < 180°$.

10.4. CHOPPERS

A chopper directly converts a fixed-voltage dc supply to a variable-voltage dc supply. The chopper can be used to control the speed of a dc motor.

10.4.1 STEP-DOWN CHOPPER (BUCK CONVERTER)

A schematic diagram of a step-down chopper with a motor load is shown in Fig. 10.34*a*. The switch S can be a conventional thyristor (i.e., SCR), a GTO thyristor, a power transistor, or a MOSFET.

When the switch S is turned on, say at $t = 0$ (Fig. 10.34*b*), the supply is connected to the load and $v_0 = V$. The load current i_0 builds up. When the switch S is turned off at $t = t_{on}$, the load current freewheels through D_{FW} and $v_0 = 0$. At $t = T$, switch S is turned on again and the cycle repeats. The waveforms of the load voltage v_0 and load current i_0 are shown in Fig. 10.34*b*. It is assumed that i_0 is continuous. Note that the output voltage v_0 is a chopped voltage derived from the supply voltage V. Hence the name chopper. The average value of the output voltage is

$$V_0 = \frac{t_{on}}{T} V \tag{10.18}$$

$$= \alpha V \tag{10.19}$$

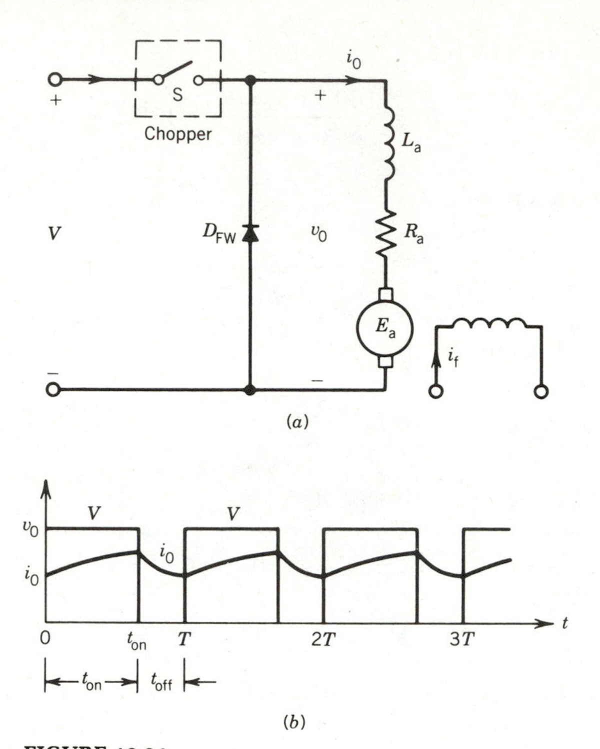

FIGURE 10.34
Step-down chopper (buck converter). (a) Circuit. (b) Waveforms.

where t_{on} is the on-time of the chopper

T is the chopping period

α is the duty ratio of the chopper

From Eq. 10.19, it is obvious that the output voltage varies linearly with the duty ratio of the chopper. The variation of V_0 with α is shown in Fig. 10.35. The output voltage can be controlled in the range $0 < V_0 < V$. This configuration of the chopper is known as a step-down chopper (or buck converter).

If the switch S is a GTO thyristor, a positive gate pulse will turn it on and a negative gate pulse will turn it off. If the switch is a transistor, the base current will control the on and off period of the switch. If the switch is an SCR, a commutation circuit is required to turn it it off. Many forms of commutation circuits[3] are used to force-commutate a thyristor. An exam-

---

[3] P. C. Sen, *Thyristor DC Drives*, Wiley–Interscience, New York, 1981.

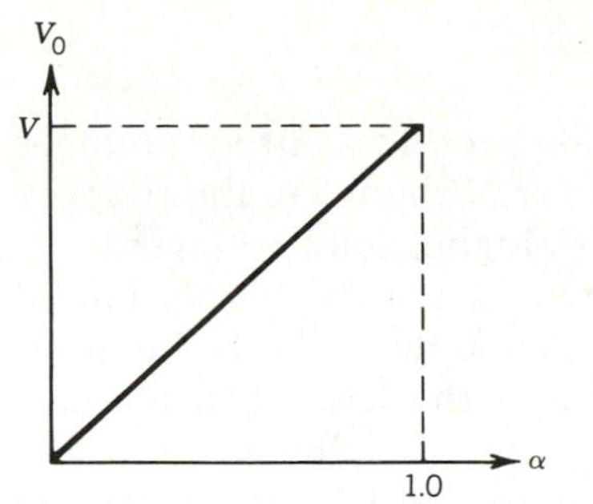

FIGURE 10.35
V_0 versus α characteristic of a step-down chopper.

ple of a commutation circuit is shown in Fig. 10.36. This circuit is sometimes known as the Hitachi commutation circuit.[3] It is used in the chopper controller for Toronto subway cars. When the main thyristor S is on, the capacitor remains charged with the polarity shown in Fig. 10.36. To turn off the thyristor S the auxiliary thyristor S_A is fired. An oscillatory current impulse i_c is generated. The positive half-cycle of i_c flows through C, S_A, and L. The negative half-cycle of i_c flows through L, D_1, S, and C. This negative commutation current i_c flows in a direction opposite to the load current i_0 already flowing through S. When $|i_c| = |i_0|$, the net current through S is zero and S will turn off. Details of the operation and design of this commutation circuit are given in the literature.[3]

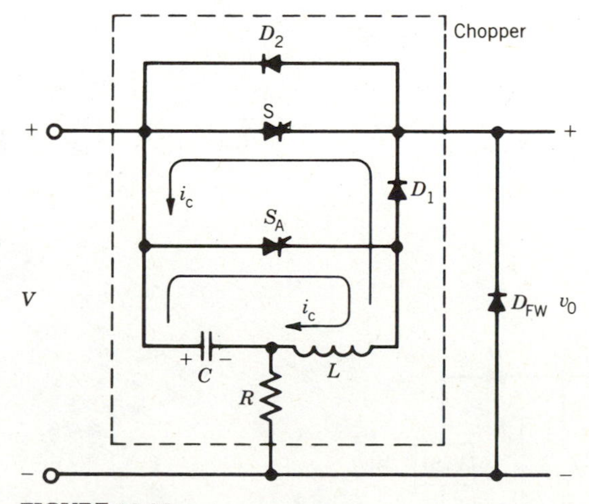

FIGURE 10.36
Commutation circuit to turn off the thyristor S.

10.4.2 STEP-UP CHOPPER (BOOST CONVERTER)

The chopper configuration shown in Fig. 10.34*a* produces output voltages less than the input voltage (i.e., $V_0 < V$). However, a change in the chopper configuration, as shown in Fig. 10.37, provides higher load voltages.

When the chopper is on, the inductor is connected to the supply V, and energy from the supply is stored in it. When the chopper is off, the inductor current is forced to flow through the diode and the load. The induced voltage v_L across the inductor is negative. The inductor voltage adds to the source voltage to force the inductor current into the load. Thus, the energy stored in the inductor is released to the load.

If the ripple in the source current is neglected, then during the time the chopper is on (t_{on}) the energy input to the inductor from the source is

$$W_i = VIt_{on} \tag{10.20}$$

During the time the chopper is off (t_{off}) the energy released by the inductor to the load is

$$W_0 = (V_0 - V)It_{off} \tag{10.21}$$

For a lossless system in the steady state, these two energies will be the same.

$$VIt_{on} = (V_0 - V)It_{off} \tag{10.22}$$

from which

$$\begin{aligned} V_0 &= V\frac{t_{on} + t_{off}}{t_{off}} \\ &= V\frac{T}{T - t_{on}} \\ &= \frac{V}{1 - \alpha} \end{aligned} \tag{10.23}$$

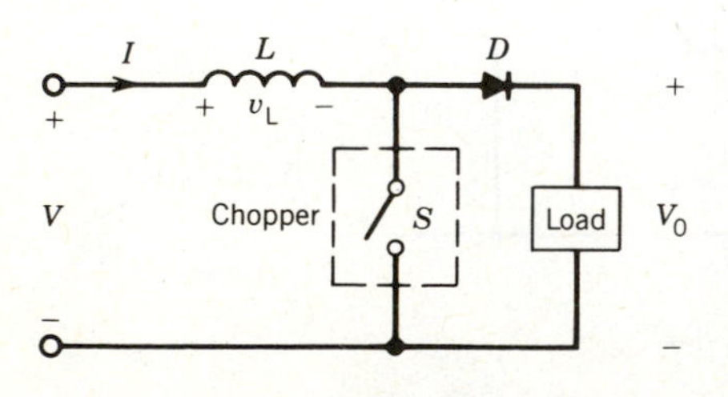

FIGURE 10.37
Step-up chopper (boost converter) configuration.

Thus, for a variation of α in the range $0 < \alpha < 1$, the voltage V_0 varies in the range $V < V_0 < \infty$. This principle of operation is utilized in the regenerative braking of a dc motor. In Fig. 10.37, if V represents the armature of the dc machine and V_0 represents the dc supply, power can be fed back from the decreasing motor voltage V to the fixed supply voltage V_0 by proper adjustment of the duty cycle (α). In Fig. 10.34*a*, if the positions of the chopper and diode are interchanged, the configuration becomes a step-up chopper with E_a as the source and V as the load (i.e., the receiver of power or sink).

10.4.3 TWO-QUADRANT CHOPPER

A combination of the step-up and step-down configurations can form a two-quadrant chopper. This circuit (or arrangement) is shown schematically in Fig. 10.38*a*. If the chopper S_1 and the diode D_1 are operated, the system operates as a step-down chopper and the dc machine operates as a motor. The output voltage V_0 is either V (when S_1 is on) or zero (when S_1 is off and D_1 conducts). The average value of the output voltage is positive and the output current i_0 flows in the positive direction (direction shown by the arrow in Fig. 10.38*a*). The chopper, therefore, operates in the first quadrant, as shown in Fig. 10.38*b*. If, however, the chopper S_2 and the diode D_2 are operated, the system operates as a step-up chopper with E_a as source and the dc machine operates in the regenerative braking mode. The output voltage V_0 is either zero (when S_2 is on) or V (when S_2 is off and D_2 conducts). The average value of the output voltage is positive, but the output current now flows in the negative direction. The chopper then operates in the fourth quadrant, as shown in Fig. 10.38*b*.

The chopper shown in Fig. 10.38*a* can thus be operated in either the first or fourth quadrant and hence is known as a two-quadrant chopper.

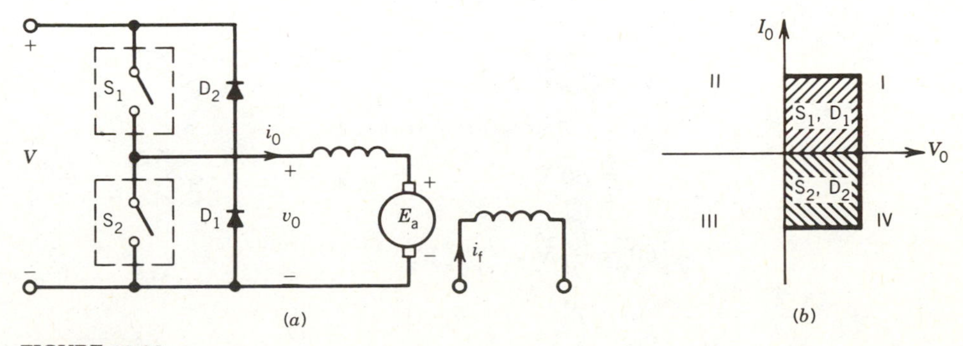

FIGURE 10.38
Two-quadrant chopper. (*a*) Circuit. (*b*) Quadrant operation.

EXAMPLE 10.5

The two-quadrant chopper shown in Fig. 10.38*a* is used to control the speed of the dc motor and also for regenerative braking of the motor. The motor constant is $K\Phi = 0.1$ V/rpm ($E_a = K\Phi n$). The chopping frequency is $f_c = 250$ Hz and the motor armature resistance is $R_a = 0.2\ \Omega$. The inductance L_a is sufficiently large and the motor current i_0 can be assumed to be ripple-free. The supply voltage is 120 V.

(a) Chopper S_1 and diode D_1 are operated to control the speed of the motor. At $n = 400$ rpm and $i_0 = 100$ A (ripple-free),

(i) Draw waveforms of v_0, i_0, and i_s.

(ii) Determine the turn-on time (t_{on}) of the chopper.

(iii) Determine the power developed by the motor, power absorbed by R_a, and power from the source.

(b) In the two-quadrant chopper S_2 and diode D_2 are operated for regenerative braking of the motor. At $n = 350$ rpm and $i_0 = -100$ A (ripple-free),

(i) Draw waveforms of v_0, i_0, and i_s.

(ii) Determine the turn-on time (t_{on}) of the chopper.

(iii) Determine the power developed (and delivered) by the motor, power absorbed by R_a, and power to the source.

Solution

(a) (i) The waveforms are shown in Fig. E10.5*a*.

(ii) From Fig. 10.38*a*

$$\begin{aligned} V_0 &= E_a + I_a R_a \\ &= 0.1 \times 400 + 100 \times 0.2 \\ &= 60\ \text{V} \end{aligned}$$

$$60 = \frac{t_{on}}{T}\,\text{V} = \frac{t_{on}}{T}\,120$$

$$t_{on} = \frac{T}{2}$$

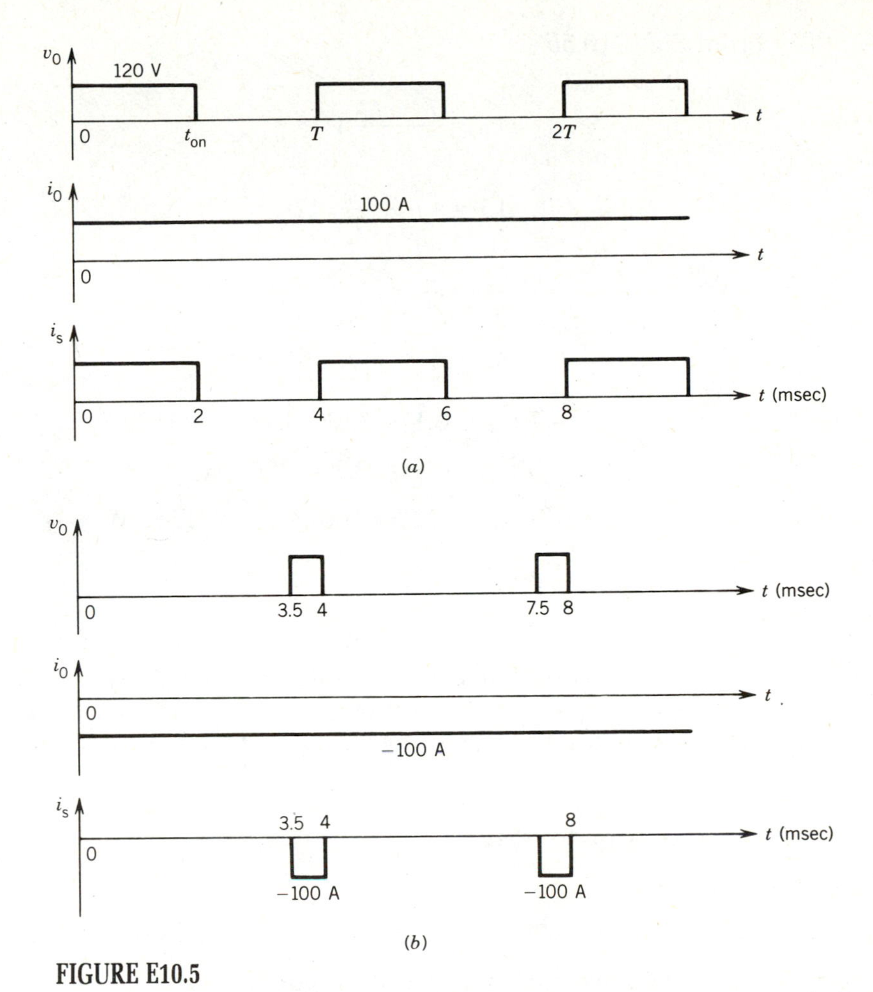

FIGURE E10.5

(iii)

$$P_{\text{motor}} = E_a I_0 = 0.1 \times 400 \times 100 = 4000 \text{ W}$$

$$P_R = (i_0)^2_{\text{rms}} R_a = 100^2 \times 0.2 = 2000 \text{ W}$$

$$P_s = V(i_s)_{\text{avg}} = 120 \times 100 \times \tfrac{2}{4} = 6000 \text{ W}$$

(b) (i) The waveforms are shown in Fig. E10.5*b*.

(ii)

$$\begin{aligned} V_0 &= E_a + (-I_0 R_a) \\ &= 0.1 \times 350 - 100 \times 0.2 \\ &= 15 \text{ V} \end{aligned}$$

From Fig. E10.5*b*

$$V_0 = \frac{T - t_{on}}{T} V$$

$$15 = \left(1 - \frac{t_{on}}{T}\right) 120$$

$$\frac{t_{on}}{T} = \frac{7}{8}$$

$$t_{on} = \tfrac{7}{8} \times 4 = 3.5 \text{ msec}$$

(iii)

$$P_{motor} = E_a I_0 = 0.1 \times 350(-100) = -3500 \text{ W}$$

$$P_R = 100^2 \times 0.2 = 2000 \text{ W}$$

$$P_s = V(i_s)_{avg} = 120(-100 \times \tfrac{1}{8}) = -1500 \text{ W}$$

10.5 INVERTERS

Inverters are static circuits that convert power from a dc source to ac power at a specified output voltage and frequency. Inverters are used in many industrial applications. The following are some of their important applications.

1. Variable-speed ac motor drives.
2. Induction heating.
3. Aircraft power supplies.
4. Uninterruptible power supplies (UPS) for computers.

In general, there are two types of inverters: voltage source inverters (VSI) and current source inverters (CSI). In the voltage source inverter, the input is a dc voltage supply and the inverter converts the input dc voltage into a square-wave ac output voltage source as shown in Fig. 10.39*a*. In the current source inverter the input is a dc current source and the inverter converts the input dc current into a square-wave ac output current as shown in Fig. 10.39*b*.

10.5.1 VOLTAGE SOURCE INVERTERS (VSI)

The input of a voltage source inverter is a stiff dc voltage supply, which can be a battery or the output of a controlled rectifier. Both single-phase and three-phase voltage source inverters are used in industry and will be

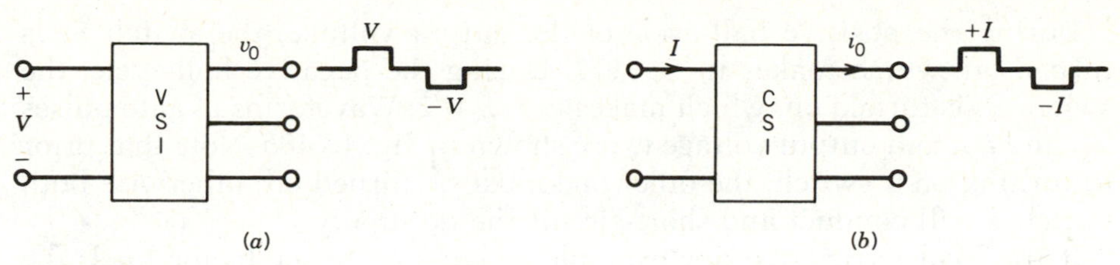

FIGURE 10.39
Inverter configurations. (a) Voltage source inverter (VSI). (b) Current source inverter (CSI).

discussed here. The switching device can be a conventional thyristor (with its commutation circuit), a GTO thyristor, or a power transistor. Here a thyristor symbol enclosed in a circle is used to represent the on/off switch.

Single-Phase VSI

The half-bridge configuration of the single-phase voltage source inverter is shown schematically in Fig. 10.40*a*. The dc supply is center-tapped. Switches S_1 and S_2 are on/off solid-state switches (SCRs or GTO thyristors, BJTs, or MOSFETs). Diodes D_1 and D_2 are known as feedback diodes because they can feed back load reactive energy.

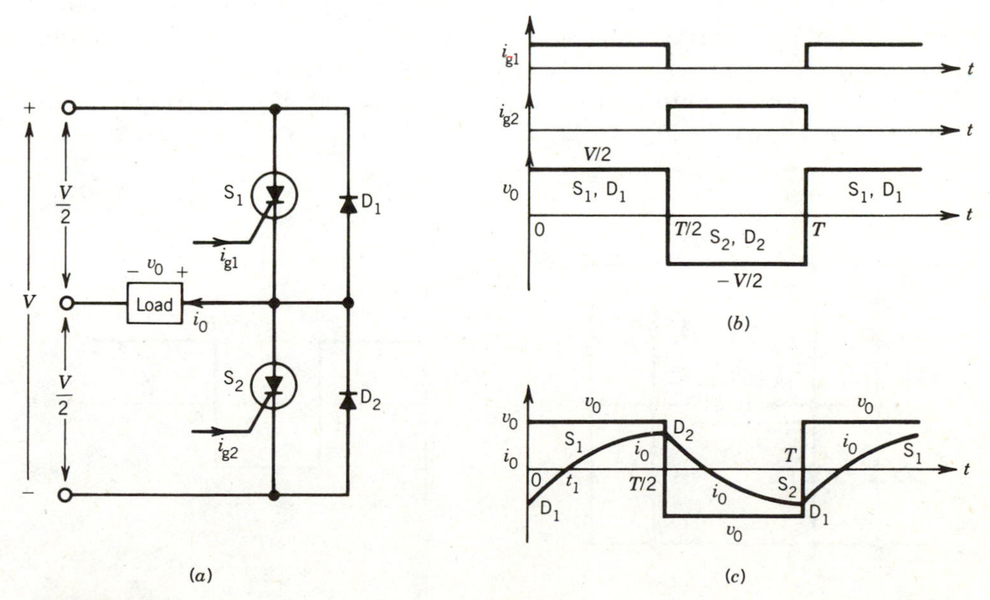

FIGURE 10.40
Half-bridge voltage source inverter. (a) Circuit. (b, c) Waveforms.

During the positive half-cycle of the output voltage, the switch S_1 is turned on, which makes $v_0 = +V/2$. During the negative half-cycle, the switch S_2 is turned on, which makes $v_0 = -V/2$. Waveforms of gate pulses (i_{g1} and i_{g2}) and output voltage v_0 are shown in Fig. 10.40*b*. Note that prior to turning on a switch, the other one must be turned off, otherwise both switches will conduct and short-circuit the dc supply.

If the load is reactive, for example, a lagging power factor load, the output current i_0 lags the output voltage v_0, as shown in Fig. 10.40*c*. Note that during $0 < t < T/2$, v_0 is positive; that is, either S_1 or D_1 is conducting during this interval. However, i_0 is negative during $0 < t < t_1$; therefore D_1 must be conducting during this interval. The load current i_0 is positive during $t_1 < t < T/2$ and therefore S_1 must be conducting during this interval. The devices conducting during various intervals of time are shown in Fig. 10.40*c*. The feedback diodes conduct when the voltage and current are of opposite polarities.

The full-bridge configuration of the single-phase voltage source inverter is shown in Fig. 10.41*a*. Switches S_1 and S_2 are fired during the first half-cycle and switches S_3 and S_4 are fired during the second half-cycle of the output voltage. The output voltage is a square wave of amplitude V, as shown in Fig. 10.41*b*. Note that the frequency of the firing pulses decides the output frequency of the inverter.

Commutation Circuits

If the switch used is a conventional thyristor (SCR), commutation circuits are required to turn it off. Many forms of commutation circuits are used to force-commutate a thyristor. One type of commutation circuit is shown in Fig. 10.36. Another commutation circuit that has been extensively used in inverters is shown in Fig. 10.42. This circuit is known as the McMurray

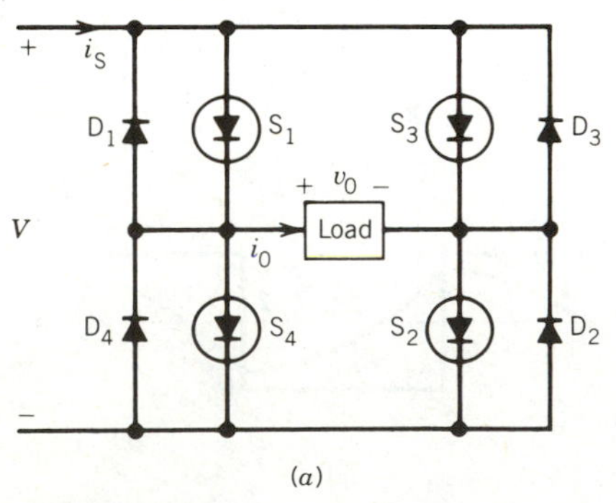

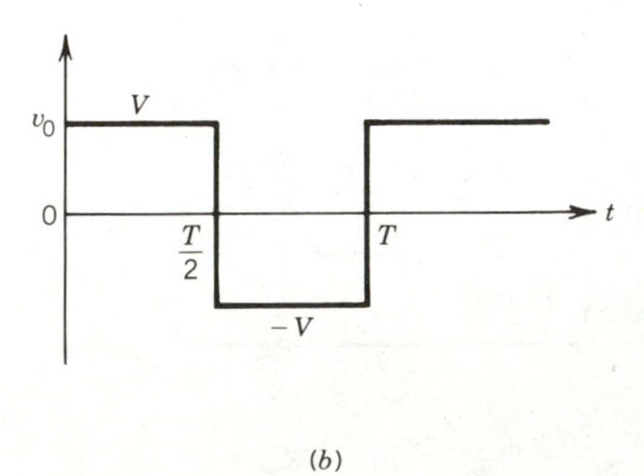

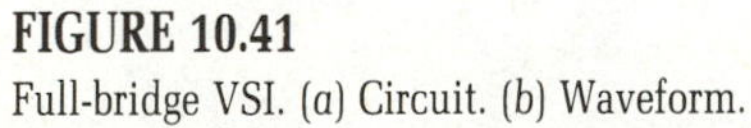
FIGURE 10.41
Full-bridge VSI. (*a*) Circuit. (*b*) Waveform.

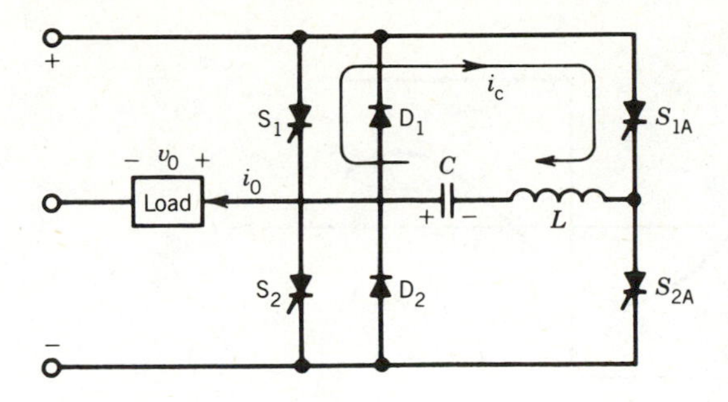

FIGURE 10.42
McMurray commutation circuit.

inverter. The elements S_{1A}, S_{2A}, L, and C form the commutation circuit, and these can be operated to turn off the main thyristors S_1 and S_2. For example, to turn off the main thyristor S_1 at instant $T/2$ (prior to turning on the other main thyristor S_2), the auxiliary thyristor S_{1A} is fired. As a result, an oscillatory current impulse i_c flows in the circuit consisting of L, C, S_1, and S_{1A}. This commutation current i_c flows opposite to the load current i_0 already flowing through S_1. When $i_c = i_0$, the net current through S_1 is zero and S_1 turns off. Details of the operation and design of this commutation circuit are given in the literature.[4]

EXAMPLE 10.6

In the single-phase bridge inverter of Fig. 10.41*a*, the load current is

$$I_0 = 540 \sin(\omega t - 45°)$$

The dc supply voltage is $V = 300$ volts.

(a) Draw waveforms of v_0, i_0, and i_s. Indicate on the waveforms of i_0 and i_s the devices that are conducting during various intervals of time.

(b) Determine the average value of the supply current and the power from the dc supply.

(c) Determine the power delivered to the load.

Solution

(a) The waveforms are shown in Fig. E10.6.

---

[4] B. D. Bedford and R. G. Hoft, *Principles of Inverter Circuits*, Wiley, New York, 1964.

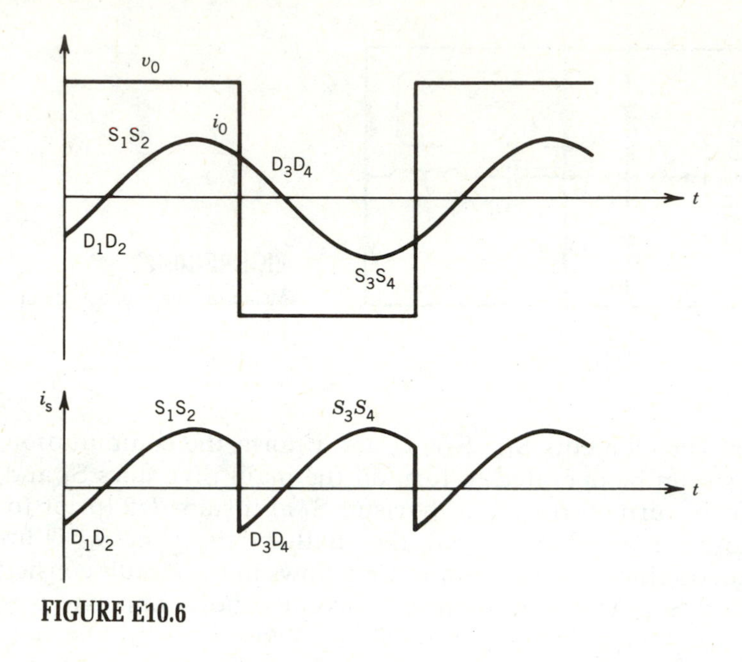

FIGURE E10.6

(b)

$$i_s|_{avg} = I_s = \frac{1}{\pi}\int_0^{\pi} 540 \sin(\omega t - 45°)\, d(\omega t)$$

$$= 243.1 \text{ A}$$

$$P_s = 300 \times 243.1 = 72.93 \text{ kW}$$

(c) From the Fourier analysis of v_0 (square wave), the rms value of the fundamental output voltage is

$$V_{01} = \frac{4V}{\pi\sqrt{2}} = \frac{4 \times 300}{\pi\sqrt{2}} = 270.14 \text{ V}$$

$$P_{out} = V_{01} I_0 \cos\theta$$

$$= 270.14 \times \frac{540}{\sqrt{2}} \cos 45°$$

$$= 72.93 \text{ kW}$$

Three-Phase Bridge Inverter

Using single-phase half-bridge inverter as a building block, a three-phase inverter can be constructed, as shown in Fig. 10.43a. The load is shown as connected in star. The firings (and hence the operation) of the three half-

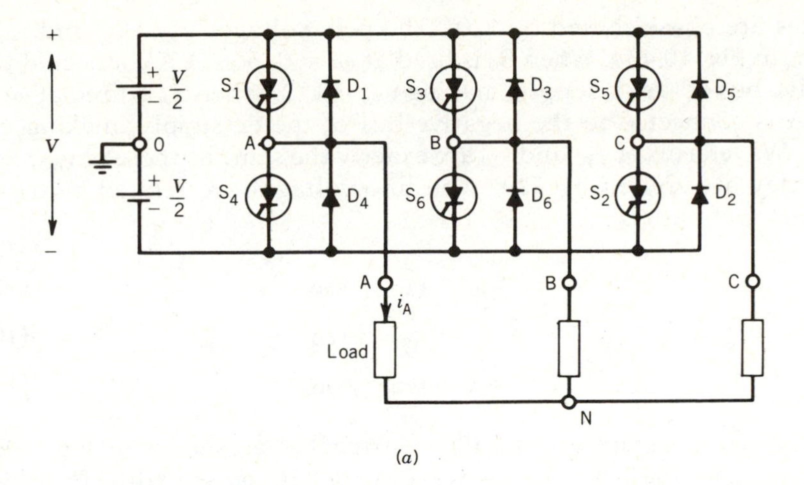

(a)

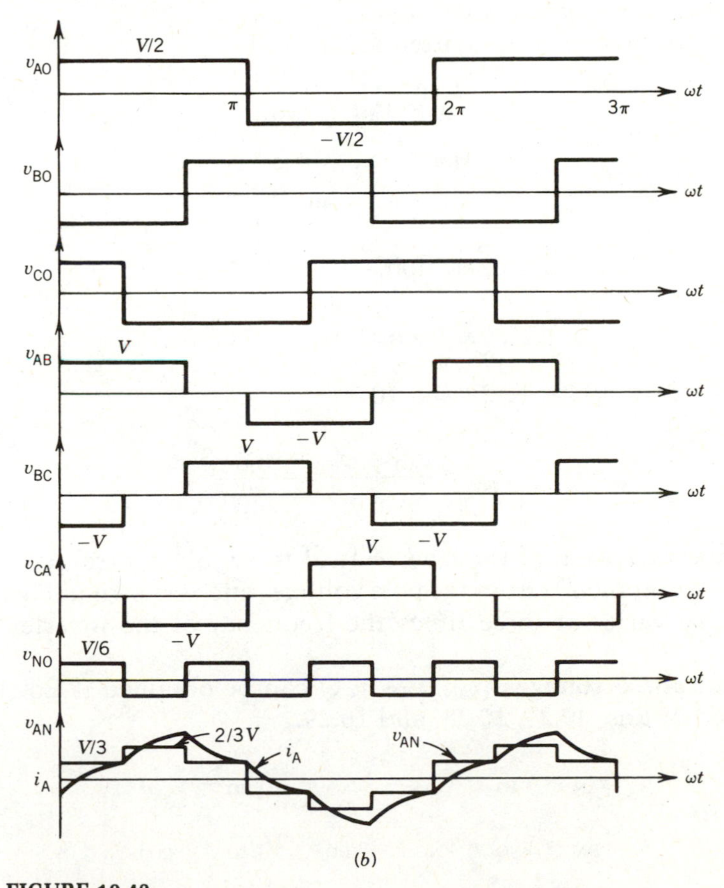

(b)

FIGURE 10.43
Three-phase bridge inverter. (a) Circuit. (b) Waveforms.

bridges are phase-shifted by 120°. The pole voltages v_{AO}, v_{BO}, and v_{CO} are shown in Fig. 10.43*b*. When S_1 is fired at $\omega t = 0$, pole A is connected to the positive bus of the dc supply, making $v_{AO} = V/2$. When S_4 is fired at $\omega t = \pi$, pole A is connected to the negative bus of the dc supply, making $v_{AO} = -V/2$. Waveforms of v_{BO} and v_{c0} are exactly the same as those of v_{AO}, except that they are shifted by 120°. The line voltages are related to the pole voltages as follows:

$$v_{AB} = v_{AO} - v_{BO} \tag{10.24}$$

$$v_{BC} = v_{BO} - v_{CO} \tag{10.25}$$

$$v_{CA} = v_{CO} - v_{AO} \tag{10.26}$$

The line voltages are graphically constructed as shown in Fig. 10.43*b*. These voltages are quasi-square waves with 120° pulse width. They have a characteristic six-stepped wave shape.

The pole voltages can be written as

$$v_{AO} = v_{AN} + v_{NO} \tag{10.27}$$

$$v_{BO} = v_{BN} + v_{NO} \tag{10.28}$$

$$v_{CO} = v_{CN} + v_{NO} \tag{10.29}$$

For balanced three-phase operation,

$$v_{AN} + v_{BN} + v_{CN} = 0 \tag{10.30}$$

From Eqs. 10.27, 10.28, 10.29 and 10.30

$$v_{NO} = \frac{v_{AO} + v_{BO} + v_{CO}}{3} \tag{10.31}$$

The voltage waveform of the load neutral to supply neutral, v_{NO}, can be constructed graphically from the pole voltages and is shown in Fig. 10.43*b*. This voltage varies at three times the frequency of the inverter output voltage.

The load phase voltages (v_{AN}, v_{BN}, v_{CN}) can be obtained if Eq. 10.31 is substituted in Eqs. 10.27, 10.28, and 10.29.

$$v_{AN} = v_{AO} - v_{NO} = \tfrac{2}{3}v_{AO} - \tfrac{1}{3}(v_{BO} + v_{CO}) \tag{10.32}$$

$$v_{BN} = v_{BO} - v_{NO} = \tfrac{2}{3}v_{BO} - \tfrac{1}{3}(v_{AO} + v_{CO}) \tag{10.33}$$

$$v_{CN} = v_{CO} - v_{NO} = \tfrac{2}{3}v_{CO} - \tfrac{1}{3}(v_{AO} + v_{BO}) \tag{10.34}$$

The load phase voltage (such as v_{AN}) can be constructed graphically as shown in Fig. 10.43*b*. It also has a six-stepped wave shape. The other phase voltages v_{BN} and v_{CN} will have the same wave shape as v_{AN} except that they are phase-shifted by 120°. A typical load current i_A for an inductive load is also shown with the wave of v_{AN} in Fig. 10.43*b*.

EXAMPLE 10.7

In the three-phase bridge inverter of Fig. 10.43*a*, the dc supply voltage is 600 V. Determine the rms value of the load line-to-line voltage and load phase voltage.

Solution

From the waveforms shown in Fig. 10.43*b*,

$$V_L = \sqrt{\tfrac{2}{3}} \times 600 = 0.8165 \times 600 = 489.9 \text{ V}$$

$$V_p = \left\{\frac{1}{\pi}\left[\left(\frac{V}{3}\right)^2 \times \frac{\pi}{3} + (\tfrac{2}{3}V)^2 \times \frac{\pi}{3} + \left(\frac{V}{3}\right)^2 \times \frac{\pi}{3}\right]\right\}^{1/2}$$

$$= \frac{V}{3}(\tfrac{1}{3} + \tfrac{4}{3} + \tfrac{1}{3})^{1/2}$$

$$= \frac{V\sqrt{2}}{3}$$

$$= \frac{600\sqrt{2}}{3}$$

$$= 282.84 \text{ V}$$

Pulse Width-Modulated (PWM) Inverters

Many applications require a variable-voltage source. For example, in the speed control of induction motors, the voltage should be changed with frequency to keep the flux level constant. The output voltage of an inverter can be varied by changing the pulse width of each half-cycle of the inverter output voltage. Consider the single-phase bridge inverter in Fig. 10.44*a*. The firings of the switches S_3 and S_2 of the right leg of the inverter are shifted by an angle $\gamma°$ with respect to the firing of S_1 and S_4 of the left leg of the inverter. This produces the pole voltages v_{AO} and v_{BO} as shown in Fig. 10.44*b*. The resulting load voltage v_{AB} has a pulse width of $\gamma°$. By changing the shift angle $\gamma°$ the inverter output voltage can be changed.

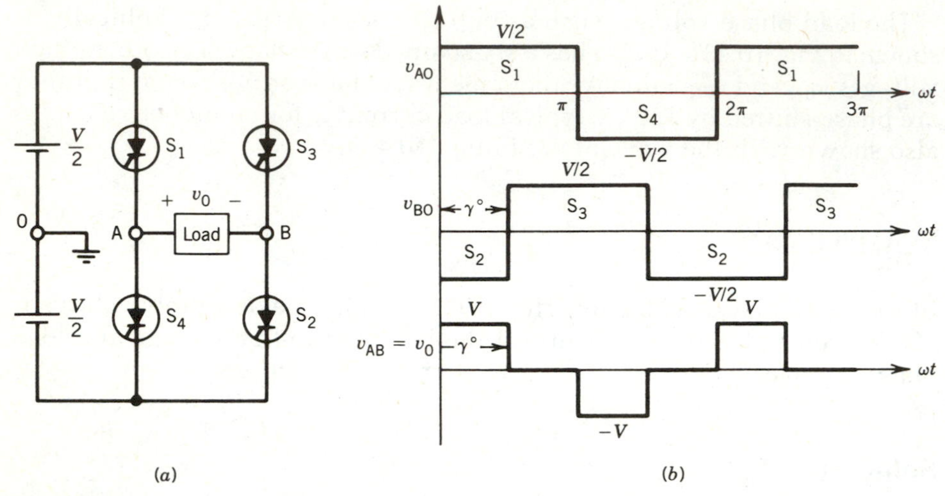

FIGURE 10.44
Pulse width modulation. (single pulse).

Note that at lower values of the pulse width $\gamma°$ (i.e., lower output voltage) the inverter output voltage will be rich in harmonic content. PWM inverters with multiple pulses in each half-cycle of the inverter output voltages can reduce the harmonic content. Various methods[5] have been used to achieve this feature. One method popular in industrial applications, is known as the *sinusoidal PWM* technique. This method will now be described.

In the sinusoidal PWM method a triangular carrier wave of frequency f_c and a modulating wave of frequency f_m (the same frequency as that of the inverter output) are used to modulate the pole voltage. Consider the single-phase inverter circuit of Fig. 10.44*a*. The triangular carrier wave (f_c) and the sinusoidal modulating waves ϕ_A and ϕ_B are shown in Fig. 10.45. The pole voltage v_{AO} is switched between positive and negative buses at the intersections of the carrier wave and the modulating wave ϕ_A. Similarly, the pole voltage v_{BO} is modulated by the carrier wave and the modulating wave ϕ_B. These pole voltages v_{AO} and v_{BO} are shown in Fig. 10.45, and the load voltage v_{AB} ($= v_{AO} - v_{BO}$) is constructed graphically and shown in Fig. 10.45.

Note that the pulses in each half-cycle have different widths. The central pulse is wider than the side pulses. Fourier analysis of this inverter voltage

---

[5] P. C. Sen and S. D. Gupta, Modulation Strategies of Three Phase PWM Inverters, *Canadian Electrical Engineering Journal*, vol. 4, no. 2, 1979.

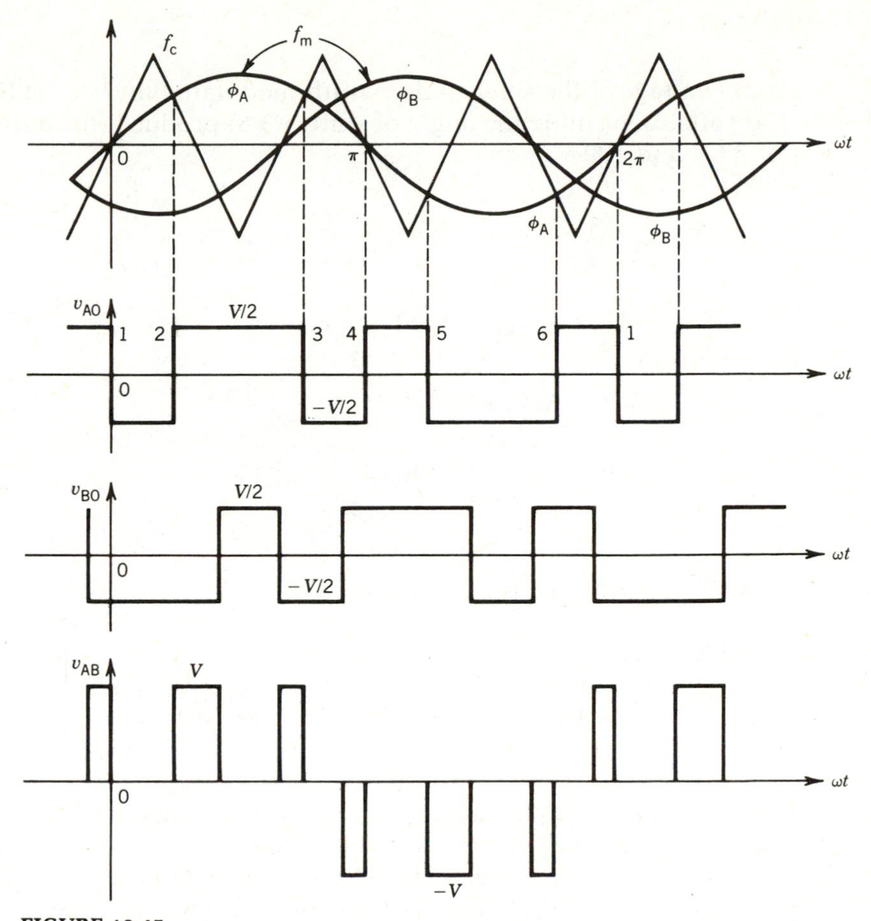

FIGURE 10.45
Sinusoidal pulse width modulation (PWM).

waveform reveals that it has less harmonic content than a single pulse per half-cycle inverter voltage (v_{AB} of Fig. 10.44).

The frequency ratio f_c/f_m is called the *carrier ratio* and the amplitude ratio A_m/A_c is called the *modulation index*. Note that the carrier ratio determines the number of pulses in each half-cycle of the inverter output voltage and the modulation index determines the width of the pulses and hence the rms value of the inverter output voltage.

For a 3ϕ inverter (Fig. 10.43*a*), using modulating waves phase-shifted from each other by 120°, a three-phase sinusoidal PWM inverter can be obtained.

EXAMPLE 10.8

The dc supply voltage of the single pulse width-modulated inverter of Fig. 10.44 is 120 volts. Determine the angle of shift ($\gamma°$) to produce rms output voltages of 50 V and 100 V.

Solution

$$V_0 = \sqrt{\frac{\gamma}{180°}} \times 120$$

$$50 = \sqrt{\frac{\gamma}{180°}} \times 120 \rightarrow \gamma = 31.25°$$

$$100 = \sqrt{\frac{\gamma}{180°}} \times 120 \rightarrow \gamma = 125°$$

10.5.2 CURRENT SOURCE INVERTERS (CSI)

A current source inverter requires a stiff dc current source at the input, as opposed to the stiff dc voltage source required in a voltage source inverter. A series inductor is present in the input to provide the stiff current source. A three-phase current source inverter[6] widely used in medium- to large-horsepower ac motor drives is shown in Fig. 10.46*a*. The idealized waveforms of the input and output currents are shown in Fig. 10.46*b*. Each thyristor conducts for a 120° interval. The capacitors (C_1–C_6) and diodes (D_1–D_6) are the commutating elements. The thyristors (S_1–S_6) are numbered in accordance with the sequence in which they are fired. When a thyristor is fired, it immediately commutates the conducting thyristor of the same group (upper group S_1–S_3–S_5, lower group S_4–S_6–S_2). For example, assume that S_1 and S_2 are conducting. Input current I will flow through S_1, D_1, phase A load, phase C load, D_2, S_2, and back to the input source. Capacitor C_1 will now be charged with the polarity shown in Fig. 10.46*a*. If S_3 is now fired, C_1 will be connected across S_1 and will reverse-bias S_1 and turn it off. The current from S_1 will be transferred to S_3. Eventually, the input current will flow through S_3, D_3, phase B load, phase C load, D_2, S_2, and back to the source. The diodes cause the charge to be held on the commutating capacitors. Without these diodes a capacitor would discharge through two phase loads.

A variable-current source can be obtained from a variable-voltage source as shown in Fig. 10.47. The current loop adjusts the output voltage

---

[6] K. P. Phillips, Current Source Inverter for AC Motor Drives, *IEEE Transactions on Industry Applications,* vol. IA-8, no. 6, pp. 679–683, 1972.

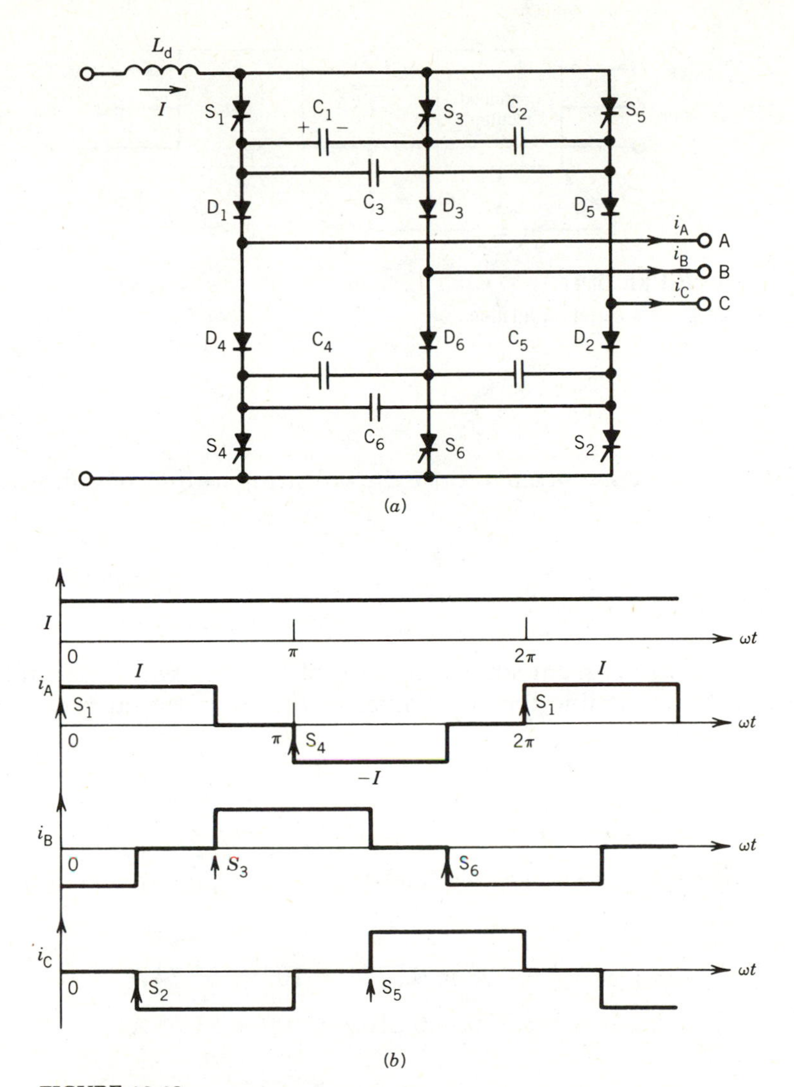

FIGURE 10.46
Three-phase CSI (current source inverter). (*a*) Circuit. (*b*) Waveforms.

of the controlled rectifier to maintain the dc link current at the desired set value (i.e., $I_{set} = I$).

Current source inverters are rugged and reliable. Even if the output terminals are shorted, current does not increase because it is regulated by the input current loop. If the load is an ac machine, such as an induction machine, the quasi-square wave current produces pulsating torque and this may make the motor rotate in jerks at low speed. Pulse width modula-

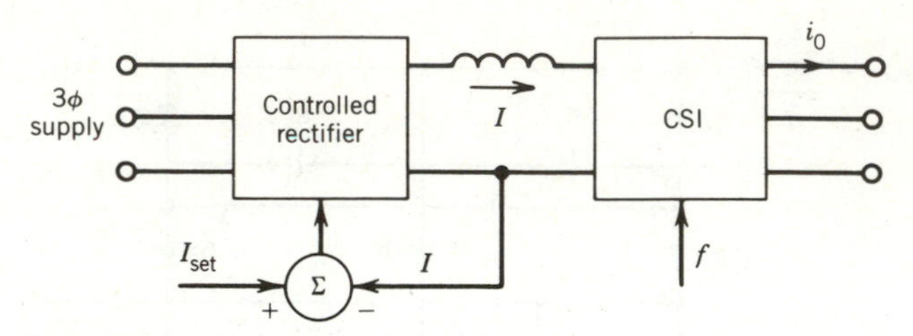

FIGURE 10.47
CSI with current regulation.

tion (PWM) of the current can reduce the pulsating torque and improve the motor performance.[7,8]

EXAMPLE 10.9

In the three-phase current source inverter of Fig. 10.46*a*, the dc link current is 100 A. Determine the rms values of the thyristor current and the output current.

Solution

From the waveforms shown in Fig. 10.46*b*, each thyristor conducts for 120°.

$$I_{SCR} = \sqrt{\tfrac{1}{3}} \times 100 = 0.5774 \times 100 = 57.74 \text{ A}$$

$$I_{out} = \sqrt{\tfrac{2}{3}} \times 100 = 0.8165 \times 100 = 81.65 \text{ A}$$

10.6 CYCLOCONVERTERS

A cycloconverter directly converts ac power at one input frequency to output power at a different (normally lower) frequency. It is essentially a dual converter as described in Section 10.2.2, operated in such a way as to produce an alternating output voltage.

---

[7] C. Namuduri and P. C. Sen, Optimal Pulse Width Modulation for Current Source Inverters, *IEEE-IA Transactions*, vol. IA-22, no. 6, pp. 1052–1072, 1986.
[8] T. A. Lipo, Analysis and Control of Torque Pulsations in Current-Fed Asynchronous Machine, *IEEE-PESC Proceedings*, pp. 89–96, 1978.

10.6.1 SINGLE-PHASE TO SINGLE-PHASE CYCLOCONVERTER

Consider the dual-converter circuit of Fig. 10.48*a* with a resistive load. Converters P and N are the positive and negative controlled rectifiers, respectively. If only converter P is operated, the output voltage is positive. If converter N is operated, the output voltage is negative.

Let the polarity of the control voltage v_c represent the polarity of the output voltage v_0 and the amplitude of v_c represent the desired average output voltage. The frequency of v_c represents the fundamental output frequency of v_0.

The supply voltage v is shown in Fig. 10.48*b*. Let the amplitude of v_c be such that the output voltage v_0 is maximum. This means that the firing angles of the two converters are zero, that is, $\alpha_p = 0$, $\alpha_n = 0$. During the positive half-cycle of v_c converter P is fired, and during the negative half-cycle converter N is fired. The output voltage waveform v_0 is shown in Fig. 10.48*b*. Note that the fundamental output frequency is one-third the input frequency. The waveform of the output voltage v_0 at a reduced value of the control voltage v_c is shown in Fig. 10.48*c*. If the control voltage v_c varies with time during each half-cycle (instead of remaining constant), the firing angles change during the half-cycle. This reduces the harmonic content in the output voltage v_0. This feature will be discussed further in the following section on the three-phase cycloconverter. For example, if the side pulses of voltages are obtained at higher values of the firing angles and the middle pulse is obtained at a lower value of firing angle, as shown in Fig. 10.48*d*, the harmonic content in the output voltage will be less than that in the output voltage shown in Fig. 10.48*c*, where all pulses are obtained at a fixed firing angle.

10.6.2 THREE-PHASE CYCLOCONVERTER

Many applications require sinusoidal ac voltage. The cycloconverter circuit shown in Fig. 10.48*a* is not a practical circuit and is seldom used because of its nonsinusoidal output voltage. An essentially sinusoidal output voltage can be synthesized from three-phase input voltages by using three-phase controlled rectifiers. Consider the controlled rectifier circuit in Fig. 10.49*a*. The average output voltage of this controlled rectifier varies as the cosine of the firing angle (Eq. 10.10). The successive firing angles can be changed so that the average output voltage changes sinusoidally. Fabrication of such an output voltage from the three-phase input voltages is illustrated in Fig. 10.49*b*. The sinusoidally varying average voltage is shown as $v_{0(avg)}$. This represents the desired output voltage. Zero average voltage is required at $t = t_0$, and therefore the firing angle at this instant is $\alpha = 90°$ ($\cos 90° = 0$), as shown in Fig. 10.49*b*. As $v_{0(avg)}$ increases, the firing angle decreases. At the peak value of $v_{0(avg)}$, the firing angle α is minimum

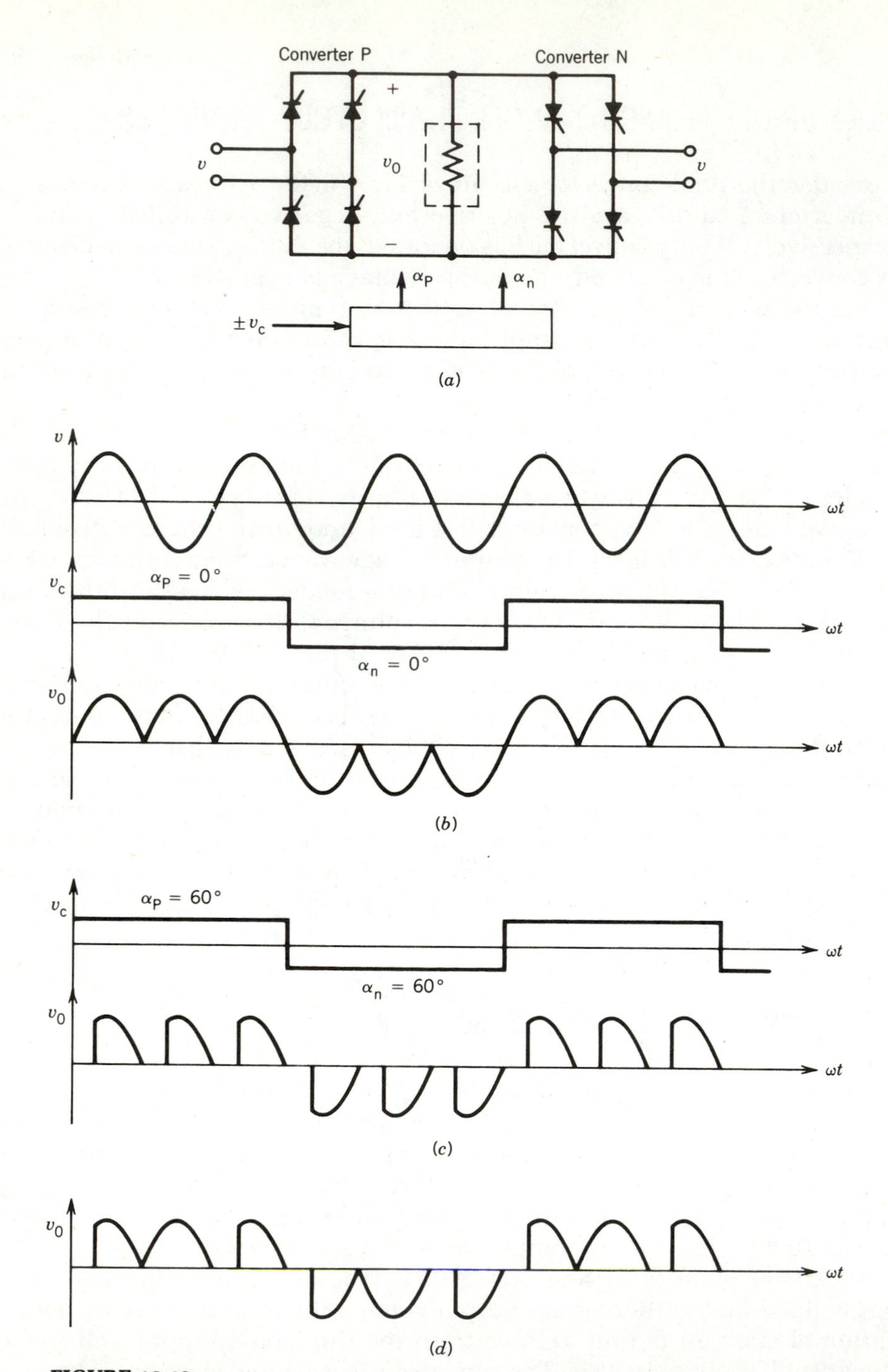

FIGURE 10.48

Single-phase to single-phase cycloconverter. (a) Circuit. (b) Waveforms for $\alpha_p = \alpha_n = 0°$. (c) Waveforms for $\alpha_P = \alpha_n = 60°$. (d) Waveforms at different values of α.

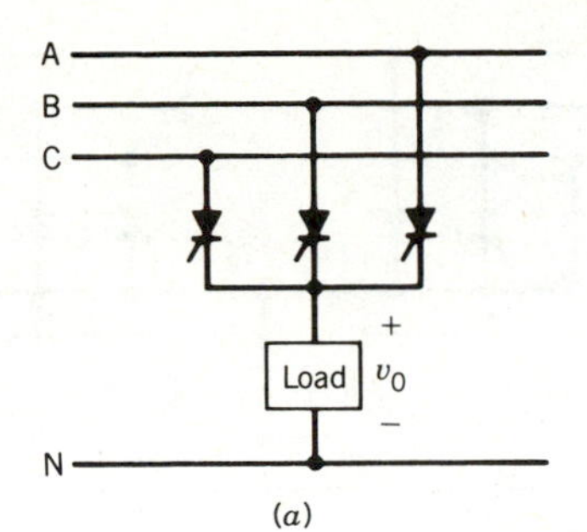

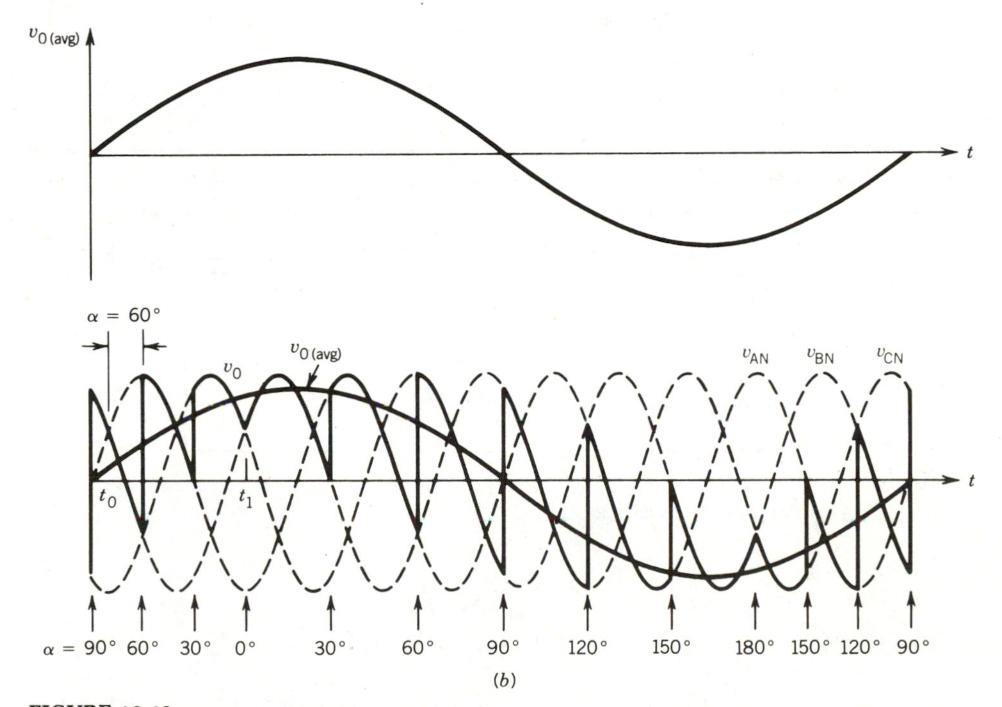

FIGURE 10.49
Synthesis of sinusoidal output voltage. (a) Controlled rectifier circuit. (b) Output voltage synthesis.

($\alpha = 0$ at $t = t_1$ in Fig. 10.49*b*). The firing angle of successive pulses is changed from 90° to 0° and back to 90° and then from 90° to 180° and back again to 90° in appropriate steps. The actual output voltage v_0 is shown by the thick line in Fig. 10.49*b* and the average of v_0 is shown as $v_{0(avg)}$.

The three-phase to single-phase cycloconverter is shown in Fig. 10.50. The converter P conducts during the positive half-cycle and the converter N conducts during the negative half-cycle of the cycloconverter output current i_0. The three-phase to three-phase cycloconverter is shown in Fig. 10.51.

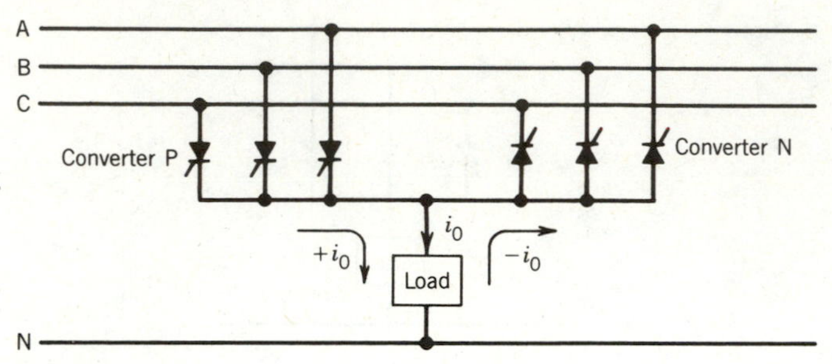

FIGURE 10.50
Three-phase to single-phase cycloconverter.

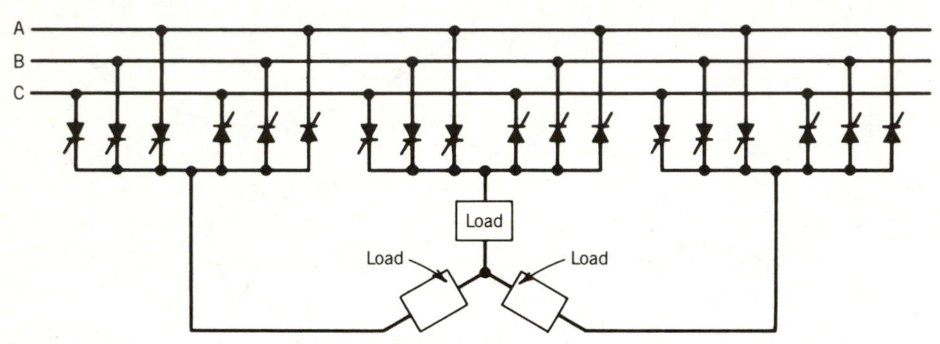

FIGURE 10.51
Three-phase to three-phase cycloconverter.

PROBLEMS

10.1 For the speed control system shown in Fig P10.1, SCRs are fired at $\alpha = 60°$. The motor current is 15 A and is assumed to be ripple-free. For this operating condition,

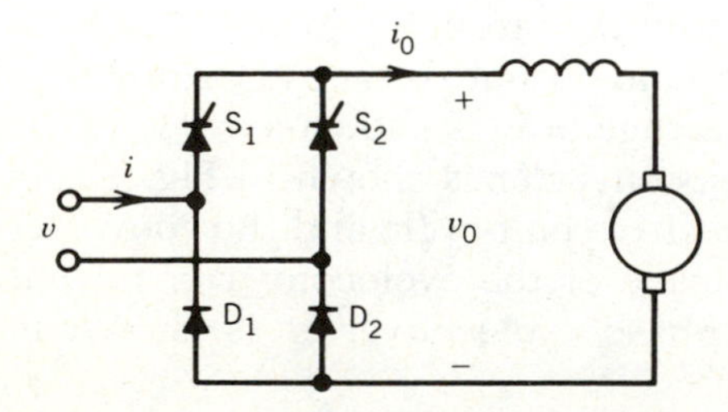

FIGURE P10.1

(a) Draw the waveforms of v, v_0, i_0, i_{S1}, i_{D1}, and i.

(b) Determine the power taken by the dc motor.

(c) Determine the supply volt–amperes and supply power factor. Assume that the converter is lossless.

(d) Determine the rms value of the SCR current and the diode current.

10.2 The speed of a 110 V, 5 hp dc motor is controlled by a 1ϕ ac/dc full converter (i.e., controlled rectifier). The ac supply is 120 V, 60 Hz. Consider the dc motor and converter to be ideal and lossless. Assume a very large inductance in series with the motor. The motor voltage constant is 0.055 V/rpm.

(a) Draw the power circuit.

(b) For a speed of 1000 rpm and rated motor current,

i. Determine the firing angle of the converter.

ii. Determine the rms value of the supply current and thyristor current.

iii. Determine the supply power factor.

iv. Draw waveforms of the supply voltage, supply current, converter output voltage, and converter output current.

10.3 The speed of a 10 kW, 250 V dc motor is controlled by a 3ϕ ac/dc semiconverter. The ac supply is 3ϕ, 208 V, 60 Hz. Assume a very large inductance in series with the dc motor.

(a) Draw the power circuit.

(b) At $\alpha = 120°$ draw the waveform of the output voltage.

(c) Determine the average output voltage at $\alpha = 120°$.

(d) Draw the waveform of a supply phase voltage and current in the same phase.

(e) Determine the width of the supply current pulse and the supply power factor (assume that the converter is lossless and motor current is 15 A).

10.4 The three-phase full converter of Fig. 10.27 is used to control the speed of a dc motor. The motor back emf constant is 0.1 V/rpm. The supply line-to-line voltage is 110 V. The motor armature resistance is 0.2 Ω. For $\alpha = 50°$, the motor speed is 900 rpm.

(a) Determine the average value of the motor current, assuming it to be ripple-free.

(b) Determine the rms value of the thyristor current and supply line current.

(c) Determine the supply power factor.

10.5 In the light dimmer circuit shown in Fig. P10.5 determine the load power at the triggering angles $\alpha = 0°, 30°, 60°, 90°, 120°, 150°$, and $180°$ and plot the load power as a function of firing angle.

10.6 The speed of a 1ϕ, 1 hp, 120 V, 60 Hz, 1750 rpm induction motor is controlled by a 1ϕ ac voltage controller connected to a 1ϕ, 120 V, 60 Hz supply.

(a) At $\alpha = 90°$, the conduction angle (γ) is $135°$ and speed is 1200 rpm.

i. Draw qualitative waveforms of the motor terminal voltage and motor current.

ii. Determine the voltage (rms) across the motor terminals.

(b) At full-load output, the motor voltage is 120 V (rms) and the motor operates at 0.7 power factor and 75% efficiency.

i. Determine the maximum value of the firing angle.

ii. If the firing angle is $\alpha = 15°$, determine the rms value of the thyristor current.

10.7 Consider the delta-connected 3ϕ ac controller of Fig. 10.33*b*. The load is purely resistive.

(a) What is the range of α for current control in the load?

(b) Draw phase currents and line currents for $\alpha = 60°$.

(c) Determine the peak voltage across a thyristor and the maximum rms current through a thyristor.

10.8 A one-quadrant chopper, such as that shown in Fig. 10.34*a*, is used to control the speed of a dc motor.

Supply dc voltage = 120 V

$R_a = 0.15\ \Omega$

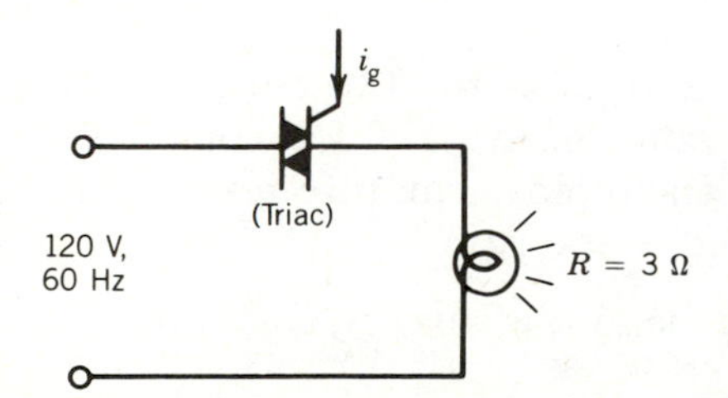

FIGURE P10.5

Motor back emf constant = 0.05 V/rpm

Chopper frequency = 250 Hz

At a speed of 120 rpm, the motor current is 125 A. The motor current can be assumed to be ripple-free.

(a) Determine the duty ratio (α) of the chopper and the chopper on time t_{on}.

(b) Draw waveforms of v_0, i_0, and i_s.

(c) Determine the torque developed by the armature, power taken by the motor, and power drawn from the supply.

10.9 The power circuit configuration during regenerative braking of a subway car is shown in Fig. P10.9. The dc motor voltage constant is 0.3 V/rpm and the dc bus voltage is 600 V. Assume the motor current to be ripple-free. At a motor speed of 800 rpm and motor current of 300 A,

(a) Draw the waveforms of v_0, i_a, and i_s for a particular value of the duty cycle α (= t_{on}/T).

(b) Determine the duty ratio α of the chopper for the operating condition.

(c) Determine the power fed back to the bus.

10.10 Consider the two-quadrant chopper systems shown in Fig. P10.10. The two choppers S_1 and S_2 are turned on for time t_{on} and turned off for time $T - t_{on}$, where T is the chopping period.

(a) Draw the waveform of the output voltage v_0. Assume continuous output current i_0.

(b) Derive an expression for the average output voltage V_0 in terms of the supply voltage V and the duty ratio α (= t_{on}/T).

10.11 In the 1ϕ bridge inverter of Fig. 10.41a, the output frequency is 10 Hz and the load is a pure inductor of L = 100 mH. The dc supply voltage is 100 V. The thyristors are GTO thyristors. For steady-state operation, sketch the

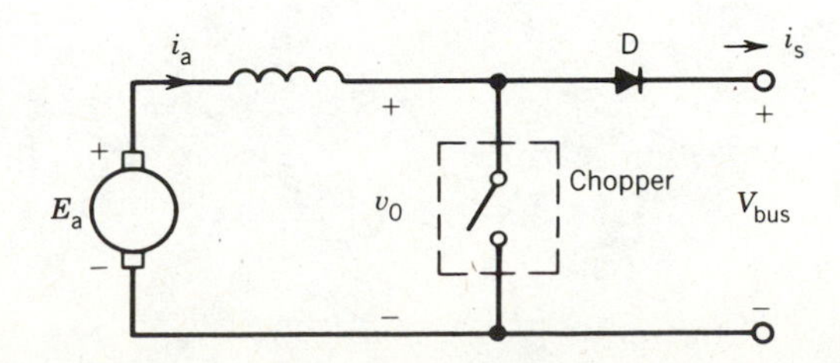

FIGURE P10.9

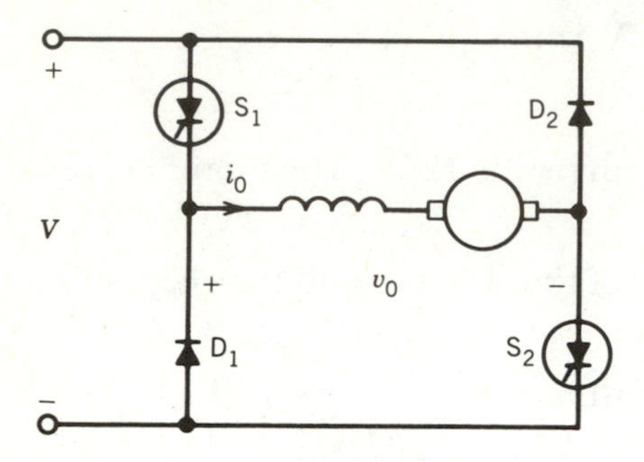

FIGURE P10.10

(a) Gate currents.

(b) Output voltage v_0.

(c) Output current i_0. Determine and show the magnitudes of maximum and minimum load current. Indicate the devices on the waveform of i_0 that conduct during various intervals of time.

10.12 Repeat Example 10.6 if $i_0 = 400 \sin(\omega t + 60°)$.

10.13 The three-phase inverter of Fig. 10.43*a* is used as a 3ϕ PWM inverter. A triangular modulation of the pole voltages is implemented; that is, the modulating waves ϕ_A, ϕ_B, and ϕ_C are triangular waves at the inverter output frequency. For a carrier ratio of three and modulation index 0.5, draw voltage waveforms for pole voltages v_{AO}, v_{BO}, and v_{CO} and line voltages v_{AB}, v_{BC}, and v_{CA}.

10.14 In the single-phase cycloconverter of Fig. 10.48*a*, the input supply is 120 V, 60 Hz. The load is a pure resistance and the output frequency is 15 Hz.

(a) Draw waveforms for input voltage v, control voltage v_c, and output voltage v_0 for $\alpha = 0°$ and $\alpha = 90°$.

(b) Determine the rms value of the output voltage at $\alpha = 0°$ and $\alpha = 90°$.

APPENDIX A

WINDINGS

The layout of a winding in an electric machine affects the mmf distribution and the performance of the machine. All the coils of a winding can be placed in two slots, and such a winding is known as a *concentrated winding*. This requires large sizes for the slots; also, a large portion of the stator or rotor is left unused. Except in a few smaller machines, concentrated windings are hardly used. The coils of a winding are usually distributed over a few slots, and such a winding is known as a *distributed winding*. Distributed windings can make better use of the stator or rotor structure and also decrease harmonics. In this appendix the properties and effects of the various types of windings are discussed.

A.1 MMF DISTRIBUTION

Consider a winding of N turns placed in two slots on the stator of a machine, as shown in Fig. A.1*a*. If a current i flows through the winding, the mmf along a path, defined by the angle θ, is given by the ampere-turns enclosed by the dotted contour:

$$F(\theta) = Ni \tag{A.1}$$

The mmf distribution in the air gap is shown in Fig. A.1*b*. Let us assume that the air gap is uniform and of length g and that the reluctance of the stator and the rotor core is neglected. Then the flux density distribution in the air gap is similar to the mmf distribution and is

$$B(\theta) = \mu_0 \frac{Ni}{2g} \tag{A.2}$$

The flux density distribution is of square-wave shape, that is, nonsinusoidal. The fundamental and the harmonic components of the flux density are as follows:

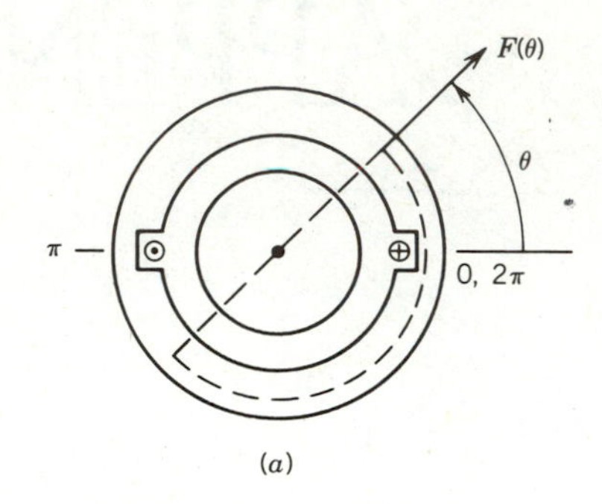

(a)

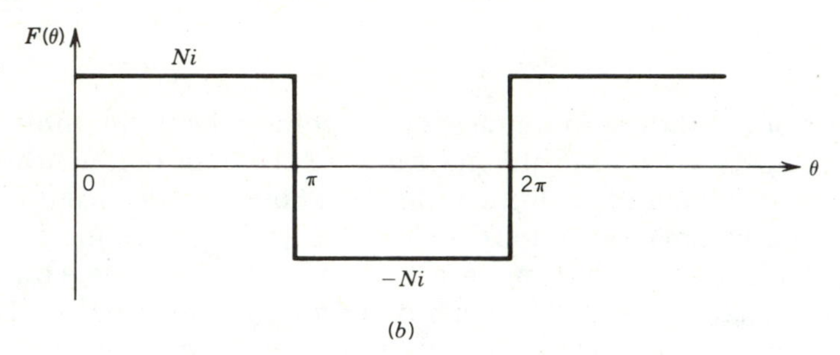

(b)

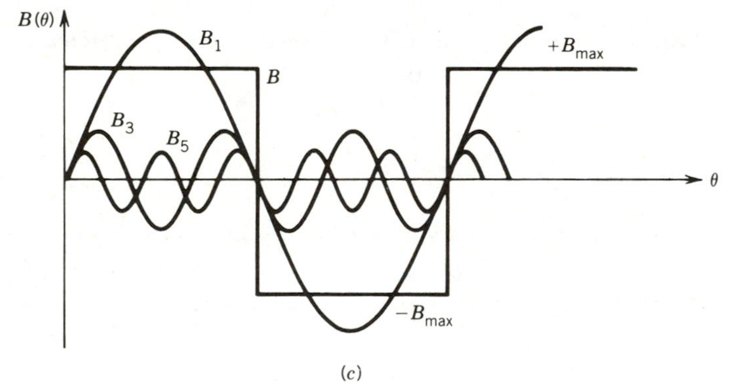

(c)

FIGURE A.1

Concentrated winding. (a) Winding in two stator slots. (b) mmf distribution in the air gap. (c) Flux density distribution in the air gap.

$$B(\theta) = B_{1(\max)} \sin \theta + B_{3(\max)} \sin 3\theta + B_{5(\max)} \sin 5\theta + \cdots \quad (A.3)$$

$$= \sum_{h=1,3,5,\ldots} \frac{4B_{\max}}{\pi h} (\sin h\theta) \quad (A.4)$$

All odd harmonics are present. In Fig. A.1c the fundamental, third harmonic, and fifth harmonic flux densities are shown. These harmonic flux densities induce harmonic voltages in the winding.

The harmonic content can be decreased if the winding is distributed over several slots. In Fig. A.2*a*, a distributed winding is placed in 12 slots. The mmf distribution resulting from the distributed winding is shown in Fig. A.2*b*. The fundamental component of the mmf is also shown in this figure. It is clear that the mmf distribution is closer to being sinusoidal as a result of distributing the winding over several slots.

If the winding could be placed in an infinitely large number of slots and the conductors in the slots were sinusoidally distributed (instead of placing the same number of conductors in each slot), the mmf distribution in the air gap would be sinusoidal. Such an ideal machine is impossible to build. Besides, it is convenient to make all the coils identical and place them in the slots. Figure A.3*a* shows three multiturn coils. The twist at the end facilitates placing it in the slot. The mmf (and hence the flux density) distribution in the air gap will contain some harmonics. Other methods are used to minimize or eliminate certain harmonics, particularly the lower-order ones. Figure A.3*b* shows the stator windings of a three-phase ac machine. In a practical electric machine, a distributed winding is placed in a finite number of slots and all coils are identical.

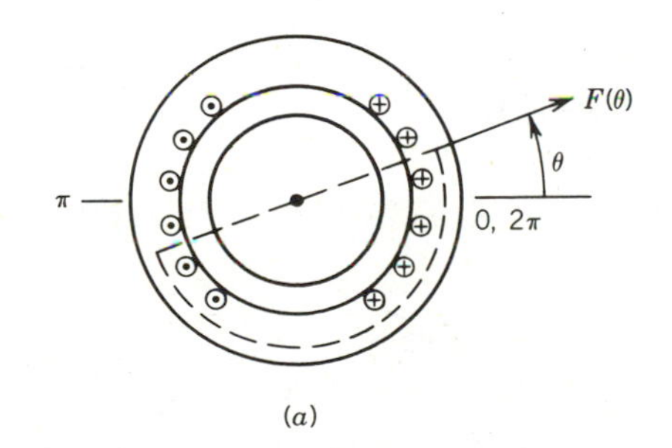

(*a*)

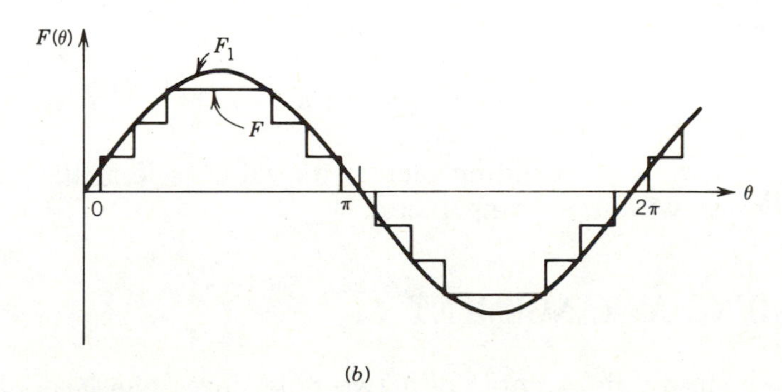

(*b*)

FIGURE A.2
Distributed winding.

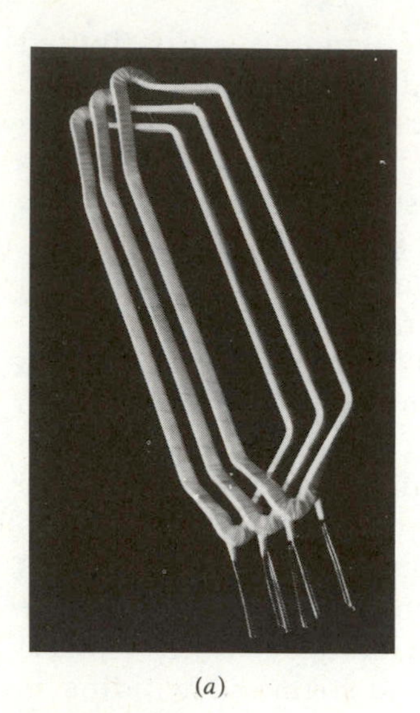

(a)

(b)

FIGURE A.3
Coil and winding in polyphase ac machine. (a) Multiturn coils. (Courtesy of Westinghouse Canada Inc.) (b) Polyphase stator winding. (Courtesy of General Electric Canada Inc.)

A.2 INDUCED VOLTAGES

For a concentrated winding of N turns per phase, the rms voltage induced in each phase is

$$E = 4.44 f N \Phi \tag{A.5}$$

where f is the frequency and Φ is the fundamental flux per pole. However, if the winding is distributed over several slots, the induced voltage is less and is given by

$$E = 4.44 f N \Phi K_w \tag{A.6}$$

where K_w is called the winding factor; its value is less than unity and depends on the winding arrangement.

A.3 WINDING ARRANGEMENT

Figure A.4*a* shows an example of a two-pole, three-phase, double-layer, full-pitch distributed winding for the stator. In each slot two coil sides are placed. The double-layer winding is used in most machines, except some smaller motors. It has the advantage of simpler end connection and it is

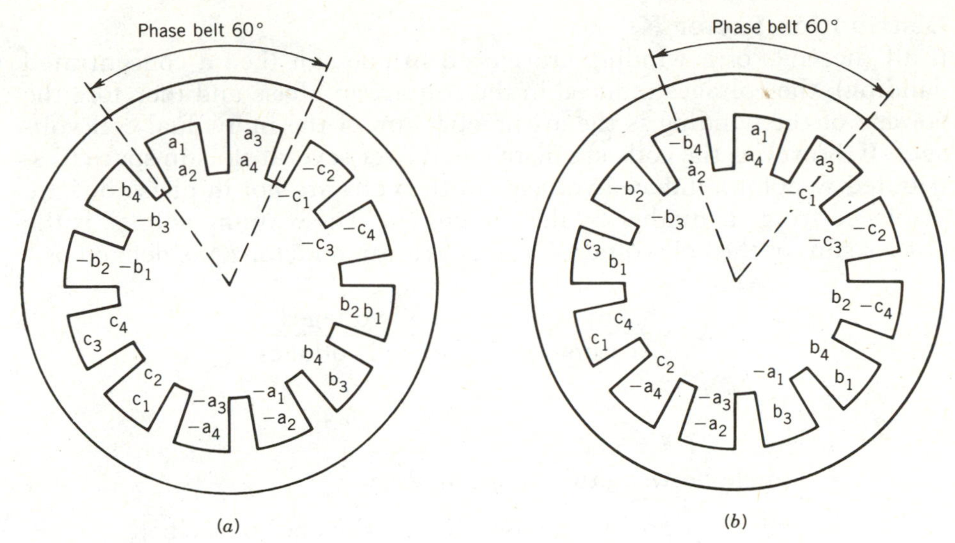

FIGURE A.4
Double-layer winding (stator). (*a*) Full-pitch coil. (*b*) Short-pitch coil.

economical to manufacture. Coil sides that are placed in adjacent slots and belong to the same phase, such as a_1, a_3 or a_2, a_4 constitute a *phase belt*. In the three-phase machine the phase belt is 60°. In the double-layer arrangement, one side of a coil, such as a_1, is placed at the bottom of a slot and the other side, $-a_1$, is placed at the top of another slot. Note that each coil in Fig. A.4*a* has a span of a full pole pitch or 180 electrical degrees; hence the winding is a full-pitch winding.

Figure A.4*b* shows a distributed winding arrangement in which the coils span less than a full pole pitch. Such a winding is called a *short-pitch*, *fractional-pitch*, or *chorded* winding. In Fig. A.4*b*, a coil such as a_1, $-a_1$ spans five-sixths of a pole pitch, that is, 150 electrical degrees. Note that the phase belts overlap. The phase belt for phase a has the coil sides of coils belonging to phases b and c.

Short-pitch windings are often used in polyphase ac machines. They reduce the length of the end connections (thereby saving copper) and reduce significantly, as we shall see, the magnitude of certain harmonics in the mmf distribution as well as voltage induced in the winding.

A.3.1 WINDING FACTORS

The distribution and pitching of the coils affects the voltages induced in the coils. Two factors are discussed here: (a) the distribution factor K_d, also known as the breadth factor, and (b) the pitch factor K_p, also known as the chord factor.

Distribution Factor K_d

If all the coils of a winding are placed in one slot (i.e., a concentrated winding), the voltages induced in the coil are in phase and therefore the voltage of the winding is the arithmetic sum of the individual coil voltages. If, however, the coils are distributed over several slots in space (distributed winding), induced voltages in the coils are not in phase but are displaced from each other by the *slot angle* α. The winding voltage is the phasor sum of the coil voltages. The distribution factor K_d is defined as

$$K_d = \frac{\text{phasor sum of coil voltages}}{\text{arithmetic sum of coil voltages}} \tag{A.7}$$

Let

α = angle between two adjacent slots

n = slots per pole per phase, that is, slots per phase belt

The distribution factor can be determined by constructing a phasor diagram for the coil voltages. Let $n = 3$. Figure A.5 shows the coil voltages as phasors RS, ST, and TU, each of which is a chord of a circle with center at 0 and subtends an angle α at 0. The phasor sum RU, representing the resultant winding voltage, subtends an angle $n\alpha$ at the center. From Eq. A.7 and Fig. A.5,

$$\begin{aligned} K_d &= \frac{RU}{n(RS)} \\ &= \frac{2Rx}{n(2Ry)} \\ &= \frac{Rx}{nRy} \\ &= \frac{0R \cdot \sin(n\alpha/2)}{n \cdot 0R \cdot \sin(\alpha/2)} \\ &= \frac{\sin(n\alpha/2)}{n \sin(\alpha/2)} \end{aligned} \tag{A.8}$$

Pitch Factor K_p

For a short-pitch coil where the coil span is less than a pole pitch, the induced voltage is less than the voltage that would be induced if the coil span were a full pole pitch. The pitch factor K_p is defined as

$$K_p = \frac{\text{voltage induced in short-pitch coil}}{\text{voltage induced in full-pitch coil}} \tag{A.9}$$

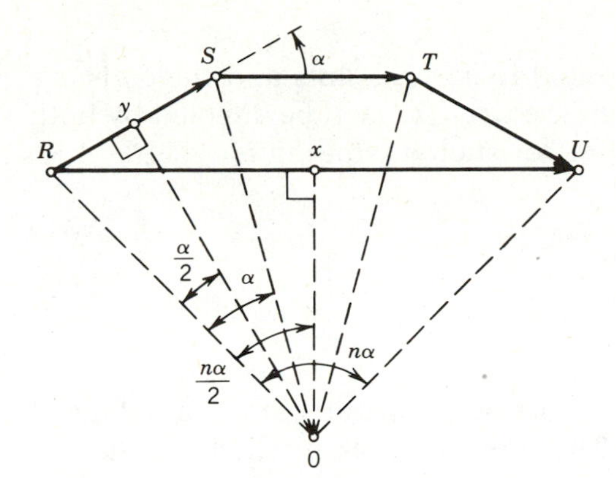

FIGURE A.5
Coil voltages in a distributed winding.

Figure A.6 shows the voltages induced in a full-pitch coil and a short-pitch coil. The coil pitch for the short-pitch coil is $180° - \gamma°$; that is, it is shorter than a full-pitch coil by $\gamma°$. The coil voltage e_c is

$$e_c = e_1 + e_2 \tag{A.10}$$

For the full-pitch coil both e_1 and e_2 are maximum at the same instant. However, for the short-pitch coil, when e_1 is maximum, e_2 is not maximum. These phenomena can be represented in the phasor diagram of these voltages, as shown in Fig. A.6*b*. For the short-pitch coil, the phasor E_2 for the voltage e_2 is phase-displaced by the angle $\gamma°$ from the phasor E_1 for the voltage e_1. From Eq. A.9 and Fig. A.6*b* and assuming $E_1 = E_2 = E$,

$$\begin{aligned} K_p &= \frac{2E \cos \gamma/2}{2E} \\ &= \cos \gamma/2 \end{aligned} \tag{A.11}$$

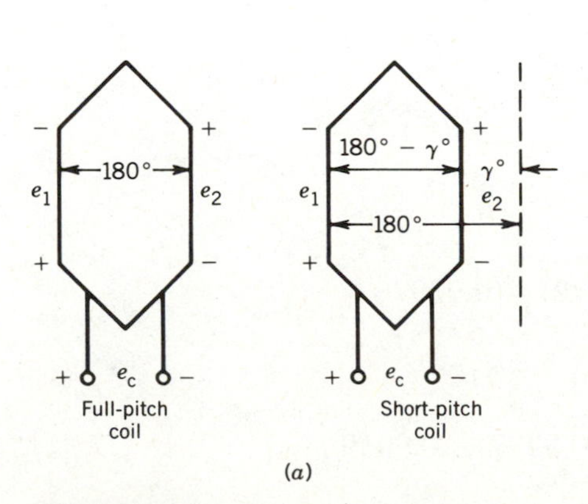

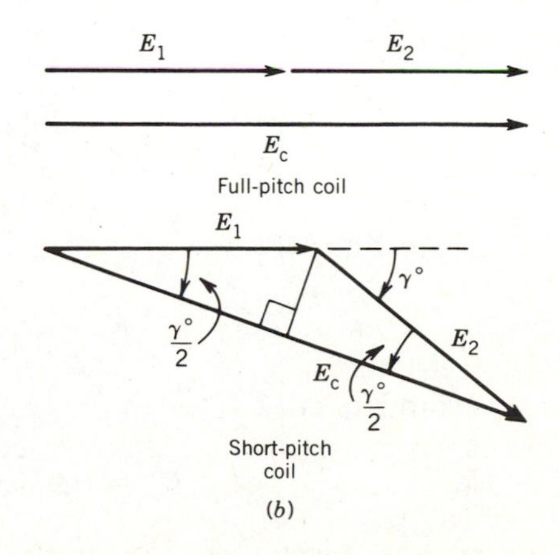

FIGURE A.6
Coil voltage in full-pitch and short-pitch coils.

Winding Factor K_w

If the coils of a winding are distributed in several slots and the coils are short-pitched, the voltage induced in the winding will be affected by both factors K_d and K_p. The winding factor for such a winding is

$$K_w = K_d \cdot K_p \tag{A.12}$$

EXAMPLE A.1

The stator of a 3ϕ machine has nine slots per pole and carries a balanced three-phase, double-layer winding. The coils are short-pitched and the coil pitch is 7/9; that is, each coil spans seven slots. Determine the winding factor.

Solution

$$\text{Slot angle} \qquad \alpha = \frac{180°}{9} = 20°$$

$$\text{Coil pitch} = \tfrac{7}{9} \times 180 = 140°$$

The coil is short-pitched by

$$\gamma = 180° - 140° = 40°$$

Number of slots per pole per phase:

$$n = \frac{9}{3} = 3$$

From Eq. A.8

$$K_d = \frac{\sin(3 \times 20°/2)}{3\sin(20°/2)} = 0.9598$$

From Eq. A.11

$$K_p = \cos(40°/2) = 0.9397$$

From Eq. A.12

$$K_w = 0.9598 \times 0.9397 = 0.9019$$

Because of distribution and short-pitching the winding voltage will be less by a factor of 0.9019.

A.4 SPACE HARMONICS AND WINDING FACTORS

In Eqs. A.5 and A.6 it is assumed that the induced voltage is sinusoidal. However, if the flux density distribution is nonsinusoidal, the induced voltage in the winding will be nonsinusoidal. The distribution factor, pitch factor, and winding factor will be different for each harmonic voltage.

The phase difference between the hth harmonic voltages of adjacent coils is $h\alpha$. Therefore, the distribution factor for the hth harmonic is

$$K_{dh} = \frac{\sin(nh\alpha/2)}{n\sin(h\alpha/2)} \tag{A.13}$$

The effect of distributing the winding over several slots is illustrated in Table A.1. For values of n (i.e., number of slots per pole per phase) ranging from 1 to 6 ($n = 1$ represents a concentrated winding), the fundamental distribution factor K_{d1} varies from 1 to 0.9561. Thus, the fundamental voltage will be lowered to some extent if the winding is distributed over several slots. However, Table A.1 shows that distributing the winding will result in a significant reduction in the harmonic content of the induced voltage in the winding.

In a short-pitch coil the phase difference between the hth harmonic voltages of the two coil sides is $h\gamma$. Therefore, the pitch factor for the hth harmonic is

$$K_{ph} = \cos(h\gamma/2) \tag{A.14}$$

The variation of harmonic pitch factors for different values of the coil pitch is shown in Table A.2. The harmonic voltages decrease in a short-pitch coil, thereby improving the waveform of the induced voltage in the

TABLE A.1
Distribution Factor in Three-Phase Machines

	Distribution Factor for Harmonics					
n[a]	$h = 1$	3	5	7	9	11
1	1.000	1.000	1.000	1.000	1.000	1.000
2	0.966	0.707	0.259	0.259	0.707	0.966
3	0.960	0.667	0.218	0.177	0.333	0.177
4	0.958	0.653	0.205	0.158	0.271	0.126
5	0.957	0.647	0.200	0.149	0.247	0.110
6	0.956	0.644	0.197	0.145	0.236	0.102
∞	0.955	0.637	0.191	0.136	0.212	0.087

[a] $n = 1$, concentrated winding; $n > 1$, distributed winding.

TABLE A.2
Pitch Factor in Three-Phase Machine

Coil Pitch	Pitch Factor for Harmonics					
$(180° - \gamma)$[a]	$h = 1$	3	5	7	9	11
120° or 2/3	0.866	0.000	0.866	0.866	0.866	0.866
144° or 4/5	0.951	0.588	0.000	0.588	0.951	0.951
150° or 5/6	0.966	0.707	0.259	0.259	0.707	0.966
154° or 6/7	0.975	0.782	0.434	0.000	0.434	0.782
160° or 8/9	0.985	0.866	0.643	0.342	0.000	0.342
180° or 1	1.000	1.000	1.000	1.000	1.000	1.000

[a] $\gamma = 0°$ for full-pitch coil.

winding. In fact, a certain harmonic can be completely eliminated from the winding voltage by choosing a pitch for the coils that makes the pitch factor zero for that harmonic. To eliminate the hth harmonic voltage,

$$\cos(h\gamma/2) = 0 \tag{A.15}$$

or

$$\frac{h\gamma}{2} = 90°$$

or

$$\gamma = \frac{180°}{h} \tag{A.16}$$

Thus, to eliminate the third harmonic, the coils are to be shorted by

$$\gamma = 180°/3 = 60°$$

The winding factor corresponding to the hth harmonic voltage is

$$K_{\mathrm{wh}} = K_{\mathrm{dh}} \cdot K_{\mathrm{ph}} \tag{A.17}$$

where K_{dh} and K_{ph} are given by Eqs. A.13 and A.14, respectively.

A.5 TIME HARMONIC VOLTAGES

The waveform of the induced voltage in a winding depends on the space distribution of the air gap flux density. This flux density distribution is not purely sinusoidal. For example, in synchronous machines the space flux density distribution of the rotor pole is nonsinusoidal. In induction machines the air gap flux is produced by currents flowing in the windings and the space flux density distribution is nonsinusoidal. The distribution of winding coils can improve the space flux density distribution (Fig. A.2) but cannot make it purely sinusoidal.

Because of a nonsinusoidal space flux density distribution, the induced voltage in a winding will contain harmonics. By distributing and chording the coils of the winding, the harmonics can be appreciably reduced. Of course, the fundamental will also be reduced, although, fortunately, by a small amount. The winding factors, discussed earlier, represent the per-unit reduction of the fundamental and each harmonic resulting from distribution and chording of the winding coils.

Figure A.7 shows a typical space flux density distribution, which can be expressed as

$$B(\theta) = B_{1(\max)} \sin \theta + B_{3(\max)} \sin 3\theta + B_{5(\max)} \sin 5\theta + B_{7(\max)} \sin 7\theta + \cdots \quad \text{(A.18)}$$

The fundamental and third harmonic flux densities are also shown. If the fundamental corresponds to a "p-pole" machine, then the third harmonic can be considered to correspond to a "$3p$-pole" machine. Each component of the flux density distribution will induce voltage in the winding.

For a concentrated winding of N turns per phase, the rms voltage induced in each phase by the fundamental component of the flux density is

$$E_1 = 4.44 f N \Phi_1 \quad \text{(A.19)}$$

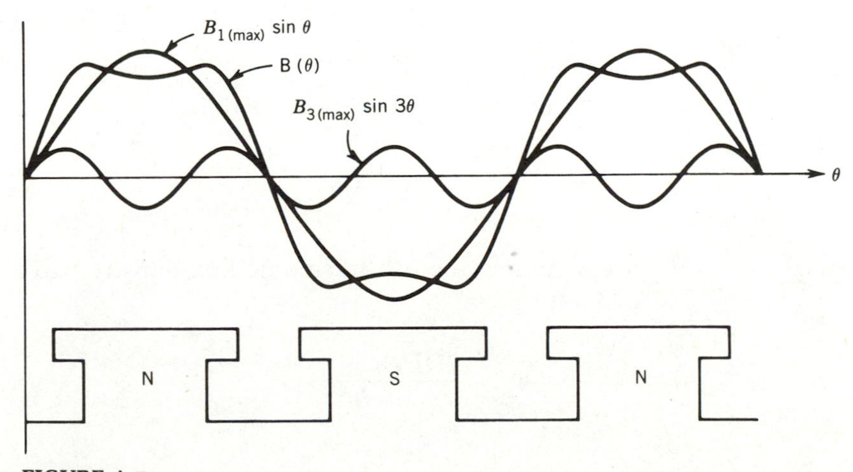

FIGURE A.7
Space flux density distribution.

where the subscript 1 denotes the fundamental, f the frequency, and Φ_1 the fundamental flux per pole.

For a distributed and chorded winding the fundamental voltage is

$$E_1 = 4.44 f N \Phi_1 K_{w1} \tag{A.20}$$

where $K_{w1} = K_{d1} \cdot K_{p1}$ is the fundamental winding factor, which is slightly less than unity for a three-phase machine.

The space harmonics in the flux density wave induce time harmonic voltages in the winding. For the hth harmonic flux, the induced voltage is

$$E_h = 4.44(hf) N \Phi_h K_{wh} \tag{A.21}$$

where Φ_h is the flux per pole corresponding to the hth harmonic of the flux density.

Let

D = diameter of the air gap

L = length of the pole

p = number of poles

Then

$$\Phi_h = A_h \cdot \overline{B}_h \tag{A.22}$$

where A_h in the area of the hth harmonic pole,

$$A_h = \frac{\pi D L}{hp} \tag{A.23}$$

and $\overline{B}_h$ is the average flux density for the hth harmonic,

$$\overline{B}_h = \frac{2 B_{h(\max)}}{\pi} \tag{A.24}$$

where $B_{h(\max)}$ is the amplitude of the hth harmonic flux density wave.

From Eqs. A.22, A.23, and A.24,

$$\Phi_h = \frac{2DL}{p} \frac{B_{h(\max)}}{h} \tag{A.25}$$

$$\propto \frac{B_{h(\max)}}{h} \tag{A.26}$$

From Eqs. A.21 and A.26,

$$E_h \propto B_{h(\max)} K_{wh} \tag{A.27}$$

The total rms voltage induced in the winding is

$$E = (E_1^2 + E_3^2 + E_5^2 + E_7^2 + \cdots)^{1/2} \tag{A.28}$$

$$E = \left(\sum_{h=1,3,5,\ldots} E_h^2 \right)^{1/2} \tag{A.29}$$

It can be shown that in a three-phase machine the triplen harmonic voltages (i.e., third and its multiples) do not appear in the line-to-line voltage. The line-to-line voltage therefore appears more sinusoidal than the line-to-neutral (i.e., winding) voltage. The rms line-to-line voltage is

$$E_{LL} = \sqrt{3}\,(E_1^2 + E_5^2 + E_7^2 + E_{11}^2 + \cdots)^{1/2} \tag{A.30}$$

EXAMPLE A.2

A 3ϕ, 60 Hz, star-connected synchronous generator has eight poles, 96 stator slots, and 9/12 chorded winding. The air gap flux density shows that third and fifth harmonics are present and are of amplitude 30% and 15% of the fundamental.

Determine the ratio of the line-to-line voltage and line-to-neutral voltage.

Solution

$$n = \frac{96}{8 \times 3} = 4 \text{ slots/pole/phase}$$

$$\alpha = \frac{180^\circ}{3 \times 4} = 15^\circ$$

$$\gamma = 3 \times 15^\circ = 45^\circ$$

The distribution factors are

$$K_{d1} = \frac{\sin(4 \times 15^\circ/2)}{4 \sin 15^\circ/2} = 0.9577$$

$$K_{d3} = \frac{\sin(4 \times 3 \times 15°/2)}{4 \sin(3 \times 15°/2)} = 0.6533$$

$$K_{d5} = \frac{\sin(4 \times 5 \times 15°/2)}{4 \sin(5 \times 15°/2)} = 0.2053$$

The pitch factors are

$$K_{p1} = \cos(45°/2) = 0.9239$$
$$K_{p3} = \cos(3 \times 45°/2) = 0.3827$$
$$K_{p5} = \cos(5 \times 45°/2) = -0.3827$$

The winding factors are

$$K_{w1} = K_{d1}K_{p1} = 0.9577 \times 0.9239 = 0.8848$$
$$K_{w3} = 0.6533 \times 0.3827 = 0.2500$$
$$K_{w5} = 0.2053 \times (-0.3827) = -0.0786$$

From Eq. A.27, the rms fundamental voltage is

$$\begin{aligned} E_1 &\propto B_{1(\max)}K_{w1} \\ &= K \times 1 \times 0.8848 \\ &= 0.8848K \end{aligned}$$

The third and fifth harmonic voltages are

$$\begin{aligned} E_3 &= KB_{3(\max)}K_{w3} \\ &= K \times 0.3 \times 0.25 \\ &= 0.075K \\ E_5 &= KB_{5(\max)}K_{w5} \\ &= K \times 0.15 \times 0.0786 \\ &= 0.0118K \end{aligned}$$

The phase voltage is

$$\begin{aligned} E_{LN} &= (E_1^2 + E_3^2 + E_5^2 + \cdots)^{1/2} \\ &= K(0.8848^2 + 0.075^2 + 0.0118^2 + \cdots)^{1/2} \\ &\simeq 0.8881K \end{aligned}$$

The line-to-line voltage is

$$\begin{aligned} E_{LL} &= \sqrt{3}\,(E_1^2 + E_5^2 + \cdots)^{1/2} \\ &= \sqrt{3}\,K(0.8848^2 + 0.0118^2 + \cdots)^{1/2} \\ &\simeq \sqrt{3}\;0.8848K \end{aligned}$$

The ratio of the line voltage to the phase voltage is

$$\frac{E_{LL}}{E_{LN}} = \frac{\sqrt{3} \times 0.8848}{0.8881} = \sqrt{3} \times 0.9963$$

Note that the line-to-line voltage is slightly lower than $\sqrt{3}$ times the phase voltage because of the absence of third harmonic voltages in the line-to-line voltage.

PROBLEMS

A.1 A 3ϕ, star-connected synchronous generator has four rotor poles, which produce a space flux density distribution as follows:

$$B(\theta) = 1.00 \sin\theta + 0.4 \sin 3\theta + 0.2 \sin 5\theta$$

The stator has 36 slots and a balanced three-phase double-layer winding. The coil pitch is 140°. Determine, in terms of the rms fundamental voltage E_1,

(a) The rms value of the phase voltage E_{LN}.

(b) The rms value of the line-to-line voltages E_{LL}.

A.2 A 3ϕ, 60 Hz, six-pole, Y-connected synchronous generator has 108 stator slots and a double-layer armature winding. The coil pitch is 150° and each coil has 30 turns. The air gap flux density due to field poles contains third and fifth harmonic components of magnitude 30 and 20%, respectively, relative to the fundamental. The fundamental flux per pole is 0.01 webers. The generator runs at the synchronous speed.

(a) Determine the rms values of the fundamental and harmonic voltages induced in the stator winding.

(b) Determine the rms value of the stator phase voltage.

(c) Determine the rms value of the stator line-to-line voltages.

A.3 A 3ϕ, eight-pole, 750 rpm, Y-connected synchronous generator has the following data:

Total stator slots = 96

Conductors in each slot = 20

Stator winding configuration is a double layer

Coil span is suitable to eliminate the third harmonic induced voltage

Fundamental flux per pole = 0.12 Wb

Analysis of the air gap flux shows third harmonic and fifth harmonic flux density amplitudes 25 and 15%, respectively, of that of the fundamental.

(a) Determine the number of turns per phase of the stator winding.

(b) Determine the coil span in electrical degrees.

(c) Determine the fundamental frequency.

(d) Determine the rms values of the phase and line voltages.

A.4 A 3ϕ, 60 Hz, synchronous generator has eight-poles and 96 stator slots. The coil span of the winding is 135°. The open-circuit phase voltage (line-to-neutral) contains fundamental, third, and fifth harmonics of relative magnitude 100, 25, and 5%, respectively. The measured rms phase voltage is 12 kV.

(a) Determine the rms values of the harmonic components of the phase voltage.

(b) Determine the total rms value of the line-to-line voltage.

(c) Determine the amplitudes of the harmonic flux densities in the air gap relative to that of the fundamental flux density.

APPENDIX B

UNITS AND CONSTANTS

B.1 UNITS

Quantity	Units	Equivalent
Length	1 meter (m)	3.281 feet (ft) 39.36 inches (in.)
Mass	1 kilogram (kg)	2.205 pounds (lb) 35.27 ounces (oz)
Time	1 second (sec)	
Force	1 newton (N)	0.2248 pounts (lbf)
Torque	1 newton-meter ($N \cdot m$)	0.738 pound-feet ($lbf \cdot ft$)
Moment of inertia	1 kilogram-meter^2 ($kg \cdot m^2$)	23.7 pound-feet^2 ($lb \cdot ft^2$)
Power	1 watt (W)	0.7376 foot-pounds/second 1.341×10^{-3} horsepower (hp)
Energy	1 joule (J)	1 watt-second 0.7376 foot-pounds ($ft \cdot lb$) 2.778×10^{-7} kilowatt-hours (kWh)
Magnetic flux	1 weber (Wb)	10^8 maxwells or lines
Magnetic flux density	1 tesla (T)	1 weber/meter^2 (Wb/m^2) 10^4 gauss
Voltage	1 volt (V)	1 watt/ampere
Current	1 ampere (A)	1 coulomb/second
Frequency	1 Hertz (Hz)	1 cycle/second
Horsepower	1 hp	746 watts
Magnetomotive force	1 ampere–turn (At)	1.257 gilberts
Magnetic field intensity	1 ampere–turn/meter (At/m)	1.257×10^{-2} oersted

B.2 CONSTANTS

Permeability of free space	$\mu_0 = 4\pi \times 10^{-7}$ H/m
Permittivity of free space	$\varepsilon_0 = 8.854 \times 10^{-12}$ F/m
Acceleration due to gravity	$g = 9.807$ m/sec^2

APPENDIX C

LAPLACE TRANSFORMS

Laplace Transform $F(s)$	Time Function $f(t)$
$\frac{1}{s}$	$U(t) \rightarrow$ unit step function
$\frac{1}{s^2}$	t
$\frac{1}{s+a}$	e^{-at}
$\frac{\omega}{s^2+\omega^2}$	$\sin \omega t$
$\frac{s}{s^2+\omega^2}$	$\cos \omega t$
$\frac{1}{(s+a)^2}$	te^{-at}
$\frac{\omega}{(s+a)^2+\omega^2}$	$e^{-at} \sin \omega t$
$\frac{s+a}{(s+a)^2+\omega^2}$	$e^{-at} \cos \omega t$
$\frac{1}{(s+a)(s+b)}$	$\frac{e^{-at}-e^{-bt}}{b-a}$
$\frac{1}{s(1+Ts)}$	$1-e^{-t/T}$
$\frac{\omega_n^2}{s(s^2+\omega_n^2)}$	$1-\cos \omega_n t$
$\frac{\omega_n^2}{s^2+2\xi\omega_n s+\omega_n^2}$	$\frac{\omega_n}{\sqrt{1-\xi^2}} e^{-\xi\omega_n t} \sin \omega_n \sqrt{1-\xi^2}\, t$
$\frac{\omega_n^2}{s(s^2+2\xi\omega_n s+\omega_n^2)}$	$1-\frac{e^{-\xi\omega_n t}}{\sqrt{1-\xi^2}} \sin(\omega_n \sqrt{1-\xi^2}\, t+\cos^{-1} \xi)$

INDEX